Florida

Preparing for FCAT Mathematics Grade 10

Second Edition

W. Kay Williams

David E. Williams

AMSCO

AMSCO SCHOOL PUBLICATIONS, INC.
315 Hudson Street, New York, N.Y. 10013

Dedication

We dedicate this Grade 10 book to the many students who have entered our classrooms: We hope that we have made a difference in your lives. We also dedicate it to our children, Jennifer and David: You are the best and we love you.

Special thanks to the teachers who have advised and encouraged us in writing this book.

W. Kay Williams

This author has taught mathematics for 28 years, 20 of which were in Palm Beach County, Florida. At present she is an adjunct professor at Allegany College of Maryland. Mrs. Williams has taught all levels of mathematics, from general mathematics to International Baccalaureate and Advanced Placement calculus. She is a Deputy Chief Examiner for the International baccalaureate program. Mrs. Williams received the William T. Dwyer Excellence in Education Award and was named Palm Beach County High School Mathematics Teacher of the Year. In 2001, she was named the Florida Region V Teacher of the Year.

David E. Williams

This author is presently an adjunct professor at Allegany College of Maryland. He has taught mathematics for 35 years, 20 years of which were in Palm Beach County. His experience includes teaching all levels of mathematics from basic math to Advanced Placement calculus, in a wide range of communities from rural to suburban to inner-city. Mr. Williams has conducted many local and state workshops on various topics. He holds the positions of table leader for grading AP Calculus exams and team leader for marking IB mathematics exams. His achievements include being named University of Miami Outstanding High School Teacher, receiving the William T. Dwyer Excellence in Education Award, and being selected as Florida Teacher of the Year in 1996. In May 1998, he was inducted into the Florida Educators Hall of Fame.

They have co-authored *Florida: Preparing for FCAT Mathematics Grade 6, Preparing for FCAT Mathematics Grade 7, Preparing for FCAT Mathematics Grade 8, Preparing for FCAT Mathematics Grade 9, and Mastering the MCAS Grade 10.*

Reviewers:

Carol Goehring
Evans High School
Orlando, Florida

Roger O'Brien
Mathematics Consultant
Lakeland, Florida

Please visit our Web site at: ***www.amscopub.com***

Text design: Merrill Haber
Cover design: Meghan Shupe
Compositor: Monotype, LLC

When ordering this book, please specify **R 6680 W** or
FLORIDA: PREPARING FOR FCAT MATHEMATICS, GRADE 10 SECOND EDITION

ISBN 978-1-56765-591-9

PRINTED IN THE UNITED STATES OF AMERICA
1 2 3 4 5 6 7 8 9 10 11 10 09 08 07

Contents

CHAPTER TWO

Measurement

CHAPTER THREE

Geometry and Spatial Sense

CHAPTER FOUR

Algebraic Thinking

CHAPTER FIVE

Data Analysis and Probability

Introduction: To the Student

About FCAT

The Sunshine State Standards, adopted by the State Board of Education in 1996, identify the skills and essential knowledge that students should learn. To judge whether students are meeting these standards for learning, the Florida Comprehensive Assessment Test (FCAT) was developed.

The grade 10 FCAT for mathematics tests five content areas:

(1) Number Sense, Concepts, and Operations

(2) Measurement

(3) Geometry and Spatial Sense

(4) Algebraic Thinking

(5) Data Analysis and Probability

The test emphasizes two categories especially: Geometry and Spatial Sense and Algebraic Thinking.

There will be three types of questions on the test:

- **Multiple-choice.** The majority of the questions will be in multiple-choice format.
- **Gridded-response.** In this format, you must first solve a problem and then write the answer in a special grid.
- **Performance tasks.** The third type of question on the test is called a "performance task." In these tasks, you must show your work and perhaps explain your answer in words. There are two types of performance tasks: **short-response** and **extended-response.** (Examples of each type will be shown later.)

The FCAT mathematics test is divided into two parts, with approximately 28 questions in each part. You will be given a total of 180 minutes to take the test.

General approach to problem solving

When you have difficulty solving math problems, the cause is often a haphazard approach to the problem. The following four-step approach can make problems more manageable:

Step 1: READ. Take a deep breath and read the problem slowly. Decide what information is given and what needs to be found. Sometimes it helps to underline the question being asked.

Step 2: PLAN. Decide what type of operation or strategy you should use to solve the problem. Does your choice of procedure make sense, considering what you are looking for?

Step 3: SOLVE. Do the math. Be very careful. Don't let careless mistakes hurt you now. If you are not comfortable doing the calculations on your own, use your calculator.

Step 4: CHECK. Go back to the problem and see if you have answered the question that was asked. Does your answer make sense? Use your calculator to check your work.

Multiple-choice problems

On the FCAT, a multiple-choice question will have four possible answers, and only *one* of them will be considered the best answer. The four choices will be labeled **A**, **B**, **C**, **D** or **F**, **G**, **H**, **I**. Each multiple-choice question is expected to take approximately 1 minute to answer. Each correct answer earns 1 point.

Here are some strategies for choosing the right answer to a multiple-choice problem:

- Work the problem out on scrap paper. Then look to see if your answer is one of the four choices.
- If the problem gives several pieces of information, organize the data in a table or a list.
- Draw a diagram or a picture.
- Try to find a pattern in the problem that will lead you to the answer.
- Make a guess and see if it works. Use the information given in the problem to set up a situation. See if your guess fits all the requirements of the problem.
- Start with each answer to the problem and work backward through the problem to get to the question.
- If a problem seems too complicated, try solving an easier problem, involving simpler numbers or a simpler situation. Then use the same procedure to solve the actual, harder problem.

Gridded-response problems

Each gridded-response problem is expected to take approximately 1.5 minutes, and each correct answer will earn 1 point. First, you work out the problem. After a problem has been worked out, you write your solution in the answer grid assigned to that problem.

A sample answer grid is shown below:

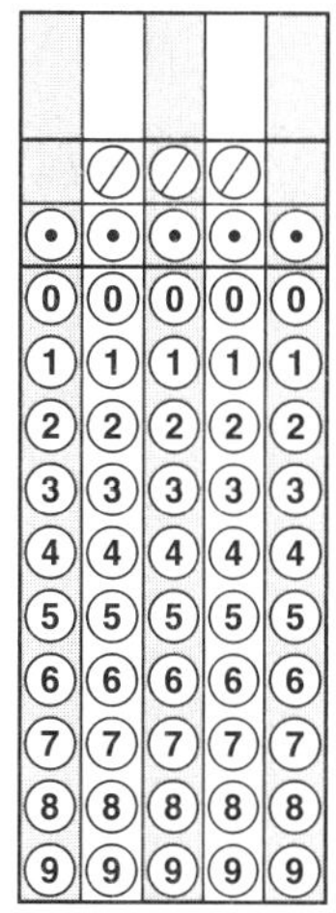

The top row of the grid is where you will write the answer. You have a choice: You may start from the left and fill in each box as needed, or you may start from the right and fill in going to the left. Do NOT leave a blank space in the middle of an answer. However, you may have blanks on the left or on the right if some spaces are not needed.

For instance, you may write the number 50 on the top row in either of these ways:

5	0			

OR

			5	0

If an answer has a decimal point (.) or a fraction bar (/), be sure to write it in the appropriate answer box.

The bottom part of the grid consists of columns of bubbles. A bubble must be filled in for every answer in the answer box. If 0 is part of the answer, be sure to bubble in the zero bubble. If a box in the top row is blank, do NOT bubble anything in that column.

Notice that the grid does not have a dollar sign ($). If an answer turns out to be in dollars and cents, simply write the amount in the answer boxes without a dollar sign. Don't forget the decimal point, to indicate cents.

If an answer turns out to be a mixed number, such as $3\frac{1}{2}$, you must change it to an improper fraction or a decimal before writing it in the answer box. For instance, $3\frac{1}{2}$ must be written as 7/2 or 3.5.

Following are some examples of how to fill in the answer grids.

EXAMPLE 1: WHOLE NUMBERS

The sum of 150 + 20 =

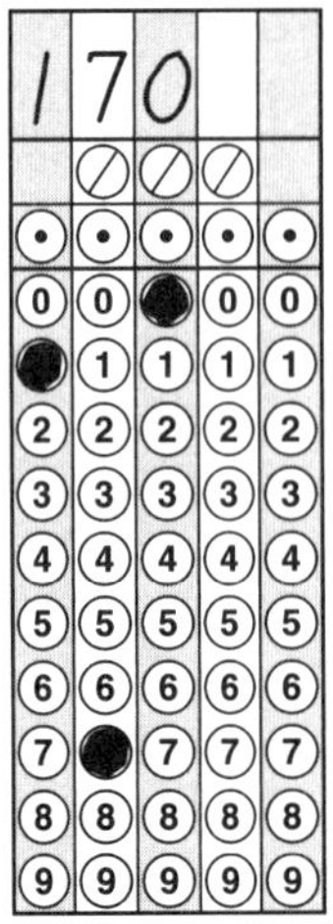

OR

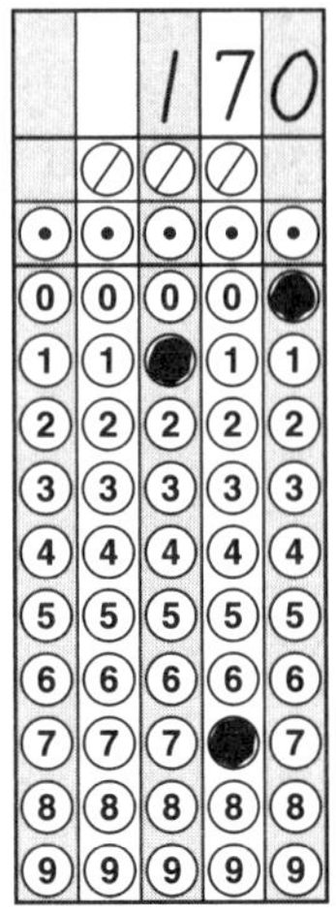

EXAMPLE 2: PERCENTS

$\frac{1}{2}$ written as a percent =

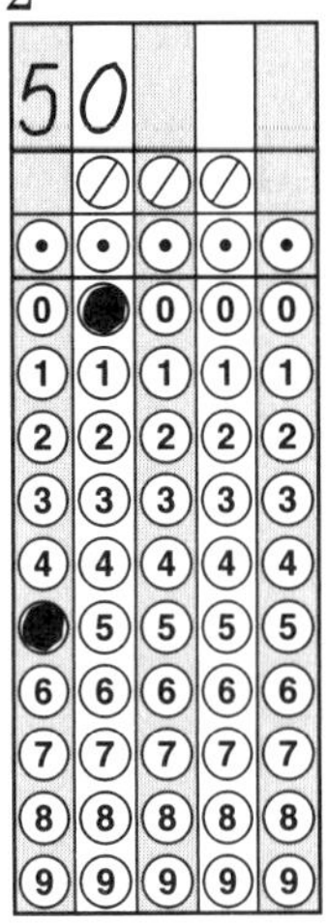

OR

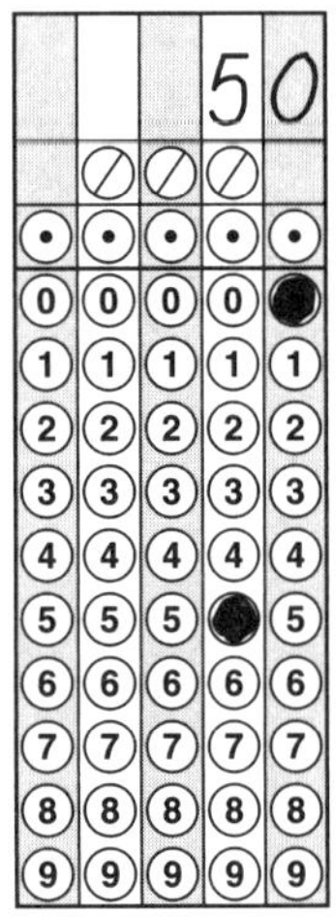

EXAMPLE 3: DECIMALS

The decimal equivalent of $2\frac{3}{100}$ =

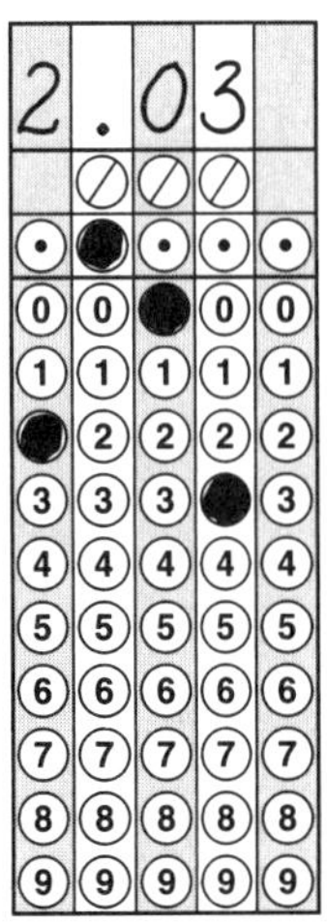

OR

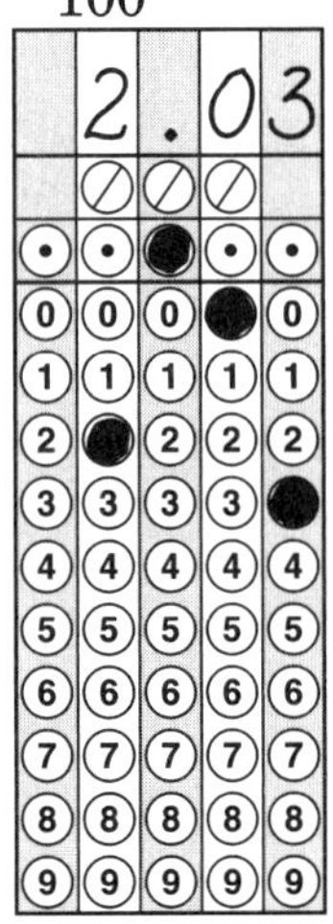

EXAMPLE 4: FRACTIONS

$$4\frac{7}{10} - 2\frac{3}{4} =$$

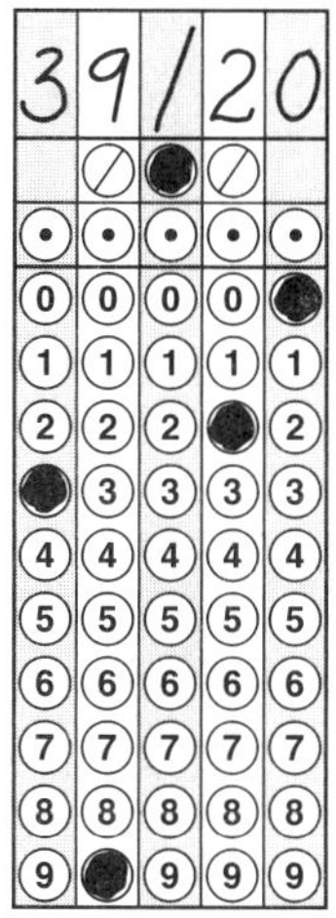

OR

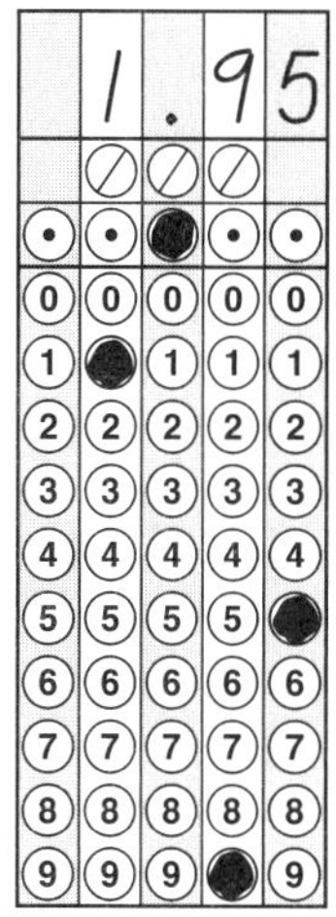

REMEMBER: Mixed numbers MUST be written as improper fractions or decimals.

Performance tasks

There are two types of performance tasks: short-response and extended-response. The answer to a performance task must be written in the proper place in the FCAT answer booklet. Be sure to show your work, even if you are using a calculator.

Your answer to a performance task will be scored with a **rubric**: a guideline that tells the scorer how many points to award. (Examples of rubrics are shown later.)

Short-response problems

On FCAT, the short-response problems are identified by a symbol that says: "Think, Solve, Respond." Each short-response problem is expected to take about 3 to 5 minutes and is worth 2 points. A complete, correct response earns the full 2 points. A response that is partly complete or partly correct may earn you 1 point. A response that is totally wrong or missing earns 0.

In this type of question, you must do more than simply solve the problem. You will also be asked to do one or more of the following: show the procedure necessary to solve the problem, explain the procedure, and justify the method used. In some cases, there are several acceptable responses.

EXAMPLE: SHORT-RESPONSE PROBLEM

Joe paddles a canoe downstream (with the current) for 6 miles, taking 40 minutes. His return trip upstream (against the current) takes 3 hours. The rate of the current is 3.5 miles per hour.

Part A Write two equations, representing the trip downstream and the trip upstream, that you could use to find the rate at which Joe is paddling.

Part B Solve one of your equations to find Joe's rate of paddling. Show your work.

<table>
<tr><th>Score</th><th>Explanation</th></tr>
<tr><td>2</td><td>

A. Use the formula $d = rt$. Let r = Joe's rate of paddling in mph. Then:

$r + 3.5$ = rate of canoe going downstream

$r - 3.5$ = rate of canoe going upstream

Equations (convert 40 minutes to $\frac{2}{3}$ hour):

$\frac{2}{3}(r + 3.5) = 6$ and $3(r - 3.5) = 6$

B. Either equation may be solved:

First equation

$\frac{2}{3}(r + 3.5) = 6$

$\frac{2}{3}r + \left(\frac{2}{3} \bullet \frac{7}{2}\right) = 6$

$\frac{2}{3}r = 6 - \frac{14}{6} = 3\frac{2}{3}$

$r = 5\frac{1}{2} = 5.5$

Second equation

$3(r - 3.5) = 6$

$3r - 10.5 = 6$

$3r = 6 + 10.5 = 16.5$

$r = 5.5$

Answer: Joe's rate of paddling is 5.5 mph.

The student shows a thorough understanding of the problem and shows complete, correct work.

</td></tr>
<tr><td>1</td><td>Both equations are set up correctly but answer is incorrect because of a minor arithmetic error.</td></tr>
<tr><td>0</td><td>Wrong equations and wrong answer OR correct answer but no work.</td></tr>
</table>

REMEMBER: You will earn partial credit if some of the work is correct. You will not earn any credit if you do not show your work when work is expected.

Extended-response problems

Extended-response problems are identified by the "Think, Solve, Explain" symbol plus several little lines (which look like ruled paper or answer blanks). Each extended-response problem is expected to take 5 to 15 minutes and is worth 4 points. You will earn the full 4 points for a complete, correct response showing correct work. You may earn partial credit (3 points, 2 points, or 1 point) for a partially correct or partially complete response. If your response is totally incorrect or missing, you will earn 0 points.

Extended-response problems often have more than one part. As with the short-response questions, you must usually do more than just give an answer for each part. You may be asked to solve the problem, show the procedure necessary to solve the problem, explain the procedure, and justify the method you have used. In most cases there are several possible responses.

EXAMPLE: EXTENDED-RESPONSE PROBLEM

A contractor needs to calculate the distance across a river. It cannot be measured directly. In the diagram below, BC is the width of the river.

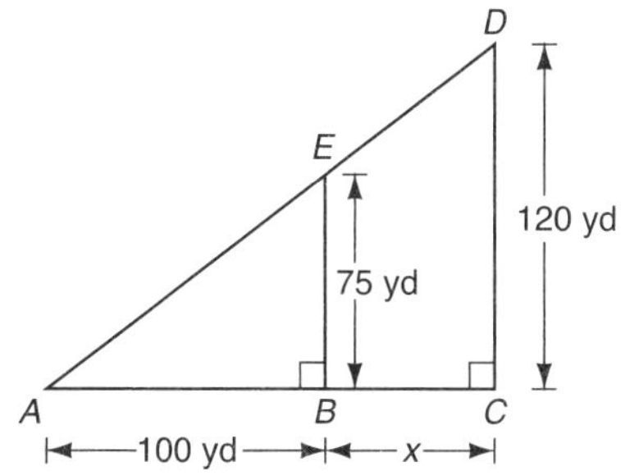

Part A Using geometric principles, explain why $\triangle AEB$ is similar to $\triangle ADC$.

Part B Set up a proportion to find the distance BC. Let x = the distance.

Part C Solve your proportion to find the approximate distance, in yards, across the river. Show your work.

Score	Explanation
4	**A.** $\triangle AEB \approx \triangle ADC$ by angle-angle (both triangles are right triangles, and they both contain $\angle A$). **B.** Proportion: $\frac{100}{x + 100} = \frac{75}{120}$ where x = distance BC in yards **C.** Solution: If $\frac{100}{x + 100} = \frac{75}{120}$, then: $75x + 7{,}500 = 12{,}000$ $75x = 4{,}500$ $x = 60$ Answer: Distance BC is 60 yards. Full credit will be awarded if a thorough understanding is demonstrated, all 3 parts (A, B, C) are answered correctly, the mathematics is correct, and the reasoning is clear and logical.
3	Essential but not thorough understanding with possible minor mathematical errors.
2	Partial understanding of the problem with only partially correct answers OR correct approach but incorrect answer OR faulty interpretation of the results OR only two of the three parts are answered correctly.
1	Error-filled, incomplete answers with faulty reasoning, but student shows some grasp of the problem.
0	Totally incorrect solutions OR uninterpretable response OR both.

IMPORTANT: In this rubric, "uninterpretable response" (which earns 0) means that the scorer cannot understand what you are trying to say or cannot read your answer at all. So be sure that your ideas—and your handwriting—are clear.

The FCAT calculator

When you arrive to take the FCAT, you will be given a simple four-function calculator that you can use during the test. You may NOT use your own calculator.

CAUTION! Your own calculator may be more sophisticated and may have the order of operations programmed into it. But the FCAT calculator is NOT programmed this way. If you put the numbers and symbols in from left to right on the FCAT calculator, it may NOT do the calculations correctly, according to the order of operations. Therefore, when using the FCAT calculator, you should do one operation at a time.

EXAMPLE

$5 + 2 \bullet 3 = ?$

According to the order of operations, you must multiply first: $2 \bullet 3 = 6$. Then you add: $6 + 5 = 11$. But if you just put the numbers into the FCAT calculator from left to right, you would get an incorrect answer: $7 \bullet 3 = 21$.

Using a calculator: points to remember

Chapter One of this book has many examples of using a calculator. Following are some things you should be aware of.

Rounding

Sometimes, the calculator gives you an answer that is not one of the choices in a multiple-choice problem. In a gridded-response problem, the answer from the calculator may not fit into the grid. In these cases, rounding may be necessary.

EXAMPLE

$\frac{2}{3}$ of 60 = ?

When you use a calculator for this problem, two different answers could be displayed, depending on whether you do division first or multiplication first.

$2 \times 60 \div 3 = 120 \div 3 = 40$ (multiplication first)

$2 \div 3 \times 60 = 0.6666\ldots \times 60 = 39.999996$ (division first)

The second answer (from division first) must be rounded to 40.

Overflow errors

The calculator display will show only 8 digits maximum. If the answer to a problem has more than 8 digits, a small E will appear. This means "error," and the answer will be incorrect.

EXAMPLE

$565656 \times 565656 = 3199.6671$ E (incorrect answer)

If an E appears when you a doing a calculation, clear the calculator and make sure that you are using the correct numbers. Redo the problem.

Memory

Sometimes it is useful to store a number while other calculations are being done. In that case, use the memory capability of the calculator.

- When a number is stored in memory, a small M will appear in the display.
- If no number is stored in memory, the key M+ will store the number that is on the display. Then, the key MR will recall that number back to the display.
- Once you have a number stored in memory, M+ will add a number in the display to the number already stored.
- The key M− will subtract the number in the display from the number that is stored in memory.
- The keys M+ and M− will NOT change what is in the display. These keys change only what is stored in memory.
- The key MR recalls a number that is presently stored in memory, but it does not change what is stored.
- The ON key clears memory and display.
- The C/AC key clears the display only.

EXAMPLE

$(52 \times 34) + (312 \div 3) = ?$

Keystrokes	Display	Memory
5 2 × 3 4 =	1768	0
M+	1768 M	1768
C/AC	0 M	1768
3 1 2 ÷ 3 =	104 M	1768
M+	104 M	1872
MR	1872 M	1872
ON	0	0

NOTE: For more calculator practice, turn to page 467.

The FCAT mathematics reference sheet

FCAT includes a "mathematics reference sheet" that you are expected to use while taking the test. This sheet provides many important formulas, such as the formulas for area, surface area, volume, and interest. It also provides a conversion table.

Reference sheets similar to FCAT's appear on pages 11 and 12 of this book.

About this book

This book will help you prepare for the mathematics portion of the grade 10 FCAT. It does not take the place of the mathematics being taught in a regular math class. Instead, the explanations and problems in this book should help you remember the mathematics you have already learned and help you make your skills stronger.

The book begins with a Diagnostic Test, so that you (and your teacher) can assess your skills. The Diagnostic Test is modeled on FCAT. It has the same types of questions as FCAT, at about the same level of difficulty. There is an answer sheet, with bubbles for the multiple-choice problems and grids for the gridded-response problems.

The book has five content chapters, corresponding to the five areas tested by FCAT: (1) Number Sense, Concepts, and Operations, (2) Measurement, (3) Geometry and Spatial Sense, (4) Algebraic Thinking, (5) Data Analysis and Probability.

Each chapter reviews what you need to know. The material is illustrated with many Examples and Model Problems showing step-by-step solutions. "Try These" exercises give you practice in applying these skills. "Assessment" exercises are provided for each major topic.

Each chapter also has "Chapter Review" exercises. Chapters 2, 3, 4, and 5 each end with "Cumulative Review" exercises that will help you keep all your skills sharp as you go along.

Following the content chapters are four sample tests: Tests 1, 2, 3, and 4. Like the Diagnostic Test, these are modeled on FCAT. Each test has an answer sheet with bubbles for the multiple-choice problems and grids for the gridded-response problems.

Finally, there are two Appendixes that review basic arithmetic and calculator skills and give you extra practice.

Good wishes and good luck!

Mathematics Reference Sheet

Area

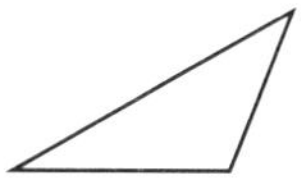	Triangle	$A = \frac{1}{2}bh$
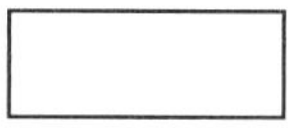	Rectangle	$A = lw$
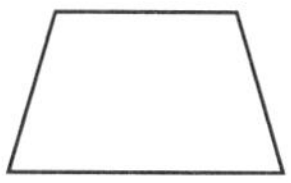	Trapezoid	$A = \frac{1}{2}h(b_1 + b_2)$
	Parallelogram	$A = bh$
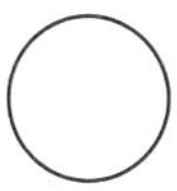	Circle	$A = \pi r^2$

Key

b = base
h = height
l = length
w = width
ℓ = slant height
d = diameter
r = radius

Use $\pi \approx 3.14$ or $\frac{22}{7}$.

Circumference of circle, $C = \pi d = 2\pi r$

		Volume	Total Surface Area
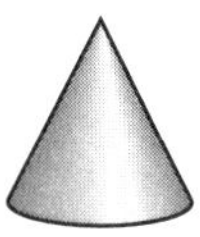	Right circular cone	$V = \frac{1}{3}\pi r^2 h$	$S.A. = \frac{1}{2}(2\pi r)\ell + \pi r^2 = \pi r\ell + \pi r^2$
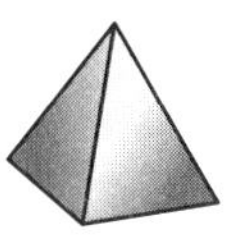	Square pyramid	$V = \frac{1}{3}lwh$	$S.A. = 4\left(\frac{1}{2}l\ell\right) + l^2 = 2l\ell + l^2$
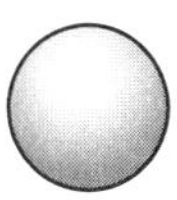	Sphere	$V = \frac{4}{3}\pi r^3$	$S.A. = 4\pi r^2$
	Right circular cylinder	$V = \pi r^2 h$	$S.A. = 2\pi rh + 2\pi r^2$
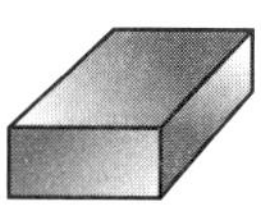	Rectangular solid	$V = lwh$	$S.A. = 2(lw) + 2(hw) + 2(lh)$

In the following formulas, n represents number of sides:

In a polygon, the sum of the measures of the interior angles is equal to $180(n - 2)$.

In a regular polygon, the measure of an interior angle is equal to $\frac{180(n - 2)}{n}$.

Mathematics Reference Sheet

Pythagorean theorem: $c^2 = a^2 + b^2$

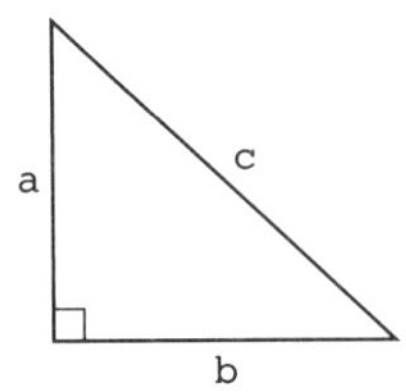

Distance between two points:

$P_1(x_1, y_1)$ and $P_2(x_2, y_2)$:

$$\sqrt{(x_2 - x_1)^2 + (y_2 - y_1)^2}$$

Slope-intercept form of an equation of a line, where m = slope and b = y-intercept:

$$y = mx + b$$

Midpoint between two points:

$P_1(x_1, y_1)$ and $P_2(x_2, y_2)$:

$$\left(\frac{x_2 + x_1}{2}, \frac{y_2 + y_1}{2}\right)$$

Distance, rate, time formula, where d = distance, r = rate, t = time:

$$d = rt$$

Simple interest, $I = prt$,

where p = principal, r = rate, t = time

CONVERSIONS

1 hour = 60 minutes
1 minute = 60 seconds

Customary System

1 yard = 3 feet = 36 inches
1 mile = 1,760 yards = 5,280 feet
1 acre = 43,560 square feet
1 cup = 8 fluid ounces
1 pint = 2 cups
1 quart = 2 pints
1 gallon = 4 quarts
1 pound = 16 ounces
1 ton = 2,000 pounds

Metric System

1 liter = 1000 milliliters = 1000 cubic centimeters
1 meter = 100 centimeters = 1000 millimeters
1 kilometer = 1000 meters
1 gram = 1000 milligrams
1 kilogram = 1000 grams

ANSWER SHEET: DIAGNOSTIC TEST

NOTE: When you answer performance tasks on your own paper, use graph paper if necessary.

PART I

1 Ⓐ Ⓑ Ⓒ Ⓓ

2 Ⓕ Ⓖ Ⓗ Ⓘ

3 Ⓐ Ⓑ Ⓒ Ⓓ

4 Ⓕ Ⓖ Ⓗ Ⓘ

5 Ⓐ Ⓑ Ⓒ Ⓓ

6 Ⓕ Ⓖ Ⓗ Ⓘ

7 Ⓐ Ⓑ Ⓒ Ⓓ

8

9

10 Use your own paper.

11 Ⓕ Ⓖ Ⓗ Ⓘ

12 Ⓐ Ⓑ Ⓒ Ⓓ

13 Ⓕ Ⓖ Ⓗ Ⓘ

14

15

16 Ⓐ Ⓑ Ⓒ Ⓓ

17 Ⓕ Ⓖ Ⓗ Ⓘ

18 Use your own paper.

19 Ⓐ Ⓑ Ⓒ Ⓓ

20 Ⓕ Ⓖ Ⓗ Ⓘ

21 Ⓐ Ⓑ Ⓒ Ⓓ

22

23

24 Use your own paper.

25

26

27

28

29 Ⓕ Ⓖ Ⓗ Ⓘ

30 Ⓐ Ⓑ Ⓒ Ⓓ

PART II

1 Ⓐ Ⓑ Ⓒ Ⓓ

2 Ⓕ Ⓖ Ⓗ Ⓘ

3 Ⓐ Ⓑ Ⓒ Ⓓ

4 Ⓕ Ⓖ Ⓗ Ⓘ

5

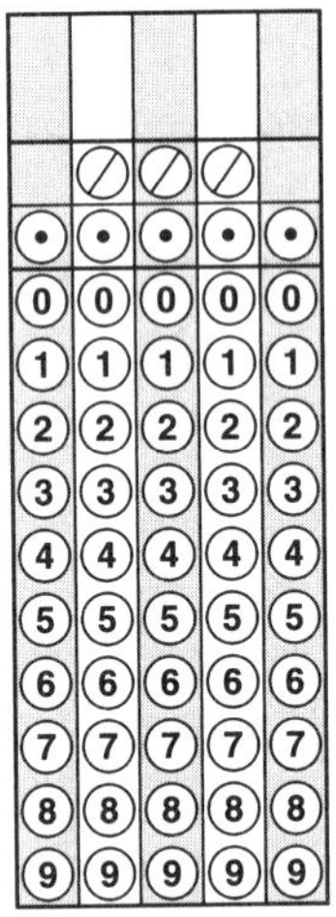

6

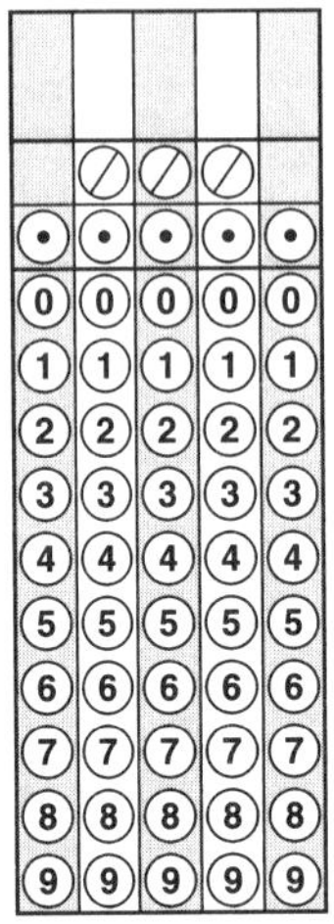

7

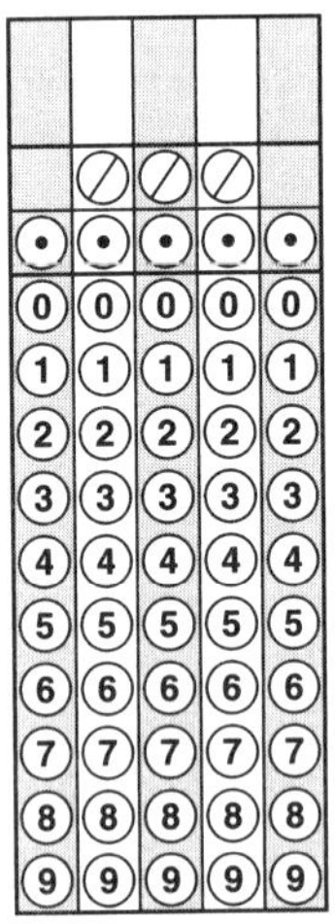

8

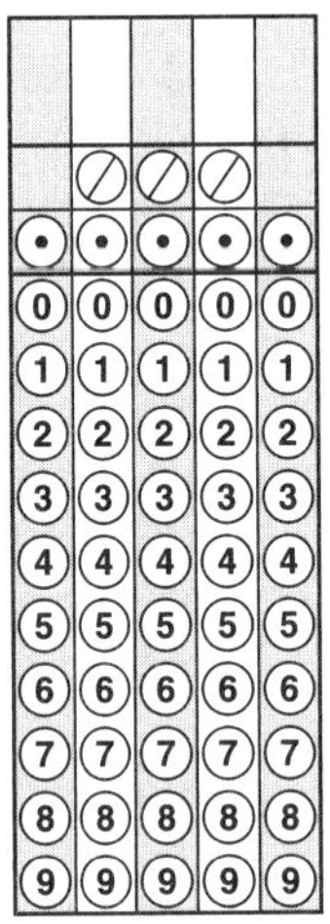

9 Use your own paper.

10 Ⓐ Ⓑ Ⓒ Ⓓ

11 Ⓕ Ⓖ Ⓗ Ⓘ

12 Ⓐ Ⓑ Ⓒ Ⓓ

13 Ⓕ Ⓖ Ⓗ Ⓘ

14 Ⓐ Ⓑ Ⓒ Ⓓ

15 Ⓕ Ⓖ Ⓗ Ⓘ

16

17

18 Use your own paper.

19 Ⓐ Ⓑ Ⓒ Ⓓ

20 Ⓕ Ⓖ Ⓗ Ⓘ

21

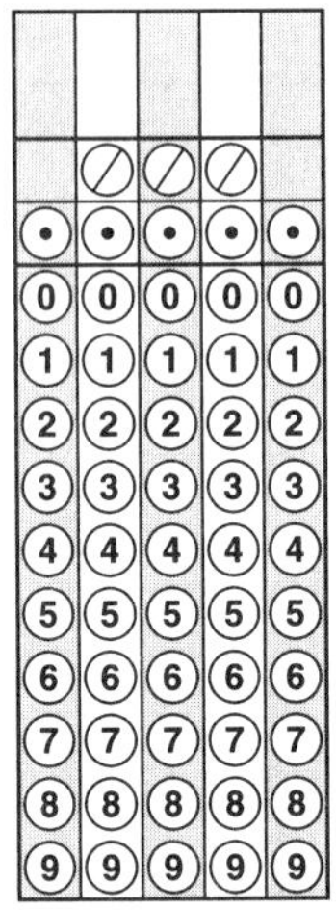

22

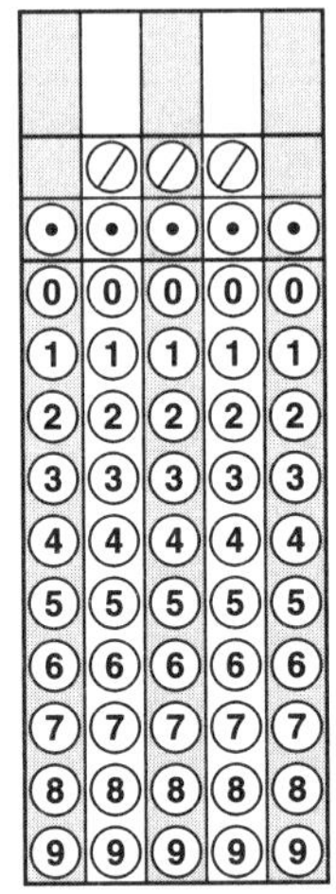

23

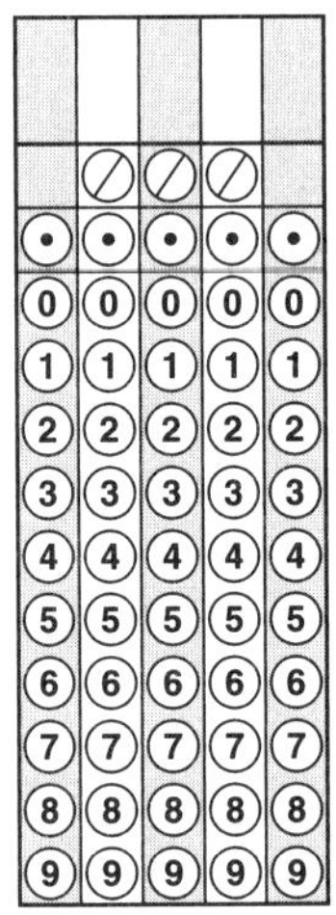

24 Ⓐ Ⓑ Ⓒ Ⓓ

25 Ⓕ Ⓖ Ⓗ Ⓘ

26 Use your own paper.

27

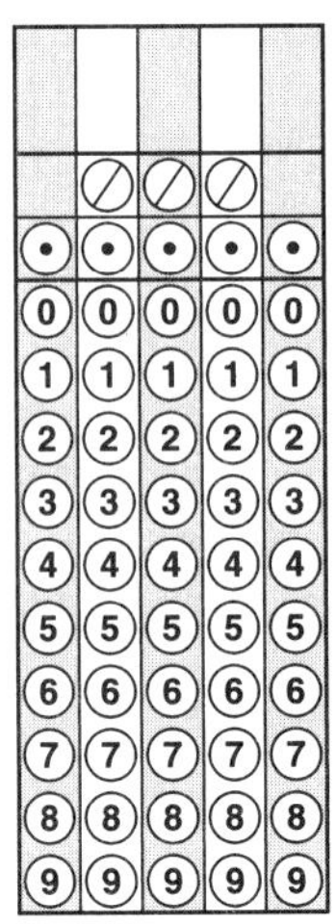

28

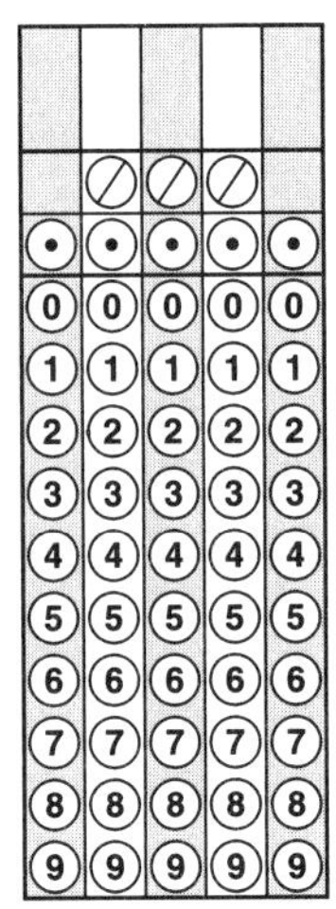

29 Ⓐ Ⓑ Ⓒ Ⓓ

30 Ⓕ Ⓖ Ⓗ Ⓘ

DIAGNOSTIC TEST

PART I

Multiple-Choice Problems

1. Which of the following describes the result when -3 is multiplied by -1?

   A. It results in a number less than -3.

   B. It results in a number greater than 3.

   C. It results in the inverse of -3.

   D. It results in a number between 3 and -3.

2. Adam receives a salary of $400 per week plus a commission of 6% on all his sales. Last week his sales totaled $1,800. What was his salary for the week?

   F. $424 **H.** $1,824

   G. $508 **I.** $2,200

3. In the equation $5y = 8x + 6$, what is the value of y when $x = 3$?

   A. 3.4

   B. 6

   C. 17

   D. 30

4. At the race-car track shown below, people are evenly distributed in the stands surrounding the track. There are approximately 600 people sitting in Section D. Which of the following is closest to the number of people sitting in the stands?

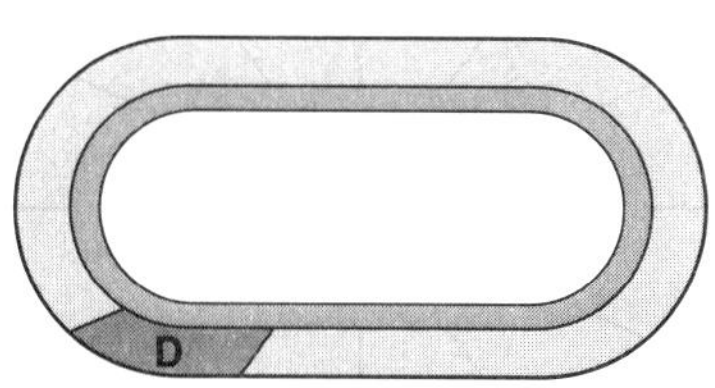

   F. 6,600

   G. 7,200

   H. 7,800

   I. 8,400

5. What is the area of the figure shown below?

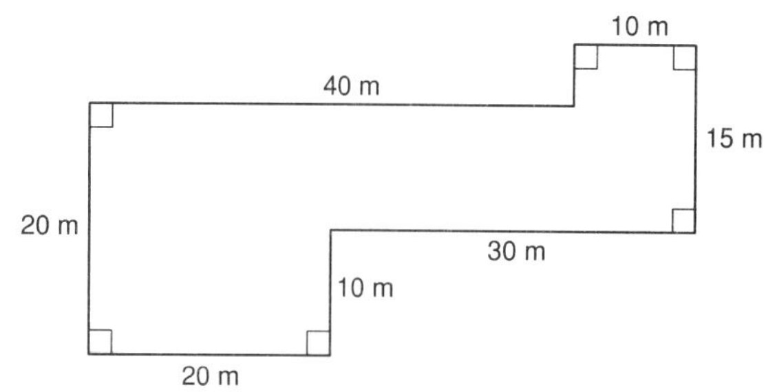

   A. 650 square meters

   B. 700 square meters

   C. 750 square meters

   D. 800 square meters

6. Adrian's lawn mower requires a mixture of 2 ounces of oil for every pint of gasoline. If Adrian fills his lawn mower with 3 gallons of gas, how many pints of oil will he need to add?

   F. 2 pints **H.** 4 pints

   G. 3 pints **I.** 5 pints

7. Maria drove 18 miles in 30 minutes to her cousin's house. What was her average speed in miles per hour?

   A. 18 mph **C.** 36 mph

   B. 27 mph **D.** 45 mph

Gridded-Response Problems

8. The speed of sound is 1.087×10^3 feet per second. How can this number be expressed in standard form?

9. In the drawing below, what is the length of PO?

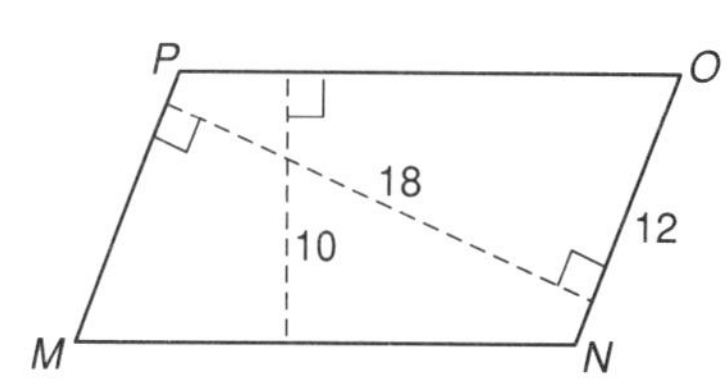

Think, Solve, Explain: Short-Response Problem

10. An inverted square pyramid has a height of 8 inches and a base edge of 12 inches. The pyramid is filled with water.

Part A What is the volume of the water in the pyramid?

Part B The water level drops 2 inches when 222 cubic inches of water are removed. What is the length of a side on the top of the water after this water is removed?

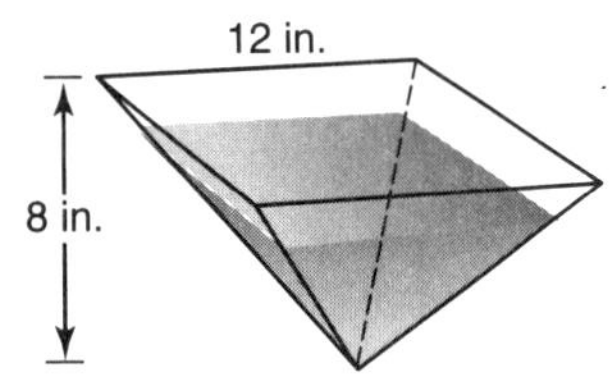

Multiple-Choice Problems

11. How many degrees are in each exterior angle of a regular pentagon?

F. 60° **H.** 108°

G. 72° **I.** 120°

12. In the figure below, $\ell_1 || \ell_2$. If the measure of $\angle 4$ is 135°, how many degrees are in $\angle 8$?

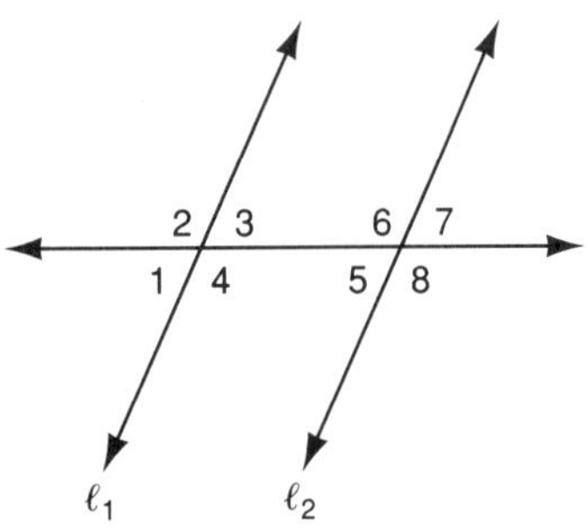

A. 45° **C.** 135°

B. 65° **D.** 165°

13. From figure 1 to figure 2, the letter "A" has gone through what transformations?

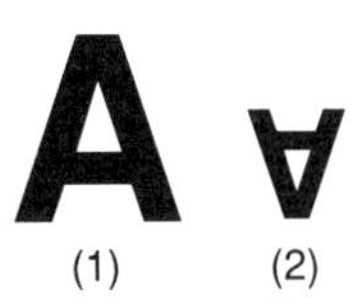

F. reflection, dilation

G. translation, dilation

H. reflection, rotation

I. translation, rotation

Gridded-Response Problems

14. Pablo has a rectangular area in his backyard that he wants to redesign. He intends to concrete half of the area and use the other half as a flower garden. Before pouring the concrete, he puts a brace in to hold the diagonal. The brace meets the diagonal so that one part is 4 feet long and the other part is 9 feet long. How long is the brace?

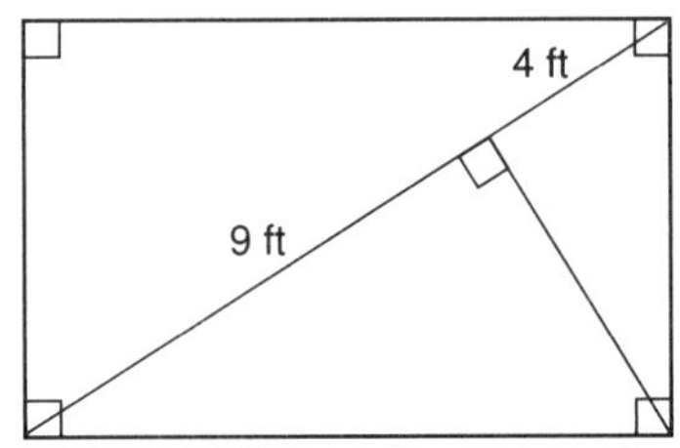

15. If box I is similar to box II, the volume of box II is how many times larger than the volume of box I?

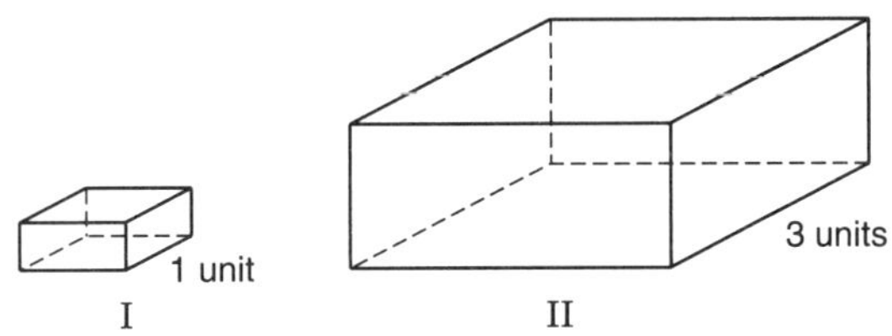

Multiple-Choice Problems

16. During an air show, an airplane flew in low past the grandstand as shown in the graph below. Billy was sitting in the stands 50 feet above the ground. During what time interval was the plane flying below him?

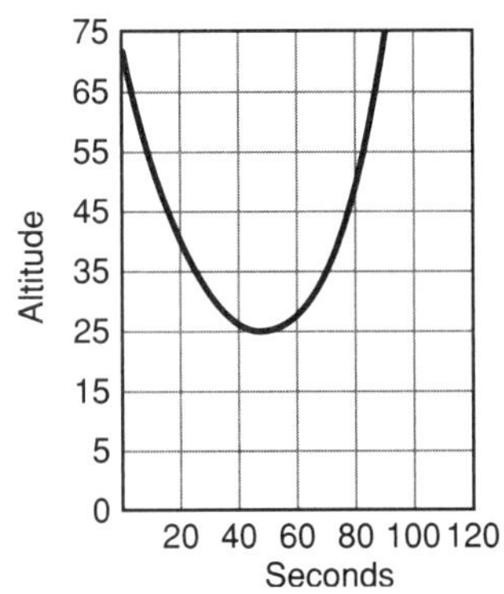

A. between 0 and 15 seconds

B. between 15 and 80 seconds

C. between 80 and 100 seconds

D. The plane never flew below 50 feet.

17. The rectangular box shown below has dimensions of 6 meters by 8 meters by 3 meters. What is the area of the shaded triangle?

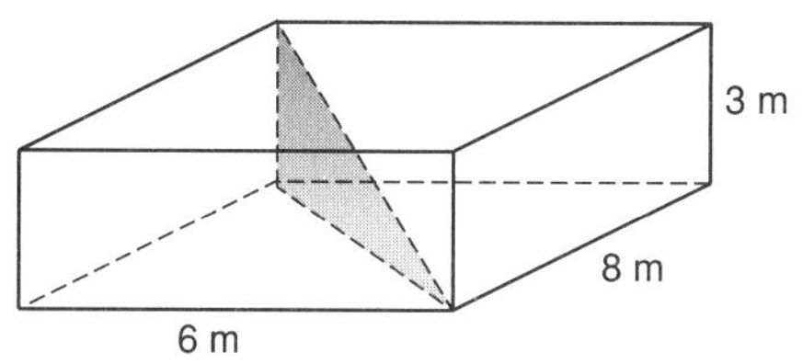

F. 10 square meters

G. 15 square meters

H. 30 square meters

I. 72 square meters

Think, Solve, Explain: Extended-Response Problem

18. Margaret wants to find the distance across a lake from point D to point B. She has drawn triangles with the distances marked and $\angle AED = \angle ACB$.

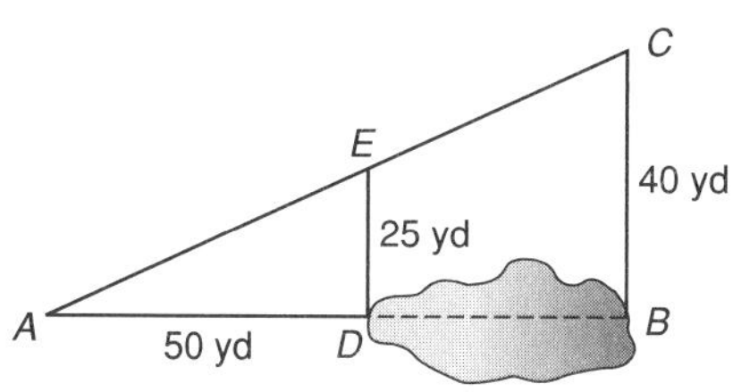

Part A Explain why triangle ADE is similar to triangle ABC.

Part B Write a proportion that can be used to find the distance between B and D.

Part C Solve the proportion showing all the mathematics necessary to determine the distance.

Multiple-Choice Problems

19. Which of the following means five less than the product of two and x?

A. $2 - 5x$

B. $5x - 2$

C. $5 - 2x$

D. $2x - 5$

20. During a routine, the dance team breaks up into small groups. The first group has 2 members, the second group has 5 members, the third group has 8 members, and so forth. If the pattern continues, how many members will be in the sixth group?

F. 11

G. 14

H. 17

I. 20

21. The quantity a is multiplied by $\frac{1}{2}$ five times. The result is what fractional part of the original?

A. $\frac{1}{64}$ **C.** 32

B. $\frac{1}{32}$ **D.** 64

Gridded-Response Problems

22. A function is described by $3c + 8 = d$. What is the missing value in the table?

c	−2	0	1	5	8
d	2	8	11		32

23. Abdul bought 2 sodas (s) and 3 candy bars (c) for \$3.75 while his friend Ferris bought 1 soda and 2 candy bars for \$2.20. Using the system of equations shown below, what is the cost of one candy bar?

$2s + 3c = 3.75$

$1s + 2c = 2.20$

Think, Solve, Explain: Short-Response Problem

24. A farmer has a right circular cylinder silo with a diameter of 5 meters and a height of 40 meters. He wants to build an additional silo 40 meters tall but with a volume 3 times the volume of the other silo. What will be the diameter of the new silo? Show all work necessary. Express your answer to the nearest hundredth.

Gridded-Response Problems

25. In the following equation, what is the value of x?

$$3(9 - 2x) + 8(x + 1) = 46$$

26. In a bag of 20 marbles, 3 are blue, 5 are red, 9 are black, 2 are green, and 1 is yellow. A person reaches into the bag and pulls out one marble. What is the probability that the marble is red?

27. The bar graph shown below displays the lengths of rivers. About how long is the Mekong River?

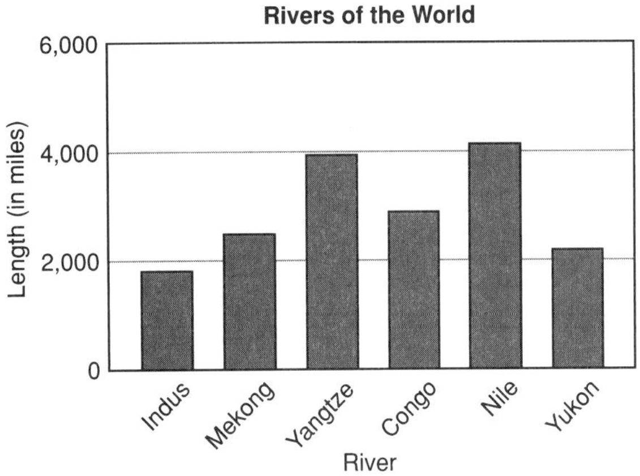

28. Mr. Thompson wants to buy his granddaughter Nina a TV for her room. He found the prices at a local store to be the following: \$98, \$125, \$147, \$163, \$95, \$151, \$134, \$118, and \$139. What is the difference between the median price and the mean price of the TVs?

Multiple-Choice Problems

29. Aaron invented a mop that can be used to clean all kinds of surfaces. Sales of the mop are displayed on the line graph in the next column. If sales continue as in the past, what amount of sales can Aaron expect to receive in 2010?

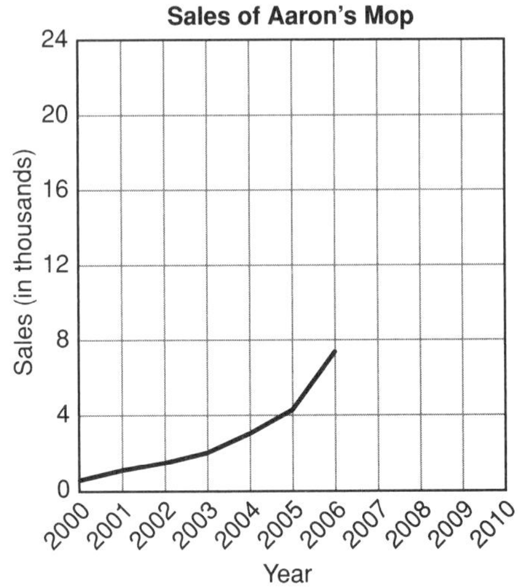

F. \$12,000

G. \$15,000

H. \$19,000

I. \$25,000

30. The table below displays the number of members in 6 different clubs at Cocoanut High School. No student is in more than one club. What percent of the students shown are in the Psychology Club? Round to the nearest percent.

Club	Tally of members
NHS	𝍸 𝍸 𝍸 𝍸 III
Interact	𝍸 𝍸 𝍸 IIII
FCA	𝍸 𝍸 𝍸 I
Psychology	𝍸 𝍸 IIII
Drama	𝍸 𝍸 𝍸 𝍸 𝍸 IIII
Dance	𝍸 𝍸 𝍸 𝍸 II

A. 11%

B. 14%

C. 18%

D. 20%

PART II

Multiple-Choice Problems

1. A 20-foot ladder is leaning against a house in order to reach a second-floor window. Using the Pythagorean theorem it is found that the base of the ladder is $3\sqrt{5}$ feet from the base of the house. Which of the following is closet to this distance?

   A. $6\frac{1}{3}$ feet **C.** $6\frac{2}{3}$ feet

   B. $6\frac{1}{2}$ feet **D.** $6\frac{4}{5}$ feet

2. Between what two numbers does $5\sqrt{6}$ lie?

   F. 5 and 6 **H.** 11 and 12

   G. 10 and 11 **I.** 12 and 13

3. What is the value of $5[2 + 5(4) - 2^2] - 4^2$?

   A. 74 **C.** 98

   B. 90 **D.** 104

4. When a block of ice, x cubic units, is placed in front of a heat source, it shrinks to $\frac{1}{5}$ of its size in 15 minutes. Which of these could be used to determine the size of the block of ice after p 15-minute periods?

   F. $\frac{1}{5}xp$ **H.** $\left(\frac{1}{5}\right)^p x$

   G. $\frac{1}{5}x^p$ **I.** $\left(\frac{1}{5}x\right)^p$

Gridded-Response Problems

5. An appliance store owner sells all stoves at 20% above the original cost. Amita bought a stove for $660. How much did the stove cost the store owner?

6. In the circle below, how many degrees are in $\angle x$?

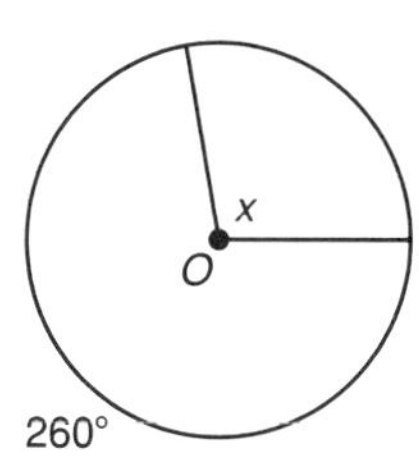

7. A cheetah can run at a speed of 60 mph. How fast is this in feet per second?

8. Barry wants to cover the rectangular box shown below with material. How much material will he need?

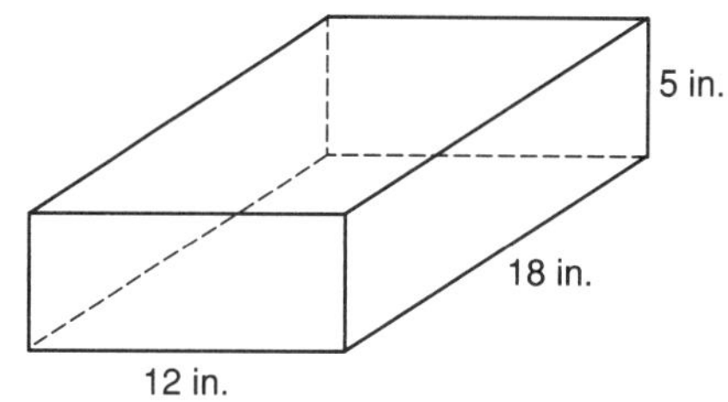

Think, Solve, Explain: Extended-Response Problem

9. The table below shows the enrollment of students in a county from 1970 to 2000.

Year	Number of students
1970	4,500
1980	5,200
1990	6,300
2000	7,200

Part A What is the difference in enrollment from 1970 to 2000?

Part B What is the percent of increase from 1970 to 2000?

Part C Draw a line graph on graph paper displaying the data from the table. Be sure to include a title for the graph, labels for both axes, consistent scales, and the line graph.

Part D Use your graph to find what 10-year interval had the greatest increase in enrollment? How does your graph itself show where the greatest increase occurs?

Part E If the trend continues, how many students would you expect to be enrolled in 2010?

Multiple-Choice Problems

10. An abstract painting has two triangles with a common base of 8 inches. The height of the top triangle is 7 inches and the height of the bottom triangle is twice as long. What is the combined area of the two triangles?

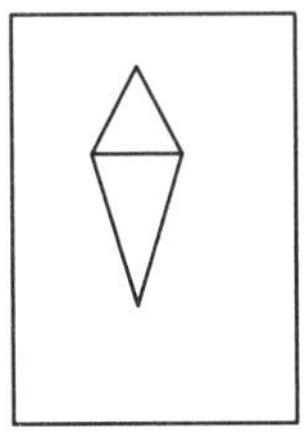

A. 36 square inches

B. 72 square inches

C. 84 square inches

D. 168 square inches

11. While looking at a map of Florida, Terry noticed that the straight-line distances between the cities of Tallahassee, Jacksonville, and Arcadia form a right triangle with the right angle at Jacksonville. The straight-line distance from Tallahassee to Jacksonville is 160 miles. From Jacksonville to Arcadia is 230 miles. What is the approximate straight-line distance between Tallahassee and Arcadia?

F. 390 miles **H.** 270 miles

G. 280 miles **I.** 240 miles

12. The letter "G" is reflected over a vertical line. Which of the following demonstrates this transformation?

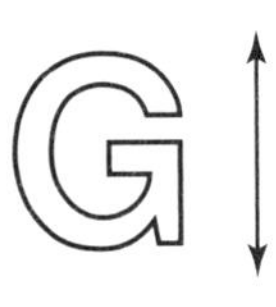

A.  **C.** 

B.  **D.** 

13. In the figure below, *BCDE* is a parallelogram with $\angle ABE = 46°$. How many degrees in $\angle BED$?

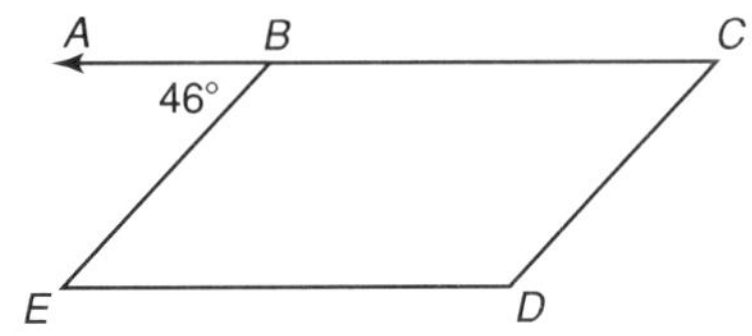

F. 44° **H.** 134°

G. 46° **I.** 154°

14. By what method are the triangles below congruent?

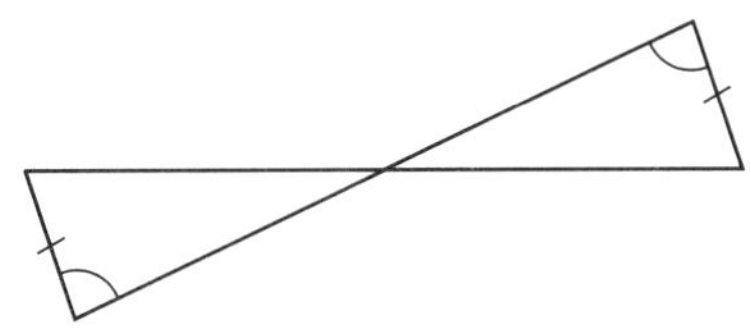

A. AAS **C.** SAS

B. ASA **D.** SSS

15. Given $\angle 1 = \angle 4$ in the figure below. The informal proof lists the reasons needed to establish what type of triangle *ABC* is.

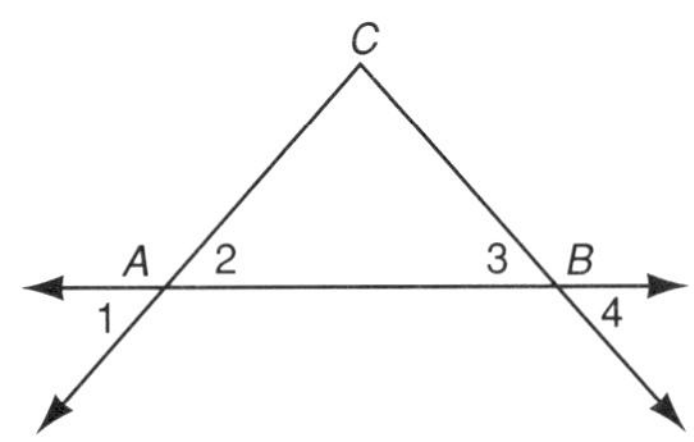

Statements	Reasons
1. $\angle 1 = \angle 4$	1. Given.
2. $\angle 1 = \angle 2$ and $\angle 3 = \angle 4$	2. _______
3. $\angle 2 = \angle 3$	3. Substitute $\angle 2$ and $\angle 3$ for $\angle 1$ and $\angle 4$.
4. $AC = CB$	4. In a triangle, the sides opposite equal angles are equal.
5. Triangle *ABC* is isosceles.	5. An isosceles triangle is a triangle with two equal sides.

Which of the following can be used in the blank space above to justify the statement in step 2?

F. $\angle 1$ and $\angle 2$, and $\angle 3$ and $\angle 4$ are supplementary angles.

G. $\angle 1$ and $\angle 2$, and $\angle 3$ and $\angle 4$ are complementary angles.

H. In an isosceles triangle, the base angles are equal.

I. Vertical angles are equal.

Gridded-Response Problems

16. In the following proportion, what is the value of x?

$$\frac{4}{5} = \frac{10}{x + 3}$$

17. In the figure below, triangle I is similar to triangle II. What is the area of triangle II?

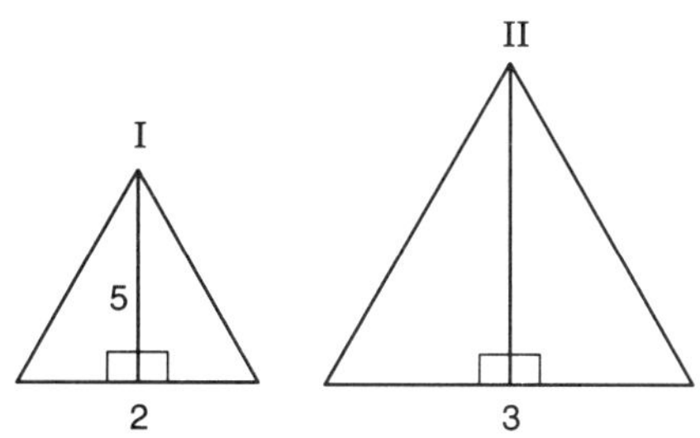

Think, Solve, Explain: Short-Response Problem

18. On the graph below, Juan found the coordinates of the midpoints D and E to be $(-6, -1)$ and $(-2, 2)$.

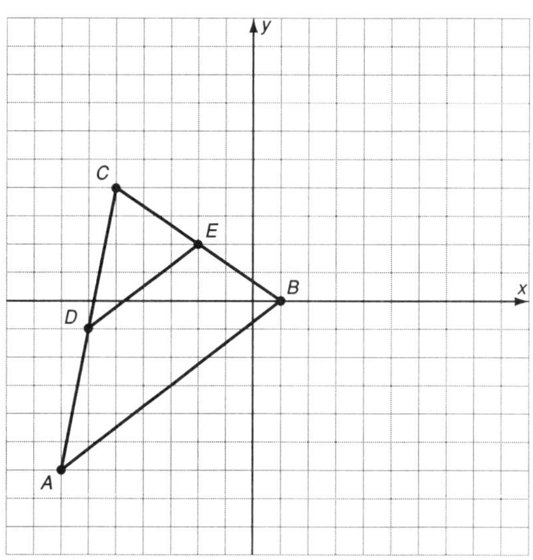

Part A What are the slopes of $\overline{DE}$ and $\overline{AB}$?

Part B What are the lengths of $\overline{DE}$ and $\overline{AB}$?

Part C Use the information obtained in Parts A and B to explain the relationship between $\overline{DE}$ and $\overline{AB}$.

Multiple-Choice Problems

19. In the geometric sequence 200, 100, 50, 25, . . . , what is the 8th term?

A. $\frac{25}{32}$

B. $\frac{25}{16}$

C. $\frac{25}{8}$

D. $\frac{25}{4}$

20. In the function $f(x) = 8x - 7$, what does $f(4)$ equal?

F. -4

G. 5

H. 25

I. 39

Gridded-Response Problems

21. The last row in the right section of the auditorium has 28 seats. As the rows continue to the front of the auditorium, each row has 2 fewer seats than the row behind it. What is the total number of seats in the last 4 rows of this section?

22. What is the largest value of x in this inequality?

$3x + 2 \leq 9$

23. Lydia has a cube with edges of 6 inches each. She makes the length and width half the size and squares the height. How many cubic inches larger is the volume of the rectangular box compared to the original cube?

Multiple-Choice Problems

24. The total weight of two items is less than 10 pounds and greater than 3 pounds. Which region, if any, represents the solution?

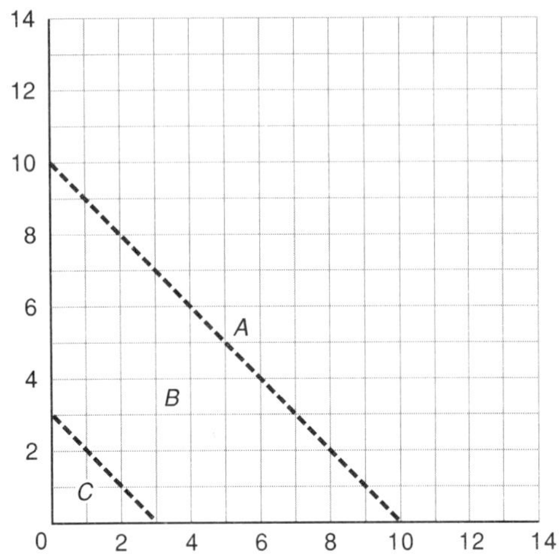

A. Region A

B. Region B

C. Region C

D. No region represents the solution.

25. Tim's Taxi Service charges a base price of \$4.90 plus \$1.00 per extra passenger. There is a charge of \$0.30 for each $\frac{1}{4}$ mile traveled. What is the maximum number of miles that three people can travel if the cost must be \$30 or less?

F. 18.5 miles **H.** 20.1 miles

G. 19.25 miles **I.** 25 miles

Think, Solve, Explain: Short-Response Problem

26. Graciella bought 3 loaves of bread (b) and 1 gallon of milk (m) for \$7.37. Last week, for the same prices, she bought 2 loaves of bread and 2 gallons of milk for \$9.58. Using the following system of equations, what is the cost of 1 loaf of bread and 1 gallon of milk?

$3b + m = 7.37$

$2b + 2m = 9.58$

Show all necessary work.

Gridded-Response Problems

27. Mr. Ibrahim gave a 15-point quiz and recorded the following results: 12, 15, 11, 14, 12, 13, 8, 9, 14, 6, 15, 12, 14. What is the median score of the quizzes?

28. Leroy has 7 dimes, 3 pennies, and 2 nickels in his pocket. He reaches in and takes out 2 coins at the same time. What is the probability that both coins will be dimes?

Multiple-Choice Problems

29. The box-and-whisker plot shown below displays the ages of the people at the Campbell family reunion.

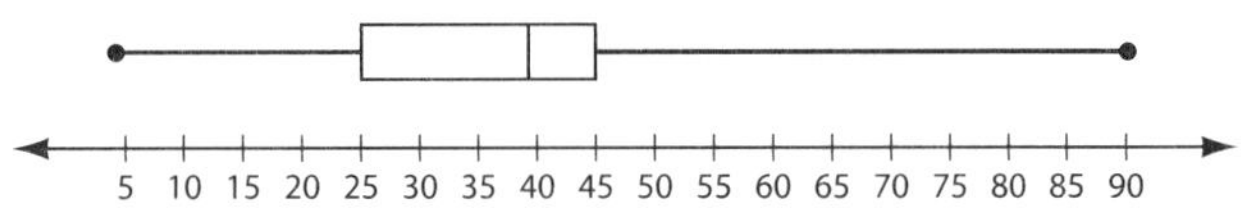

Which statement is NOT supported by the box-and-whisker plot?

A. The range of ages is over 80 years.

B. The median age is 39 years.

C. Fifty percent of the people are over the age of 45.

D. The youngest person is 4 years old.

30. Which statement about the stem-and-leaf plot shown below is true?

Stem	Leaf
2	1 2 3
3	0 2 3 3
4	1 2 2
5	0 0 0 3

Key: $2|1 = 21$

F. The stem-and-leaf plot is bimodal.

G. The median value is not one of the pieces of data found on the plot.

H. The range of data is 30.

I. There are 18 data points on the plot.

CHAPTER ONE

Number Sense, Concepts, and Operations

CHAPTER CONTENTS

Section 1 Representing numbers in the real world

A CLASSIFYING NUMBERS

Numbers are the tools used to represent measurements in our world. The expression of these numbers can take many forms.

Sets of numbers

Sets, or collections, of numbers include the following. In these examples, note that . . . means *and so on, repeating the pattern.*

- **Natural numbers** are also the counting numbers: 1, 2, 3, . . .
- **Whole numbers** are the natural numbers with zero included: 0, 1, 2, 3, . . .
- **Integers** are whole numbers and their negatives: . . . , −3, −2, −1, 0, 1, 2, 3, 4, . . .
- **Rational numbers** are defined as numbers that can be expressed as a ratio of two integers, or $\frac{a}{b}$ where $b \neq 0$. We commonly refer to these as **fractions**. For example: "I'll be home in half an hour, or $\frac{1}{2}$ hour." "The race was won by $\frac{7}{10}$ of a second." A rational number can also be expressed as a **terminating decimal** (such as $\frac{1}{2} = 0.5$) or a **repeating decimal** (such as $\frac{2}{3} = 0.666\ldots$). A repeating decimal, such as 0.777 . . . , can be written as $0.\overline{7}$.
- **Irrational numbers** are numbers that cannot be expressed as the ratio of two integers and cannot be expressed as a terminating or repeating decimal. For example: $\sqrt{2}$ or $\sqrt{7}$ or the value of π. (The symbol π, or *pi*, stands for the ratio of the circumference to the diameter in a circle. Pi is usually rounded to $\frac{22}{7}$ or to the decimal approximation 3.14.)
- **Real numbers** are defined as all rational and irrational numbers.
- **Imaginary numbers** are those based on the imaginary unit $\sqrt{-1}$, usually written as i. The imaginary numbers are i, $2i$, $3i$, . . . (We use the word *imaginary* because the square root of a negative number cannot be calculated in the real number system.)
- **Complex numbers** are defined as numbers that can be written in the form $a + i$ or $a + bi$, where a and b are real numbers and $i = \sqrt{-1}$.

Types of Numbers, with Examples

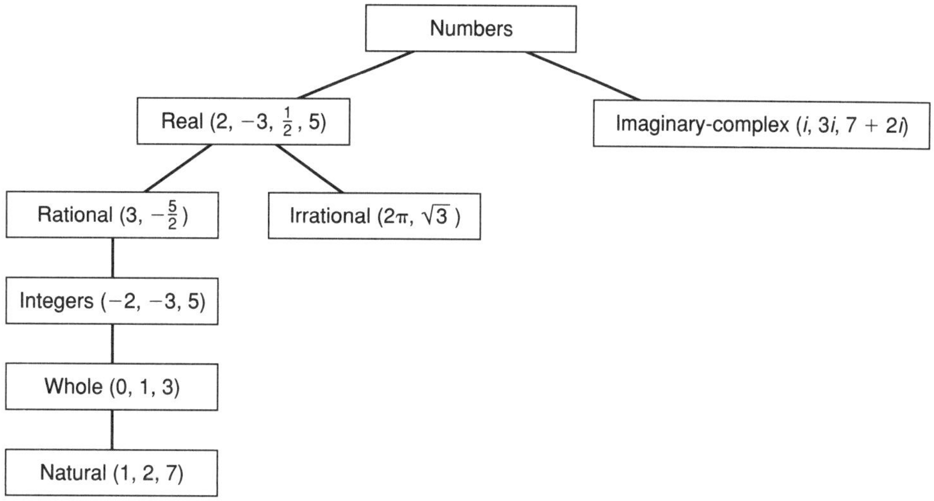

MODEL PROBLEMS

Which term from the chart best describes each of the following numbers?

		Answers:
1.	$-5i$	imaginary-complex
2.	π	irrational
3.	$3\frac{1}{2}$	rational
4.	$-\frac{16}{4}$	integer $\left(\text{because } -\frac{16}{4} = -4\right)$
5.	$\sqrt{4}$	whole or natural number (because $\sqrt{4} = 2$)
6.	$\sqrt{7}$	irrational
7.	$-2 - 3i$	imaginary-complex
8.	-7	integer
9.	$\frac{15}{3}$	whole or natural number $\left(\text{because } \frac{15}{3} = 5\right)$
10.	two-thirds	rational $\left(\frac{2}{3}\right)$
11.	5.6	rational $\left(\text{because } 5.6 = \frac{56}{10} \text{ or } 5\frac{6}{10} \text{ or } \frac{28}{5}\right)$
12.	8.333 . . .	rational $\left(\text{because } 8.333\ldots = 8\frac{1}{3} \text{ or } \frac{25}{3}\right)$

TRY THESE

Which term or terms from the number chart describe each of the following numbers?

1. eighteen-thirds

2. $-\sqrt{9}$

3. 3π

4. $2 + 4i$

5. one and three-fourths

6. $-3\sqrt{13}$

7. $\frac{1}{2}i$

8. -8

9. $\sqrt{144}$

10. $\frac{3}{9}$

11. $5.\overline{6}$

12. $2.\overline{7}$

Place values of numbers

Each digit of a number is said to have a **place value**. Use the diagram below to review the names of the places.

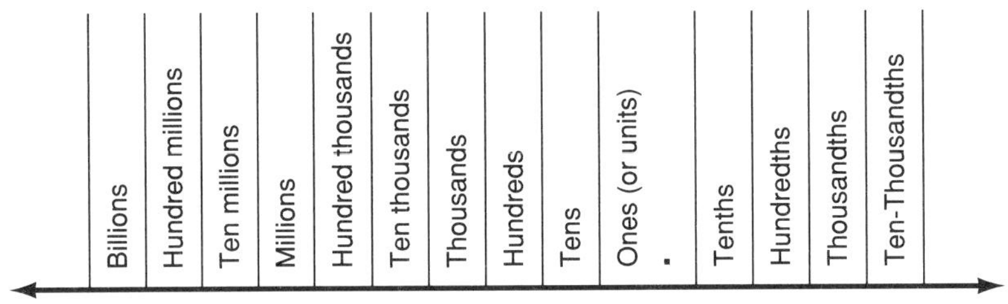

EXAMPLES

748 written in words would be seven hundred forty-eight.

4,000 would be written as four thousand or 4 thousand.

The word *thousand* takes the place of three zeros. Thus 70,000 could be written as 70 thousand.

MODEL PROBLEMS

1. Write 74,300 using the word *thousand*.

   Answer: 74.3 thousand (three places have been replaced)

2. Write 9,000,000 using the word *million*.

   Answer: 9 million (six places have been replaced)

3. Write 9,300,000 using the word *million*.

   Answer: 9.3 million

TRY THESE Write the following numbers using the given word.

1. 4,600 thousand
2. 82,300 thousand
3. 73,000,000 million
4. 6,400,000 million
5. 9,450,000 million
6. 540 hundred
7. 8,600,000,000 billion
8. 12,000,000,000 billion
9. 9,200 thousand
10. 52,800,000 million

Concrete numbers and variables

When a number is known, it can be represented in a **concrete** way. For instance: Suppose that the Simpson family took 10 hours to drive 450 miles. In this example, 10 hours and 450 miles are concrete numbers.

When a number is not known, it is usually represented with a symbol such as a letter. In that case, it is called a **variable**. For example: If d represents the distance the Simpson family travels in 1 hour, then 10 times d, or $10d$, would be the distance traveled in 10 hours. This is referred to as **symbolic representation**.

MODEL PROBLEM

The distance a plane travels can be represented by the product of its rate of speed and the time it travels (in hours). If the rate is 500 miles per hour, express the distance traveled after a certain number of hours.

Answer: Let t stand for time (hours) traveled. Then the distance would equal $500 \times t$ or $500t$.

TRY THESE

1. If a represents the present age of a student, how would the student's age 10 years from now be represented?
2. A wrestler weighs 350 pounds. How could his weight be represented if he loses p pounds?
3. Bill is twice as old as Mary. If Mary's age is represented by m, how will Bill's age be represented?
4. Felecia made n three-point shots in the basketball game. Represent the number of points she scored on n shots.
5. Sonni ran 4 laps around the track in t minutes. Represent the average length of time it took her to run one lap.
6. Amet is h inches tall. Represent his height if he grows 3 inches.
7. Raja is y years old. Represent his age 5 years ago.
8. The total rainfall for a 7-day period is t inches. Represent the average rainfall each day.
9. Beth spent d dollars for dinner and m dollars at the movie theater. Represent the total amount of money spent.
10. Jerome made g touchdowns, 6 points each, during the football season. Represent how many points he scored.

B ORDERING NUMBERS

Ordering integers

Integers are related to each other in an ordered manner. For example, three is less than four and negative seven is greater than negative nine. Using the symbols for *less than* ($<$) and *greater than* ($>$), these statements would be written as $3 < 4$ and $-7 > -9$.

Another way to represent order is to use a number line marked off in integer units. The value of the number increases as the number line is read from left to right. The value decreases as the number line is read from right to left. The numbers are said to be in **consecutive order**:

EXAMPLES

Locate −7 and −9 on the number line. Negative seven is to the right of −9. Therefore, −7 is greater than −9.

Now locate 3 and 4 on the number line. Three is to the left of 4. Therefore, 3 is less than 4.

MODEL PROBLEMS

Complete these relationships by using the symbols for *less than* (<) and *greater than* (>).

		Answers:
1.	0 ◯ 3	$0 < 3$
2.	−5 ◯ −6	$-5 > -6$
3.	2 ◯ −1	$2 > -1$

TRY THESE

Complete the following relationships, by using the symbols for *less than* (<) and *greater than* (>).

1. 9 ◯ 15

2. 8 ◯ −3

3. −5 ◯ −2

4. 0 ◯ −7

5. −3 ◯ −9

6. −4 ◯ 2

7. 1 ◯ 15

8. −10 ◯ −20

9. −5 ◯ −1

10. 7 ◯ −7

Ordering rational numbers

Rational numbers can also be ordered by using *less than* and *greater than*. Every rational number either reduces to an integer or lies between two integers. For example, the rational number $\frac{12}{3}$ reduces to the integer 4 and can be located on the number line. The rational number $\frac{11}{3}$ can be written as a **mixed number**, $3\frac{2}{3}$, and so it would lie between the integers 3 and 4 on the number line.

Using the signs < and >, we would write: $3\frac{2}{3} > 3$ and $3\frac{2}{3} < 4$.

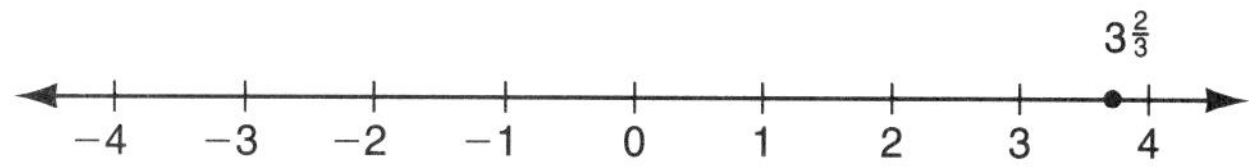

NOTE: $\frac{11}{3}$ can be written as: $3\overline{)11}$ with a quotient of 3 and a remainder of 2 = $3\frac{2}{3}$.

MODEL PROBLEMS

Between what two consecutive integers do the following rational numbers lie?

1. $\frac{4}{3}$

Answer: $\frac{4}{3} = 1\frac{1}{3}$, so it would lie between 1 and 2.

2. $-\frac{5}{2}$

Answer: $-\frac{5}{2} = -2\frac{1}{2}$, so it would lie between −3 and −2.

3. $\frac{7}{8}$

Answer: $\frac{7}{8}$ would lie between 0 and 1.

TRY THESE

Between what two consecutive integers do the following rational numbers lie?

1. $\frac{8}{5}$ **2.** $-\frac{13}{7}$ **3.** $\frac{28}{3}$ **4.** $-\frac{1}{5}$ **5.** $\frac{18}{4}$

6. $\frac{10}{3}$ **7.** $-\frac{29}{5}$ **8.** $\frac{23}{2}$ **9.** $\frac{9}{16}$ **10.** $\frac{-71}{7}$

Ordering irrational numbers

Irrational numbers can be ordered and placed on the number line. Even though a specific point cannot be identified, a close approximation can be found.

For example, $\sqrt{2}$ is irrational. Without using a calculator, we know that $\sqrt{2}$ is between $\sqrt{1}$ and $\sqrt{4}$, which are both integers: $\sqrt{1} = 1$ and $\sqrt{4} = 2$. Therefore, $\sqrt{2}$ will lie somewhere between 1 and 2. And $\sqrt{3}$ would lie to the right of $\sqrt{2}$, because $\sqrt{3} > \sqrt{2}$.

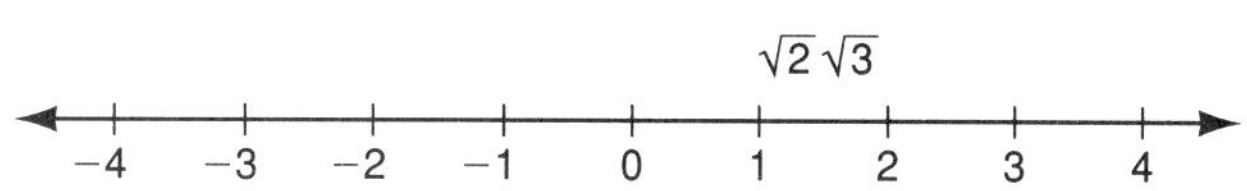

MODEL PROBLEM Between what two integers does $\sqrt{7}$ lie?

Answer: $\sqrt{7}$ lies between $\sqrt{4} = 2$ and $\sqrt{9} = 3$. Therefore, $\sqrt{7}$ lies between 2 and 3.

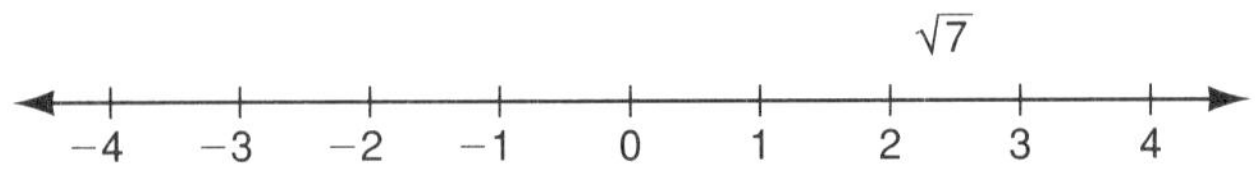

TRY THESE

1. Between what two integers does $\sqrt{19}$ lie?
2. Between what two integers does π lie?
3. Between what two integers does $\sqrt{11}$ lie?
4. Between what two integers does $-\sqrt{12}$ lie?
5. Between what two integers does $2\sqrt{2}$ lie?
6. Between what two integers does $\sqrt{13}$ lie?
7. Between what two integers does 2π lie?
8. Between what two integers does $-\sqrt{5}$ lie?
9. Between what two integers does $\frac{\pi}{2}$ lie?
10. Between what two integers does $-2\sqrt{7}$ lie?

C VARYING REPRESENTATIONS OF NUMBERS

Fractions

Rational numbers, more commonly known as **fractions**, have a numerator and a denominator. The **numerator** is the top number, and the **denominator** is the bottom number.

$$\text{A fraction consists of: } \frac{\text{numerator}}{\text{denominator}}$$

The numerator and denominator can be any integer, with one exception. The denominator can *never* be zero, because in our number system we are not able to divide by zero. For example, for $\frac{6}{0}$, there is no number n so that $n \bullet 0 = 6$. Therefore, we say that division by zero is undefined.

Decimals

Decimal fractions, or **decimals**, are like fractions except that the denominator must be 10 or a power of 10 such as 100 or 1,000. Because of this restriction, decimals are not written with a denominator. Instead, they have a notation called a **decimal point**. For example:

$$0.7 \text{ means } \frac{7}{10} \qquad 0.07 \text{ means } \frac{7}{100} \qquad 0.007 \text{ means } \frac{7}{1{,}000}$$

Places right of decimal point	Denominator	Example	Fraction
1	$10^1 = 10$	0.3	$\frac{3}{10}$
2	$10^2 = 100$	0.31	$\frac{31}{100}$
3	$10^3 = 1{,}000$	0.312	$\frac{312}{1{,}000}$
4	$10^4 = 10{,}000$	0.3124	$\frac{3{,}124}{10{,}000}$

- To change a fraction to a decimal, divide the numerator by the denominator.

EXAMPLE

$$\frac{1}{4} = 0.25$$

$$\begin{array}{r} .25 \\ 4\overline{)1.00} \\ \underline{-8} \\ 20 \\ \underline{-20} \end{array}$$

Now try this on your calculator:

Keystrokes	Display
ON	0
1 ÷	1
4 =	0.25

NOTE: The above calculator operations begin with ON, but in the rest of the calculator examples we *won't* repeat this reminder.

MODEL PROBLEMS

1. Change $2\frac{3}{5}$ to a decimal without using a calculator.

Answer: First change $2\frac{3}{5}$ to a fraction: $2\frac{3}{5} = \frac{(2 \times 5) + 3}{5} = \frac{13}{5}$.

Then divide.

$$\begin{array}{r} 2.6 \\ 5\overline{)13.0} \\ \underline{-10} \\ 30 \\ \underline{-30} \end{array}$$

$2\frac{3}{5} = \frac{13}{5} = 2.6$

2. Using a calculator, change the fraction $\frac{13}{5}$ to a decimal.

Answer:

Keystrokes	Display
1 3 ÷	13
5 =	2.6

TRY THESE

Change the following fractions to decimals (a) with a calculator and (b) without a calculator.

1. $\frac{4}{5}$

2. $-\frac{11}{2}$

3. $7\frac{1}{4}$

4. $\frac{3}{4}$

5. $-\frac{26}{5}$

6. $10\frac{1}{3}$

7. $-\frac{5}{8}$

8. $\frac{43}{9}$

9. $2\frac{2}{5}$

10. $-1\frac{9}{10}$

- To change decimals to fractions, use the appropriate power of ten for the denominator. Then reduce the fraction, if possible.

MODEL PROBLEM

Change 0.35 to a fraction.

Answer: There are two places to the right of the decimal point, so the denominator will be 100.

$0.35 = \frac{35}{100}$

$\frac{35}{100}$ will reduce to $\frac{7}{20}$

TRY THESE

Change the following decimals to fractions and reduce if possible.

1. 0.4
2. 6.75
3. −0.364
4. 0.9
5. 1.25
6. −0.216
7. 0.45
8. 8.4
9. 0.37
10. −2.05

Percents

Another way to express numbers is as percents. **Percent** means "hundredths" or "out of 100" or "per 100." The symbol % means *percent*.

MODEL PROBLEMS

1. Change $\frac{13}{100}$ to a percent.

   Answer: $\frac{13}{100}$ or 0.13 would be 13 percent or 13%

Notice that the decimal point is moved two places to the right and the word *percent* or the sign % is used.

2. Change $\frac{2}{5}$ to a percent.

   Answer: $\frac{2}{5} = \frac{40}{100}$ or 0.40, and this would be 40%

TRY THESE

Change the following decimals to percents.

1. 0.35
2. 0.04
3. 0.536
4. 0.2
5. 1.15
6. 0.63
7. 0.072
8. 10.12
9. 0.9
10. 8.25

- To change a percent to a decimal, move the decimal point two places to the left and remove the percent sign.

EXAMPLE

45% = 0.45

TRY THESE

Fill in the blanks.

Item	Fraction	Decimal	Percent
1.	$\frac{1}{8}$		
2.		0.375	
3.			50%
4.	$\frac{5}{6}$		
5.		0.625	
6.	$\frac{3}{10}$		
7.			42.5%
8.	$\frac{14}{5}$		
9.		0.400	
10.			87.5%

NOTE: For more practice, turn to page 462.

Exponents

Exponents are used to show a form of multiplication. When a number is multiplied by itself, the operation can be written with an **exponent**, and the product is called a **power**. In exponent notation, the exponent indicates the number of times a number is multiplied by itself

For example, 5×5 would be written as 5^2 and $5^2 = 25$. We say that 25 is the second power of 5, or 5 raised to the second power. Similarly, 5^3 means $5 \times 5 \times 5$ or 125. We say that 125 is the third power of 5, or 5 raised to the third power.

NOTE: If a number is raised to the second power, we say it is *squared*. If a number is raised to the third power, we say it is *cubed*.

MODEL PROBLEM Use your calculator to find 6^3.

Answer:

Keystrokes	Display
6 ×	6
6 ×	36
6 ×	216

TRY THESE

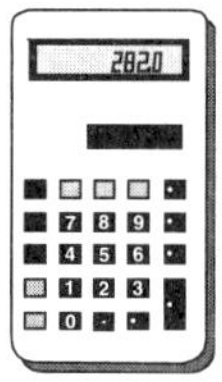

Using your calculator, find each value.

1. 12^2 **2.** 3^4 **3.** 7^3 **4.** 2^5 **5.** 16^2

6. 5^3 **7.** 8^2 **8.** 9^3 **9.** 10^4 **10.** 4^5

Square roots

When we reverse the operation of multiplying a number by itself, the result is a **square root**. The symbol $\sqrt{\ }$ stands for square root.

EXAMPLES

$\sqrt{25} = 5$ because $5^2 = 25$

$\sqrt{100} = 10$ because $10^2 = 100$

MODEL PROBLEM Using your calculator, find $\sqrt{25}$ and $\sqrt{100}$.

Answer:

Keystrokes	Display
[2] [5]	25
[√]	5
[1] [0] [0]	100
[√]	10

When a number is not a perfect square, the calculator will display a decimal approximation. For example, $\sqrt{12}$ on the calculator will be displayed as 3.4641016:

Keystrokes	Display
[1] [2]	12
[√]	3.464101

MODEL PROBLEM Using your calculator, find the approximate decimal value of $3\sqrt{11}$ rounded to three decimal places.

Answer:

Keystrokes	Display
[1] [1]	11
[√]	3.316624
[×] [3]	9.949874

$3\sqrt{11} \approx 9.950$

NOTE: You must find the $\sqrt{\ }$ first and then multiply.

TRY THESE

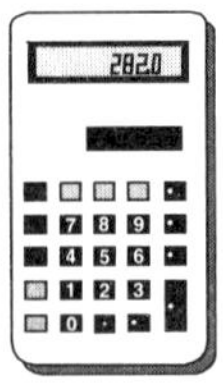

Using your calculator, find each value. Round to three decimal places when needed.

1. $\sqrt{169}$

2. $\sqrt{81}$

3. $3\sqrt{144}$

4. $\sqrt{24}$

5. $2\sqrt{18}$

6. $5\sqrt{3}$

7. $\sqrt{42}$

8. $7\sqrt{121}$

9. $\sqrt{157}$

10. $9\sqrt{2}$

Scientific notation

Scientific notation is used to represent very large or very small numbers in real-world situations. For instance, we could use scientific notation to show \$1,240,000 (a large number) or 0.000045 inch (a small number).

EXAMPLE

In the number 5,740,000, the decimal point is located to the right of the last zero: 5,740,000.

Move the decimal point to the left until it is located between the 5 and the 7: 5.740000

You have now moved the decimal point six places or six powers of ten. The number would now have to be written as: 5.74×10^6

The form above is called **scientific notation**.

In scientific notation, remember:

- When we move the decimal point left or right, only one place is to the left of the new decimal position.
- When we move the decimal point to the right, the power of ten is negative.
- When we move the decimal point to the left, the power of ten is positive.

MODEL PROBLEMS

1. Write 356,240 in scientific notation.

Answer: 356,240 would be written as 3.5624×10^5

2. Write 0.00034 in scientific notation.

Answer: 0.00034 would be written as 3.4×10^{-4}

TRY THESE

Write the following numbers in scientific notation.

1. 0.0289

2. 0.000061

3. 0.0005

4. 3,257

5. 0.0008

6. 0.0056

7. 3,000,000

8. 0.0178

9. 56.23

10. 806,000,000

Absolute value

Absolute value is used when a nonnegative number is required. The symbol for absolute value is $|\ |$. Thus $|x|$ is x if $x \geq 0$ and $-x$ if $x < 0$.

EXAMPLES

The absolute value of 5 is 5 or: $|5| = 5$

The absolute value of -5 is 5 or: $|-5| = -(-5) = 5$

MODEL PROBLEMS

		Answers:
1.	$\|3 + 4\| = \|7\| =$	7
2.	$\|3 - 4\| = \|-1\| =$	1
3.	$\|5 \times 2\| = \|10\| =$	10
4.	$\|-5 \times 2\| = \|-10\| =$	10
5.	$3\|4\| = 3 \times 4 =$	12
6.	$-3\|4\| = -3 \times 4 =$	-12
7.	$3\|-4\| = 3 \times 4 =$	12
8.	$-3\|-4\| = -3 \times 4 =$	-12

TRY THESE

Evaluate.

1. $|8|$

2. $|-8|$

3. $-|6|$

4. $-|-6|$

5. $2|3|$

6. $2|-3|$

7. $-4|4|$

8. $-4|-4|$

9. $4 + |-2|$

10. $4 - |-2|$

11. $|4 - 7|$

12. $|-6 \times 2|$

Assessment

Multiple-Choice Problems

1. The Federal Reserve raised the prime interest rates by 0.4%. How would this percent be expressed as a decimal?

   A. 0.004 C. 0.4

   B. 0.04 D. 40

2. A support for a roof truss is $2\sqrt{5}$ feet long.

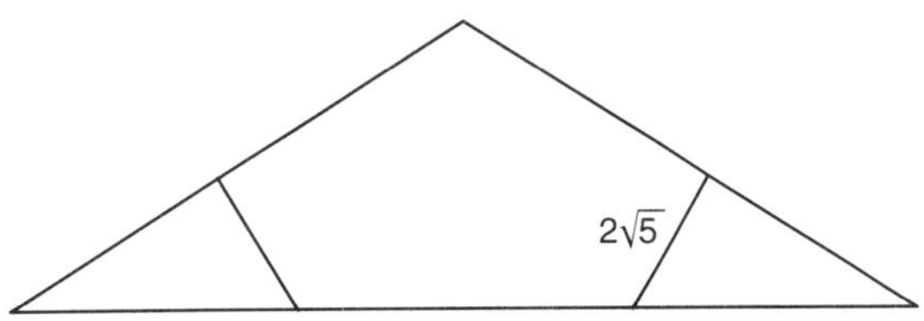

   Which of the following is the same measurement?

   A. $5\sqrt{2}$ feet C. $\sqrt{20}$ feet

   B. $\sqrt{10}$ feet D. 10 feet

3. At its farthest point of orbit, Mercury is 69 800 000 kilometers from the sun. What is the scientific notation for this distance?

   A. 69.8×10^6 C. 6.98×10^6

   B. 69.8×10^7 D. 6.98×10^7

4. To celebrate Valentine's Day, the 24 members of the dance club decided to have a party. Which of the following represents the cost per member if the expenses for the party totaled E dollars?

   A. $24E$ C. $E + 24$

   B. $24 - E$ D. $E \div 24$

5. Between what two integers does 6.23×10^3 lie?

   A. 6 and 7

   B. 62 and 63

   C. 622 and 624

   D. 6,229 and 6,231

6. A contractor is building a rectangular deck beside the house. To make sure that the sides are perpendicular, he places a diagonal brace between the two sides. He uses the Pythagorean theorem to determine that the brace is $4\sqrt{6}$ feet long.

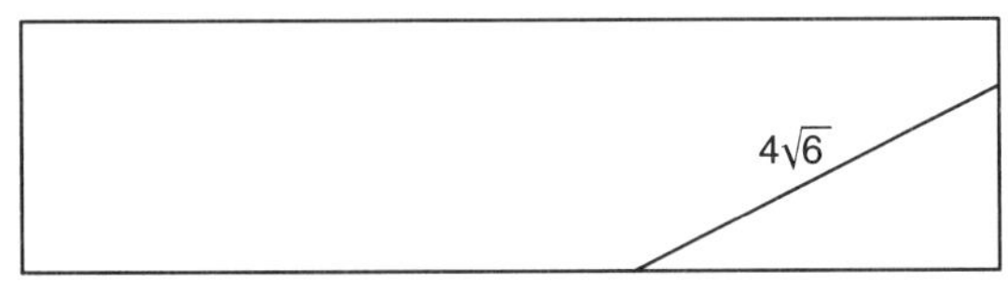

   Which of the following is closest to the length of the brace?

   A. $9\frac{3}{4}$ feet C. $9\frac{7}{9}$ feet

   B. $9\frac{4}{5}$ feet D. $9\frac{10}{13}$ feet

7. A certain number of bacteria, x, doubles in size every 3 hours. Which of these expressions could be used to determine the number of bacteria in the sample after h three-hour periods?

   A. $2^h x$ C. $2x^h$

   B. $2xh$ D. $(2x)^h$

8. There are approximately 1.4 million acres in Everglades National Park. Which of the following could be the actual size of the park?

   A. 1,400,000,000 acres

   B. 1,399,078 acres

   C. 1,321,962 acres

   D. 1,400 acres

9. If $x = \sqrt{36y}$, which value of y will give a rational value for x?

   A. $\frac{1}{2}$ C. $\frac{1}{4}$

   B. $\frac{1}{3}$ D. $\frac{1}{6}$

Gridded-Response Problems

10. The angstrom unit, used by scientists, is equivalent to 1×10^{-8} centimeter. A certain sheet of paper is 1,000,000 angstroms thick. How many centimeters, written in standard form, is the thickness of the sheet of paper?

11. What number is equivalent to the expression $2|5|-3|-2|$?

10

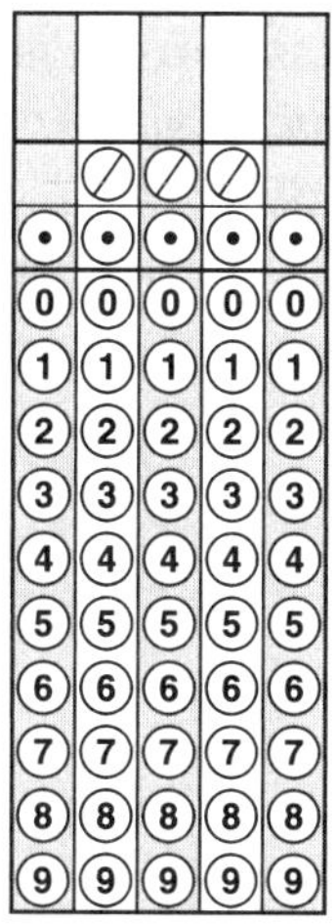

11

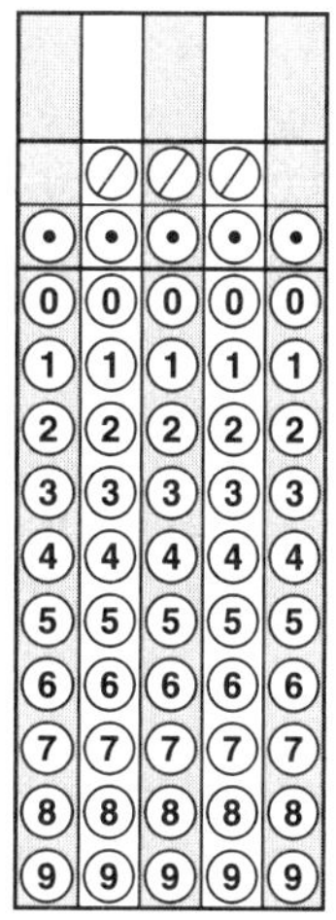

Section 2 Understanding basic operations with numbers

In our number system there are four basic operations: addition (+), subtraction (−), multiplication, (×) and division (÷). All the types of numbers reviewed in Section 1 are operated on by these four basic operations.

A OPERATIONS WITH INTEGERS

Addition is a combining of two or more numbers. The numbers being added are called **addends**, and the result is called the **sum**.

When you are adding numbers, be sure to notice whether you are adding a positive or a negative number. A negative number will move you to the left on the number line. A positive number will move you to the right.

EXAMPLES

$7 + 3 = 10$

$-5 + (-3) = -8$

$-4 + 2 = -2$

$6 + (-1) = 5$

Here are the rules for addition:

- To add two numbers with like signs, add the absolute values of the numbers. The sum will have the sign of the numbers being added.
- To add two numbers with unlike signs, subtract their absolute values. The sum will have the sign of the number with the greater absolute value.

When zero is added to any number, the result is the number itself. For example:

$7 + 0 = 7$

$-3 + 0 = -3$

$\sqrt{2} + 0 = \sqrt{2}$

$x^2 + 0 = x^2$

In other words, when zero is added to a number, the result is the identical number. Therefore, zero is called the **additive identity**.

When two numbers added together have the sum of zero, the numbers are said to be **additive inverses** of each other. For example, 4 and −4 are additive inverses of each other, and −10 and 10 are additive inverses of each other:

$4 + (-4) = 0 \qquad -10 + 10 = 0$

Subtraction is the opposite of addition. On the number line, subtraction will move you in the opposite direction from addition. The result of subtraction is called the **difference**.

EXAMPLES

$7 - 3 = 4$

$8 - (-2) = 10$

$-7 - 3 = -10$

$-6 - (-2) = -4$

To subtract signed numbers, change the sign of the number being subtracted and follow the rules for addition.

Subtracting zero (the additive identity) from a number gives you the same number. For example, $9 - 0 = 9$.

NOTE: For more practice on addition and subtraction, turn to page 459.

Multiplication is a shortcut method of doing addition. The number being multiplied is called the **multiplicand**. The number doing the multiplying is called the **multiplier**. The result of multiplication is called the **product**.

The following symbols are used to indicate multiplication:

Times sign	5×3
Raised dot	$5 \bullet 3$
Parentheses	5(3) or (5)3 or (5)(3)

When we use variables:

$5y$ means 5 times y

ab means a times b

Any variable by itself, such as k, means 1 times k or k times 1.

Here are the rules for multiplication with positive and negative numbers:

Rule	**Example**
$+ \bullet + = +$	$5 \bullet 4 = 20$
$+ \bullet - = -$	$5 \bullet (-4) = -20$
$- \bullet + = -$	$-5 \bullet 4 = -20$
$- \bullet - = +$	$-5 \bullet (-4) = 20$

When the signs are the same, the product is positive.
When the signs are different, the product is negative.

When zero is multiplied by any number, the result is zero. For example:

$3 \bullet 0 = 0$ $\quad -8 \bullet 0 = 0$ $\quad a \bullet 0 = 0$

When 1 is multiplied by any number except zero, the result is the number itself, the identical number. Therefore, 1 is called the **multiplicative identity**. For example:

$10 \bullet 1 = 10$

$-4 \bullet 1 = -4$

$2\sqrt{7} \bullet 1 = 2\sqrt{7}$

$k \bullet 1 = k$

MODEL PROBLEMS

		Answers:
1.	$-8 \bullet 5 =$	-40
2.	$3 \bullet -2 =$	-6
3.	$-9 \bullet -5 =$	45
4.	$10 \bullet 0 =$	0

Division is the operation in which a number is broken down into smaller, identical groups. It is the opposite of multiplication. The number being divided is called the **dividend**. The number doing the dividing is called the **divisor**. The result of division is called the **quotient**.

There are several symbols for division:

Division sign	$8 \div 2$
Fraction bar	$\frac{8}{2}$ or 8/2
Long division symbol	$2\overline{)8}$

The rules for division of signed numbers are the same as the rules for multiplication:

Rule	Example
$+ \div + = +$	$10 \div 2 = 5$
$+ \div - = -$	$10 \div -2 = -5$
$\frac{+}{-} = -$	$\frac{10}{-2} = -5$
$\frac{-}{+} = -$	$\frac{-10}{2} = -5$
$\frac{-}{-} = +$	$\frac{-10}{-2} = 5$

When the signs are the same, the quotient is positive.
When the signs are different, the quotient is negative.

When zero is divided by any number, except zero, the result is zero. For example:

$$\frac{0}{3} = 0$$

$$\frac{0}{-2} = 0$$

$$\frac{0}{a} = 0 \quad \text{when } a \neq 0$$

Numbers cannot be divided by zero in our number system. Any number divided by 0 is **undefined**. For example, $\frac{7}{0}$ is undefined because there is no number n such that $n \bullet 0 = 7$.

NOTE: For more practice on multiplication and division, turn to page 460.

MODEL PROBLEMS

		Answers:
1.	$-14 \div -2 =$	7
2.	$50 \div -10 =$	-5
3.	$-\frac{28}{4} =$	-7
4.	$-\frac{14}{0} =$	undefined
5.	$\frac{0}{32} =$	0

The **reciprocal** of any number except zero is the number inverted. To **invert** a number, first express the number as a fraction. If the number is an integer, the denominator is 1. Then, simply turn the fraction upside down. The reciprocal of a number (except zero) is called the **multiplicative inverse**.

MODEL PROBLEMS

Write the reciprocal of each number.

Number	*First step:* Fraction	*Answer:* Reciprocal
5	$\frac{5}{1}$	$\frac{1}{5}$
$\frac{1}{3}$	(already a fraction)	$\frac{3}{1}$ or 3
$-\frac{5}{12}$	(already a fraction)	$-\frac{12}{5}$ or $-2\frac{2}{5}$
5.3	$5\frac{3}{10} = \frac{53}{10}$	$\frac{10}{53}$

When a number is multiplied by its reciprocal, the product is 1. For example:

$$\frac{5}{1} \bullet \frac{1}{5} = \frac{5}{5} = 1 \qquad -\frac{2}{3} \bullet -\frac{3}{2} = \frac{6}{6} = 1$$

TRY THESE

1. $9 + 35 + 124$
2. $96 + (-28)$
3. $128 - 97$
4. 56×29
5. $-5{,}561 - 83$
6. $0 - 75$
7. $\frac{23}{0}$
8. What is the reciprocal of $-\frac{5}{6}$?
9. What is the reciprocal of 8?
10. $\frac{0}{-5}$

B OPERATIONS WITH FRACTIONS

Now we will apply the four basic operations to fractions.

When fractions are added, they must have the same denominator, called their **common denominator**. If the denominators are not alike, the fractions must be changed to equivalent fractions with like denominators before addition can take place. To add fractions, add the numerators and keep the common denominator the same. Reduce the sum if possible.

MODEL PROBLEMS

1. $\frac{3}{16} + \frac{4}{16} = ?$

   Answer: $\frac{7}{16}$

2. $\frac{2}{9} + \frac{4}{9} = ?$

   Answer: $\frac{6}{9}$, which reduces to $\frac{2}{3}$

3. $\frac{1}{8} + \frac{3}{16} = ?$

   Answer: Find a common denominator: $\frac{1}{8} = \frac{2}{16}$

   Add: $\frac{2}{16} + \frac{3}{16} = \frac{5}{16}$

4. $\frac{2}{3} + \frac{1}{4} = ?$

   Answer: Find a common denominator: $\frac{2}{3} = \frac{8}{12}$ and $\frac{1}{4} = \frac{3}{12}$

   Add: $\frac{8}{12} + \frac{3}{12} = \frac{11}{12}$

In subtracting fractions, the fractions must have the same denominator. Subtract the numerators and keep the common denominator.

MODEL PROBLEMS

1. $\frac{7}{8} - \frac{2}{8} = ?$

   Answer: $\frac{5}{8}$

2. $\frac{9}{11} - \frac{3}{22} = ?$

Answer: Find a common denominator: $\frac{9}{11} = \frac{18}{22}$

Subtract: $\frac{18}{22} - \frac{3}{22} = \frac{15}{22}$

NOTE: See page 461 for practice on adding and subtracting fractions.

When fractions are multiplied, there is no need for a common denominator, because you are finding a part of another fraction. Multiply numerators and multiply denominators. Reduce the product if possible.

MODEL PROBLEMS

1. $\frac{2}{3} \bullet \frac{4}{5} = ?$

Answer: $\frac{2}{3} \bullet \frac{4}{5} = \frac{2 \bullet 4}{3 \bullet 5} = \frac{8}{15}$

2. $\frac{-1}{4} \bullet \frac{8}{7} = ?$

Answer: $\frac{-1}{4} \bullet \frac{8}{7} = \frac{-1 \bullet 8}{4 \bullet 7} = \frac{-8}{28}$, which will reduce to $\frac{-2}{7}$ or $-\frac{2}{7}$

When fractions are divided, a common denominator is not necessary. Since division is the opposite of multiplication, invert the second fraction, which is the divisor. Then multiply. For example:

$$\frac{a}{b} \div \frac{e}{d} = \frac{a}{b} \bullet \frac{d}{e} = \frac{a \bullet d}{b \bullet e} \quad \text{or} \quad \frac{ad}{be}$$

After you multiply, reduce the product if possible.

MODEL PROBLEMS

1. $\frac{5}{6} \div \frac{3}{2} = ?$

Answer: $\frac{5}{6} \div \frac{3}{2} = \frac{5}{6} \bullet \frac{2}{3} = \frac{10}{18}$, which will reduce to $\frac{5}{9}$

2. $8 \div \frac{1}{2} = ?$

Answer: $8 \div \frac{1}{2} = 8 \bullet \frac{2}{1} = 16$

3. $\frac{7}{9} \div 2 = ?$

Answer: $\frac{7}{9} \div 2 = \frac{7}{9} \bullet \frac{1}{2} = \frac{7}{18}$

NOTE: For more practice on multiplying and dividing fractions, turn to page 461.

TRY THESE

1. $\frac{2}{3} + \frac{5}{3}$
2. $\frac{3}{7} - \frac{2}{3}$
3. $\frac{2}{3} \bullet \frac{3}{5} \bullet \frac{4}{7} \bullet \frac{1}{2}$
4. $-2\frac{3}{5} - (-3)$
5. $\frac{14}{25} \div \frac{7}{5}$
6. $\frac{4}{5} + \frac{3}{7}$
7. $-1\frac{2}{3} - \frac{1}{6}$
8. $2\frac{2}{3} \times 3\frac{5}{7}$
9. $\frac{3}{4} \div \frac{5}{8}$
10. $3\frac{2}{3} \div 1\frac{1}{2}$

C OPERATIONS WITH DECIMALS

Now we'll apply the basic operations to decimal fractions, or decimals. In the addition or subtraction of decimals, the decimal points must be lined up.

MODEL PROBLEMS

1. 5.32 + 76.4 = ?

Answer:
$$\begin{array}{r} 5.32 \\ +\ \underline{76.40} \\ 81.72 \end{array}$$

NOTE: 76.4 = 76.40

2. 7 + 32.9 = ?

Answer:
$$\begin{array}{r} 32.9 \\ +\ \underline{7.0} \\ 39.9 \end{array}$$

3. 85.326 − 4.1 = ?

Answer:
$$\begin{array}{r} 85.326 \\ -\ \underline{4.100} \\ 81.226 \end{array}$$

4. 63 − 4.28 = ?

Answer:
$$\begin{array}{r} 63.00 \\ -\ \underline{4.28} \\ 58.72 \end{array}$$

When decimals are multiplied, it is *not* necessary to line up the decimal points. Follow these steps:

- *First step.* Set up the problem as you would with whole numbers, and then multiply.
- *Second step.* Count the total number of decimal places to the right of the decimal point in each part of the problem (that is, in each number you are multiplying).

- *Third step*. Starting from the right in the answer (the product), count the total number of decimal places you found in the second step. This is where you put the decimal point in the answer.

MODEL PROBLEM $43.27 \bullet 5.2 = ?$

Answer:

$$\begin{array}{rl} 43.27 & \quad 2 \text{ places} \\ \times\ \underline{5.2} & \quad +\ \underline{1 \text{ place}} \\ 8654 & \\ \underline{21635} & \\ 225.004 & \quad 3 \text{ places} \end{array}$$

In dividing decimals, first move the decimal point in the divisor to the right until the divisor is a whole number. Next, because you have changed the divisor, you must move the decimal point in the dividend to the right the same number of places as you moved it in the divisor. Then, divide.

MODEL PROBLEM $13.44 \div 0.12 = ?$

Answer: $.12.\overline{)13.44.}$

move 2 places   move 2 places

$$\begin{array}{r} 112. \\ 12\overline{)1344.} \\ \underline{12} \\ 14 \\ \underline{12} \\ 24 \\ \underline{24} \\ 0 \end{array}$$

TRY THESE

1. $-12.63 + (-8.2) + 5.9$
2. $7.34 - 12$
3. $-318.1 - (-493.02)$
4. $9.3 \bullet 14.8$
5. $52.48 \div 16.4$
6. $9.4 - 5.23 + 7.1$
7. $21.35 - 42$
8. $-61.2 - (-12.3)$
9. $7.2 \bullet 8.23$
10. $22.36 \div 5.2$

Your calculator can be a great help with operations on decimals. Here are some problems done on the calculator.

Problem	Keystrokes	Display
$5.623 + 1.84 = ?$	5 . 6 2 3 +	5.623
	1 . 8 4 =	7.463
$0.932 - 0.04 = ?$	0 . 9 3 2 −	.932
	0 . 0 4 =	.892
$53.6 \bullet 15.72 = ?$	5 3 . 6 ×	53.6
	1 5 . 7 2 =	842.592
$12.63 \div 1.5 = ?$	1 2 . 6 3 ÷	12.63
	1 . 5 =	8.42

TRY THESE

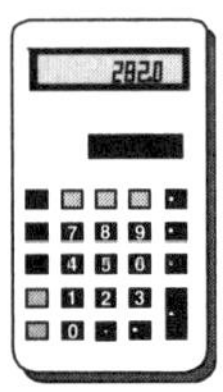

Use your calculator:

1. $17.3 + 12.25 + 26 + 3.59$
2. $4.35 - (-16.2)$
3. $81.7 - 59.87$
4. $3.24 \bullet 91.5$
5. $593.64 \div 87.3$
6. $14.17 + (-3.2) - 4.63$
7. $12.92 - 56$
8. $2.7 \bullet 32.5$
9. $212.52 \div 25.3$
10. $6.375 \div 0.125$

D OPERATIONS WITH PERCENTS

The following section gives examples of various kinds of percentage problems.

REMEMBER: **Percent** means "per hundred" or "out of 100."

Simple interest

One of the most common ways to save money is to open a savings account in a bank. Money is deposited in the account and left there to accumulate **interest**—the money the bank pays you for the use of the money you deposited. The rate of interest is usually expressed as a percent.

In problems involving interest, three important terms are used:

- **Principal (*p*)** is money deposited.
- **Rate (*r*)** of interest is a percentage determined by the bank.
- **Time (*t*)** is the period of the investment (the deposit), measured in years. If time is given as months, we need to convert to years.

You can use a **formula** to find the amount of simple interest:
Interest = principal × rate × time
$I = prt$

NOTE: Money is rounded to the nearest cent.

MODEL PROBLEMS

1. Lucas deposits $200 in a savings account at 3.5 percent interest for 1 year. At the end of 1 year, how much interest has he earned?

   Answer: Use the formula for simple interest. Convert the interest rate from a percent to a decimal.

   $I = prt$

   $I = \$200 \times 3.5\% \times 1$

   $I = \$200 \times 0.035 \times 1$

   $I = \$7.00$

2. How much interest does Lucas earn in 3 months?

   Answer: Use the formula, converting 3 months to $\frac{3}{12} = 0.25$ year.

   $I = prt$

   $I = \$200 \times 0.035 \times 0.25$

   $I = \$1.75$

Borrowing money

We can apply the same formula, $I = prt$, when someone is borrowing money at a simple rate of interest. The borrower owes the lender the amount of the loan plus the interest.

MODEL PROBLEM

Deandra borrows $3,000 from the bank for a down payment on a car. The bank charges 7 percent simple interest annually. How much interest will Deandra owe at the end of 2 years? What is the total amount she will owe the bank?

Answer: First, find the interest.

$I = prt$

$I = \$3{,}000 \times 0.07 \times 2$

$I = \$420$ interest

Next, to find the total amount Deandra will owe, add the interest and the principal:

Total owed = $3,000 + $420 = $3,420

TRY THESE

1. Mr. Lopez deposits $225 in an account that pays 4.5 percent simple interest. How much interest does he receive in 1 year?
2. Lucia deposits $675 in a savings account that earns 3.75 percent simple interest. After 2 years, what will be the total amount in her account?

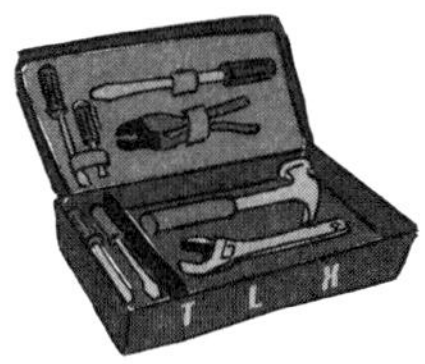

3. The Marquettes want to borrow $8,000 from a bank to do some repair work on their home. The bank will charge them 7 percent interest for a period of 3 years. How much will they have to pay the bank at the end of this period?
4. Preston invests $15,000 at a rate of 5.5%. What is the value of his investment at the end of 9 months?
5. Sylvia puts $1,000 in her savings account for 3 months. The bank is paying 4% simple interest on her account. How much interest will she have at the end of the 3 months?
6. Sonar Bank is offering a special interest rate of 6%. Mrs. Ramos invests $500 for 4 years. How much interest will she receive at the end of this time?

7. In order to pay for a new TV, the Chins borrow $600 from the bank at 5% simple interest. After a year, how much interest do they owe?
8. Mr. Rialto borrows $800 from the bank to pay for car repairs. The bank charges 7.5% simple interest. After 1 year, how much interest is due on the loan? How much does he owe the bank?

Three cases of percent

Most percent problems have three elements. In other words, a typical percent problem involves three numbers. Usually you are given two of the three numbers, and asked to find the third.

Case 1. **Find percent** when we are given part and whole.

Case 2. **Find part** when we are given whole and percent.

Case 3. **Find whole** when we are given part and percent.

Remember that percent means "per 100." A percent compares a part and the whole amount using a fraction with a denominator of 100.

Most percent problems can be set up in the following way:

$$\frac{\text{part}}{\text{whole}} = \text{percent}$$

A percent written as a fraction has a denominator of 100, so:

$$\frac{\text{part}}{\text{whole}} = \frac{\text{number}}{100} \text{ or } \frac{\%}{100}$$

Even though there are three different cases of percent problems, we can rewrite each case so that the problem can be done in the same way, no matter which case is presented.

Here is how:

Rewrite each problem so that it fits in these blanks:

_______% of _______ is _________.

In this way, the first blank has the percent amount, the second blank contains the whole amount, and the last blank contains the part. Using this method, the statement could be set up as

$$\frac{\text{is}}{\text{of}} = \frac{\text{percent amount}}{100}$$

Then cross multiply and divide to get the answer.

MODEL PROBLEMS

1. What percent of 60 is 12? (This problem is already written in the form shown above.)

Solution: Part (12) and whole (60) are given, and we need to find the percent.

$$\frac{\text{part}}{\text{whole}} = \frac{\%}{100}$$

$$\frac{12}{60} = \frac{?}{100}$$

Cross multiply: 12 • 100 = 1,200

Divide: 1,200 ÷ 60 = 20

Answer: 20% of 60 is 12.

2. 40 percent of 45 = ?

Solution: We are given the percent and the whole. We need to find the part.

$$\frac{\text{part}}{\text{whole}} = \frac{\%}{100}$$

$$\frac{?}{45} = \frac{40}{100}$$

Cross multiply: 45 • 40 = 1,800

Divide: 1,800 ÷ 100 = 18

Answer: 40% of 45 is 18.

3. 35 is 20 percent of what?

Solution: Rewrite 20% of _________ is 35. We are given the part and the percent, and we need to find the whole.

$$\frac{\text{part}}{\text{whole}} = \frac{\%}{100}$$

$$\frac{35}{?} = \frac{20}{100}$$

Cross multiply: $35 \bullet 100 = 3{,}500$

Divide: $3{,}500 \div 20 = 175$

Answer: 35 is 20% of 175.

CALCULATOR CLUE: You can use your calculator to find the missing amount in these three cases of percent.

Problem	Keystrokes	Display
Case 1: ____% of 300 is 75.	7 5 ÷ 3 0 0 %	25
Case 2: 25% of 520 is ____.	5 2 0 × 2 5 %	130
Case 3: 15% of ____ is 180.	1 8 0 ÷ 1 5 %	1200

TRY THESE

1. The down payment for a house on your street is 12 percent of the selling price. If the selling price is $80,000, how much is the down payment?
2. The trip to West Palm Beach from Orlando is 140 miles. What percent of the trip is completed after driving 35 miles?
3. George takes out a bank loan for minor home repairs. He pays the bank $80, which is 20 percent of the amount he owes. How much did he originally borrow from the bank?
4. Linc pays $150 for a new stereo. The sales tax comes to $9. What is the tax rate?

5. In a class of 36 students, 25% ride the bus. How many students ride the bus?

6. Eight students received an A on the math test. This is 20% of the class. How many students are in the class?
7. If the quarterback completes 12 out of 32 passes, what percent of the passes did he complete?
8. The distance from the school to the mall is 24 miles. If Mai has driven 3 miles, what percent of the distance has she driven?
9. Jamie borrowed $260 from his mother. So far, he has repaid 14% of the loan. How much money has he paid his mother?
10. Four members of the football team have turned in their uniforms. This represents 8% of the team. How many members are on the team?

"On" and "off" problems

When you come across a problem that involves a discount or a sale, you can think of the problem as ON or OFF:

- ON is the amount or part that you pay or receive.
- OFF is the amount or part that you do *not* have to pay or do *not* receive.

Think of the whole amount as 100 percent, or 1. Then:

ON + OFF = whole
ON + OFF = 100%
ON + OFF = 1

Therefore, we can also subtract:

whole − OFF = ON
100% − OFF = ON
1 − OFF = ON

MODEL PROBLEMS

1. Dina wants to buy a picture that normally sells for $30. Today there is a special sale and the picture is offered at 15 percent off. If Dina buys it, what will the cost be, *not* including tax?

Solution: This is either an ON or an OFF problem.

First, you must decide whether the question is asking you for the amount Dina will pay (ON) or the amount she will not pay (OFF). In this case, you are being asked to find the amount she will pay, the cost, or the ON amount.

Whole amount	OFF	Calculation: whole − OFF = ON	ON
100%	15%	100% − 15% = 85%	<u>85%</u>

Next, change 85% to a decimal:

85% = 0.85

Finally, multiply the original price by 0.85:

$30 × 0.85 = $25.50

Answer: Dina will pay $25.50.

2. How much will Dina save?

Solution: When the problem is in this form, you use the OFF amount, 15 percent. Change 15% to a decimal, 0.15, and multiply the original amount by this decimal.

$30 × 0.15 = $4.50

Answer: Dina will save $4.50.

3. At a bookstore there is a special table with a sign saying: ALL BOOKS ON THIS TABLE ARE ONE-THIRD OFF THE ORIGINAL PRICE. Emit finds a book about his favorite hobby that was originally priced at $24. If he decides to buy the book, how much will it cost without the sales tax?

Solution:

Step 1. Ask yourself: Is this an ON-OFF problem? Yes. Therefore, break up 1 whole into ON and OFF parts. The sale is $\frac{1}{3}$ OFF, so:

whole = ON + OFF

$$1 = \frac{2}{3} + \frac{1}{3}$$

Step 2. Which part are you being asked to find? ON. Therefore, multiply the original price by the ON part:

$$24 \bullet \frac{2}{3} = \frac{48}{3} = 16$$

Answer: Emit will pay $16 for the book.

4. How much will Emit save?

Solution: If you are asked to find the amount Emit saves, you will multiply $24 by the OFF amount:

$$24 \bullet \frac{1}{3} = \frac{24}{3} = 8$$

Answer: Emit will save $8.

Check by adding the ON and OFF amounts. The sum should equal the original amount.

TRY THESE

1. A dress that normally sells for $50 is marked 20 percent off. How much will the dress cost if you buy it at the sale price?

2. A department store in Margaret's town is having a GOING OUT OF BUSINESS sale. Everything in the store is marked down 25 percent. How much will Margaret save if she buys a $400 sofa?

3. Every Wednesday, a bakery has a one-fourth-off sale on doughnuts. If a dozen doughnuts usually cost $3, what will one dozen cost on Wednesday?

4. All tapes at a music store are being sold for a third off. If a tape normally sells for $9, how much would you save by purchasing it during the sale?

5. The regular price of a jacket is $44.40. If it is being sold at 25 percent off, what is the sale price?

6. A car battery is marked $\frac{1}{5}$ off. If the regular price of the battery is $59.20, how much will be saved by buying it on sale?
7. Best Buy Insurance gives a 10% discount for students with a B average or above. Karim normally pays $50 per month for car insurance. How much will he save each month if he brings home a report card with five A's and two B's?

8. What will be the new cost of a $45 bracelet that is marked 10% off?

"One-plus" percentage problems

In some problems, we must add a percentage to an original amount. Examples include problems involving sales tax, commission, tips, inflation, and interest. Any problem of this type can be solved with a strategy called **one-plus percentage**:

- The original amount is **one** whole, so it is 100 percent, or 1.
- The percentage of the whole is the **plus** part.

Not every problem involving percent is a one-plus problem. Before you try to answer a percentage question, be sure to ask yourself what type of problem it is.

MODEL PROBLEMS

1. Jaquet buys a shirt for $15. A sales tax of 6 percent is added to the price. How much is the sales tax? How much does she pay for the shirt?

   Solution: Is this a one-plus problem? Yes, because you must find the tax and the total she pays.

   "one" whole = $15

   "plus" part = 6 percent

   To find the tax, change 6 percent to 0.06 and multiply (or use the % key on your calculator):

   sales tax = $15 • 0.06 = $0.90

   To find the amount she pays, add:

   total cost = "one" whole + "plus" part
   = price + tax
   = $15 + $0.90
   = $15.90

   Answer: The sales tax is $0.90. Jaquet pays $15.90 for the shirt.

2. At James Restaurant, Jennifer's bill comes to $13.60. She wants to leave a 20 percent tip. How much money will she spend in all?

   Solution: Is this a one-plus problem? Yes, because you must add a percentage (the tip) to a given amount (the bill).

This problem does *not* ask you to find the amount of the tip. The key words *in all* indicate that you need to find just the total. In this situation, we can use the following method:

"one" whole = 100 percent = 1

"plus" part = 20 percent = 0.20

one-plus percentage = 1 + 0.20 = 1.20

Now multiply:

total = \$13.60 × (1 + 0.20)

= \$13.60 × 1.20 = \$16.32

Answer: Jennifer will spend \$16.32.

3. Marcus is a salesman in a furniture store. He makes 7 percent commission on the furniture he sells. On Monday, Marcus had \$4,500 in sales. How much commission did he make that day?

Solution: Is this a one-plus problem? No, because you must find *just* the commission. In other words, you must find only the "plus" part, *not* the "one" part. Therefore, simply multiply, changing 7 percent to 0.07:

\$4,500 • 0.07 – \$315.00

Answer: Marcus's commission was \$315.00.

TRY THESE

1. Maria deposits \$10,000 in a bank that pays 5 percent simple interest if she leaves the money in her account for 1 year. How much interest will she earn at the end of 1 year?
2. Enrique wants to make his 100-foot driveway 20 percent longer. How long will the driveway be when it is finished?
3. Rolanda buys a pair of hiking boots for \$50. If the sales tax is 6 percent, what is the total bill?
4. Serena leaves a 15 percent tip for a dinner bill that totals \$45.20. How much is the tip?
5. A sidewalk is 12 feet long. If it is lengthened by 25 percent, how long will it be?
6. Erik works in an electronics store. He gets a 6 percent commission on all merchandise he sells. Last week his total sales were \$4,250. How much commission did he receive?
7. Tony received a 5% raise in his salary this year. His salary last year was \$25 per hour. What will his salary be this year?
8. Inna's insurance company raised its rates by 15% after a hurricane. If Inna's premium was \$956 before the storm, how much does she pay now?

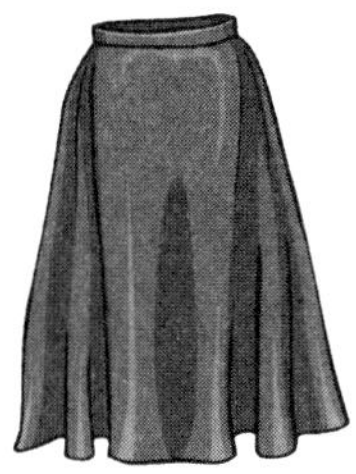

9. Luz buys a new skirt for \$27.95. If the sales tax is 6.5%, what is the total cost of the skirt?

10. Guy takes out a loan to buy a new car that costs \$22,500. He makes a down payment of \$3,000. If the rate on the loan is 7%, how much will he owe the bank?

Percent of increase and decrease

Percent can be used as a measure of how much a quantity has increased or decreased. To find the **percent of increase or decrease**, we follow these three steps:

Step 1. Subtract the original amount from the new amount.

Step 2. Compare the difference and the *original* amount by writing a fraction. This is the ratio of the difference to the original value.

Step 3. Change the fraction, or ratio, to a percent.

MODEL PROBLEMS

1. The price of a \$200 TV is increased to \$250. What is the percent of increase?

   Solution: Follow the three steps.

   Step 1. new amount minus original amount $= \$250 - \$200 = \$50$

   Step 2. ratio of difference to original amount $= \dfrac{50}{200}$

   Step 3. percent $= \dfrac{50}{200} = \dfrac{1}{4} = 25\%$

   Answer: Percent of increase = 25%

2. The price of a lawn mower is reduced from \$150 to \$120. What is the percent of decrease?

   Solution: Follow the steps.

   Step 1. original amount minus new amount $= \$150 - \$120 = \$30$

   Step 2. ratio of difference to original amount $= \dfrac{30}{150} = \dfrac{1}{5}$

   Step 3. percent $= \dfrac{1}{5} = \dfrac{20}{100} = 0.2 = 20\%$

   Answer: Percent of decrease = 20%

NOTE: Here is a useful **formula** to remember:

$$\text{percent of increase or decrease} = \frac{\text{amount of change}}{\text{original amount}} = \frac{\%}{100}$$

TRY THESE

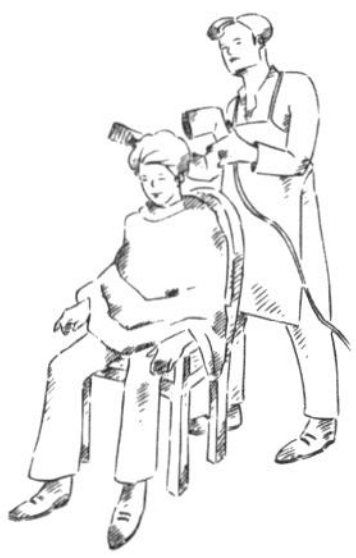

1. The cost of towing a car increased from \$40 to \$52. What is the percent of increase?
2. The price of a DVD player is reduced from \$100 to \$90. What is the percent of decrease?
3. The price of a chair is reduced from \$200 to \$180. What is the percent of decrease?
4. The price of a haircut increases from \$15 to \$18. What is the percent of increase?
5. This year, the population of Mayfield decreased from 2,500 to 2,200. What is the percent of decrease?
6. Last year, there were 1,500 seats in the auditorium. After a remodeling this year, there are 1,800 seats. What is the percent of increase?
7. Blue jeans, once priced at \$35 a pair, are on sale at 2 pairs for \$60. What is the percent of decrease of the price?
8. The average class grade for the first math exam was 85. The average class grade for the second exam was 91. What was the percent of increase?

9. Last week the Landers family spent \$150 on groceries. Their bill for groceries this week was \$90. What is the percent of decrease?
10. The Pawlak family bought a house for \$98,000 25 years ago. The house is now worth \$350,000. How much has the house increased in value?

E OPERATIONS WITH EXPONENTS

We now review the basic operations on numbers with exponents.

To add or subtract numbers with exponents, first evaluate the numbers, then do the addition or subtraction. If the numbers are variables, the symbols *and* the exponents must be the same before the numbers can be added or subtracted.

MODEL PROBLEMS

1. $5^2 + 3^2 = ?$

   Answer: $25 + 9 = 34$

2. $8^2 - 3^3 = ?$

   Answer: $64 - 27 = 37$

3. $3x^2 + 2x^2 = ?$

   Answer: $5x^2$

4. $6x^2 + x^2 = ?$

   Answer: $7x^2$

5. $3x^2 + 2y^2$

Answer: This cannot be added because the variables x and y are not the same.

6. $3x^2 + 2x^3 = ?$

Answer: This cannot be added because the exponents of x are not the same.

7. $8a^2 - 3a^2 = ?$

Answer: $5a^2$

8. $8a^2 - 3b^2 = ?$

Answer: This cannot be subtracted because the variables a and b are not the same.

9. $8a^2 - 3a^3 = ?$

Answer: This cannot be subtracted because the exponents of a are not the same.

To multiply numbers with exponents, evaluate the numbers first, then do the multiplication. For example:

$3^2 \bullet 4^2 = ?$

$9 \bullet 16 = 144$

Variables with exponents can be multiplied if the variable is the same. To multiply, *add* the exponents. For example:

$x^2 \bullet x^3 = x^{2+3} = x^5$ because $(x \bullet x) \bullet (x \bullet x \bullet x) = x^5$

But variables that are different cannot be multiplied, or simplified. For example:

$x^2 \bullet y^3 = x^2y^3$

To divide numbers with exponents, evaluate the numbers first, and then do the division. For example:

$3^2 \div 4^2 = ?$

$9 \div 16 = \frac{9}{16}$ or 0.5625

Variables with exponents can be divided if the variable is the same. To divide, find the *difference* of the exponents. For example:

$x^3 \div x^2 = x^{3-2} = x^1$ or x because $\frac{x^3}{x^2} = \frac{x \bullet x \bullet x}{x \bullet x} = x$

But variables that are different cannot be divided, or simplified. For example:

$$x^2 \div y^3 = \frac{x^2}{y^3}$$

MODEL PROBLEMS

1. $2x^3 \bullet 4x^4 = ?$

Answer: $8x^7$

Notice that the exponents have been added.

2. $8x^2 \bullet 4y^3 \bullet 2y^8 = ?$

Answer: $64x^2y^{11}$

Notice that an exponent shows the number of repetitions of a variable in the problem.

3. $4x^5 \div 2x^3 = ?$

Answer: $\frac{4x^5}{2x^3} = \frac{4x \bullet x \bullet x \bullet x \bullet x}{2x \bullet x \bullet x} = \frac{4x^2}{2} = 2x^2$

Notice that in division, the exponent in the divisor is subtracted from the exponent in the dividend.

4. $16x^2 \div 32x^5 = ?$

Answer: $\frac{16x^2}{32x^5} = \frac{16x \bullet x}{32x \bullet x \bullet x \bullet x \bullet x} = \frac{1}{2x^3}$ or $2x^{-3}$

Notice the subtraction of the exponents: $2 - 5 = -3$.

TRY THESE

1. $42x + 12x - 6x + 3x$

2. $7x^2 - 12x^2 - 6x$

3. $\frac{1}{2}x + \frac{2}{3}x^2 + \frac{1}{4}x + \frac{2}{5}y$

4. $8x^3 \bullet 12x^2 \bullet x$

5. $\frac{54x^4y}{14x^3y^3}$

6. $12a^2 + 7a^2 - 4a - a + 5$

7. $-4b^3 - 8b^3 - b^3 + b^2$

8. $6c^2 \bullet 2c^3 \bullet c^5$

9. $\frac{3d^2e^3}{6de}$

10. $\frac{24f^3g^2}{8f^5g^2}$

The calculator can be a great help in evaluating numbers with exponents. Make sure that the number with the exponent is evaluated first. Look at the following examples.

Problem	Keystrokes	Display
$3^2 + 4$	[3] [×] [3] [+]	9
	[4] [=]	13
$5^2 + 2^3$	[5] [×] [5] [=]	25
	[2] [×] [2] [×] [2] [=]	8
	[2] [5] [+]	25
	[8] [=]	33
$28 - 4^2$	[4] [×] [4] [=]	16
	[2] [8] [−]	28
	[1] [6] [=]	12

NOTE: You must evaluate each number and exponent separately and then combine.

REMEMBER: Always evaluate the numbers with the exponents *first*.

TRY THESE

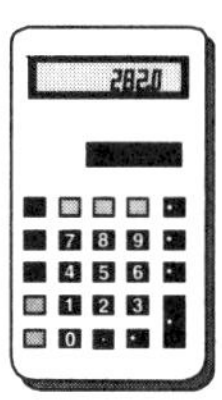

Use your calculator.

1. $3^2 + 12^2$
2. $(-7)^2 - 5^3$
3. $3.1^2 \bullet \left(\frac{2}{3}\right)^2$
4. $4^2 \bullet 7^2$
5. $3^3 \div 5^3$
6. $7^3 - 5^2$
7. $2^5 \bullet 4^3$
8. $6^3 \div 3^5$
9. $4.2^3 \bullet \left(\frac{1}{2}\right)^3$
10. $3^6 + 15^2$

F OPERATIONS WITH RADICALS

Finally, we'll apply the basic operations to radicals, or square roots.

In a square root, the number under the radical sign is called the **radicand**. A number in front of the radical sign is the **coefficient**. In $3\sqrt{10}$, for example, 10 is the radicand and 3 is the coefficient.

Square roots can be added or subtracted in radical form if the radicand is the same. For addition, combine the coefficients and keep the same radicand. For subtraction, subtract the coefficients and keep the same radicand.

NOTE: If the radicands are not the same, square roots can be evaluated and then added or subtracted. The results may be approximate.

MODEL PROBLEMS

1. $2\sqrt{3} + 1\sqrt{3} = ?$

Answer: $3\sqrt{3}$

2. $2\sqrt{3} + 3\sqrt{2}$

Answer: $2\sqrt{3} + 3\sqrt{2}$ cannot be added in radical form, because the radicands are different.

3. $5\sqrt{7} + 3\sqrt{7} + \sqrt{5} = ?$

Answer: $8\sqrt{7} + \sqrt{5}$

4. $8 + 5\sqrt{2} = ?$

Answer: This cannot be added in radical form.

5. $12\sqrt{3} - 3\sqrt{3} = ?$

Answer: $9\sqrt{3}$

6. $12\sqrt{3} - 3\sqrt{2} = ?$

Answer: This cannot be subtracted in radical form.

To multiply square roots, multiply positive numbers under the radical and then simplify if possible. For example:

$\sqrt{2} \bullet \sqrt{8} = \sqrt{16} = 4$

$\sqrt{3} \bullet \sqrt{5} = \sqrt{15}$, which will not simplify

Square roots can be "split apart" using the same concept. For example:

$\sqrt{8} = \sqrt{4} \bullet \sqrt{2}$. Since $\sqrt{4} = 2$, this number can be written as $2\sqrt{2}$.

$\sqrt{27} = \sqrt{9} \bullet \sqrt{3}$. This would become $3\sqrt{3}$.

Notice that for both $\sqrt{8}$ and $\sqrt{27}$ the radicand was split into factors (numbers multiplied) so that one of the factors was a perfect square and could be simplified to a whole number. Listed below are the perfect squares through 100. This list will help you simplify radicals.

Perfect square	Square root	Perfect square	Square root
1	1	36	6
4	2	49	7
9	3	64	8
16	4	81	9
25	5	100	10

MODEL PROBLEMS

1. $\sqrt{3} \bullet \sqrt{6} = ?$

Answer: $\sqrt{3} \bullet \sqrt{6} = \sqrt{18} = \sqrt{9} \bullet \sqrt{2} = 3\sqrt{2}$

2. $\sqrt{24} = ?$

Answer: $\sqrt{24} = \sqrt{6} \bullet \sqrt{4} = 2\sqrt{6}$

This procedure helps us add and subtract square roots. For example, the expression $\sqrt{8} + 3\sqrt{2}$ looks as if it cannot be added. But if we simplify $\sqrt{8}$ to $\sqrt{4} \bullet \sqrt{2}$ or $2\sqrt{2}$, then we can add: $2\sqrt{2} + 3\sqrt{2} = 5\sqrt{2}$.

MODEL PROBLEMS

1. $\sqrt{32} + 2\sqrt{2} = ?$

Answer: $\sqrt{32} = \sqrt{16} \bullet \sqrt{2} = 4\sqrt{2}$ and $4\sqrt{2} + 2\sqrt{2} = 6\sqrt{2}$

2. $\sqrt{3} - \sqrt{27} = ?$

Answer: $\sqrt{27} = \sqrt{9} \bullet \sqrt{3} = 3\sqrt{3}$ and $\sqrt{3} - 3\sqrt{3} = -2\sqrt{3}$

Two square roots can be divided as follows. First, rewrite the positive numbers under the two radicals as one fraction under one radical. Next, simplify.

EXAMPLES

$$\frac{\sqrt{32}}{\sqrt{2}} = \sqrt{\frac{32}{2}} = \sqrt{16} = 4$$

$$\frac{\sqrt{7}}{\sqrt{2}} = \sqrt{\frac{7}{2}} = \sqrt{3.5}$$

The square root of a fraction can be found by taking the square root of the numerator and the denominator separately.

EXAMPLES

$$\sqrt{\frac{1}{4}} = \frac{\sqrt{1}}{\sqrt{4}} = \frac{1}{2}$$

$$\sqrt{\frac{7}{4}} = \frac{\sqrt{7}}{\sqrt{4}} = \frac{\sqrt{7}}{2}$$

A procedure called **rationalizing** will eliminate a radical in the denominator of a fraction. To rationalize, multiply the numerator and the denominator by the radical you wish to eliminate.

EXAMPLE

$$\frac{3}{\sqrt{7}} = \frac{3}{\sqrt{7}} \bullet \frac{\sqrt{7}}{\sqrt{7}} = \frac{3\sqrt{7}}{\sqrt{49}} = \frac{3\sqrt{7}}{7}$$

TRY THESE

1. $4\sqrt{26} + 7\sqrt{26} - 2\sqrt{26}$
2. $\sqrt{48}$
3. $14\sqrt{2} - (-\sqrt{128})$
4. $3\sqrt{5} \bullet 2\sqrt{2}$
5. $\sqrt{128} \div \sqrt{24}$
6. Rationalize $\frac{2\sqrt{5}}{\sqrt{3}}$
7. Rationalize $\frac{3\sqrt{7}}{\sqrt{2}}$
8. $\sqrt{72}$
9. $8\sqrt{2} - 3\sqrt{2}$
10. $4\sqrt{6} \bullet 7\sqrt{5}$

In some problems, you need to find the decimal approximation of a square root. This can be done with your calculator. Remember that if the square root is an irrational number, the decimal will be only an approximation. Look at the following calculator problems.

Problem	Keystrokes	Display
$\sqrt{10}$	1 0 √	3.162277
$\sqrt{5} + 3$	5 √ +	2.236067
	3 =	5.236067
$9 - \sqrt{2}$	2 √	1.414213
	9 − 1 . 4 1 4 2 1 3 5 =	7.585786

REMEMBER: Just like exponents, square roots must be done first.

TRY THESE

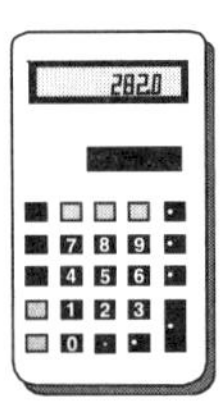

Use your calculator to find the decimal approximation of the following. Round to 4 decimal places. (Remember to do operations within parentheses first.)

1. $\sqrt{37} + 3\sqrt{5}$
2. $5\sqrt{15} - 2\sqrt{7}$
3. $2\sqrt{3} - 3\sqrt{8} - \left(-\frac{1}{2}\sqrt{2}\right)$
4. $(3\sqrt{3})^3 \bullet \sqrt{2}$
5. $\sqrt{135} \div \sqrt{7}$
6. $8\sqrt{3} + 4\sqrt{7}$
7. $6\sqrt{5} - 9\sqrt{2}$
8. $(7\sqrt{3})^2 \bullet \sqrt{5}$
9. $\sqrt{87} \div \sqrt{6}$
10. $2\sqrt{3} \div 4\sqrt{11}$

Assessment

Multiple-Choice Problems

1. S, T, V, and W are points on the number line below. Which point on the number line represents a number that when squared will result in a number greater than itself?

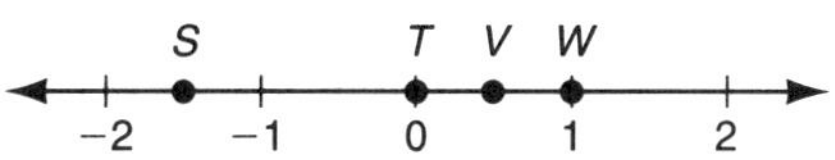

   A. S

   B. T

   C. V

   D. W

2. Which of the following is equivalent to 2^8?

   A. $2^2 \bullet 4^4$

   B. 8^2

   C. 16^2

   D. $4^3 \bullet 2^4$

3. Kim saw an $87 dress that was marked 30% off. How much will the dress cost before the sales tax is added?

   A. $26.10

   B. $37.28

   C. $60.90

   D. $124.28

4. The trip from Tallahassee to Key West is 640 miles. Approximately what percent of the trip is completed after driving 350 miles?

   A. 57%

   B. 55%

   C. 53%

   D. 51%

Items 5 and 6: The guidance counselor at Star Cove Beach High School wants to see each of the 265 tenth-grade students some time during the first month of school. By the end of the second week, she had seen 159 of the students.

5. What part of the total number of students had she seen? Express in decimal form.

   A. 0.60

   B. 0.55

   C. 0.50

   D. 0.40

6. What percent of the students had she seen?

   A. 60%

   B. 55%

   C. 50%

   D. 40%

7. One day in January, the Midwest of the country was having very cold weather. At 7 A.M. it was 9° above zero. At noon the temperature had dropped 16 degrees. What was the temperature at noon?

   A. 7°

   B. 0°

   C. −7°

   D. −25°

Gridded-Response Problems

8. In one week, the price of gasoline rose from \$2.35 per gallon to \$2.50 per gallon. What is the percent of increase in the price of gasoline during the week? Round your answer to the nearest integer.

9. Laurel Hill Sports Shop is having a 20% off sale. Jamie has a coupon for an additional $\frac{1}{4}$ off. He buys a salt water fishing outfit that was originally priced at \$120. If he uses his discount coupon, what is the cost of the fishing outfit before sales tax is added?

10. The formula for compound interest is $A = p\left(1 + \frac{r}{n}\right)^{nt}$ where A is the amount in the account, p is the principal, r is the annual interest rate, t is the time in years, and n is the number of compound periods per year. Five hundred dollars is deposited in the bank at 4% interest, where it is compounded twice a year. How much will be in the account at the end of one year? Express your answer to the nearest dollar.

8

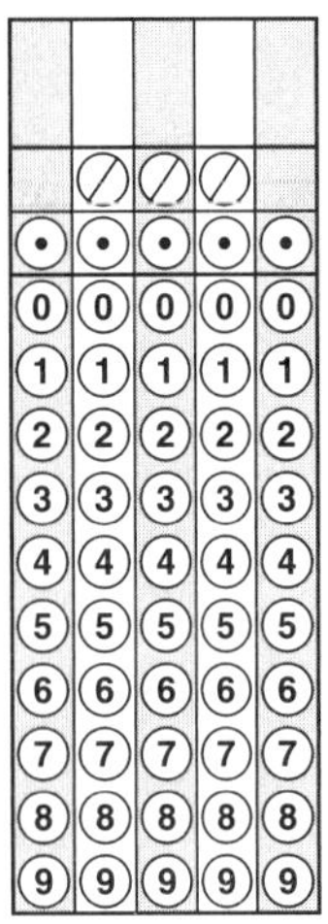

9

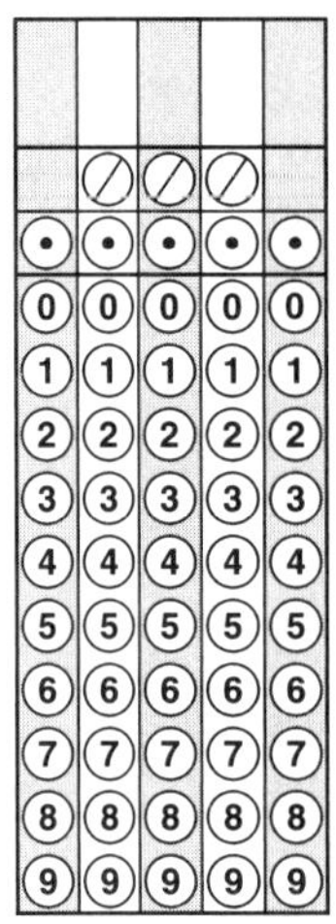

10

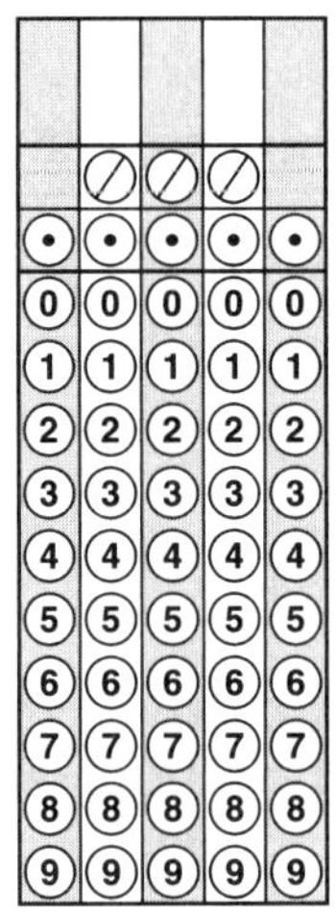

Section 3 Selecting properties of numbers and using them in operational shortcuts

A PROPERTIES OF NUMBERS

Numbers have some properties that allow operational shortcuts in mathematic problems.

One of these properties is the **inverse property**. Recall that the additive inverse of a number is its opposite. For example, 2 and -2 are additive inverses, since their sum is 0. The multiplicative inverse of a number (except zero) is its reciprocal. For example, the fractions $\frac{2}{3}$ and $\frac{3}{2}$ are multiplicative inverses, since their product is 1.

Another useful property is the **identity property**. The additive identity is 0. The multiplicative identity is 1. When 0 is added to a number or 1 is multiplied by a number, the result is the same number (the "identical" number).

The **distributive property** is used in multiplying a number or variable by a sum or difference.

EXAMPLE

$5(3 + 4) = 5 \bullet 7 = 35$

Using the distributive property gives us the same answer:

$5(3 + 4) = (5 \bullet 3) + (5 \bullet 4) = 15 + 20 = 35$

MODEL PROBLEMS

1. $7(9 - 3) = ?$

Answer: $7(9 - 3) = (7 \bullet 9) - (7 \bullet 3) = 63 - 21 = 42$

2. $5(x + y) = ?$

Answer: $5(x + y) = 5x + 5y$

TRY THESE

Multiply using the distributive property.

1. $14(6 - 8)$

2. $-9(x - 5)$

3. $3(8 - 4x)$

4. $6(-10 + 9x)$

5. $(2x + 3y)5$

6. $2a(b + c)$

7. $7(-5 + 2)$

8. $x(3m + p)$

9. $(6r - 7s)4$

10. $-8y(x - 9)$

Another helpful property is the **commutative property of addition**. This means that the sum of two or more numbers can be obtained by adding the numbers in any order. For example:

$3 + 2 + 5 = 10$ and $2 + 5 + 3 = 10$ and $5 + 3 + 2 = 10$.
The order does not matter.

The **commutative property of multiplication** means that the product of two or more numbers can be obtained by multiplying the numbers in any order. For example:

$5 \times 3 \times 2 = 30$ and $5 \times 2 \times 3 = 30$ and $2 \times 3 \times 5 = 30$.
The order does not matter.

The **associative property** is also helpful. If numbers are grouped in a certain way in a problem, the associative property allows us to group the numbers differently. For example, $5 + 2 + 3$ can be added in different ways. Five plus 2 would be 7, and then 7 plus 3 would result in 10. Or we can group 2 and 3 together to get 5 as the sum, and then add 5 to get the final answer, 10:

$(5 + 2) + 3 = 10$ and $5 + (2 + 3) = 10$

MODEL PROBLEM

Combine $5 + (-2) + 4 = 5 - 2 + 4 = 7$

Answer: All the positive numbers can be grouped together: $5 + 4 = 9$. Then -2 can be combined with this sum: $9 + (-2) = 7$

TRY THESE

1. $8 + 9 - 3 + 4$
2. $26 - 18 - 10 + 28$
3. $-43 - 21 + 16 - 33$
4. $37 + 18 - 54 - 42$
5. $85 - 58 + 29 - 33$
6. $21 - 18 - 25 + 7$
7. $-46 - 29 - 32 + 17$
8. $62 - 51 + 10 + 21$
9. $123 - 42 - 55 + 8$
10. $-8 + 12 - 16 + 21$

The **transitive property** is also useful. Here are two examples of the transitive property:

If $7 + 3 = 10$ and $10 = 8 + 2$, then $7 + 3 = 8 + 2$.

If $x = 5$ and $5 = y$ then $x = y$.

The transitive property can be expressed in symbolic form:

If $a = b$ and $b = c$ then $a = c$.

The property of **substitution** involves replacing a symbol with an equivalent symbol or a value.

EXAMPLE

If $x = 3$, find the value of $x + 4$ by substituting 3 for x:

$x + 4 = 3 + 4 = 7$

MODEL PROBLEM If $y = x - 4$, find the value of $2y + 5$ in terms of x.

Answer:

$2y + 5$	
$2(x - 4) + 5$	substitution
$2x - (2 \bullet 4) + 5$	distributive property
$2x - 8 + 5$	
$2x - 3$	associative property

TRY THESE

1. If $x = 3$ and $y = -4$, find the value of $8x - 10y$.
2. If $x = -7$ and $y = \frac{3}{4}$, find the value of $2.3x - \frac{1}{2}y$.
3. If $y = 7(x + 1)$, find the value of $3 - 5y$ in terms of x.
4. If $y = 9(x - 6)$, find the value of $\frac{3}{4}y$ in terms of x.
5. If $3y = 4x - 1$, find the value of $y - 2$ in terms of x.
6. If $a = 2$ and $b = -3$, find the value of $5a^2 - 6b$.
7. If $p = 6$ and $r = -4$, find the value of $(-2p + 3r)^2$.
8. If $d = 6(2 - x)$, find the value of $7 - 2d$ in terms of x.
9. If $b = 2a + 4$ and $c = 3a + 7$, find the value of $4b - 3c$ in terms of a.
10. If $2p = 5r - 9$, find the value of $p + 6$ in terms of r.

B OPERATIONAL SHORTCUTS

In solving problems, it is sometimes helpful to use shortcuts. A **shortcut** is also called **mental mathematics**.

One familiar shortcut is **multiplying by a power of 10**. Recall that $10^1 = 10$, $10^2 = 100$, $10^3 = 1{,}000$, $10^4 = 10{,}000$, and so on. Notice that the exponent (the small numeral at the right) indicates the number of zeros.

To multiply an integer by a power of ten, put in the same number of zeros as the exponent. If the number contains a decimal, move the decimal point to the right the same number of places as the exponent. To move the decimal point, you may need to insert a zero or zeros.

MODEL PROBLEMS

		Answers:
1.	$53 \bullet 10 =$	530
2.	$8.32 \bullet 10 =$	83.2
3.	$53 \bullet 100 =$	5,300
4.	$53 \bullet 10^3 =$	53,000
5.	$8.32 \bullet 100 =$	832
6.	$8.32 \bullet 10^3 =$	8,320

To **divide by a power of ten**, move the decimal point to the left as many places as the power of 10. You may need to insert a zero or zeros when you move the decimal point.

MODEL PROBLEMS

1. $75 \div 10 = ?$

Answer: 7.5 (Move the decimal point 1 place to the left, since 10 is 10 to the first power.)

2. $894.3 \div 100 = ?$

Answer: 8.943 (Move the decimal point 2 places to the left, since 100 is 10 to the second power.)

3. $0.062 \div 10 = ?$

Answer: 0.0062 (Move the decimal point 1 place to the left, since 10 is 10 to the first power.)

4. $54.32 \div 10^3 = ?$

Answer: 0.05432 (Move the decimal point 3 places to the left, since you are dividing by 10 to the third power.)

TRY THESE

Use powers of 10 as shortcuts.

1. $29.07 \bullet 10^2$

2. $0.005 \div 10$

3. $135 \div 10^3$

4. $56.24 \bullet 10^3$

5. $0.0348 \bullet 10^2$

6. $1.35 \bullet 10^4$

7. $0.078 \div 10^3$

8. $0.986 \div 10^5$

9. $16.195 \bullet 10^4$

10. $4.325 \div 10^3$

When you are combining several numbers, it helps to **group numbers into tens**. This shortcut can be used in each column of numbers with more than one digit.

EXAMPLES

$3 + 2 + 7 + 9 + 8 = ?$

10 10

$10 + 10 + 9 = 29$

$$\begin{array}{r} 24 \\ 32 \\ 56 \\ \hline 112 \end{array} \quad (4 + 6) = 10$$

If you are combining positive and negative numbers, look for **additive inverses** that you can combine to get zero or zeros.

EXAMPLE

$$5 - 2 + 8 - 5 + 2 = ?$$

$$\qquad 0 \quad 0$$

$$0 + 0 + 8 = 8$$

You can use this same shortcut when multiplying. Look for multiplicative inverses that give the product 1.

EXAMPLE

$$7 \bullet \frac{3}{4} \bullet \frac{1}{7} \bullet 5 \bullet \frac{4}{3} = ?$$

$$1 \qquad 1$$

$$1 \bullet 1 \bullet 5 = 5$$

Look for the possibility of **multiplying a nonzero number by zero** or **dividing zero by a nonzero number**. In both cases, the result is zero, and this can make a problem much easier.

EXAMPLES

$$5 \bullet 2 \bullet \frac{1}{2} \bullet 7 + \frac{8}{3} \bullet \frac{4}{5} \bullet 0 \bullet \frac{9}{2}$$

$$1$$

$$5 \bullet 1 \bullet 7 + 0 = 35$$

Remember that multiplication must be done first.

$$\frac{16 \bullet 0 \bullet -5 \bullet 14}{5 \bullet 3 \bullet \frac{1}{4} \bullet -52} = \frac{0}{-195} = 0$$

Remember that the numerator and denominator must be simplified first.

TRY THESE

1. $56 + 31 + 24 + 72$

2. $39 - 24 - 59 + 16$

3. $9 \bullet \frac{2}{3} \bullet \frac{1}{2} \bullet 6 \bullet \frac{1}{2}$

4. $\frac{1}{2} \bullet \frac{2}{3} \bullet \frac{3}{4} \bullet \frac{4}{5} \bullet \frac{5}{6} \bullet \frac{6}{7}$

5. $3 \bullet 8 \bullet \frac{2}{3} \bullet \frac{1}{4} - \frac{1}{2} \bullet 5 \bullet \frac{3}{5} \bullet 4$

6. $8 \bullet \frac{4}{9} \bullet 82.3 \bullet 0 + 23 \bullet 42$

7. $\dfrac{6 \bullet 0 \bullet 1.3 \bullet 4}{5 \bullet 6 \bullet 4 \bullet 0.5}$

8. $18 + 14 - 32 + 9 \bullet \frac{1}{9} \bullet 42$

9. $\frac{3}{7} \bullet \frac{4}{9} \bullet \frac{7}{11} \bullet \frac{1}{3} \bullet \frac{9}{4} \bullet \frac{1}{2}$

10. $\dfrac{3 \bullet 5 \bullet 0 - 4 \bullet \frac{1}{4} \bullet 9}{6 \bullet 3 \bullet \frac{1}{6}}$

C ORDER OF OPERATIONS

In simplifying numerical expressions, a specific order must be followed.

- *First:* Perform operations within parentheses (or within braces or brackets).
- *Second:* Simplify exponents. That is, raise bases to powers.
- *Third:* Perform multiplication and division from left to right.
- *Fourth:* Perform addition and subtraction from left to right.

MODEL PROBLEMS

1. Simplify $6(1 + 2)^2 - 15 \div 5$.

Answer:

$6 \bullet 3^2 - 15 \div 5$

$6 \bullet 9 - 15 \div 5$

$54 - 3$

51

2. Simplify $\frac{9 - 5(3 - 2)}{3 + 5}$.

Answer:

$\frac{9 - 5 \bullet 1}{3 + 5}$

$\frac{9 - 5}{8}$

$\frac{4}{8} = \frac{1}{2}$

TRY THESE

Simplify.

1. $5(3 + 1)^2 \div 4 + 3$

2. $\frac{4(5 + 2) - 1}{4 + 9}$

3. $12 + (7 - 3) + 9^2$

4. $9[4(6 + 5) - 7]$

5. $8 \div 4 + 7^2 - 12$

6. $\frac{4 + |7 - 9| + 5^2}{10 - 4}$

7. $3 \bullet 2 - 4 \div 2 + 6 \div 8$

8. $\frac{15 - |12 - 16| + 3^2}{12 - 4}$

9. $2(-5)^2 + 8\left(3 \bullet \frac{1}{3}\right) - 24 \div 12$

10. $\frac{1}{4}[8 - 3 \bullet 6 \div 9 - 4^2]$

NOTE: For more practice, turn to page 463.

Assessment

Multiple-Choice Problems

1. Which of the following expressions is equivalent to $7(6x + 3) + 5(4x - 3)$?

   A. $62x + 36$

   B. $62x + 6$

   C. $22x + 8$

   D. $22x + 2$

2. Which of the following expressions is equivalent to $9(a + b) - 7(2a + b)$?

   A. $-5a$

   B. $16b$

   C. $-5a + 2b$

   D. $-5a + 16b$

3. Which of the following expressions is equivalent to $4(x + 3y + 2) + 7(y - 1)$?

   A. $4x + 14y - 1$

   B. $4x + 14y + 13$

   C. $4x + 19y + 15$

   D. $4x + 19y + 1$

4. Which of the following expressions is equivalent to $8(2x + 3) - 5(3x - 4)$?

   A. $x + 44$

   B. $x - 1$

   C. $x + 1$

   D. $31x + 44$

5. Which of the following expressions is equivalent to $15 + 3 \bullet 2 - 8$?

   A. 28

   B. 13

   C. 5

   D. -3

6. Which of the following expressions is equivalent to $3(6 - 8)^3 + (-2)^2(-3)$?

   A. 36 **C.** -12

   B. 12 **D.** -36

7. Which of the following expressions is equivalent to $4(7 - 8)^3 - 2^4 \div 8$?

   A. -6

   B. -2

   C. 2

   D. 6

8. Given the values $a = \frac{1}{2}$, $b = 5$, $x = -1$, $y = -3$, and $z = 12$, what is the value of the expression $\frac{4a - 3x}{2y + z}$?

   A. $-\frac{5}{18}$

   B. $-\frac{1}{6}$

   C. $-\frac{1}{18}$

   D. $\frac{5}{6}$

9. If $6y = 2(x - 9)$, which expression is equivalent to $y - 5$ in terms of x?

   A. $\frac{1}{3}x - 23$

   B. $\frac{1}{3}x - 13$

   C. $\frac{1}{3}x - 8$

   D. $\frac{1}{3}x - \frac{3}{2}$

10. The volume of a right circular cone is $V = \frac{1}{3}\pi r^2 h$ where r is its radius and h is its height. If $r = 4$ and $h = 5$, which of the following is the volume in terms of π?

    A. $\frac{100\pi}{3}$

    B. 30π

    C. $\frac{80\pi}{3}$

    D. $\frac{40\pi}{3}$

Section 4 Using estimation in problem solving and computation

It is important to be able to estimate. We need to estimate so that we can predict the outcome of a real-world problem. After the problem is calculated mathematically, we need to determine whether or not the result is reasonable.

A USING ESTIMATION STRATEGIES

Rounding numbers is one common method of estimating in order to predict. When you round a number, you are changing an exact value to an approximate value. A number is rounded to a specific **place**, depending on the problem and on how much accuracy is needed.

To round a number, follow these steps:

Step 1. Decide to what place you will round the number.

Step 2. Underline the number that is in this place.

Step 3. Look at the number to the right of the underlined number.

- If the number to the right is 5 or greater, add 1 to the underlined number and substitute zeros for all the numbers to the right of the underlined number.
- If the number to the right of the underlined number is less than 5, leave the underlined number as is and substitute zeros for all the numbers to the right of the underlined number.

MODEL PROBLEMS

1. Round 362 to the nearest hundreds place.

Answer:

362 Underline 3, since that is in the hundreds place.
The number to the right of 3 is 6, and 6 > 5.

400 Therefore, add 1 to the 3, which makes 4.
Then substitute zeros for all the numbers that were to the right of 3.

2. Round 2,809 to the nearest hundreds place.

Answer:

2,809 Underline 8, since that is in the hundreds place.
The number to the right of 8 is 0, and 0 < 5.

2,800 Therefore, leave 8 as is.
Substitute zeros for all the numbers that were to the right of 8.

3. Round 57.246 to the nearest tenth (one-tenths place).

Answer:

$57.\underline{2}46$ Underline $\underline{2}$, since that is in the one-tenths place. The number to the right of $\underline{2}$ is 4, and $4 < 5$.

57.2 Therefore, leave $\underline{2}$ as is. Substitute zeros for all the numbers that were to the right of $\underline{2}$. $57.200 = 57.2$.

4. Round 6.897 to the nearest tenth.

Answer:

$6.\underline{8}97$ Underline $\underline{8}$, since that is in the one-tenths place. The number to the right of $\underline{8}$ is 9, and $9 > 5$.

6.9 Therefore, add 1 to the $\underline{8}$, which makes 9. Substitute zeros for all the numbers that were to the right of $\underline{8}$. $6.900 = 6.9$.

NOTE: For more practice with rounding, turn to page 463.

TRY THESE

Round as indicated.

1. 9,754 to the nearest tens
2. 5.296 to the nearest tenths
3. 107,277 to the nearest thousands
4. 28.934 to the nearest hundredths
5. 3.1337 to the nearest thousandths
6. 624,999 to the nearest ten-thousands
7. 981 to the nearest hundreds
8. 12.54 to the nearest tenths
9. 87.6451 to the nearest hundredths
10. 12,463,635 to the nearest millions

Rounding can be used to estimate, or predict, answers.

MODEL PROBLEM

What is the approximate area enclosed by a rectangle that measures 7.24 miles by 4.72 miles?

Answer: Round 7.24 miles to 7 miles. Round 4.72 miles to 5 miles. Then multiply: 7 times 5 = 35. So the estimated area is approximately 35 square miles.

Check: The actual square mileage would be $7.24 \times 4.72 =$ 34.1728. The estimate of 35 square miles is a fair approximation.

TRY THESE

1. The diameter of a metal rod is 0.679 inch. What would this measurement be given to the nearest hundredth?
2. A basketball player scored 890 points in 53 games. What is the approximate number of points scored per game?
3. If the county tax rate is \$4.21 for each \$1,000 of assessed value, what is the approximate county tax on a house valued at \$98,000?
4. What is the approximate cost of 230 gallons of gasoline at \$2.45 per gallon?
5. You buy the following items at the supermarket: 2 loaves of bread at \$0.89 each, 3 bottles of soda for \$0.99 each, and 4 bags of pretzels for \$0.55 each. What is your approximate bill?
6. What is the approximate area of a square that has a side that measures 3.8 feet?
7. David went to the lumber store to buy 40 feet of molding. The molding costs \$0.24 a foot. David estimates that his bill will be approximately \$10. Explain why this is or is not a good estimate.
8. At the county academic awards ceremony, 2 students from each of the 50 schools are being honored. It takes approximately $1\frac{3}{4}$ minutes to honor each student. Explain why the program should take approximately 3 hours.
9. The diameter of a circle is 10 inches. Explain why 75 square inches is a good approximation of the area. ($A = \pi r^2$)
10. Explain why 18 meters is a good estimation for the perimeter of a rectangle with a length of 5.3 meters and a width of 3.9 meters.

In order to estimate, you must first use the information that you are given.

MODEL PROBLEM

It takes Meridith 3 hours to clean the house by herself. In other words, she can clean $\frac{1}{3}$ of the house in 1 hour. It takes Brad 4 hours to clean the same house by himself. That is, he can clean $\frac{1}{4}$ of the house in 1 hour. Working together, they can clean $\frac{7}{12}$ of the house in 1 hour, because $\frac{1}{3} + \frac{1}{4} = \frac{4}{12} + \frac{3}{12} = \frac{7}{12}$. Estimate how long it will take them to clean the house working together.

Answer: It should take them a little less than 2 hours to clean the house, since they can clean $\frac{7}{12}$ in 1 hour, and $\frac{7}{12}$ is slightly more than half $\frac{1}{2} = \frac{6}{12}$.

TRY THIS ONE

A large pipe and a small pipe are used to fill a swimming pool. When the larger pipe is used alone, the pool can be filled in 4 hours. When the smaller pipe is used alone, it takes 7 hours to fill the pool. Approximately how long will it take to fill the pool if both pipes are used?

MODEL PROBLEM

In a forest, an area measuring 1,000 square yards contains 700 trees. How many trees would you estimate are contained in an area measuring 1,400 square yards?

> *Answer:* Since 1,000 square yards contains 700 trees, 2,000 square yards would contain twice that number: 1,400 trees. Halfway between 1,000 and 2,000 would be 1,500 square yards, and halfway between 700 and 1,400 would be 1,050 trees. Since 1,400 square yards is a little less than 1,500 square yards, you can estimate that there would be a little less than 1,000 trees in 1,400 square yards.

NOTE: In the above model problem, we found the value halfway between 1,000 and 2,000 and the value halfway between 700 and 1,400. Here is the method. First add the two numbers: $1{,}000 + 2{,}000 = 3{,}000$ and $700 + 1{,}400 = 2{,}100$. Then divide the sums by 2: $\frac{3{,}000}{2} = 1{,}500$ and $\frac{2{,}100}{2} = 1{,}050$.

TRY THIS ONE

If a poster with an area of 80 square inches can contain a printed text of 300 words, approximately how many words could be used on a poster with an area of 150 square inches?

MODEL PROBLEM

Two adults decide to take a one-week trip by car to visit St. Augustine. They estimate the cost of gas, food, and motel to be $150 a day. If two other adults join them, what would the cost be for all four per day?

> *Answer:* The cost would be less than $300 per day, since the cost of the gas would be the same. It would have to be more than $150 per day even if they shared the lodging, since the cost of food would increase. An estimate of $225 per day would be a fair approximation.

TRY THIS ONE

Mr. Sanchez has offered to take his family of four to Animal Kingdom for the day. The cost of admission is $29 per car. He estimates that food at the park will cost $8 per person. If each of the Sanchezes' two children brings a friend along, what will be the approximate cost for the day? Justify your answer.

B CHECKING THE REASONABLENESS OF RESULTS

Once you have calculated an answer, or if you are looking at possible answers, be sure to check to see if the answer is reasonable.

For example, if the time it takes to wash 10 windows in a house turns out to be 3 days, something is probably wrong. If the answer to a problem indicates that a man would be 13 feet tall or that a customer would pay $2.50 for a 10-carat diamond ring, something is wrong. When this happens, redo the problem, looking for a mistake.

MODEL PROBLEM

Michael is 50 years younger than his uncle Paul. Which of the following could NOT be Uncle Paul's age? Explain your answer.

A. 47 years

B. 61 years

C. 80 years

D. 103 years

Answer: Uncle Paul could not be 47 years old, since that would make Michael −3 years old, which is impossible.

TRY THESE

1. Three workers can dig a drainage ditch in 4 days. You are asked to find the amount of time required to complete the job if more workers are hired to help dig the ditch. Which could NOT be a correct answer? Explain your reasoning.

   A. 1 day

   B. 2 days

   C. 3 days

   D. 5 days

2. Jennifer inherited $2,500 from her uncle. She invested it all at 9 percent per year. At the end of the year, she calculated that there was $2,509 in her account. Was she right? Why or why not?

3. The Spillanes buy a new rug to cover part of their living room floor. The rug measures 108 square feet. Which could NOT be the measurements of their living room? Explain.

   A. 12 feet by 10 feet

   B. 8 feet by 10 feet

   C. 13 feet by 15 feet

   D. 14 feet by 12 feet

4. A gallon of paint covers between 350 and 400 square feet of surface. Hector wants to paint the 4 walls of his bedroom. Two of the walls are 9 feet by 10 feet and the other two are 9 feet by 15 feet. He calculates that he needs to buy one gallon of paint. Are his calculations correct? Why or why not?

5. In 1896, the first Olympic games, in which 311 men participated, were held in Greece. A movie director wants to make a film about this famous time. He wants to depict the group of men, divided evenly, standing in the 5 Olympic rings. The director is allowing each man 4 square feet on which to stand. He estimates that the radius of each circle (ring) should be 8 feet. Is his estimate correct? Explain why or why not.

Assessment

Multiple-Choice Problems

1. The band director is estimating the number of people who are attending the "Battle of the Bands" show. The people are evenly distributed in the stands on both sides of the field. Approximately 150 people are sitting in the shaded section shown. What is the approximate number of people in attendance?

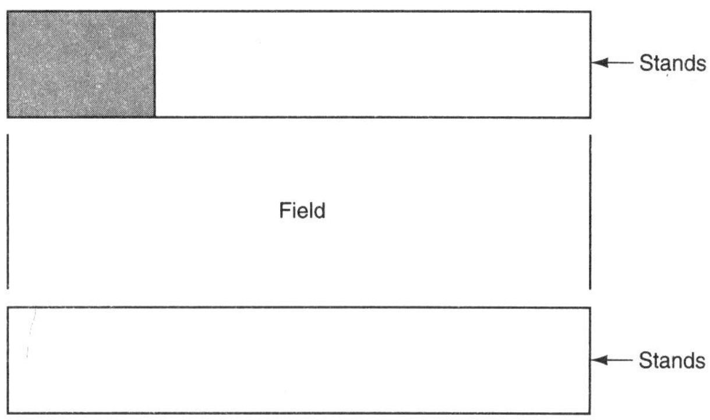

   A. 900 people

   B. 1,000 people

   C. 1,200 people

   D. 1,400 people

2. The diagram below is the layout of the first floor of a building.

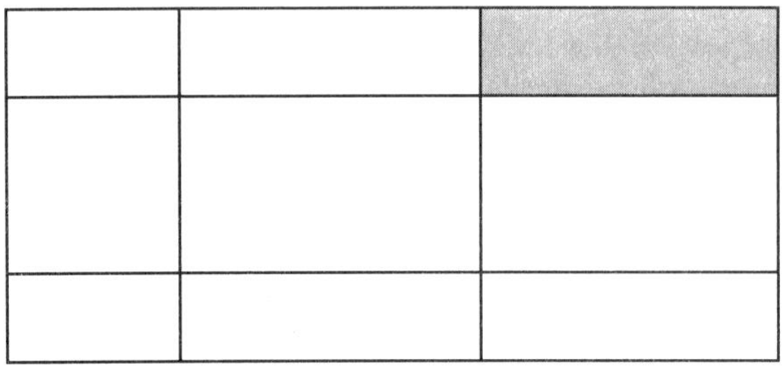

It is estimated that 90 people can comfortably occupy the room that is shaded. With that number in mind, what is the estimate of the occupancy of the entire first floor?

A. 680 people **C.** 1,080 people

B. 880 people **D.** 1,280 people

3. Phillippi wants to put a lace trim around her 8 place mats. Each place mat has a perimeter of 4 feet. The trim costs $0.26 a foot. She estimates that the lace trim will cost $8.00. Which statement best describes her estimate?

A. The trim costs exactly $8.00.

B. She has overestimated the cost.

C. She has underestimated the cost by less than $1.00.

D. She has underestimated the cost by less than $2.00.

4. Minh wants to paint a storage shed. The surface area is approximately 39 square meters. A gallon of paint covers $13\frac{1}{2}$ square meters and costs $16 per gallon. What is the cost of the paint for the shed?

A. $16

B. $32

C. $48

D. $64

5. The director of the Theater in the Round is estimating the number of people in attendance at the play. The people are evenly distributed throughout the theater. There are approximately 60 people sitting in Section C. What is the approximate number of people in attendance?

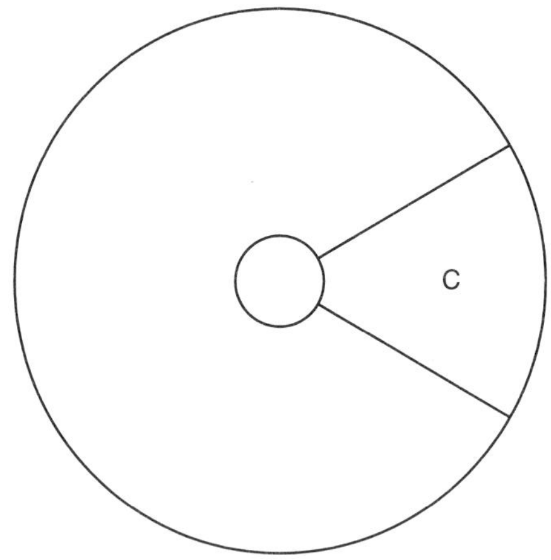

A. 290 people

B. 350 people

C. 410 people

D. 470 people

6. The Friendly Place restaurant has 4 dining rooms. Each room can hold 15 tables and each table can seat 4 people. What is the best approximation of the total number of people the restaurant can seat?

A. 189 people

B. 216 people

C. 234 people

D. 300 people

CHAPTER REVIEW

Multiple-Choice Problems

1. Points S, T, V, and W are located on the number line below.

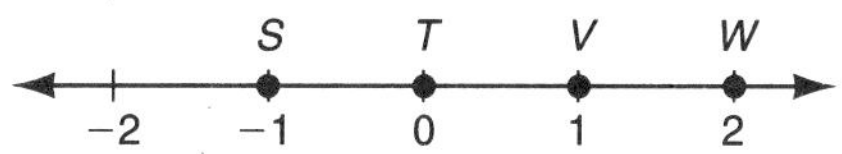

Which point on the number line represents a number that is the product of any number and its reciprocal?

A. S

B. T

C. V

D. W

2. Which of the following numbers is closest to $5\sqrt{6}$?

A. $12\frac{1}{3}$ **C.** $12\frac{1}{4}$

B. $12\frac{2}{7}$ **D.** $12\frac{1}{5}$

3. A micrometer is an instrument that can be used to measure the thickness of metal. A particular piece of metal is 0.0057 centimeter thick. What is this measurement in scientific notation?

A. 57×10^{-4} centimeters

B. 5.7×10^{-3} centimeters

C. 57×10^{4} centimeters

D. 5.7×10^{3} centimeters

4. Chrystal left home and drove 153 miles the first day and 190 miles each of the next 4 days. What would be a good estimate of the number of hours that it would take for her to return home, averaging 60 miles per hour?

A. 8 hours

B. 12 hours

C. 14 hours

D. 16 hours

5. Vincent has answered 132 requests each day for the last 18 days. His goal is to answer 4,000 requests. What is the best estimate of how many more days he will need to reach his goal?

A. 9 days

B. 11 days

C. 14 days

D. 16 days

6. If $x = 3$, $y = 2$, and $z = \frac{1}{3}$, what is the value of $\frac{2xy - 3z}{4} + 7$?

A. $\frac{11}{4}$

B. $\frac{13}{4}$

C. $\frac{39}{4}$

D. $\frac{41}{4}$

7. Which of the following is equivalent to $3(2x - 4y + 7) - 2(5x - 3y - 1)$?

A. $-4x - 6y + 23$

B. $-4x - 18y + 23$

C. $-4x - 18y + 19$

D. $-4x - 6y + 19$

Gridded-Response Problems

8. As part of a physics experiment, Abdul must evaluate the following expression: $(-3.2)^2 - (-0.2)^2$. What is its value?

9. Peaches sell for 3 for $0.99. How much will $2\frac{1}{2}$ dozen cost?

10. A recipe for 3 dozen cookies requires $1\frac{1}{3}$ cups of sugar. Gerrie wants to make 12 dozen cookies. How much sugar will she need?

11. After delivering 25 newspapers, Art had fewer than 60 left. At most, how many newspapers did he have originally?

12. In the diagram below, what is the length of the longest side? Round to the nearest integer.

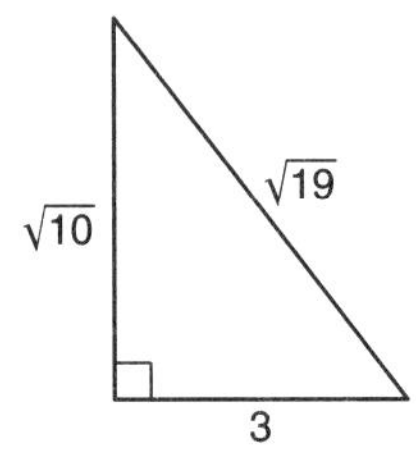

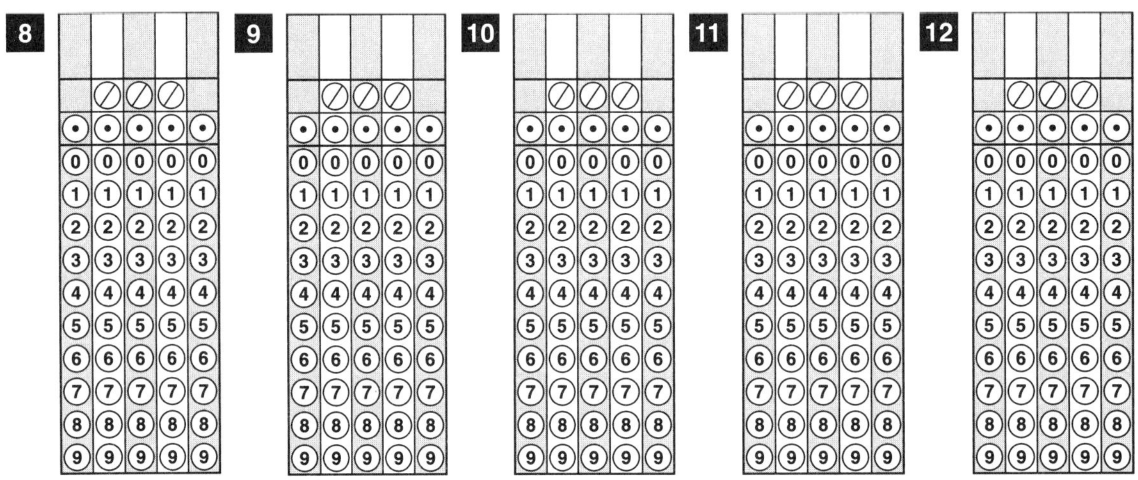

CHAPTER TWO

Measurement

CHAPTER CONTENTS

Section 1

Measuring and using quantities to solve real-world problems

A USING CONCRETE AND GRAPHIC MODELS TO DERIVE AND APPLY FORMULAS FOR RATE, DISTANCE, TIME, ANGLE MEASURE, AND ARC LENGTH

Rate, distance, and time

A ratio is a way to compare two like quantities. A **rate** is a way to compare two *un*like quantities.

For example, if a worker completes three projects every two days, the worker's rate can be expressed as 3:2 or 3/2 or $\frac{3}{2}$ projects per day. If a student can eat 5 apples in 10 minutes, the student's rate would be 5:10 or 5/10 or $\frac{5}{10}$ apples per minute, which is the same as 1:2 or 1/2 or $\frac{1}{2}$ apple per minute.

A familiar rate is miles per hour or miles/hour, abbreviated mph. This rate compares distance with time to find out how fast something is moving. For instance, if a car is moving at an average rate of 50 mph, it means that the car would cover 50 miles in one hour.

A formula used in this kind of situation is *distance equals rate multiplied by time*, or $d = rt$. If we know any two parts of the formula $d = rt$, we can find the missing part: $r = \frac{d}{t}$ and $t = \frac{d}{r}$.

EXAMPLES

If a truck travels at 60 miles per hour for 3 hours, what is the distance (d) covered?

$d = rt$

$d = 60 \bullet 3$

$d = 180$ miles

If a car covers 150 miles in 3 hours, what is its rate (r)?

$r = \frac{d}{t}$

$r = \frac{150}{3}$

$r = 50$ miles per hour

How much time (t) would it take a plane traveling at 450 miles per hour to travel 2,025 miles?

$$t = \frac{d}{r}$$

$$t = \frac{2{,}025}{450}$$

$$t = 4.5 \text{ hours}$$

MODEL PROBLEMS

1. Shari traveled 130 miles in 2.5 hours. What was her rate?

Answer:

$$r = \frac{d}{t}$$

$$r = \frac{130}{2.5}$$

$$r = 52 \text{ miles per hour}$$

2. A car averaged 60 miles per hour for 3 hours and 15 minutes. How far did it travel?

Answer: 3 hours and 15 minutes = 3.25 hours, since 15 minutes is $\frac{1}{4}$ or 0.25 hour. So:

$$d = rt$$

$$d = 60 \bullet 3.25$$

$$d = 195 \text{ miles}$$

3. How long will a 250-mile trip take if a truck averages 55 miles per hour?

Answer:

$$t = \frac{d}{r}$$

$$t = \frac{250}{55}$$

$$t \approx 4.5 \text{ hours}$$

4. If Virgil types 1,500 words in 30 minutes, what is his rate?

Answer: Rate $= \frac{1{,}500}{30} = \frac{150}{3} = 50$ words per minute

5. Jacques read 6 books in 3 months. What was his rate?

Answer: Rate $= \frac{6}{3} = 2$ books per month

TRY THESE

1. Amira worked as a receptionist for $2\frac{1}{2}$ hours on Saturday morning. She was paid \$13.75. What was her rate of pay?
2. If Yolanda scored 175 points in 7 basketball games, what was her rate per game?
3. Carlos can run 300 meters in 40 seconds. What is his rate per second?
4. How long will it take a boat to go a distance of 75 nautical miles at a speed of 30 knots? (A **knot** is a measure of a boat's speed and is equal to 1 nautical mile per hour.)
5. How many miles will a plane fly in 4 hours if it averages 600 miles per hour?
6. At the Civil War enactment, the blacksmith can make 4 horseshoes in 6 hours. What is his rate per hour?
7. Oskar, the shirt inspector, examined 400 shirts in 5 hours. What is his hourly rate?
8. Roger drove 67 miles per hour for 6 hours and 45 minutes. How far did he travel?
9. Sally wants to drive from Tallahassee to Orlando, a distance of 257 miles. What speed does she have to average in order to arrive in 4 hours?
10. Whitney was hired as a temporary employee to do some filing. For 5 days, she filed papers 8 hours each day. At the end of the 5 days, she had filed 1,440 papers. How many papers per hour did she file?

Angles

To define an angle, we must first define a ray. A **ray** is a set of an infinite number of points extending indefinitely in one direction from one endpoint. The symbol for a ray is $\rightarrow$ and a ray is named by its endpoint and one other point; for instance, ray $\overrightarrow{AB}$.

An **angle** is the union of two rays having the same endpoint. The two rays are called the **sides** of the angle. The common endpoint is called the **vertex** of the angle. An angle is named by the letter at its vertex or by three letters: the vertex and a point from each side. When a three-letter name is used, the middle letter is the vertex. The symbol for an angle is $\angle$.

$\angle A$ or $\angle ABC$ or $\angle CBA$

The "opening" of the angle is its measurement. An angle is measured in **degrees**, and the measuring tool is the **protractor**. There are three ways to indicate the measure of an angle:

$\angle A = 40$ degrees

$\angle A = 40°$

$m\angle A = 40$

Angles are named according to their measure in degrees.

- **Acute angle**: angle that measures less than 90°
- **Right angle**: angle that measures exactly 90°
- **Obtuse angle**: angle that measures greater than 90° but less than 180°
- **Straight angle**: angle that measures 180°

A degree can be divided into 60 parts, with each part called a **minute** (′). Each minute can be divided into 60 parts, with each part called a **second** (″).

NOTE: When we are talking about angles, the words *minute* and *second* do *not* refer to time.

Here are some useful rules to remember:

- To change degrees to minutes, multiply by 60:

  4 degrees = ? minutes

  $4 \bullet 60 = 240$ minutes
- To change minutes to seconds, multiply by 60:

  10 minutes = ? seconds

  $10 \bullet 60 = 600$ seconds
- To change minutes to degrees, divide by 60:

  300 minutes = ? degrees

  $\frac{300}{60} = 5$ degrees
- To change seconds to minutes, divide by 60:

  2,400 seconds = ? minutes

  $\frac{2{,}400}{60} = 40$ minutes
- To change seconds to degrees, divide by 3,600:

  7,200 seconds = ? degrees

  $\frac{7{,}200}{3{,}600} = 2$ degrees

MODEL PROBLEMS

1. Change 12° 15′ to degrees.

Answer: To change 12° 15′ to degrees, divide 15 by 60 and add it to 12.

$$12 + \frac{15}{60} = 12 + 0.25 \text{ or } 12.25°$$

2. Change 35° 15′ to minutes.

Answer: To change 35° 15′ to minutes, multiply 35 by 60 and add 15 to the product.

$35 \bullet 60 + 15 = 2{,}100 + 15$ or 2,115 minutes

TRY THESE

1. Change 21° to minutes.

2. Change 480′ to seconds.

3. Change 36,000″ to minutes.

4. Change 40° 45′ to degrees.

5. Change 5° 23′ to minutes.

6. Change 56° to seconds.

7. Change 900′ to degrees.

8. Change 48,000″ to degrees.

9. Change 85° to seconds.

10. Change 8° 15′ to minutes.

Arcs

If an angle is rotated all the way around its vertex, the entire rotation will measure 360 degrees. This rotation represents a circle. If the circle is divided into 360 parts, each part will measure 1 degree. Here is another way to describe this: If lines are drawn from these divisions on the circle to the center of the circle, 360 angles will be formed, each angle measuring 1 degree.

The center of a circle is often called O. Angles that are formed at the center of a circle are called **central angles**.

An **arc** is identified by its endpoints: for example, arc AB, or $\overset{\frown}{AB}$. The **measure of an arc** is given in degrees and is determined by the central angle that intercepts the arc. For example, in the drawing below, the angle intercepts $\overset{\frown}{AB}$. Since the angle measures 60 degrees, the measure of $\overset{\frown}{AB}$ is 60 degrees, or m $\overset{\frown}{AB} = 60$.

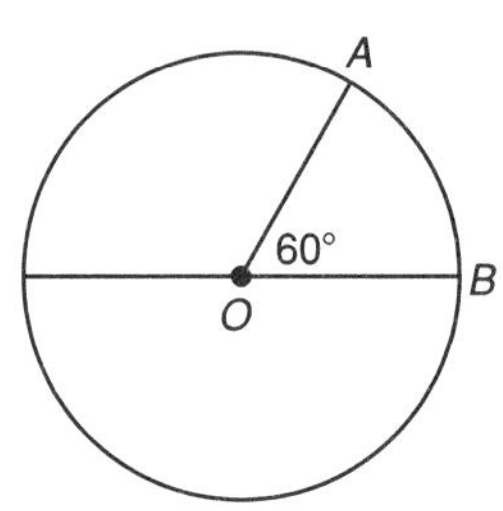

MODEL PROBLEMS

1. How many degrees are in $\overset{\frown}{AB}$?

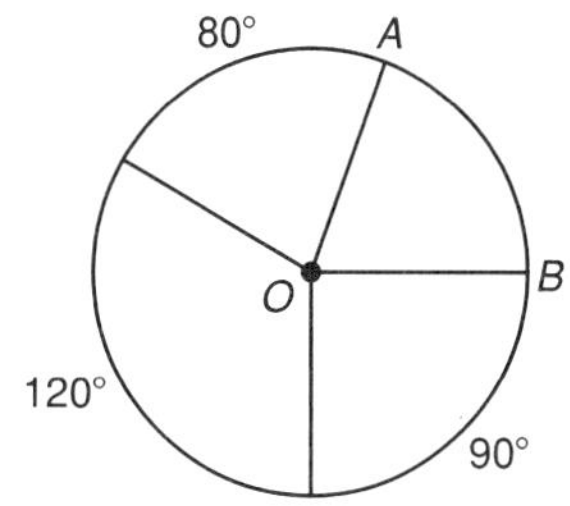

Answer: There are 360° in the complete circle.

$120° + 80° + 90° + \text{m}\,\overset{\frown}{AB} = 360$

$\text{m}\,\overset{\frown}{AB} = 360 - 290 = 70°$

2. How many degrees are in arc x?

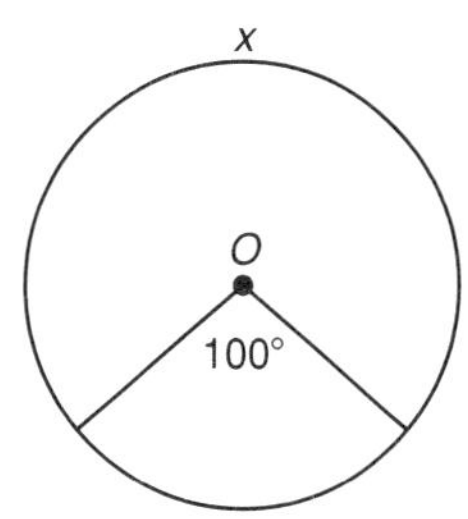

Answer: The measure of the arc intercepted by angle O equals 100°.

The measure of arc x is $360° - 100° = 260°$.

TRY THESE

Find the measure of arc x in each diagram.

1.

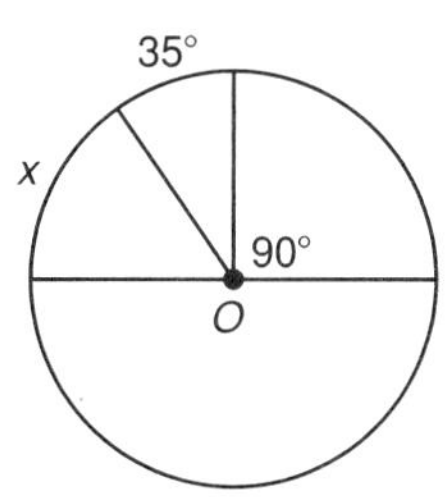

2.

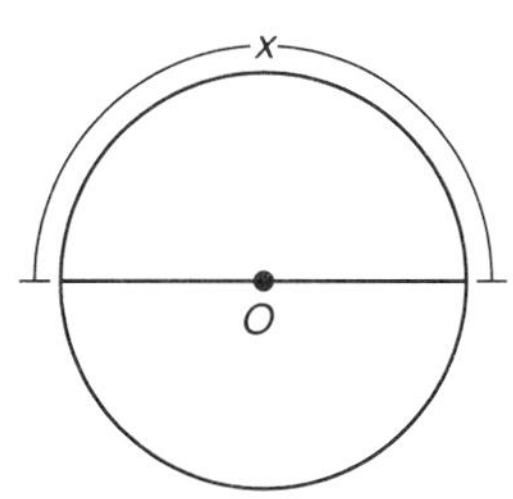

3.

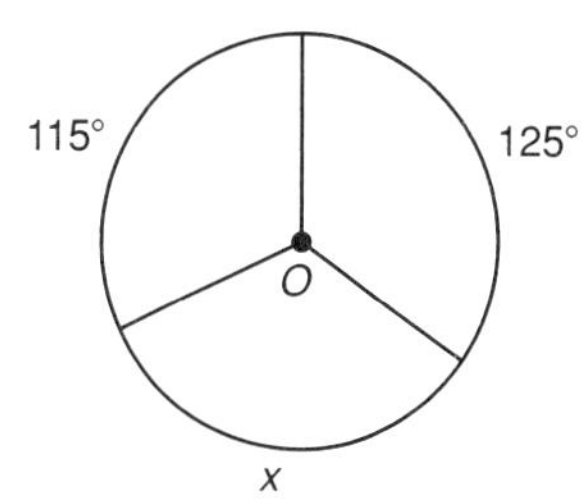

4.

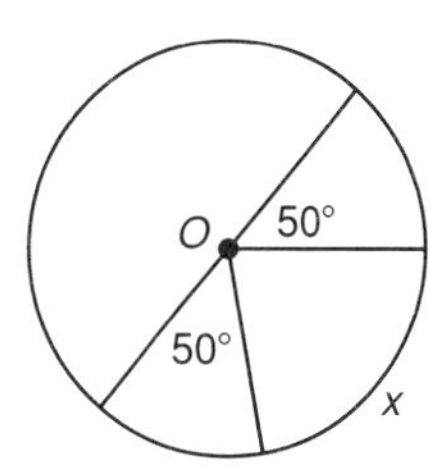

5.

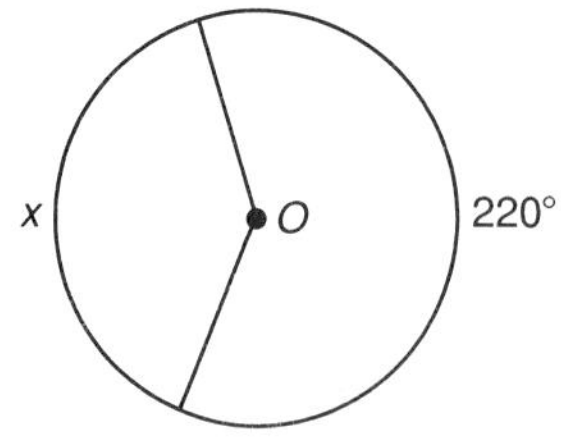

6.

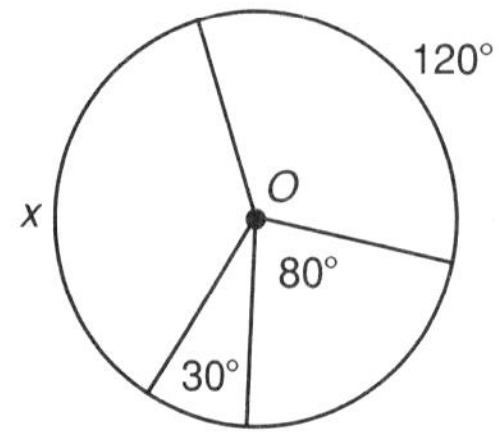

7.

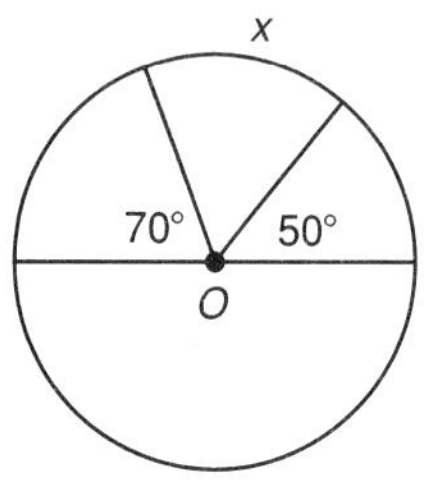

8.

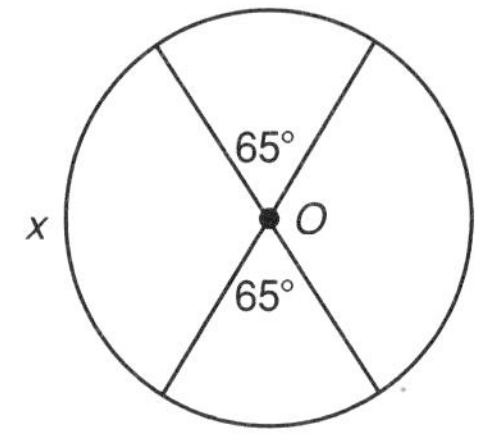

9.

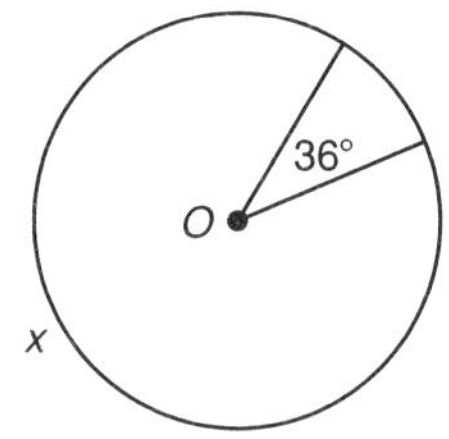

10.

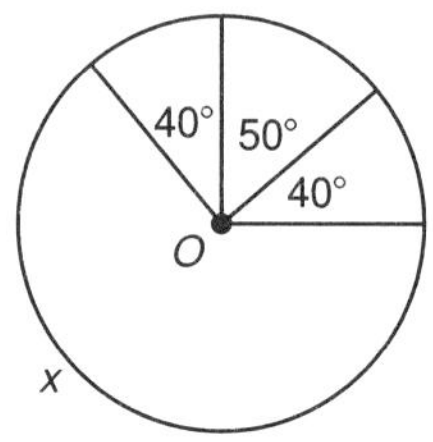

The distance around a circle is called the **circumference**. The **length of an arc** is a fractional part of the circumference.

For example, if the circumference of a circle is 16 inches and an arc on the circle measures 90 degrees, the arc is $\frac{90}{360}$, or $\frac{1}{4}$, of the circumference. Therefore, the arc length $= \frac{1}{4} \bullet 16$ inches $= 4$ inches.

If another arc in the same circle measures 45 degrees, that arc is $\frac{45}{360}$, or $\frac{1}{8}$, of the circumference. To find the length of this arc, multiply: $\frac{1}{8} \bullet 16$ inches $= 2$ inches.

A formula for the length of an arc is $\frac{n}{360} \bullet$ circumference, where n is the measure of the arc in degrees.

NOTE: Finding the circumference will be covered later in this chapter.

MODEL PROBLEMS

1. A circular track has a circumference of 125 yards. When Tom has run a 30° arc, approximately what distance has he run?

   Answer: $\frac{30}{360} \bullet 125 \approx 10.42$ yards

2. The members of a sewing club are adding lace to the edge of a circular tablecloth. So far, they have covered a 200° arc. If the circumference of the circle is 16 feet, how many more inches of lace do they still need to add?

Answer: Arc still to be covered = 360° − 200° = 160°

Length of this arc = $\frac{160}{360} \bullet 16$ feet = $7\frac{1}{9}$ feet

Change feet to inches: $7\frac{1}{9} \bullet 12 = 85\frac{1}{3}$ inches

TRY THESE

1. In a circle with circumference of 24 inches, find the length of an arc intercepted by a 30° central angle.
2. Find the length of an arc intercepted by a 240° central angle in a circle with a circumference of 48 centimeters.
3. An arc of a central angle contains 72° and is 10 inches long. What is the circumference of the circle?
4. The circumference of a circle is 84 meters. What is the degree measure of the central angle that intercepts an arc of 21 meters?

Items 5 and 6: Find the measure of arc x in each diagram.

5.

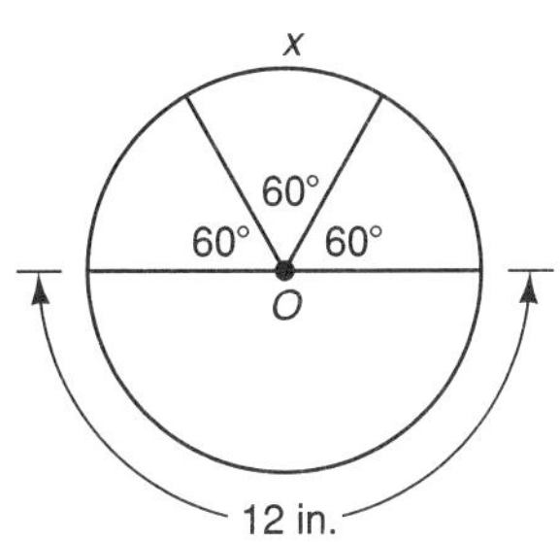

6.

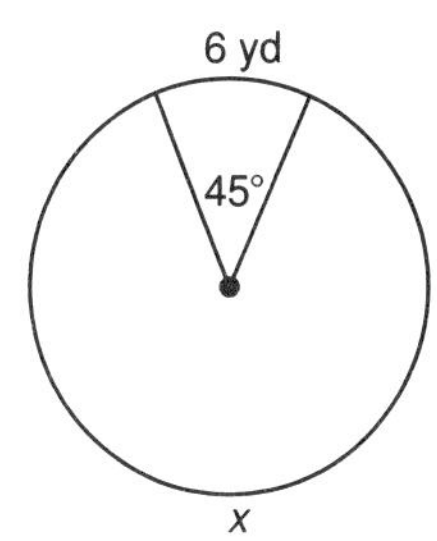

7. The diagram shows a circular pattern designed by a student. The circumference of the pattern will be 12 inches, and the pattern will have a string around the edge. How much string will be needed to cover the arc indicated by the arrow?

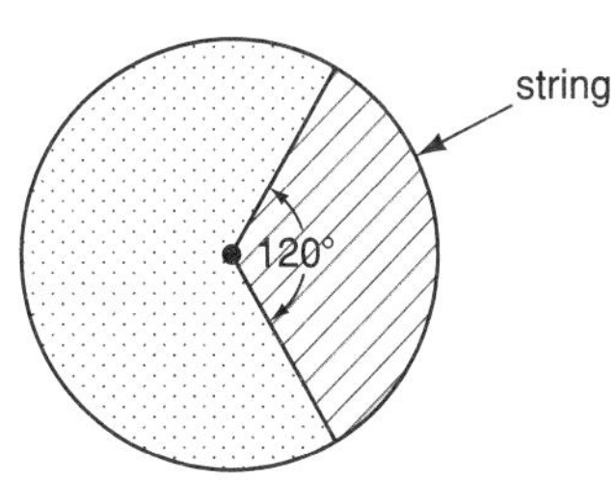

8. A central angle intercepts an arc that has a length of 30 centimeters. The entire circle has a circumference of 120 centimeters. What is the measure of the central angle?

9. There is a corner cupboard in Willa's kitchen that rotates out for easy access. If the shelf had been a circle, it would have a circumference of 36 inches. How long is the arc from *A* to *B*?

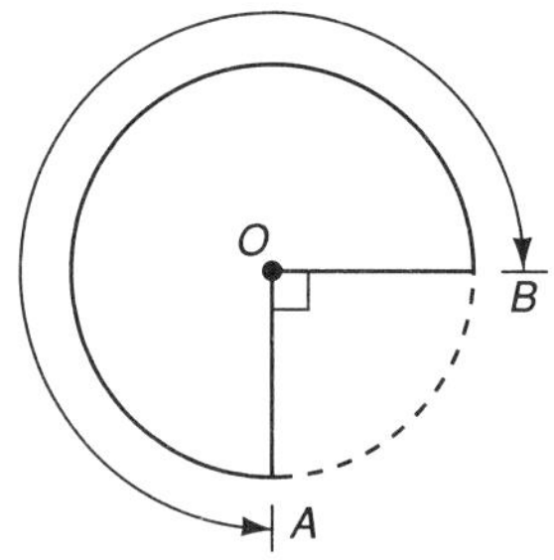

10. The tire on Martin's bike has a circumference of 180 centimeters. If a mark on the tire travels an arc of 40°, how far has Martin traveled?

Assessment

Multiple-Choice Problems

1. In the drawing below, what type of angle is $\angle x$?

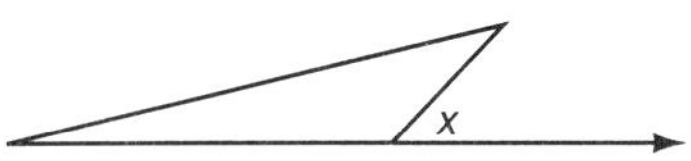

A. acute **C.** obtuse

B. right **D.** straight

2. Marson runs the 60-yard dash in 12 seconds. What is his rate in yards per second?

A. 0.1 yard per second

B. 0.5 yard per second

C. 2 yards per second

D. 5 yards per second

3. In preparation for the opening of a new Super Center, Judith spent 8 hours stocking shelves. She stocked 116 shelves during her shift. What is her rate for stocking shelves?

A. 0.069 shelf per hour

B. 6.9 shelves per hour

C. 14.5 shelves per hour

D. 20.2 shelves per hour

4. During the school year, Lydia swims an average of 10 hours per week. During the months of June, July, and August, she swims a total of 143 hours. During these three months, how close is she to her average school year swimming time?

A. more than 20 hours above her average

B. less than 20 hours above her average

C. more than 20 hours below her average

D. exactly 20 hours below her average

5. Hector flew his plane 1284 kilometers in 6 hours. What was his rate in kilometers per hour?

A. 204 kilometers per hour

B. 214 kilometers per hour

C. 223 kilometers per hour

D. 224 kilometers per hour

Gridded-Response Problems

6. The blade on a windmill turns through an angle of 60° 48′. Through how many degrees does it turn?

7. In Britain, the symbol shown below is the logo for the Campaign for Nuclear Disarmament. Many regard it as the Peace Symbol. The two equal smaller central angles are 60 degrees each. The other two central angles are equal in measure.

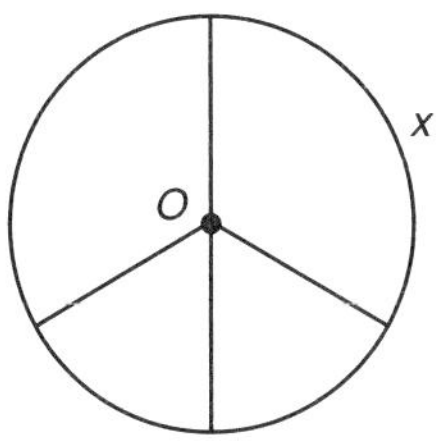

If the circumference is 243 centimeters, how long is the arc labeled x?

8. The London Eye Ferris Wheel has a circumference of 424 meters. If a rider travels over an arc of 45°, how far has the rider gone?

9. Wayne drove his car 567 miles using 18 gallons of gasoline. How many miles per gallon did he average?

10. How many hours longer will it take a car averaging 55 miles per hour than a car averaging 70 miles per hour to go a distance of 770 miles?

6

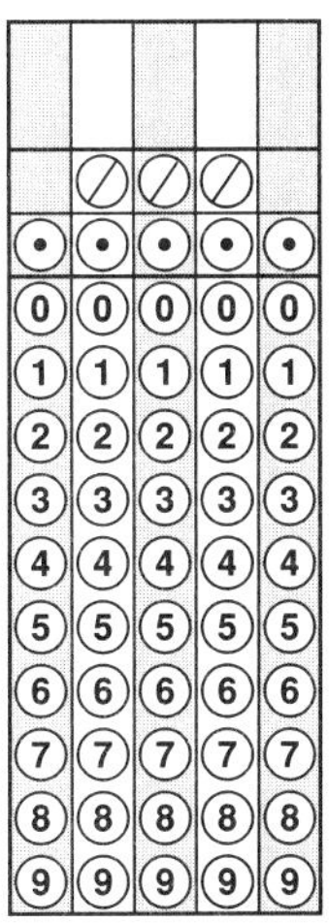

7

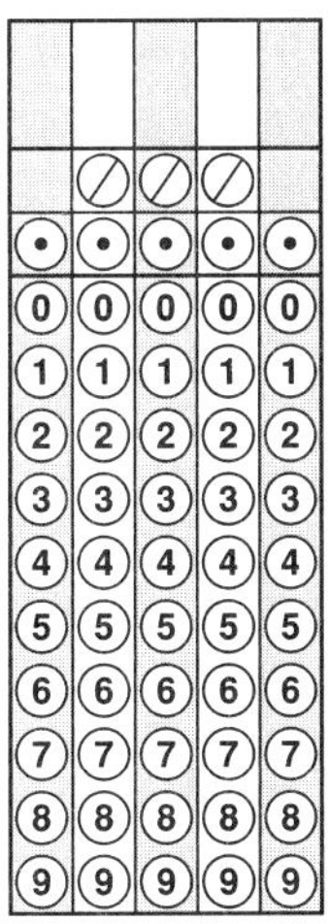

8

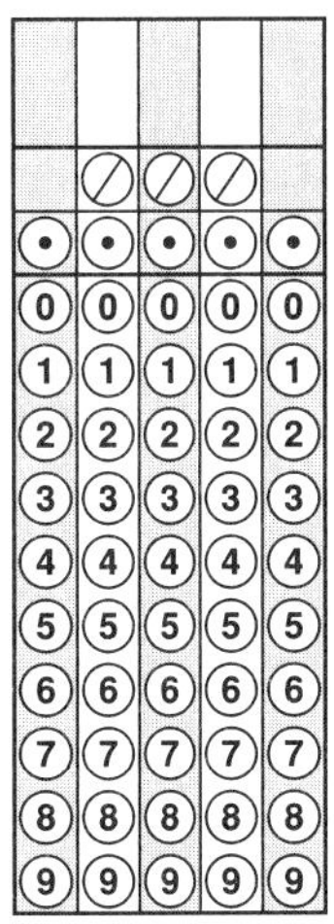

9

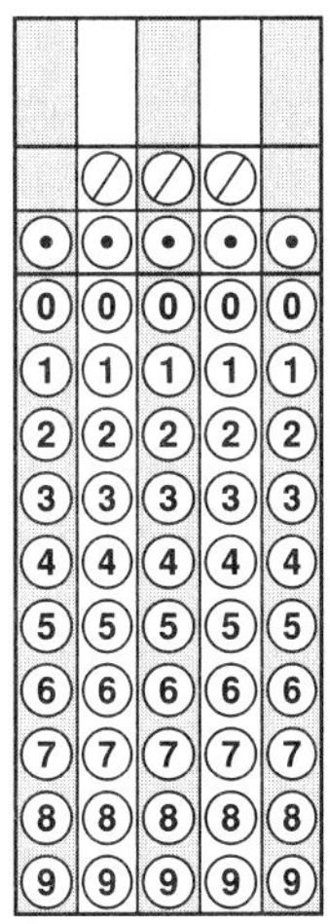

10

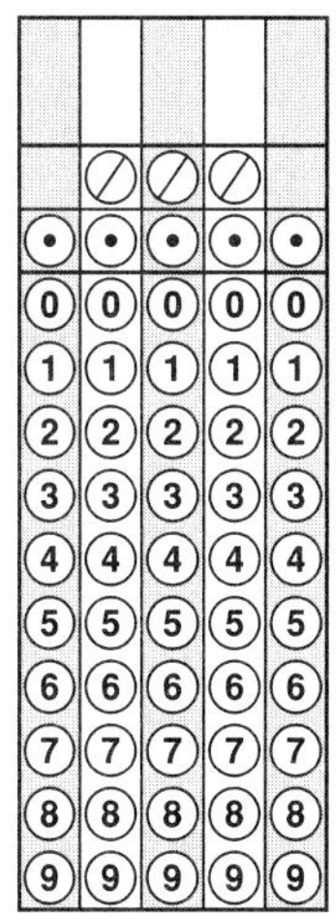

B USING CONCRETE AND GRAPHIC MODELS TO DERIVE AND APPLY FORMULAS FOR PERIMETER, CIRCUMFERENCE, AREA, SURFACE AREA, AND VOLUME

A group of students met at the park to play volleyball on Saturday afternoon. They wanted to tie the net to two trees. Could they tie the net to one tree and stretch it to reach the other tree? Were the trees too far apart or too close together? They had a measurement problem. Did they go home and get a measuring device such as a measuring tape or a yardstick? No—they simply stretched the net out between the trees to see if it would fit.

Measurement is a mathematical tool that we use in almost every aspect of our lives, from everyday situations like this one to intricate engineering problems at NASA. This chapter will review how, when, and why different things are measured. We will start with **length**.

In the example of the volleyball game, the unit of length used was one volleyball net. Walking distances are sometimes measured in paces a person takes when walking normally—the unit is one pace. Other units of measure are a thumb length and a hand span. All of these are measurements of length, even though they may differ greatly from person to person. Units of measurement like these vary, so they are not exact. They are called **nonstandard**.

By contrast, an exact system of measurement does not vary and is called a **standard** system. Two standard systems of measurement have been adopted for general use: the metric system and the English system.

Metric units of measure

The **metric system** is based on the decimal system. Units increase and decrease by factors of ten. The basic unit is called a **meter**, abbreviated m. For other units, the prefix of the unit name indicates how that unit is related to the meter.

Here is a table of prefixes for the metric system:

Prefix	Meaning	Value relative to meter
kilo (km)	1000	10^3 m
hecto (hm)	100	10^2 m
deka (dkm)	10	10 m
unit (meter)	1	1 m
deci (dm)	0.1	10^{-1} m
centi (cm)	0.01	10^{-2} m
milli (mm)	0.001	10^{-3} m

In the metric system, a bicycle might be 110 centimeters (cm) high, and the distance between two cities might be 10 kilometers (km).

For conversions from one metric unit to another, follow these rules:

- To convert from a smaller to a larger unit, divide by a power of 10 (the **conversion factor**).
- To convert from a larger to a smaller unit, multiply by a power of 10 (the conversion factor).

EXAMPLES

Change 900 centimeters to meters: $\frac{900}{100} = 9$ meters

Change 9 meters to centimeters: $9 \times 100 = 900$ centimeters

TRY THESE ✓

Change:

1. 17 centimeters to millimeters
2. 3.5 decimeters to centimeters
3. 87 meters to decimeters
4. 704 centimeters to meters
5. 8 hectometers to kilometers
6. 325 millimeters to centimeters
7. 950 meters to kilometers
8. 1.25 centimeters to millimeters
9. 640 millimeters to meters
10. 8 kilometers to meters

Customary units of measure

The second standard system of measurement is the **English system**, or **customary system**. This is based on standard units, but it is not a decimal system—it does not increase and decrease by powers of 10. Length is measured by four common units: inches, feet, yards, and miles.

NOTE: Inches can be abbreviated as in. or ″. Feet can be abbreviated as ft or ′.

Below are conversions in the English system:

1 mile = 5,280 feet	1 yard = 36 inches
1 yard = 3 feet	1 foot = 12 inches

The rules for conversion in the English system are:

- To convert from a smaller unit to a larger unit, divide by a conversion factor.
- To convert from a larger unit to a smaller unit, multiply by a conversion factor.

EXAMPLES

Change 48 inches to feet: $\frac{48}{12} = 4$ feet

Change 13 feet to inches: $13 \times 12 = 156$ inches

TRY THESE

Change:

1. 11 feet to inches
2. 15 yards to feet
3. 10,560 feet to miles
4. 60 inches to feet
5. 4 feet 7 inches to inches
6. $2\frac{1}{2}$ miles to feet
7. 18 feet to inches
8. $8\frac{1}{3}$ yards to feet
9. 6 miles to yards
10. 90 inches to feet

Now fill in the following table:

Customary system		Metric system	
15 inches	feet	16 m	cm
3 miles	feet	146 dkm	m
1.5 yards	feet	4 hm	km
8 feet	inches	9 km	cm
7,040 yards	miles	18.3 cm	dkm
27 inches	yards	24 dm	hm
$2\frac{1}{4}$ miles	feet	57 km	m
80 feet	yards	0.32 dkm	dm
$\frac{1}{4}$ mile	inches	123 m	km
56 inches	yards	76 hm	dm

NOTE: For more practice, turn to page 464.

Measurement of plane figures

Many real-world measurements involve figures that are made up of lines. These **plane** (flat) geometric figures that are bounded by three or more sides are called **polygons**. The point where any two sides meet is called the **vertex**. If all sides are equal in length and all angles have an equal measure, the figure is called a **regular polygon**. The most common polygons are listed below.

Polygon	Number of sides
triangle	3
quadrilateral	4
pentagon	5
hexagon	6
octagon	8
decagon	10
dodecagon	12

Triangles can be classified according to their sides and angles.

Classification by sides

- **Equilateral triangle**: All three sides are equal in length.
- **Isosceles triangle**: At least two sides (legs) are equal in length.
- **Scalene triangle**: No sides are equal in length.

Classification by angles

- **Equiangular triangle**: All three angles are equal in measure.
- **Right triangle**: Has one **right angle** (a 90-degree angle formed by perpendicular lines).
- **Obtuse triangle**: Has one **obtuse angle** (an angle between 90 and 180 degrees).
- **Acute triangle**: Has three **acute angles** (angles less than 90 degrees).

Quadrilaterals are also classified according to specific properties.

- **Parallelogram**: Opposite sides have equal length and are **parallel** (the same distance apart at all points, so that they will never intersect).

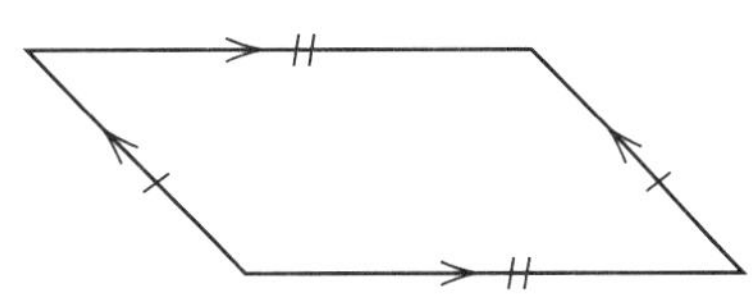

Parallelogram

- **Rectangle**: Opposite sides have equal length and are parallel, and all four angles are right angles.
- **Rhombus**: All four sides have equal length and opposite sides are parallel.
- **Square**: All four sides have equal length, opposite sides are parallel, and the four angles are right angles.
- **Trapezoid**: Exactly one pair of opposite sides are parallel.

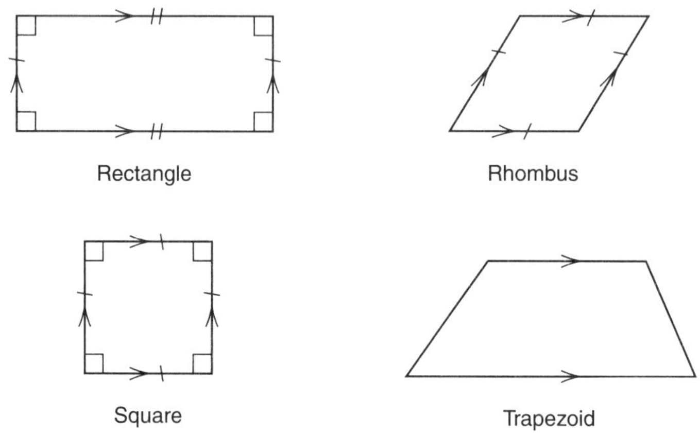

Diagrams are marked with these symbols: Perpendicular Parallel Equal in length

PERIMETER

Perimeter is the measurement of the distance around any figure. To remember this, think of peRIMeter as the measurement of the RIM of a figure. To get a perimeter measurement, simply add up the length of all the sides.

MODEL PROBLEM

Below is a drawing of a park. The community wants to fence in the entire park. How many feet of fencing, including the gate, will be necessary?

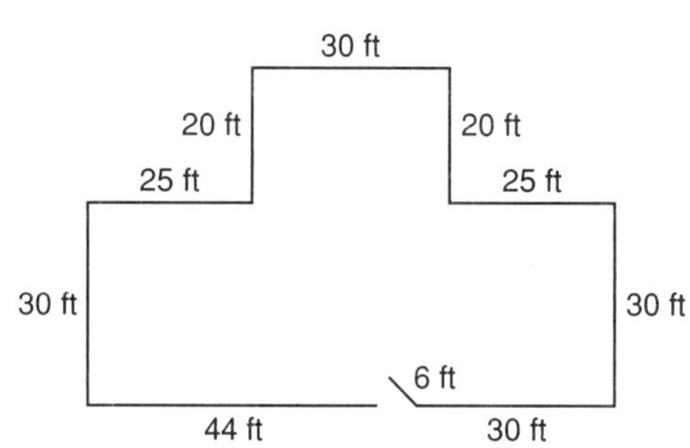

Answer: Add up all the sides plus the gate:

44 + 6 + 30 + 30 + 25 + 20 + 30 + 20 + 25 + 30 = 260 feet

The park will require 260 feet of fencing.

TRY THESE ✔

1. Jennifer is designing a flower garden in her back yard, as shown below. She wants to put a decorative edging all around the garden. How many feet of edging will she need to buy?

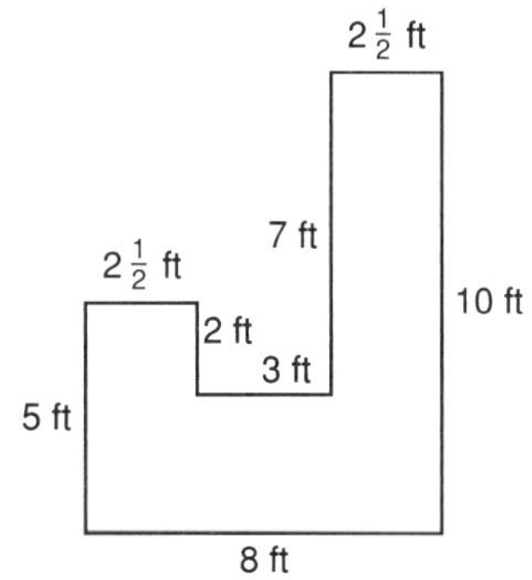

2. Marion is working on a banner. She wants to sew the letter **M** on the banner. Using the diagram below, find the perimeter of the **M**.

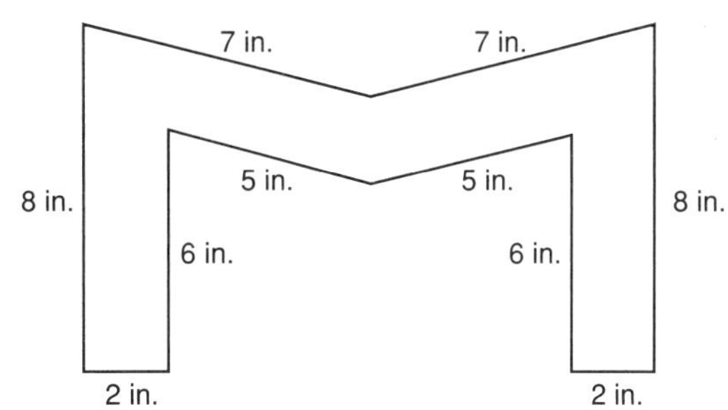

3. Arlo looked at the drawing of his house shown below. What is the perimeter of the house?

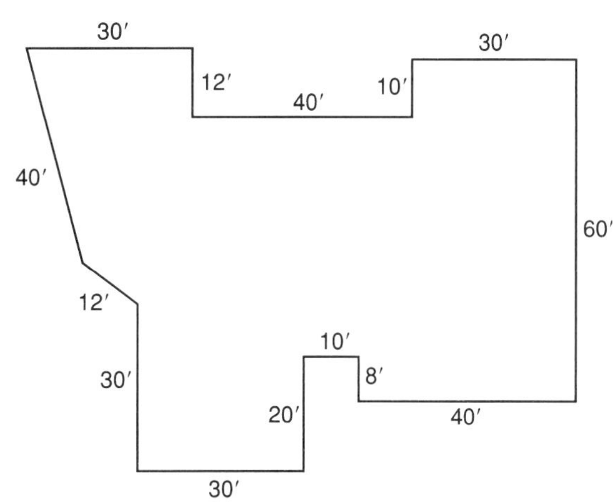

Perimeter of Triangles When the perimeter measure of a polygon is required, there are some formulas that may make the computation easier.

If the figure is a triangle, the formula for its perimeter is $P = a + b + c$ where a, b, and c represent the lengths of the three sides. In an equilateral triangle, all three sides are the same length, so the formula for its perimeter is simply $P = 3s$ where s is the length of any side.

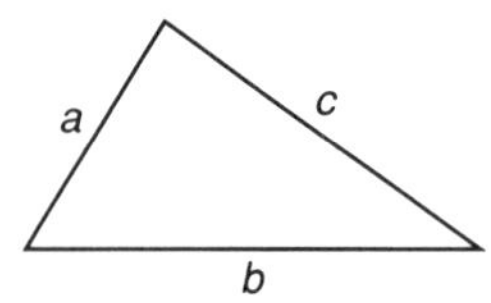

$P = a + b + c$

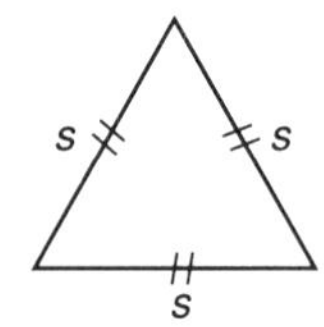

$P = 3s$

MODEL PROBLEMS

1. Find the perimeter of a triangle with sides that measure 17 yards, 8 yards, and 14 yards.

Answer:

$P = a + b + c$

$P = 17 + 8 + 14$

$P = 39$ yards

2. A design is in the shape of an equilateral triangle, each side measuring 2 centimeters. What is the perimeter of the design?

Answer:

$P = 3s$

$P = 3 \cdot 2$

$P = 6$ centimeters

TRY THESE

Fill in the missing measurements.

Triangle	Side *a*	Side *b*	Side *c*	Perimeter
1.	15 in.	12 in.	10 in.	
2.	22 m	48 m	32 m	
3.	11 yd	20 yd		47 yd
4.		10 ft	9 ft	23 ft
5.	64 mm		76 mm	170 mm

Fill in the missing measurements.

Equilateral triangle	Side	Perimeter
6.	14 km	
7.		183 miles
8.		52 cm

9. How many feet of hedge are needed to enclose a triangular lot with sides measuring 190 feet, 203 feet, and 177 feet?

10. A toy is in the shape of an equilateral triangle, 7 centimeters on a side. What is the perimeter of the toy?

Perimeter of Rectangles and Squares In a rectangle, opposite sides have equal length. Therefore, a formula for the perimeter of a rectangle is $P = 2l + 2w$ where l represents the length of the rectangle and w represents the width.

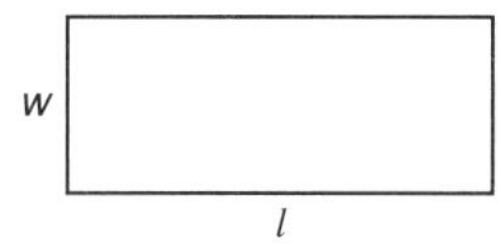

$P = 2l + 2w$

MODEL PROBLEMS

1. Find the perimeter of a rectangle with a length of 14 inches and a width of 5 inches.

Answer:

$P = 2l + 2w$

$P = (2 \bullet 14) + (2 \bullet 5)$

$P = 38$ inches

2. How many centimeters of border material would be needed to enclose a rectangle that is 35 centimeters by 12 centimeters?

Answer:

$P = 2l + 2w$

$P = (2 \bullet 35) + (2 \bullet 12)$

$P = 94$ centimeters

3. Find the length of a rectangle if its perimeter is 36 centimeters and its width is 8 centimeters.

Answer:

$P = 2l + 2w$

$36 = 2l + (2 \bullet 8)$

$36 = 2l + 16$

$36 - 16 = 2l + 16 - 16$

$\frac{20}{2} = \frac{2l}{2}$

$10 = l$

$l = 10$ centimeters

TRY THESE

Fill in the following table.

Rectangles

	Length	Width	Perimeter
1.	24 feet	10 feet	
2.	7.5 yards	2.3 yards	
3.	12 km		32 km
4.	82.5 cm	31.7 cm	
5.	12 in.		56 in.
6.	$3\frac{1}{4}$ feet	$1\frac{1}{2}$ feet	
7.	4 in.		20 in.
8.		9.3 mm	28.2 mm

A square is a rectangle in which all sides are equal. Therefore, a formula for the perimeter of a square is $P = 4s$ where s represents the length of a side.

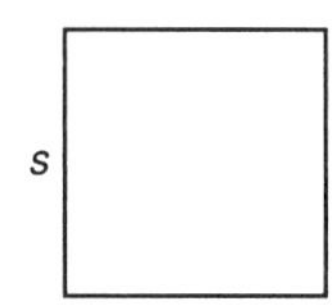

$P = 4s$

MODEL PROBLEMS

1. The town square in Sunnyville is truly a square. If a side of the town square measures 60 yards, what is its perimeter?

   Answer:

   $P = 4s$

   $P = 4 \bullet 60$

   $P = 240$ yards

2. A baseball diamond is actually in the shape of a square. If one side of the square is 90 feet, and a player runs all the way around the diamond, how far does the player go?

   Answer:

   $P = 4s$

   $P = 4 \bullet 90$

   $P = 360$ feet

TRY THESE

Items 1–3: Complete the table.

Square	Side	Perimeter
1.	21 m	
2.	$3\frac{1}{2}$ in.	
3.		42 ft

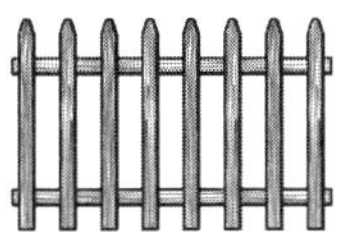

4. A square 4 feet on a side will be marked off with a fence. How many feet of fence will be needed?

5. You want to make a square banner for a homecoming game. Each side will be 2.5 meters long. How much trim will you need to go around the edges of the banner?

Perimeter of Irregular Figures Irregular figures, such as the one below, are sometimes made up of two or more polygons.

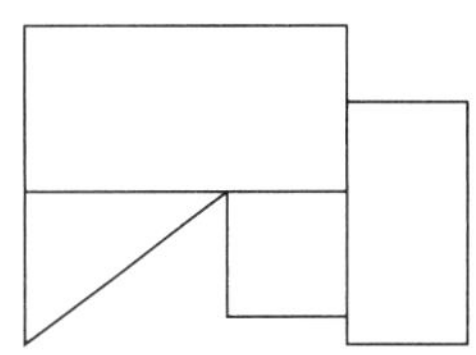

To find the perimeter of an irregular figure made up of polygons, you need to calculate the sum of the sides. The measures of the sides depend on the properties of the particular figures. For instance, in a rectangle opposite sides are equal, and in a square all sides are equal.

MODEL PROBLEM

This figure consists of a rectangle and a square. Find its perimeter.

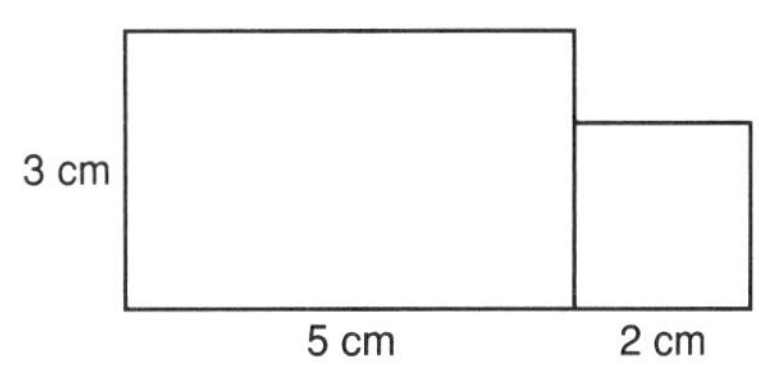

Answer: Use the properties of the rectangle and the square to find the other measurements.

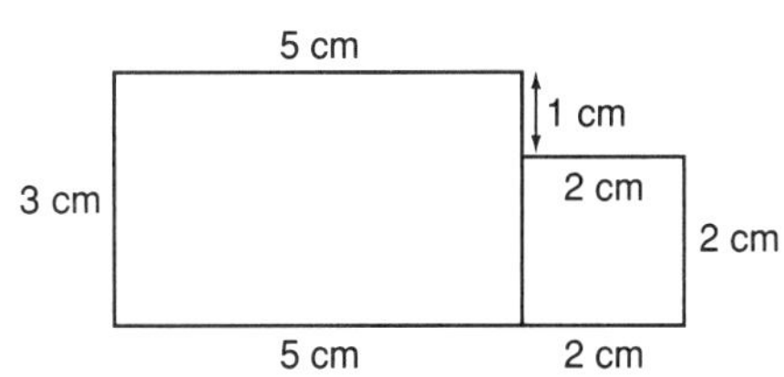

$3 + 5 + 2 + 2 + 2 + 1 + 5 = 20$

The perimeter of this irregular figure is 20 centimeters.

TRY THESE

1. This figure is made up of two squares and a rectangle. Find its perimeter.

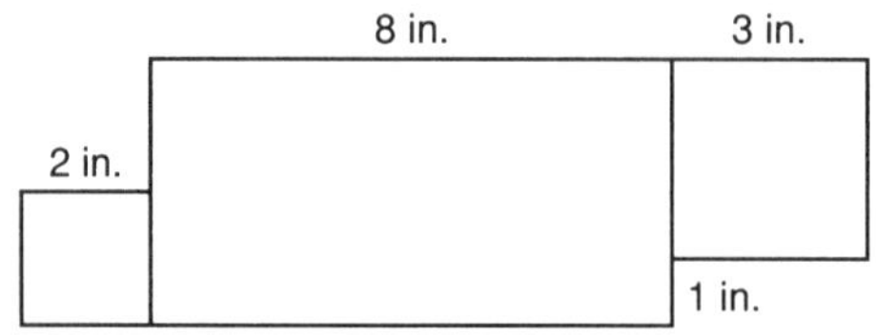

2. The figure below is made up of four squares and one rectangle. Find its perimeter.

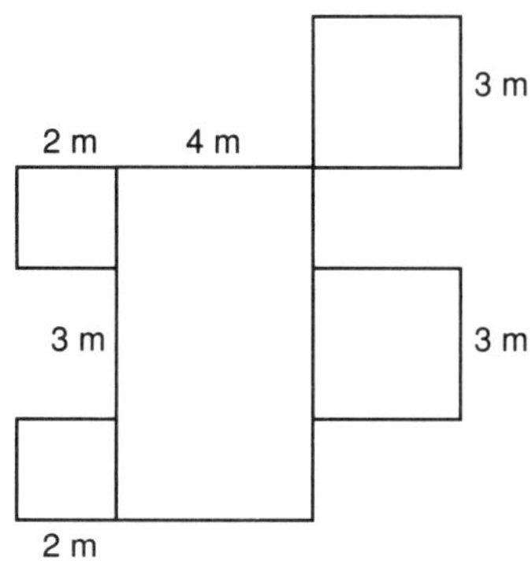

3. The figure below is made up of one rectangle, two squares, one equilateral triangle, and one isosceles triangle. Find its perimeter.

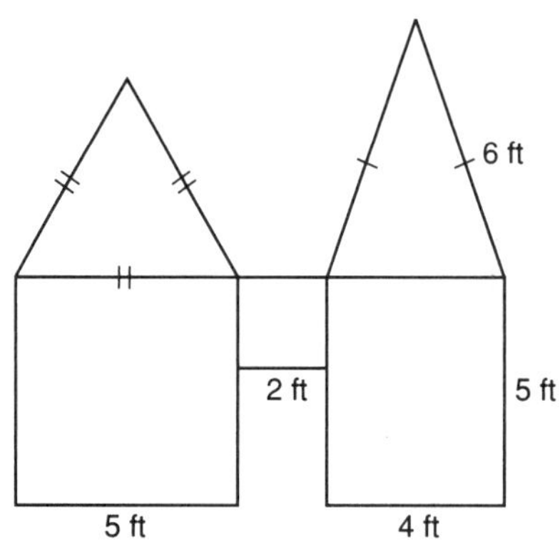

CIRCUMFERENCE

A **circle** is defined as closed curved line in a plane, with all points of the circle the same distance from a point within the circle called the **center** (often labeled O).

A line segment drawn from the center of the circle to any point on the circle is called the **radius** (r). A line segment drawn from any point on the circle through the center to another point on the circle is called a **diameter** (d). In any circle, the length of the diameter is twice the length of the radius: $d = 2r$. Therefore, $r = \frac{d}{2}$.

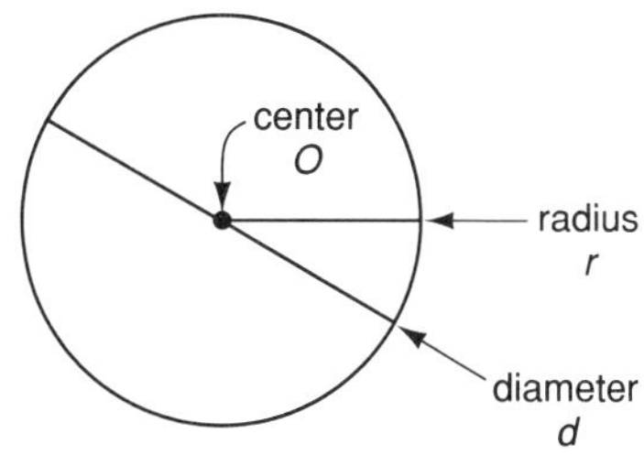

The distance around a circle is called the **circumference** (C). The circumference is equal to pi (π) times the diameter. The formula for circumference is $C = \pi d$. Since $d = 2r$, this formula can also be written as $C = 2\pi r$.

The symbol π represents an irrational number. For computational purposes, the approximation 3.14 or $\frac{22}{7}$ is used. To show that it is an approximation, the symbol $\approx$ is used instead of $=$.

MODEL PROBLEMS

1. Find the circumference of a circular table that has a diameter of 5 feet. Use $\pi \approx 3.14$.

Answer:

$C = \pi d$

$C \approx 3.14 \bullet 5$

$C \approx 15.7$ feet

2. Find the circumference of a hoop if the radius of the hoop is 14 centimeters. Use $\pi \approx \frac{22}{7}$.

Answer:

$C = 2\pi r$

$C \approx 2 \bullet \frac{22}{7} \bullet 14$

$C \approx 88$ centimeters

TRY THESE

Items 1–5: Complete the following table.

	Radius	Diameter	Approximation of π	Circumference
1.	10 in.		3.14	
2.		21 cm	$\frac{22}{7}$	
3.	6.4 ft		3.14	
4.		1.4 m	$\frac{22}{7}$	
5.	$\frac{5}{2}$ yd		3.14	

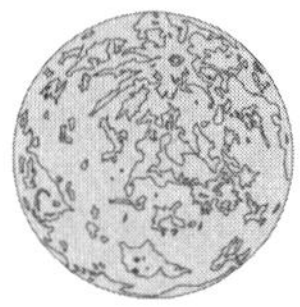

6. Find the circumference of a circular pattern that has a diameter of 3 millimeters. Use $\pi \approx 3.14$.

7. What is the circumference of the moon if its radius is 1,050 miles? Use $\pi \approx \frac{22}{7}$.

AREA

The measurement of the area of any surface is calculated in **square units**. For example, if a square measures 1 inch on each side, its area would be 1 square inch. The same is true of a square foot, a square mile, a square meter, and so forth.

Square units can be abbreviated as 1 sq in., 1 sq ft, 1 sq m, and so on. Another way to write these is 1 in.^{2}, 1 ft^2, 1 m^2, and so on.

Conversion factors for inches, feet, and yards are shown below.

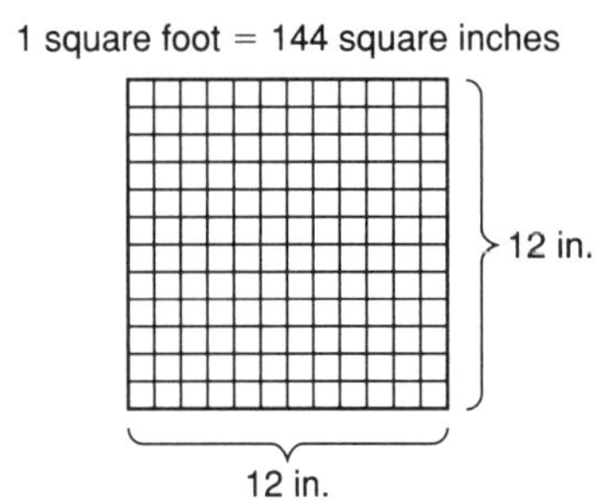

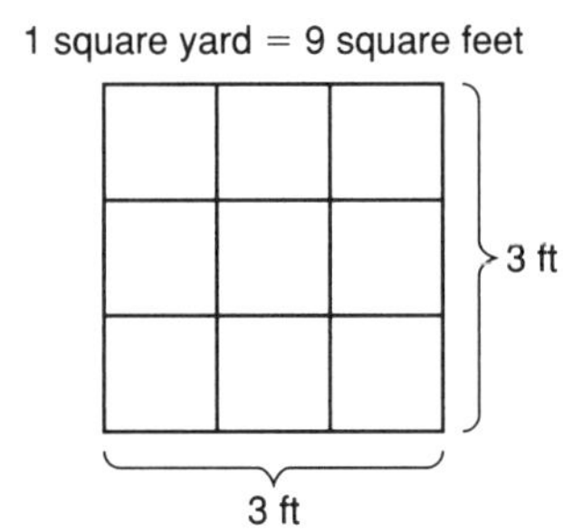

MODEL PROBLEMS

1. 45 square feet = ? square yards

   Answer: To convert from a smaller square unit to a larger square unit, divide by the conversion factor.

   $\frac{45}{9} = 5$ square yards

2. 10 square feet = ? square inches

   Answer: To convert from a larger square unit to a smaller square unit, multiply by the conversion factor.

   $10 \bullet 144 = 1{,}440$ square inches

Conversions are done in a similar way in the metric system. For example, 1 square meter = 100 • 100 or 10 000 square centimeters. This can be abbreviated as 1 sq m = 10 000 sq cm or as 1 m^2 = 10 000 cm^2.

TRY THESE

Fill in the chart with equivalent measures.

Customary system		Metric system	
0.5 sq mile	sq ft	12 sq cm	sq m
36 sq ft	sq yd	900 sq m	sq dm
8 sq yd	sq ft	63 sq dkm	sq hm
10 sq ft	sq in.	47 sq hm	sq m
1,760 sq yd	sq ft	28 sq hm	sq km

Area of Rectangles and Squares The area (A) of a rectangular surface can be found by multiplying the length of the rectangle by the width. The formula is $A = lw$ where l represents the length and w represents the width.

Since the length and width of a square are the same, the formula for the area of a square is $A = s^2$ where s is the length of a side.

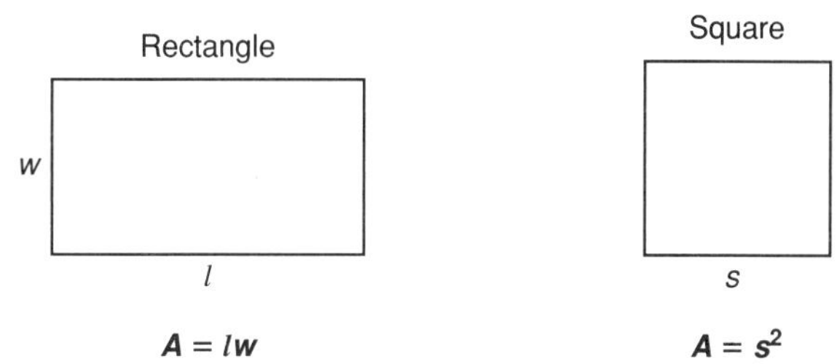

MODEL PROBLEMS

1. A rectangle measures 15 feet by 4 feet. What is its area?

Answer:

$A = lw$

$A = 15 \bullet 4$

$A = 60$ square feet

2. How many square yards are in a football field that is 120 yards long by 160 feet wide?

Answer: Before this problem can be calculated, the units of measure must be the same. Convert 160 feet to yards: $\frac{160}{3} = 53\frac{1}{3}$ yards. Then find the area.

$A = lw$

$A = 120 \bullet 53\frac{1}{3}$

$A = 6{,}400$ square yards

3. What is the area of a square plot of land that is 34 feet on a side?

Answer:

$A = s^2$

$A = 34^2$

$A = 1{,}156$ square feet

TRY THESE

Items 1–7: Complete the tables.

Rectangle	Length	Width	Area
1.	22 yd	9 yd	
2.	56 ft	32 ft	
3.	$8\frac{1}{2}$ cm	$5\frac{2}{3}$ cm	
4.	7.4 m	6.2 m	

Square	Side	Area
5.	9 m	
6.	1.2 in.	
7.	$3\frac{1}{5}$ ft	

8. The dimensions of a rectangular playground are 18 yards by 10 yards. What is the area of the playground?

9. A piece of fabric measures 18 inches by 4.5 yards. What is the square yardage of the fabric?

10. Hana is making a quilt from pieces of fabric. Each piece is a square 6 inches on a side. How many square inches would be in 4 pieces?

Area of Parallelograms To find the area of a parallelogram, multiply the base of the parallelogram by its height. The **height**, or **altitude, of a parallelogram** is defined as the perpendicular distance between the **base** and the side that is parallel to the base. The formula for the area is $A = bh$.

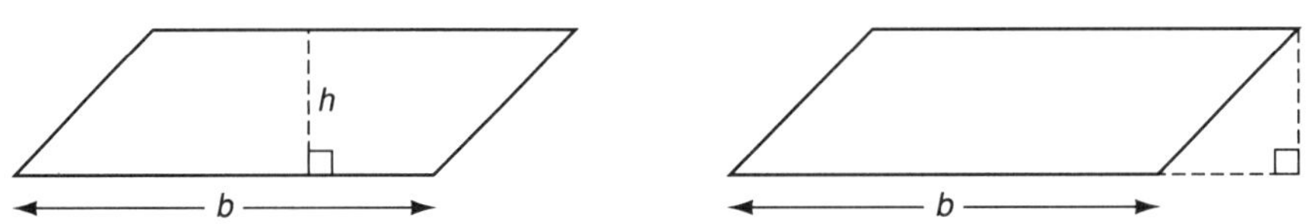

MODEL PROBLEMS

1. A parallelogram has a base of 25 inches and a height of 4 inches. Find its area.

Answer:

$A = bh$

$A = 25 \bullet 4$

$A = 100$ square inches

2. If the area of a parallelogram is 30 square feet and its base is 10 feet, what is its height?

Answer:

$A = bh$

$30 = 10h$

$h = 3$ feet

TRY THESE

Items 1–3: Find the area of the parallelogram.

1.

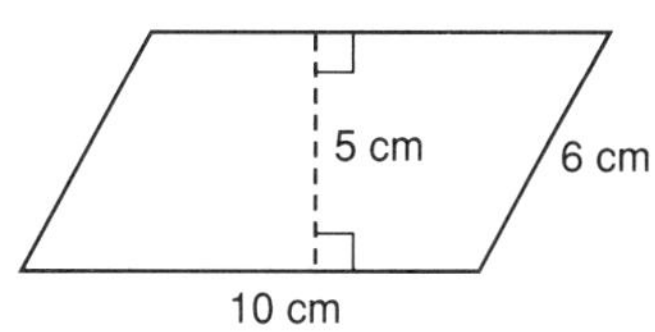

2.

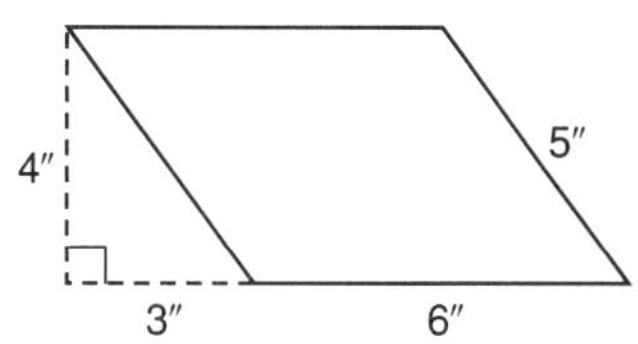

3.

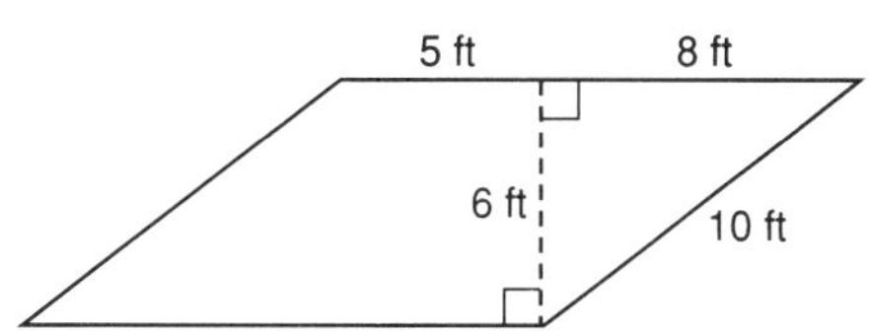

Items 4 and 5: Use the following diagram.

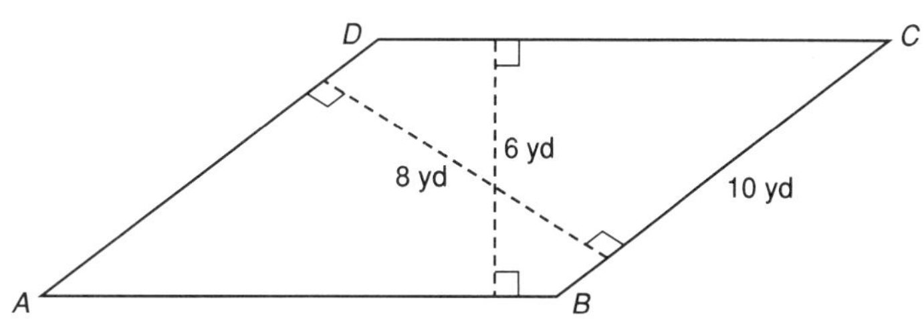

4. What is the area of parallelogram *ABCD*?

5. Find the length of *AB*.

6. In the diagram below, what is the length of OP?

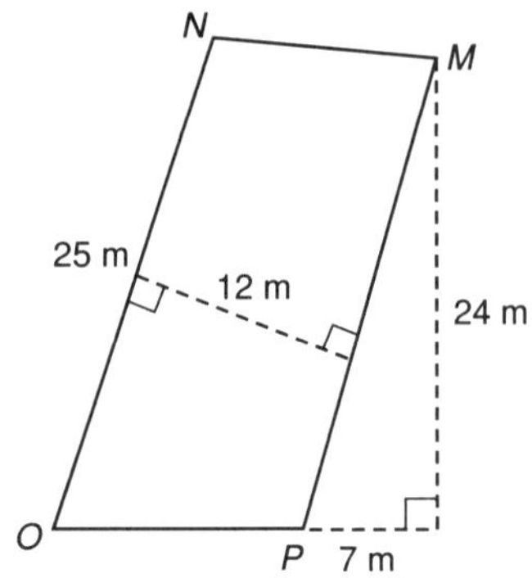

7. Find the area of a parallelogram with a base of 4.7 meters and an altitude of 2.1 meters.

8. A parallelogram has a height of 14 feet and an area of 112 square feet. What is its base?

Area of Triangles The area of a triangle is half the area of a parallelogram: one-half times base times height. The formula is $A = \frac{1}{2}bh$. The **height of a triangle** is measured as a line segment drawn from the vertex of the triangle perpendicular to (forming right angles with) the opposite side. (The height is also called the **altitude**.)

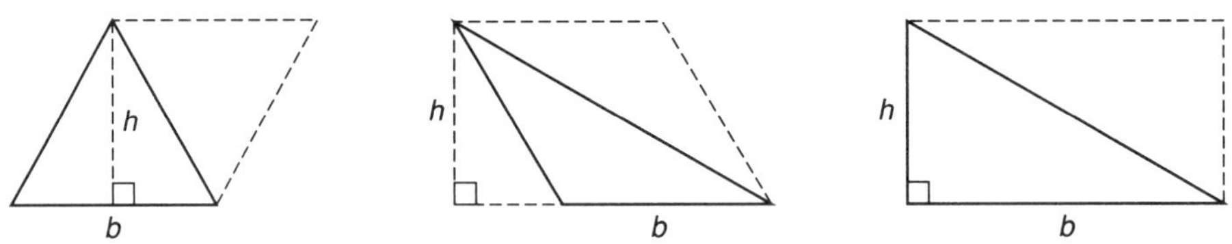

In a right triangle, the sides have specific names. The two sides that form the right angle of the triangle are called **legs**. The side opposite the right angle is called the **hypotenuse** and is the longest side. There is a theorem called the **Pythagorean theorem** that relates these three sides:

$$\text{hypotenuse}^2 = (\text{one leg})^2 + (\text{other leg})^2$$

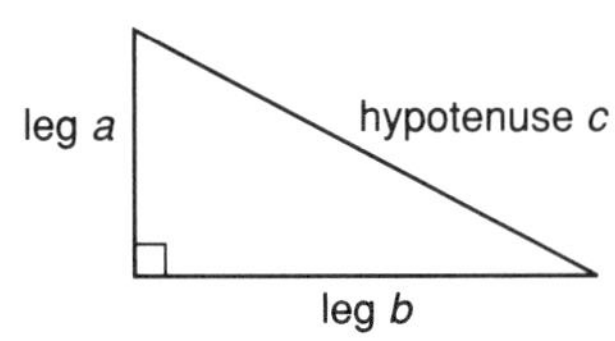

This is usually written as $c^2 = a^2 + b^2$, where c is the length of the hypotenuse and a and b are the lengths of the legs. Then $c = \sqrt{a^2 + b^2}$.

An isosceles triangle has two equal sides. The angles opposite these sides are called the **base angles**. The third angle is called the **vertex angle**. Here, $\angle A$ is the vertex angle and $\angle B$ and $\angle C$ are the base angles.

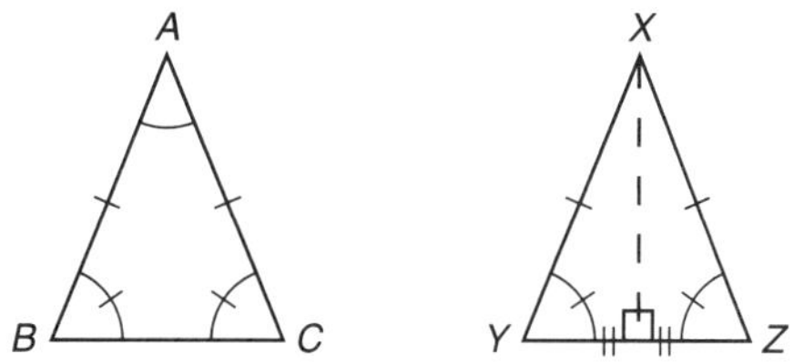

In an isosceles triangle, if a line is drawn from the vertex angle perpendicular to the base, it divides the base ($\overline{YZ}$) into two equal parts. In that case, the line is said to **bisect** the base, and the base is said to be **bisected**.

MODEL PROBLEMS

1. If a square block is 300 yards on each side, what is the length of the diagonal?

   Answer: The diagonal of a square divides the square into two **congruent** right triangles (triangles having the same measurements). The diagonal is the hypotenuse of each triangle. Therefore:

   $c^2 = a^2 + b^2$

   $c^2 = 300^2 + 300^2$

   $c^2 = 180{,}000$

   $c \approx 424$

   The length of the diagonal is approximately 424 yards.

2. You are holding a kite by a string. The length of the string is 130 feet, and the horizontal distance between you and the kite is 50 feet. How far off the ground is the kite?

NOTE: Sometimes it is helpful to draw a diagram and label the parts you know.

   Answer: The kite's height would be measured as a line perpendicular to the ground. Thus its height is one leg of a right triangle. The string is the hypotenuse, and the horizontal distance is the other leg. Therefore:

   $c^2 = a^2 + b^2$

   $130^2 = 50^2 + b^2$

   $b^2 = 16{,}900 - 2{,}500$

   $b^2 = 14{,}400$

   $b = 120$

   The height of the kite is 120 feet.

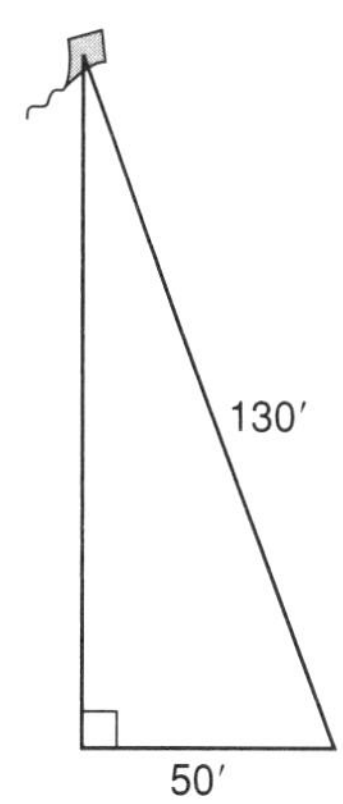

3. A design is in the shape of an isosceles triangle. The base of the triangle is 14 inches long, and each leg is 17 inches long. To change the design, a line is drawn from the vertex angle perpendicular to the base. Find the lengths of the sides of the two triangles that are now formed.

Answer: The two triangles formed are congruent right triangles. The hypotenuse of each is 17 inches. Each leg that was originally part of the base of the isosceles triangle is 7 inches long, since the altitude of an isosceles triangle bisects the base. The other leg of the right triangle is the altitude of the isosceles triangle. The altitude can be found by using $c^2 = a^2 + b^2$:

$17^2 = 7^2 + b^2$

$289 - 49 = b^2$

$b^2 = 240$

$b \approx 15.5$

The sides of the two new triangles are 7 inches, 17 inches, and 15.5 inches.

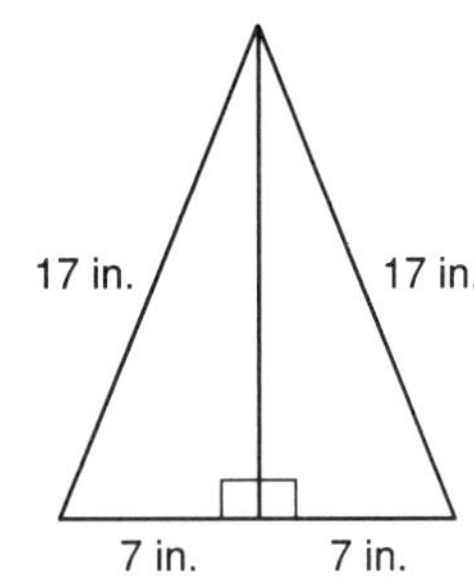

4. Find the area of a triangular flower garden that has a base of 4 feet and a height of 7 feet.

Answer:

$A = \frac{1}{2}bh$

$A = \frac{1}{2} \cdot 4 \cdot 7$

$A = 14$ square feet

5. Calculate the area of this triangle:

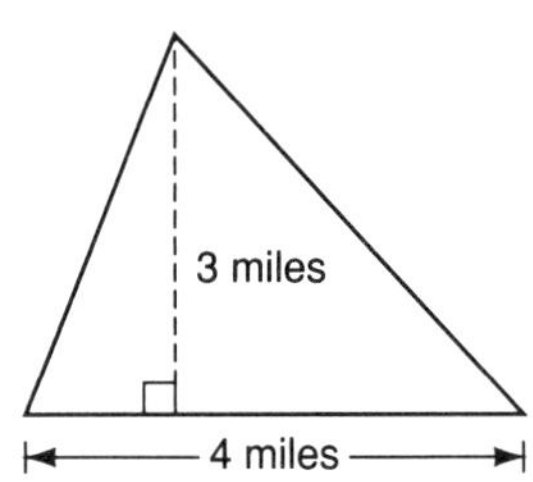

Answer:

$A = \frac{1}{2}bh$

$A = \frac{1}{2} \cdot 4 \cdot 3$

$A = 6$ square miles

6. The cost of seed is \$2.25 per square yard. You want to seed a triangular area with a base that measures 12 yards and a height that measures 5 yards. How much will the seed cost?

Answer:

$A = \frac{1}{2}bh$

$A = \frac{1}{2} \bullet 12 \bullet 5$

$A = 30$ square yards

$30 \bullet \$2.25 = \$67.50$

The seed will cost \$67.50.

TRY THESE

Items 1–3: Fill in the table.

Triangle	Base	Height	Area
1.	3m	8m	
2.		4 yd	10 sq yd
3.	22cm		22 sq cm

Items 4–6: Find the area.

4.

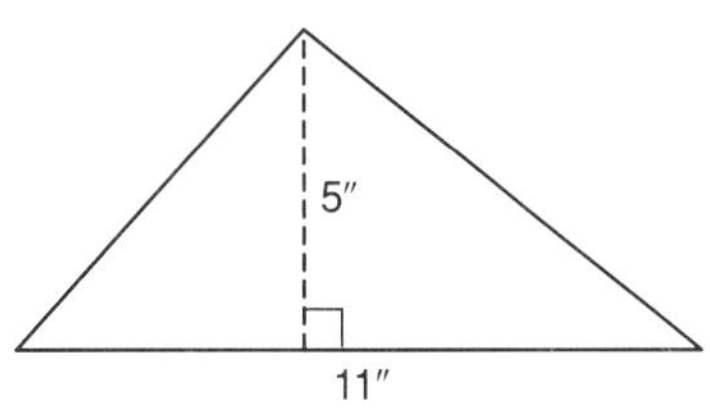

5.

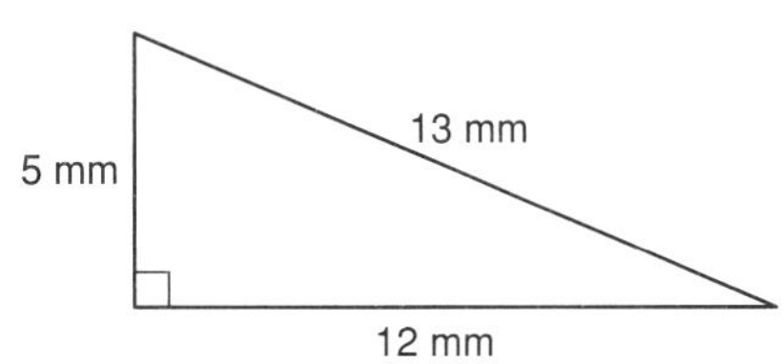

6.

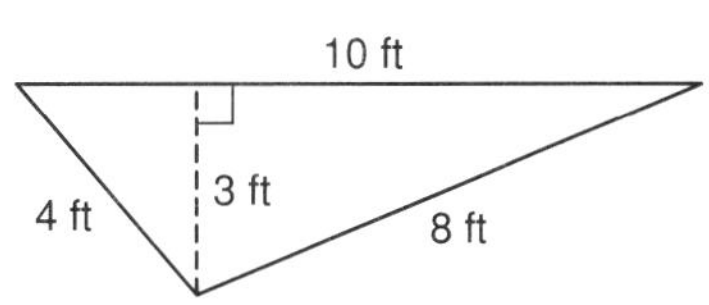

7. What is the area of a triangular flag that has a base of 2 feet and a height of 4.5 feet?

8. Dyane wants to make a stained-glass window for her new apartment. The measurements are shown in the diagram.

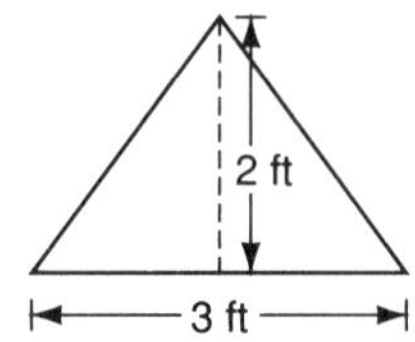

If the glass costs $8 per square foot, what will the window cost?

9. A 25-foot ladder is leaning against a garage wall. Its bottom is 7 feet from the wall. How high up the wall does it reach?

10. Two poles are 33 feet and 48 feet high, respectively. A wire is strung from the top of one pole to the top of the other. How long is the wire if the poles are 36 feet apart?

Area of Trapezoids The **height**, or **altitude**, **of a trapezoid** is the perpendicular distance between the bases (the one pair of parallel sides). To find the area of a trapezoid, multiply half of the height times the sum of the bases. The formula is $A = \frac{1}{2}h(b_1 + b_2)$ where h represents the height and b_1 and b_2 represent the lengths of the bases.

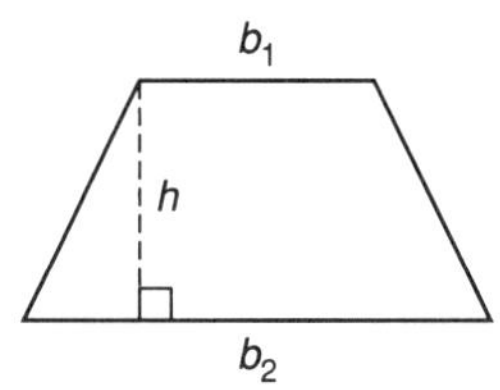

MODEL PROBLEMS

1. What is the area of a trapezoid that has a height of 6 inches and bases (parallel sides) of 8 and 14 inches?

Answer:

$A = \frac{1}{2}h(b_1 + b_2)$

$A = \frac{1}{2} \bullet 6 \bullet (8 + 14)$

$A = 66$ square inches

2. A trapezoid has an area of 56 square centimeters, an upper base of 4 centimeters, and a lower base of 10 centimeters. Find its height.

Answer:

$A = \frac{1}{2}h(b_1 + b_2)$

$56 = \frac{1}{2}h(4 + 10)$

$56 = 7h$

$h = 8$ centimeters

3. Find the area of this trapezoid:

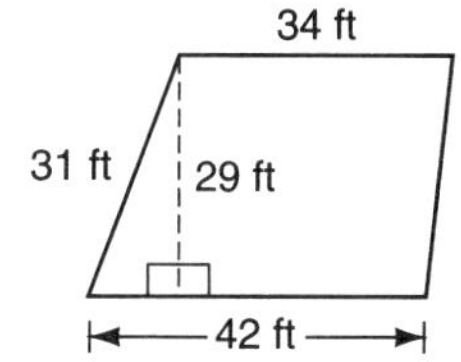

Answer:

$A = \frac{1}{2}h(b_1 + b_2)$

$A = \frac{1}{2} \bullet 29 \bullet (34 + 42)$

$A = 1{,}102$ square feet

TRY THESE

Items 1–3: Fill in the table.

Trapezoid	Height	$Base_1$	$Base_2$	Area
1.	4 in.	12 in.	10 in.	
2.	5 ft	20 ft	16 ft	
3.		11 m	14 m	15 sq m

Items 4–6: Find the area.

4.

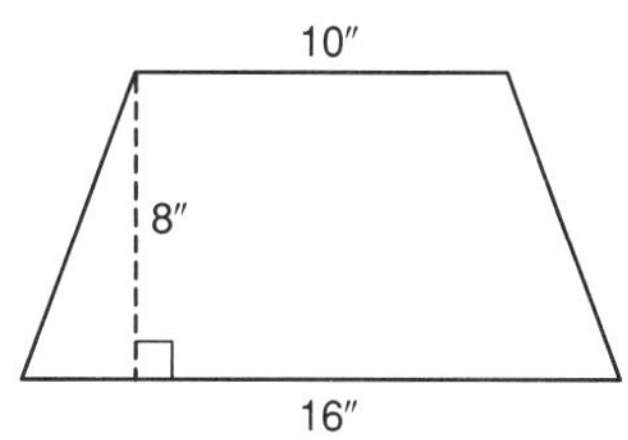

5.

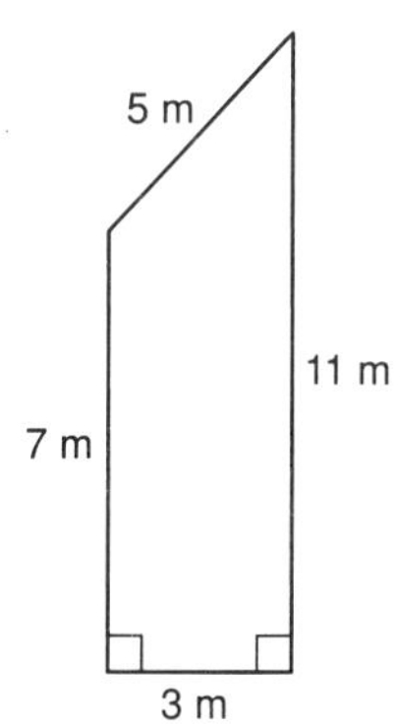

6.

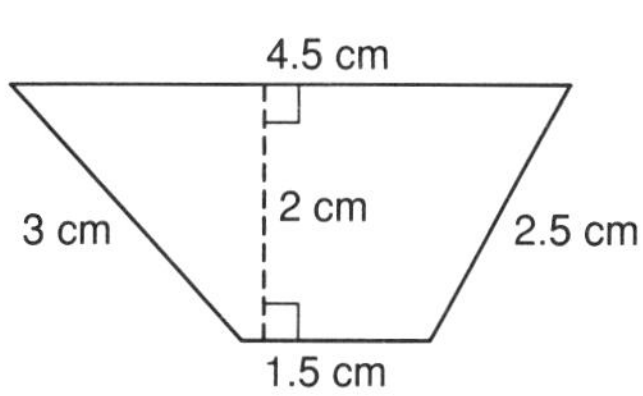

7. In town, Main Street and Center Street are parallel. The town park is in the shape of a trapezoid and is bordered by Main, Center, 1st, and 2nd streets. The length of the park that borders on Main Street is 50 yards. The length of the park that borders on Center Street is 35 yards. The perpendicular distance between Main Street and Center Street is 20 yards. What is the area of the park?

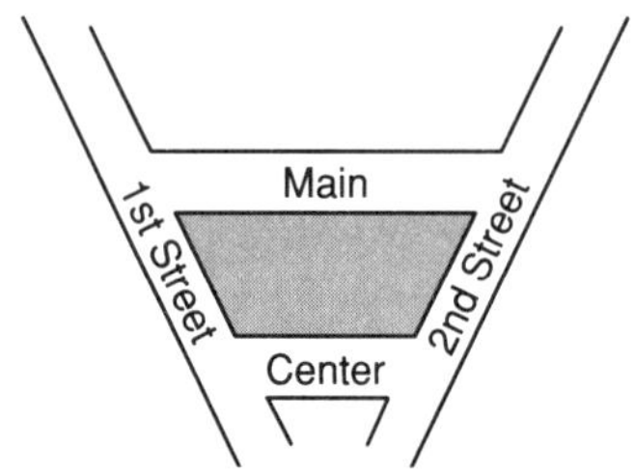

8. A trapezoid has bases that measure 11.75 yards and 14.5 yards and a height of 6 yards. Find its area.

9. The area of a trapezoid is 63 sq km. Its bases measure 8 km and 13 km. What is the height of the trapezoid?

10. A section of an airplane wing is shaped like a trapezoid. The parallel sides measure 3.5 feet and 5 feet and are 18 feet apart. What is the area of this wing section?

Area of Compound Shapes If a region is made up of several shapes that do not overlap, first find the area of each shape. Then, add those areas to get the area of the entire region.

MODEL PROBLEMS

1. Find the area of the entire region. Measurements are in feet.

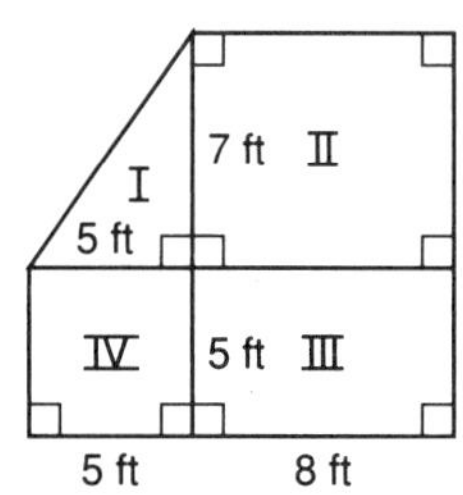

Answer: Add area I + area II + area III + area IV.

Area I is a triangle: $A = \frac{1}{2}bh$, so $A = \frac{1}{2} \bullet 5 \bullet 7 = 17.5$.

Area II is a rectangle: $A = bh$, so $A = 8 \bullet 7 = 56$.

Area III is a rectangle: $A = bh$, so $A = 8 \bullet 5 = 40$.

Area IV is a square: $A = s^2$, so $A = 5^2 = 25$

Total area = 17.5 + 56 + 40 + 25 = 138.5 square feet

2. This diagram shows the floor of a tent. Find the area of the floor.

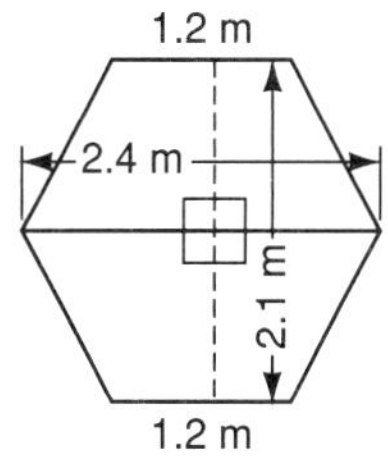

Answer: The tent floor can be divided into two equal trapezoids. To get the total area, find the area of one trapezoid and multiply that by 2.

$$A = \frac{1}{2}h(b_1 + b_2)$$

$$A = \frac{1}{2} \bullet 1.05 \bullet (2.4 + 1.2)$$

$$A = 1.89 \text{ sq m}$$

$$1.89 \bullet 2 = 3.78 \text{ sq m}$$

TRY THESE ✔

1. Find the entire area of the region shown in the diagram below.

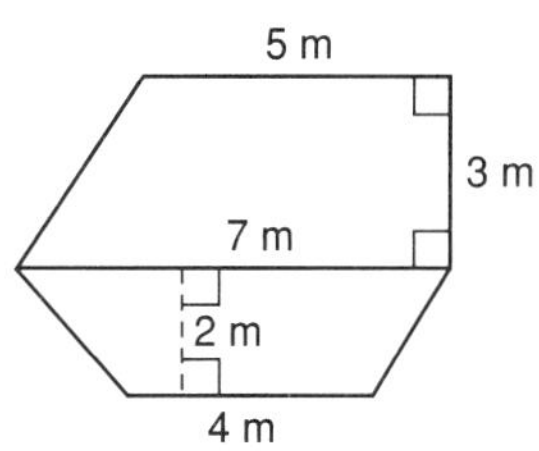

2. Find the entire area of the region shown in the diagram below.

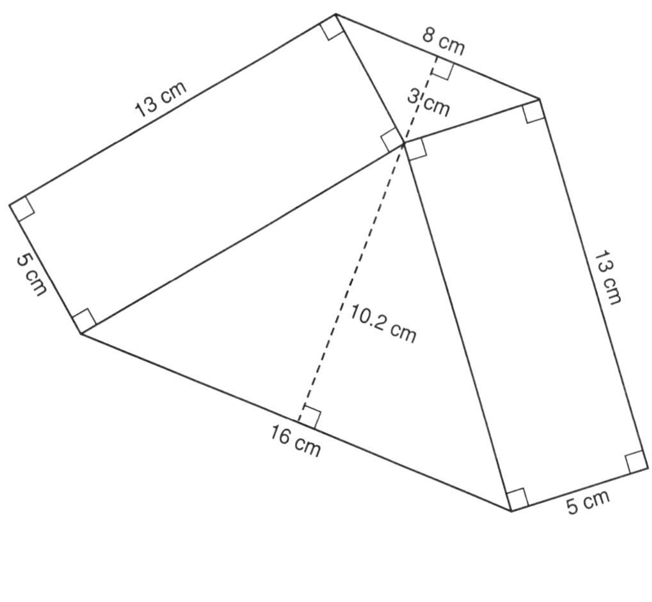

3. Find the entire area of the region shown in the diagram below.

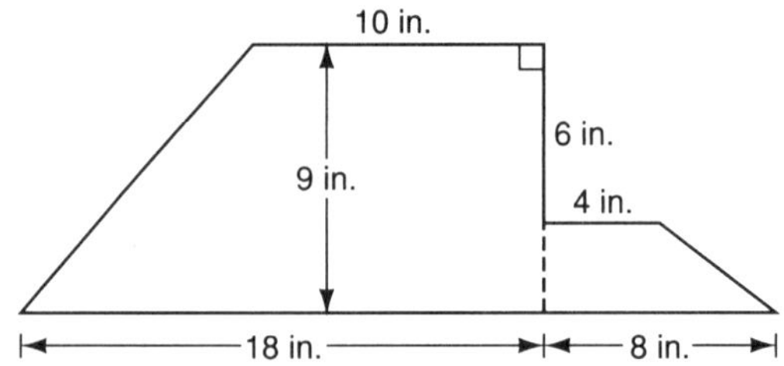

4. The diagram below is a floor plan of a cabin that Nadia is considering buying. How many square feet are shown in the plan?

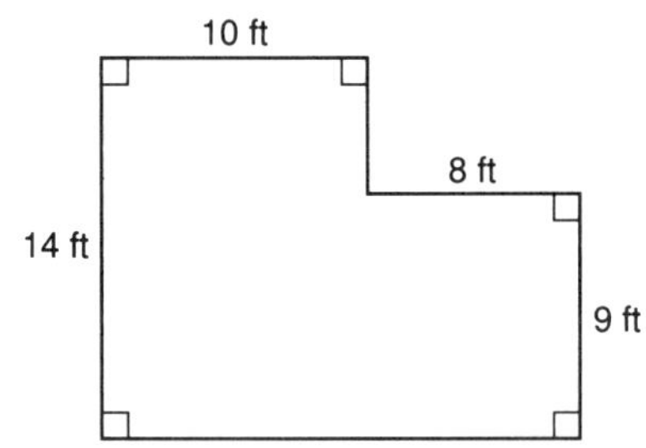

Area of Circles To find the area of a circle, multiply the square of the radius r by pi (π). The formula is $A = \pi r^2$.

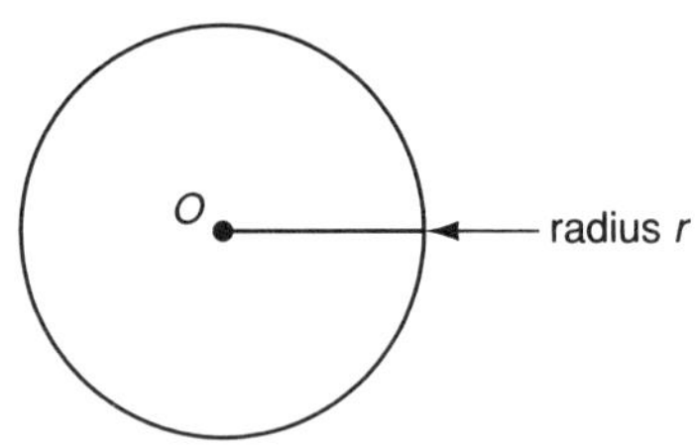

MODEL PROBLEMS

1. Find the area of a circle with a radius of 14 centimeters. Use $\pi \approx \frac{22}{7}$.

Answer:

$A = \pi r^2$

$A \approx \frac{22}{7} \bullet 14 \bullet 14$

$A \approx 616$ sq centimeters

2. A circle is to be cut out of a piece of wood. The diameter of the circle must be 10.5 inches. What will be the area of the circle? Use $\pi \approx 3.14$.

Answer:

$A = \pi r^2$

$r = \frac{d}{2} = \frac{10.5}{2} = 5.25$

$A \approx 3.14 \bullet 5.25 \bullet 5.25$

$A \approx 86.55$ square inches

TRY THESE

Items 1–5: Fill in the table. Use $\pi \approx 3.14$.

Circle	Radius	Diameter	Area
1.	2 ft		
2.		12 cm	
3.	9 m		
4.		8 yd	
5.	1.4 in.		

6. What is the area of a circular tablecloth that measures 5 feet in diameter? Use $\pi \approx 3.14$.

7. A revolving sprinkler waters a lawn in a circular motion in a radius of 28 feet. How many square feet does the sprinkler water in one revolution? Use $\pi \approx \frac{22}{7}$.

8. Hakim bought a circular swimming pool for his children. The pool has a circumference of 44 feet. What is the area of the base of the pool? Use $\pi \approx \frac{22}{7}$.

9. Wilma's white cat, Pi, has two, non-overlapping black circles on her back. The diameter of each circle is 3 inches. What area of Pi's fur is black? Use $\pi \approx 3.14$.

10. Tamara jogs three times around a circular lake. The total distance that she jogs is 0.35 mile. What is the area of the lake in square feet? Use $\pi \approx \frac{22}{7}$.

Measurement of three-dimensional figures

Until now, we have been discussing plane, or two-dimensional, geometric figures. But the real world, of course, is not entirely made up of plane figures. We also need to be able to measure **solid**, or **three-dimensional**, figures such as rectangular solids, cylinders, cones, pyramids, and spheres. Two types of measurement are needed for solids: surface area and volume.

Surface area (*S.A.*) is just that, the area of the surface of a three-dimensional figure. It is usually found by totaling the areas of all sides, or **faces**, of the figure. The unit of measure of surface area is **square units**, just as it is for the area of plane figures.

Volume is a measurement of capacity: it tells us how many three-dimensional units are contained in a given space. Two units of measure for volume are **cubic units** and **liters**. The words *volume* and *capacity* mean the same thing, but in the metric system capacity is often measured in liters rather than cubic units. (A liter is actually 1 cubic decimeter.)

NOTE: The formulas given in this section can all be found in the formula sheet on pages 11 and 12.

SURFACE AREA AND VOLUME OF RECTANGLE SOLIDS

A **rectangular solid** is the three-dimensional figure that we commonly think of as a box. It has a bottom base, a top base, and four faces (or sides). The faces meet at **edges**, and the edges meet at **vertices**. This book has the shape of a rectangular solid.

Dimensions of a rectangular solid are normally given in this order: length by width by height.

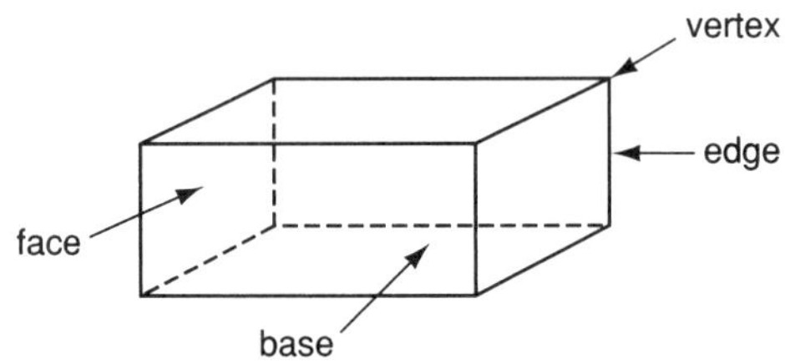

Rectangular solid

To find the surface area (*S.A.*) of a rectangular solid, find the area of each base and each face. Then total the results:

$$S.A. = (2 \bullet \text{area of the base}) + (2 \bullet \text{area of face}) + (2 \bullet \text{area of other face})$$

Or:

$$S.A. = [2(\text{length} \bullet \text{width})] + [2(\text{length} \bullet \text{height})] + [2(\text{height} \bullet \text{width})]$$

$$S.A. = 2lw + 2lh + 2hw$$

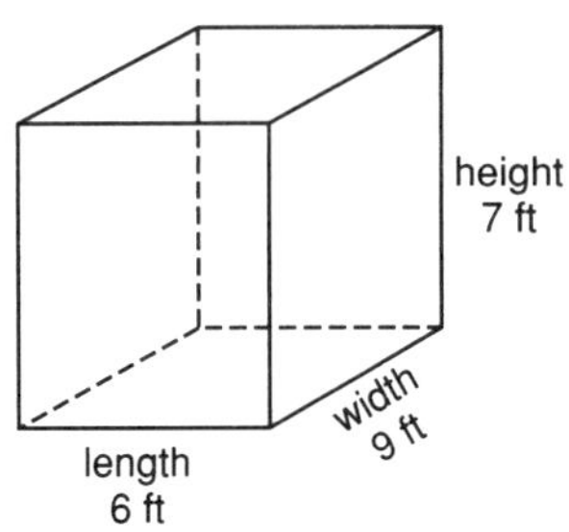

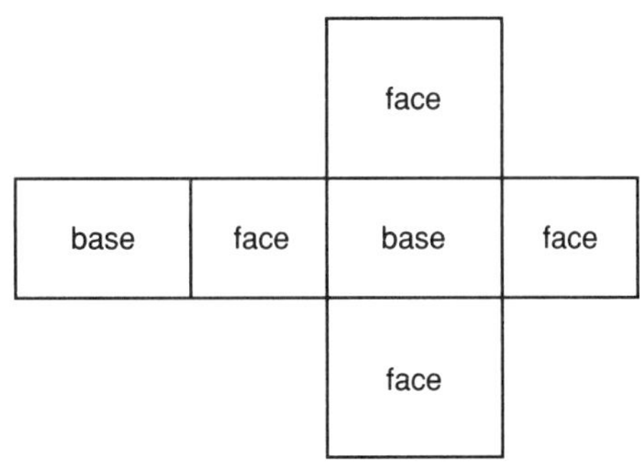

In the figure above, for example:

$$S.A. = (2 \bullet 54) + (2 \bullet 42) + (2 \bullet 63)$$

$$S.A. = 318 \text{ square feet}$$

If a rectangular solid has the same measure for length, width, and height, it is called a **cube**. In a cube, every face and base has the same area, and every edge has the same measure. Therefore, the formula for surface area of a cube is:

$S.A. = 6s^2$ where s is the measure of any edge

The **volume** (V) of a rectangular solid is found by multiplying length times width times height:

$$V = lwh$$

For example, the volume of the rectangular solid above would be $V = 6 \bullet 9 \bullet 7$ or 378 cubic feet.

Since all edges of a cube have the same measure, the volume of a cube can be found with this formula:

$V = s^3$ where s is the measure of any edge

MODEL PROBLEMS

1. The dimensions of a small packing crate are 10 inches by 12 inches by 18 inches. How many square inches of wrapping paper are needed to cover the entire crate?

   Answer: This problem is asking for surface area (*S.A.*).

   $S.A. = [2(10 \bullet 12)] + [2(10 \bullet 18)] + [2(12 \bullet 18)]$

   $S.A. = 1{,}032$ square inches

2. What is the volume of the crate in model problem 1?

   Answer:

   $V = 10 \bullet 12 \bullet 18$

   $V = 2{,}160$ cubic inches

3. The dimensions of a box are 3 feet by 5 feet by 2 feet. The box has no lid. What is its surface area?

   Answer: Calculate only one base, since there is no top:

   $S.A. = (3 \bullet 5) + [2(3 \bullet 2)] + [2(2 \bullet 5)]$

   $S.A. = 47$ square feet

TRY THESE

Items 1–4: Fill in the table.

Rectangular solid	Length	Width	Height	Surface area	Volume
1.	4 yd	8 yd	2 yd		
2.	11 cm	5 cm	10 cm		
3.	3 in.	7 in.			63 cu in.
4.	9 ft		5 ft		315 cu ft

Items 5–7: Fill in the table.

Cube	Side	Surface area	Volume
5.	2 m		
6.		54 sq cm	
7.			216 cu ft

8. You want to make a wooden packing box that has these dimensions: 6 feet by 4 feet by 5 feet. How many square feet of wood will you need?

9. What will be the volume of the box in problem 8?

10. A cube measures 15 inches on a side. What is its surface area? What is its volume?

SURFACE AREA AND VOLUME OF CYLINDERS

A **cylinder** has two circular bases and a curved outside surface. If the curved surface, called the **lateral** area, is flattened out, it forms a rectangle.

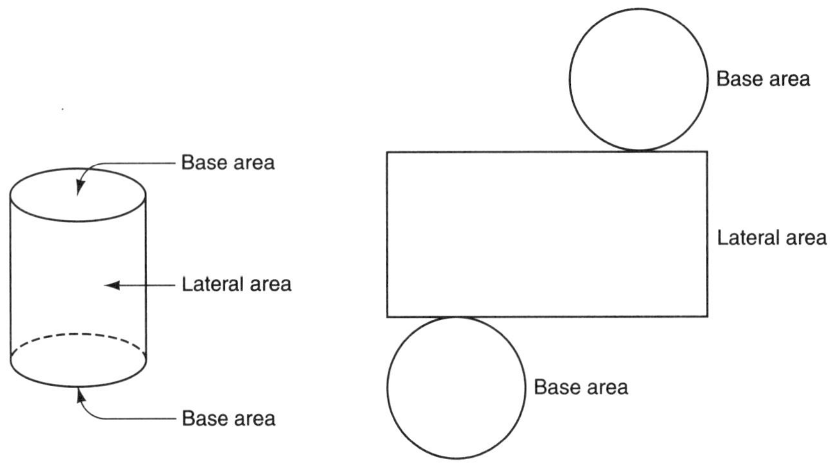

Therefore, to find the **total surface area** of a cylinder, think of it as two circles and a rectangle. The area of each circle is πr^2 and since there are two circles, this result must be doubled.

The area of the rectangle is length times width. The length of the rectangle is the circumference of the base ($2\pi r$). The height of the rectangle is the height (h) of the cylinder. So the area of the rectangle is $2\pi rh$.

The formula for the total surface area of a cylinder is:

$$S.A. = 2\pi r^2 + 2\pi rh$$

Sometimes, we need to find only the **lateral surface area** ($L.A.$) of a cylinder: just the curved area, without the bases. For instance, a drainage pipe has no ends, so its surface area is only the lateral area. The formula for lateral surface area of a cylinder is $L.A. = 2\pi rh$.

To find the **volume** of a cylinder, calculate the area of the base and multiply this by the height. Since the base is a circle, the formula for volume is:

$$V = \pi r^2 h$$

MODEL PROBLEMS

1. A cylindrical can is 5 inches tall and has a radius of 1 inch. Its label covers the entire curved surface. How many square inches are there in the label?

   Answer: We need to find lateral area (*L.A.*). Here, we'll use $\pi \approx 3.14$.

   $L.A. = 2\pi rh$

   $L.A. \approx 2 \bullet 3.14 \bullet 1 \bullet 5$

   $L.A. \approx 31.4$

   The label has approximately 31.4 square inches.

2. Terah is wrapping a large toy that comes in a cylindrical box 3 feet tall, with a radius of 2 feet. How many square feet will she wrap?

   Answer: We need to find total surface area. We'll use $\pi \approx 3.14$.

   $S.A. = 2\pi r^2 + 2\pi rh$

   $S.A. \approx (2 \bullet 3.14 \bullet 2 \bullet 2) + (2 \bullet 3.14 \bullet 2 \bullet 3)$

   $S.A. \approx 25.12 + 37.68$

   $S.A. \approx 62.8$

   She will wrap approximately 62.8 square feet.

3. What is the volume (V) of the can in model problem 1?

   Answer: Use $\pi \approx 3.14$.

   $V = \pi r^2 h$

   $V \approx 3.14 \bullet 1 \bullet 1 \bullet 5$

   $V \approx 15.7$

   The volume of the can in model problem 1 is approximately 15.7 cubic inches.

TRY THESE

Items 1–3: Fill in the table using $\pi \approx \frac{22}{7}$.

Cylinder	Radius	Diameter	Height	Surface area	Volume
1.	7 cm		10 cm		
2.		28 in.	5 in.		
3.	21 mm		30 mm		

Items 4–7: Fill in the table using $\pi \approx 3.14$.

Cylinder	Radius	Diameter	Height	Surface area	Volume
4.	3 m		8 m		
5.		20 ft	15 ft		
6.	4.5 in.		12 in.		
7.		3 m	2 m		

8. A hot water tank has a diameter of 14 inches and is 60 inches tall. How much insulation is needed to cover it, including the two ends? $\left(\text{Remember that } d = 2r \text{ and use } \pi \approx \frac{22}{7}.\right)$

9. How much water will the tank in problem 8 hold? $\left(\text{Use } \pi \approx \frac{22}{7}.\right)$

10. A cylinder is 30 feet high and has a circumference of 60 feet. What is its lateral surface area?

SURFACE AREA AND VOLUME OF CONES

A **cone** (like a cylinder) has a base that is a circle. But the cone has only one base, and its curved surface does not go straight up. Instead, the curved surface slants into a point, the **vertex**. A cone has two kinds of "height." The distance from the center of the base to the vertex is called the **height** of the cone and is abbreviated h. The measurement from a point on the circumference of the base to the vertex is called the **slant height**. Slant height is represented by the symbol ℓ.

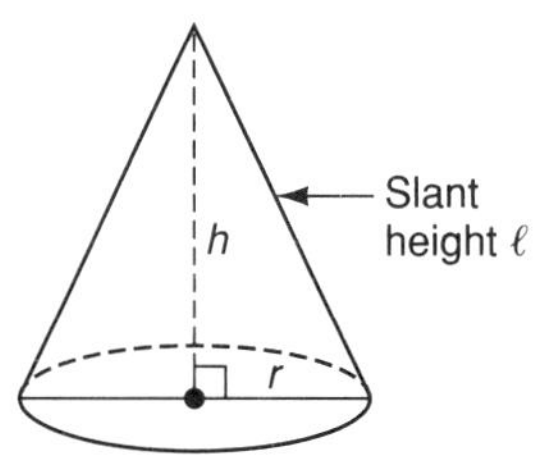

Slant height is the hypotenuse of a right triangle. The legs of this right triangle are the radius of the base and the height of the cone. Therefore, to find slant height we can use the Pythagorean theorem: $c^2 = a^2 + b^2$ and $c = \sqrt{a^2 + b^2}$.

EXAMPLE

If the height of a cylinder is 12 inches and the radius of its base is 5 inches, its slant height ℓ is 13 inches:

$5^2 + 12^2 = ?$

$25 + 144 = 169$

$\sqrt{169} = 13$

TRY THESE

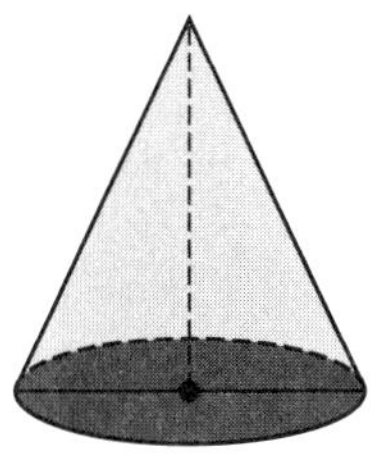

For each cone, find the missing part.

1. Radius = 12 inches, height = 9 inches, slant height = ?
2. Radius = 28 m, height = ?, slant height = 35 m
3. Radius = ?, height = 63 feet, slant height = 87 feet
4. Radius = 25 mm, height = 60 mm, slant height = ?
5. Radius = 3 yards, height = ?, slant height = 5 yards
6. Radius = ?, height = 32 inches, slant height = 40 inches
7. Radius = 6 cm, height = 8 cm, slant height = ?
8. Radius = 15 feet, height = ?, slant height = 39 feet

The formula for the **total surface area** of a cone with radius r is:

$S.A.$ = lateral surface area + surface area of base = $\pi r\ell + \pi r^2$

The lateral surface area of a cone is just the curved surface, without the base, so the formula for its lateral area is $L.A. = \pi r\ell$.

The formula for the **volume** of a cone is:

$$V = \frac{1}{3}\pi r^2 h$$

MODEL PROBLEMS

1. A cone has a base with a diameter of 6 inches and a slant height of 5 inches. Find its total surface area. Use $\pi \approx 3.14$.

   Answer:

   $S.A. = \pi r\ell + \pi r^2$

   $S.A. \approx (3.14 \bullet 3 \bullet 5) + (3.14 \bullet 3 \bullet 3)$

   $S.A. \approx 47.1 + 28.26$

   $S.A. \approx 75.36$ square inches

2. Find the volume of the cone in model problem 1 if the height of the cone is 4 inches.

   Answer:

   $V = \frac{1}{3}\pi r^2 h$

   $V \approx \frac{1}{3} \bullet 3.14 \bullet 3 \bullet 3 \bullet 4$

   $V \approx 37.68$ cubic inches

TRY THESE ✔

Items 1–5: Complete the table using $\pi \approx 3.14$. When necessary, round answers to the nearest hundredth.

Cone	Radius	Diameter	Height	Slant height	Total surface area	Volume
1.	8 cm		6 cm			
2.		10 ft	12 ft			
3.	9 in.		9 in.			
4.		22 mm	19 mm			
5.	3.5 yd		10 yd			

6. You want to paint the outside of a cone-shaped toy. Its base has a diameter of 10 centimeters and its slant height is 13 centimeters. What is the total surface area that you will cover? Give your answer in terms of π.

7. Your little brother fills the toy in problem 1 with clay. If the height of the toy is 12 centimeters, how many cubic centimeters of clay will fit into it? Give your answer in terms of π.

8. A large tent is to be made in the shape of a cone. The radius of the tent is 21 feet and the height of the tent is 28 feet. How much material is needed to construct the tent, not including the base? Use $\pi \approx \frac{22}{7}$.

9. How many cubic feet of grain are in a conical pile that is 16 feet in diameter and 5 feet high? Use $\pi \approx 3.14$. Round your answer to the nearest hundredth.

10. A pile of sand is in the shape of a cone. The pile is 8 feet in diameter and 3 feet high. How many cubic yards of sand are in the pile? Use $\pi \approx 3.14$. Round your answer to the nearest hundredth.

SURFACE AREA AND VOLUME OF PYRAMIDS

A **pyramid** is like a rectangular solid because its base is a polygon. But the faces of a pyramid are triangles instead of rectangles. These faces, called the **lateral faces**, meet at a point called the **vertex**. A pyramid can have any polygon as a base, but the kind of pyramid that is studied for FCAT has a square base.

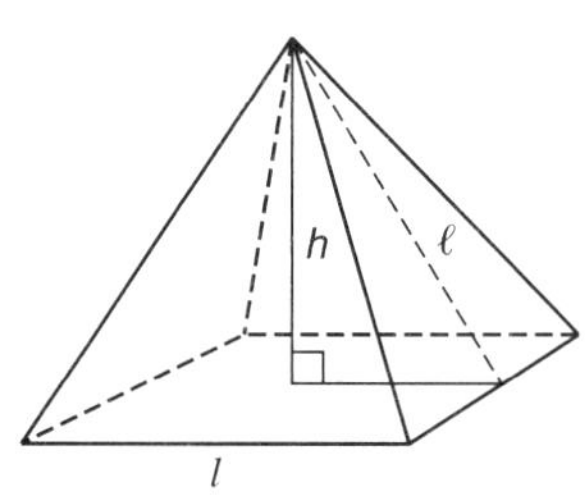

The **height** of a pyramid is measured from the vertex of the pyramid to the center of the base. For a pyramid with a square base, the height is the distance from the vertex to the center of the square.

Like a cone, a pyramid has a **slant height**. Since each lateral face of a pyramid is a triangle, the slant height is the altitude of a triangular face. We can use the Pythagorean theorem to find the slant height. The height and half the length of the base of the pyramid are the legs of a right triangle, and the hypotenuse is the slant height.

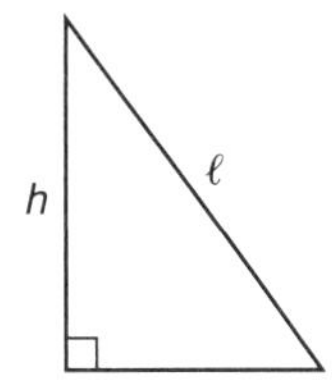

The formula for the total surface area of a pyramid is:

$$S.A. = 2l\ell + l^2$$

where l = length of a side of the square base and ℓ = slant height.

NOTE: The symbols l and ℓ look somewhat alike. Also, l can look somewhat like a one (1). Therefore, it is important to read carefully when these symbols are used in a problem. In your own calculations for square pyramids, you may find it clearer simply to use s for side of base, instead of l. But remember: l and ℓ are used on the FCAT Mathematics Reference Sheet, so be prepared to deal with them.

The formula for the **volume** of the pyramid is:

$$V = \frac{1}{3}lwh$$

where l = length of base, w = width of base, and h = height of pyramid. In a square pyramid, l and w are equal.

MODEL PROBLEMS

1. A square pyramid has a base that measures 16 feet on a side and an altitude (height) of 15 feet. Find the total surface area.

   Answer: The formula for total surface area is $S.A. = 2l\ell + l^2$. We know that length l of a side of the base = 16, but we must find the slant height ℓ. Since $l = 16$, one-half of l is 8. Now use the Pythagorean theorem:

   $8^2 + 15^2 = 289$

   $\sqrt{289} = 17$

   Therefore:

   $S.A. = 2 \bullet 16 \bullet 17 + 16^2 = 800$ square feet

2. Find the volume of the pyramid in model problem 1.

   Answer: The formula for volume of a pyramid is $V = \frac{1}{3}lwh$. Therefore:

   $V = \frac{1}{3} \bullet 16 \bullet 16 \bullet 15 = 1{,}280$ cubic feet

TRY THESE

Items 1–5: Complete the table.

Square pyramid	Edge of base	Height	Slant height	Total surface area	Volume
1.	6 in.	4 in.			
2.	24 yd	5 yd			
3.		15 m	17 m		
4.	48 ft		25 ft		
5.		8 cm	10 cm		

6. A tent is in the shape of a square pyramid. Its base is 10 feet on each side, and its height is 12 feet. What is the total outside surface area of the tent, including the base?

7. How many cubic feet of space are there inside the tent in problem 6?

Items 8–10: The right square pyramid shown below has a height of 6 meters and a volume of 128 cubic meters.

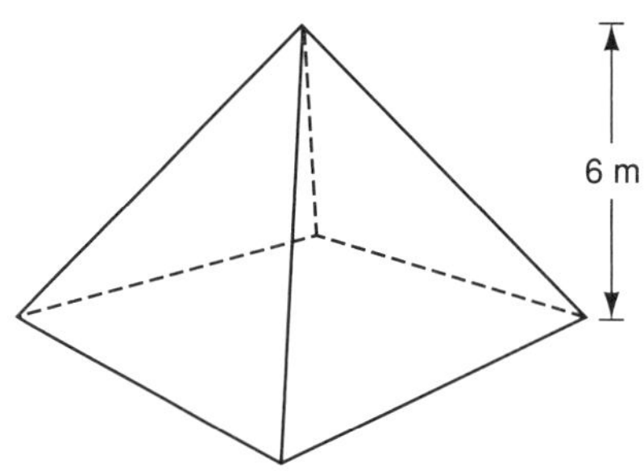

8. What is the length of one side of the base of the pyramid?

9. Find the slant height of the pyramid. Round your answer to the nearest tenth.

10. Find the total surface area.

SURFACE AREA AND VOLUME OF SPHERES

A **sphere** is the geometric figure that we commonly think of as a ball. The radius (r) of a sphere is measured from the center of the sphere to the surface.

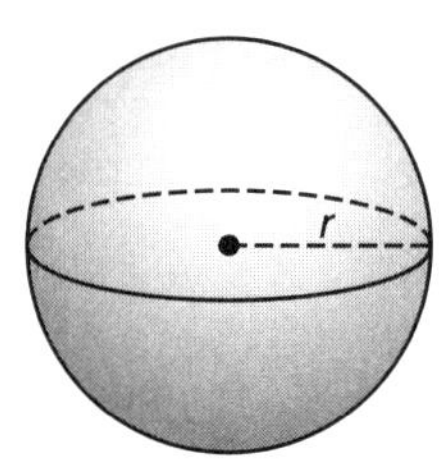

The formula for the **total surface area** of a sphere is

$S.A. = 4\pi r^2$

The formula for the **volume** of a sphere is:

$V = \frac{4}{3}\pi r^3$

MODEL PROBLEMS

1. If the radius of a sphere is 5 inches, find its total surface area and its volume. Use $\pi \approx 3.14$.

   Answer:

   Surface area $S.A. = 4\pi r^2 = 4 \bullet \pi \bullet 5^2 = 100\pi \approx 314$ square inches

   Volume $V = \frac{4}{3}\pi r^3 = \frac{4}{3} \bullet \pi \bullet 125 = \frac{500}{3}\pi \approx 523.33$ cubic inches

2. A sphere has a diameter of 2 meters. What is its total surface area? What is its volume? Use $\pi \approx 3.14$.

   Answer: Since $d = 2r$, the radius of this sphere = 1 meter. Therefore:

   $S.A. = 4\pi \bullet 1^2 = 4\pi \approx 12.56$ square meters

   $V = \frac{4}{3}\pi \bullet 1^3 = \frac{4}{3}\pi \approx 4.19$ cubic meters

TRY THESE

Items 1–5: Complete the table. When necessary, express your answer in terms of π and round to the nearest hundredth.

Sphere	Radius	Diameter	Total surface area	Volume
1.	6 in.			
2.		8 ft		
3.	9 m			
4.			576π sq yd	
5.			2.56π sq cm	

6. The radius of the earth is approximately 4,000 miles. What is the earth's total surface area? Give your answer in terms of π.
7. What is the volume of a spherical container that has a diameter of 14 inches? Give your answer in terms of π.
8. Dominic has a ball that has a circumference of 4.4 inches at its center. What is the total surface area and volume of the ball? Use $\pi \approx \frac{22}{7}$ and round the volume to the thousandths place.

Assessment

Gridded-Response Problems

1. Shown below is the floor plan of a house. All angles in the plan are right angles. Find the perimeter of the house.

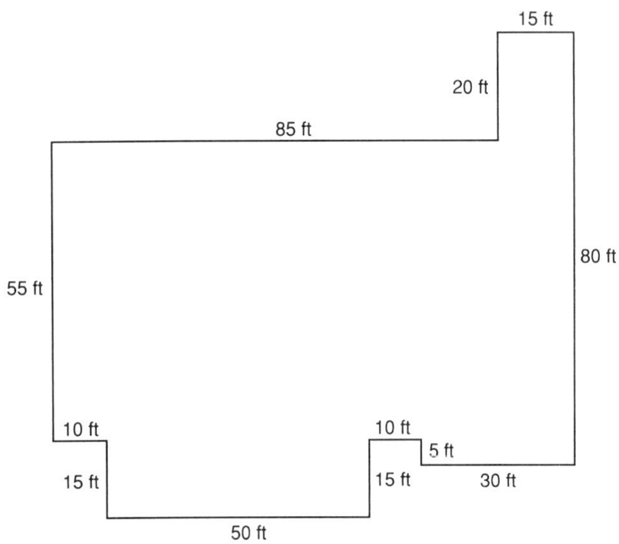

2. Refer to the floor plan in problem 1. Find the area of the floor plan of the house.

3. Marko has an entertainment center, shown as the shaded region in the diagram below, that is used as a divider to separate the living room and the dining room. He plans to put carpet in both rooms, but not on the floor under the entertainment center. How many square feet of carpet should he buy?

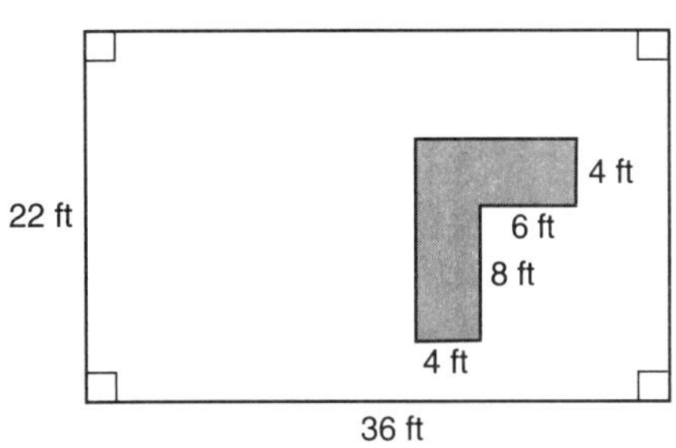

4. The area of a square room is 144 feet. What is the perimeter of the room?

5. A belt is tightly placed on two wheels. Each wheel has a radius of 6 inches. The centers of the wheels are 14 inches apart. Find the length of the belt. Use $\pi \approx 3.14$.

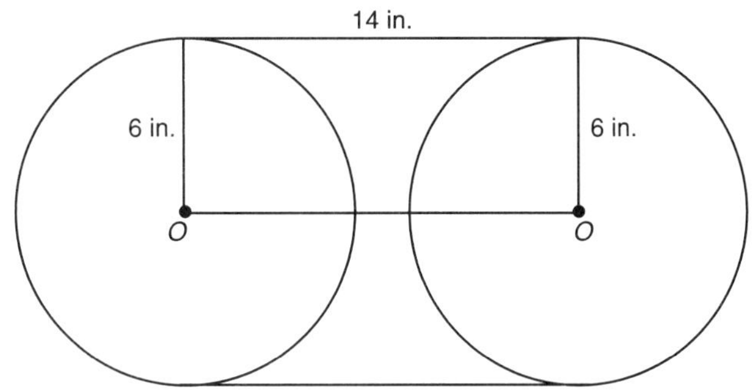

Items 6 and 7: A soup can has a base radius of 3 centimeters and a height of 10 centimeters.

6. What is the volume of the can? Use $\pi \approx$ 3.14.

7. What is the total surface area of the can? Use $\pi \approx 3.14$. Round your answer to the nearest integer.

1 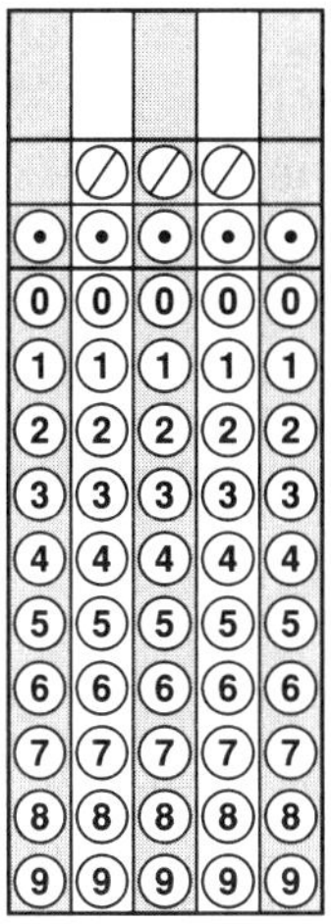

2 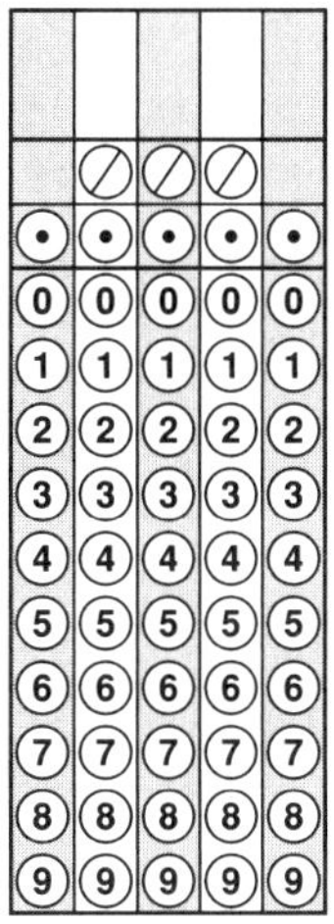

3 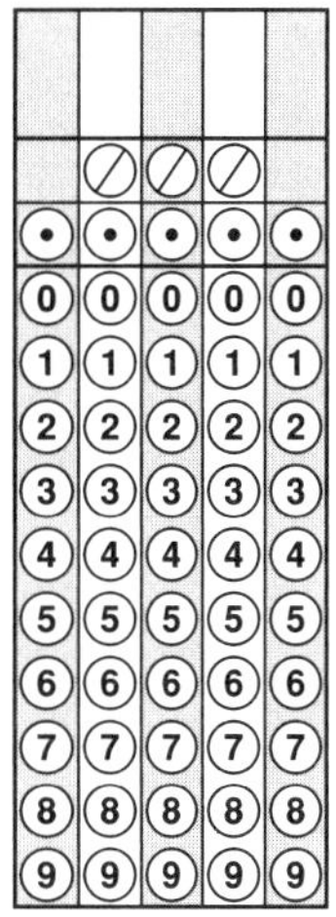

4 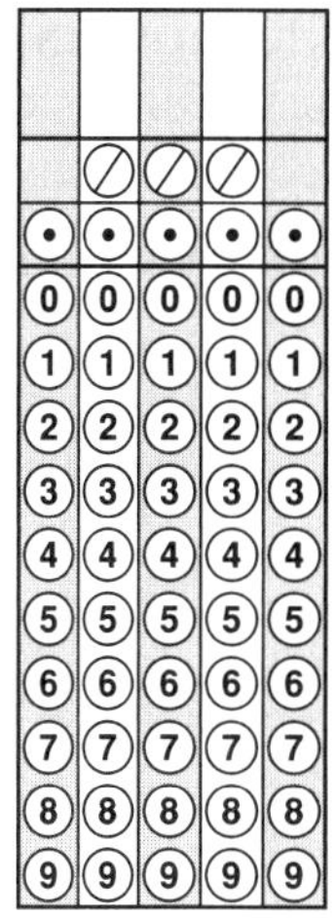

5 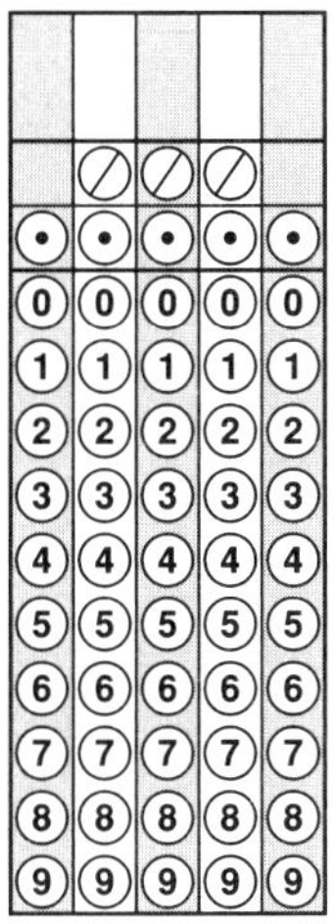

6 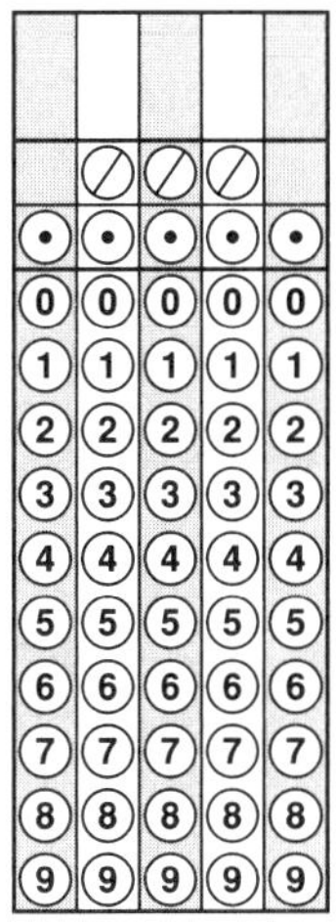

7 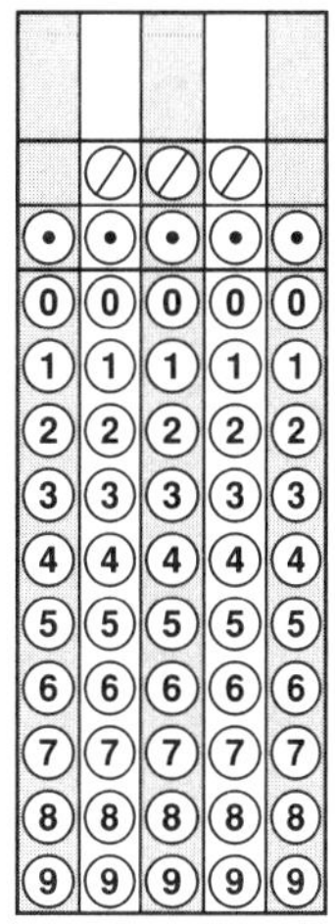

Multiple-Choice Problems

8. What is the length of a side of a square that has the same number of units in its area as in its perimeter?

A. 2 units

B. 3 units

C. 4 units

D. 5 units

9. What is the height of a trapezoid that has parallel sides of 12 feet and 18 feet and an area of 126 square feet?

A. 10.5 feet

B. 8.4 feet

C. 4.2 feet

D. 2.1 feet

10. The circumference of a circle is 22 inches. What is its area? Use $\pi \approx \frac{22}{7}$.

A. 3.5 square inches

B. 7 square inches

C. 38.5 square inches

D. 154 square inches

11. A 20-foot ladder is leaning against a house. If the base of the ladder is 8 feet from the base of the house, how high up the side of the house will the ladder reach? Round your answer to the nearest foot.

A. 12 feet

B. 18 feet

C. 19 feet

D. 22 feet

12. **Concentric circles** have the same center but different radii. If two concentric circles have radii of 6 inches and 10 inches, what is the difference in their areas?

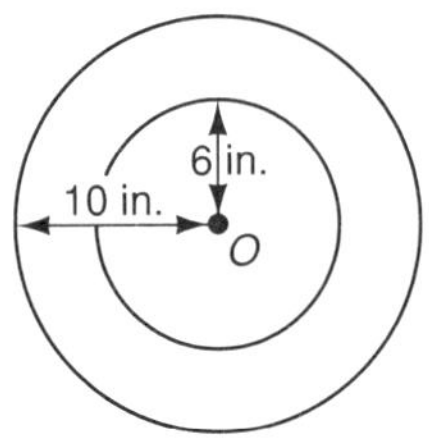

A. 4π

B. 16π

C. 64π

D. 256π

13. Sonni's dad had a truckload of mulch delivered to their house. When dumped, the mulch formed a cone-shaped pile with a diameter of 16 feet and a height of 9 feet. How many cubic feet of mulch are in the pile? Use $\pi \approx 3.14$.

A. 7,234.56 cubic feet

B. 2,411.52 cubic feet

C. 1,808.64 cubic feet

D. 602.88 cubic feet

Think, Solve, Explain: Short-Response Problems

14. Norma is going to have gravel put on her driveway. The driveway measures 150 feet long by 10 feet wide.

Part A If the gravel is to be 4 inches deep, how many cubic feet of gravel will be needed? Show your work.

Part B If she is charged $1.25 per cubic foot, what will it cost Norma for the driveway? Show your work.

15. The size of a bicycle is given as the diameter of its wheels. For example, a 20-inch bicycle has a wheel diameter of 20 inches.

Part A How far does a 20-inch bicycle move with one revolution of the wheel? Show your work.

Part B How much farther does a 24-inch bicycle move in one revolution of the wheel than a 20-inch bicycle? Show your work.

Items 16 and 17: The distance around the jogging track shown below is one-third mile, when measured along the inner edge of lane 1. The inside edge of the track is a 566-foot by 200-foot rectangle with semicircles on both ends. The track is 30 feet wide and has 6 running lanes of equal width.

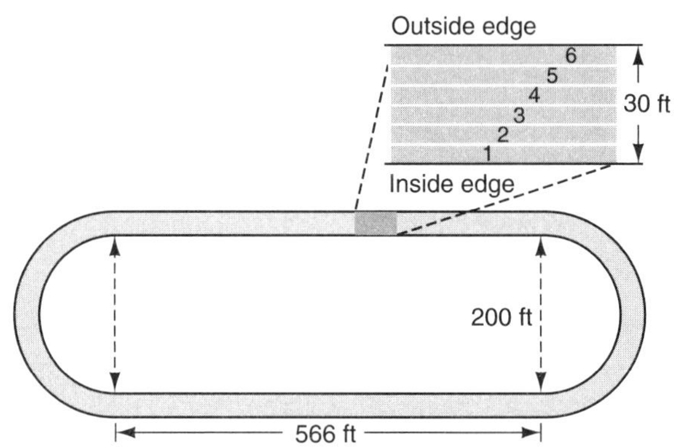

16. *Part A* Write an expression that can be used to find the distance around the track when measured along the inside edge of lane 3.

Part B What is the distance, in feet, around the track when measured along the inside edge of lane 3? Show all work necessary to justify your answer.

17. *Part A* What is the distance, in feet, around the track when measured along the inside edge of lane 6? Show all work necessary to justify your answer.

Part B How much farther is it along the inside edge of lane 6 than along the inside edge of lane 3? Show all work necessary to justify your answer.

18. It is estimated that about 25 percent of the earth's surface is dry land and the rest is covered by water. Using 4,000 miles as the radius of the earth, find the amount of surface that is covered by water. Give your answer in millions of square miles. Show your work.

19. The rectangle shown below is divided into 15 squares.

The shaded region covers $6\frac{1}{2}$ squares.

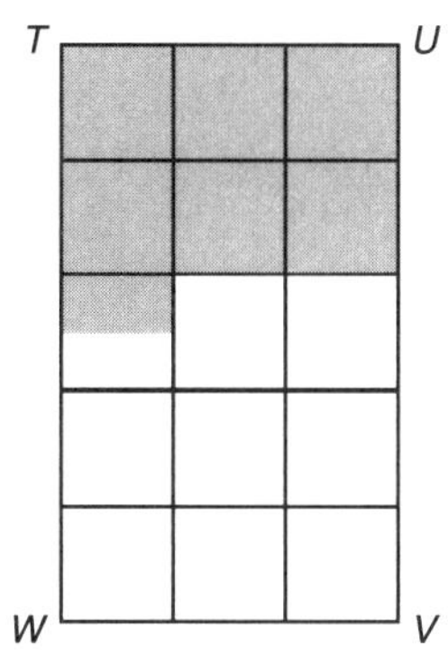

If the area of the shaded region is 104 square meters, what is the length of side *TW*? Show your work.

20. A cylindrical can has a radius of 5 inches and a height of 10 inches. The contents of this can are to be poured into a rectangular container whose dimensions are $11 \times 4 \times 18$ inches. Will the container be too small, too big, or exactly the right size? To support your answer, show the mathematical computations.

21. In a circle with a radius of 14 inches, how much longer is an arc that is intercepted by a central angle of 72° than an arc that is intercepted by a central angle of 45°? Use $\pi \approx \frac{22}{7}$. Show your work.

22. A container is in the shape of a cylinder on top of a right circular cone. How much water will be needed to fill the container? Show your work and give your answer to the nearest cubic inch.

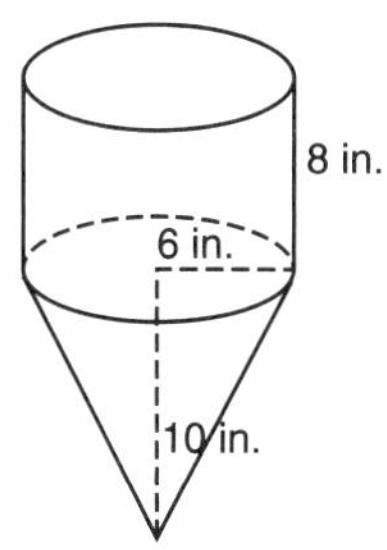

Section 2

Comparing, contrasting, and converting within systems of measurement—standard-nonstandard and metric-customary

A SELECTING AND USING DIRECT AND INDIRECT METHODS OF MEASUREMENT

After accidentally dropping a book out of her second-floor window, Jennifer wondered how far it was from the window to the ground, but she had no measuring device long enough to let her measure this directly. Then she remembered that her dad had a 22-foot ladder that was higher than the window. She got the ladder, leaned it against the house so that its end touched her windowsill, and measured the distance from the bottom of the ladder to the base of the house. That distance was 14 feet. Using the Pythagorean theorem, $c^2 = a^2 + b^2$, she found that her window was about 17 feet off the ground.

$$c^2 = a^2 + b^2$$
$$22^2 = 14^2 + b^2$$
$$22^2 - 14^2 = b^2$$
$$b^2 = 288$$
$$b \approx 17 \text{ feet}$$

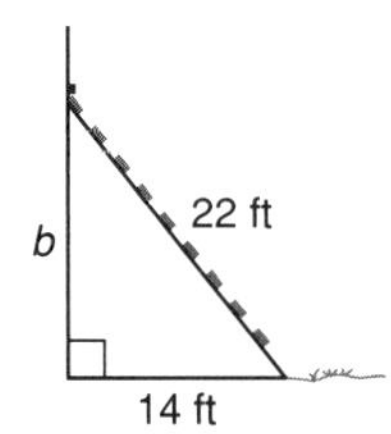

This is an example of **indirect measurement**. The following model problems show just some of the many ways to make indirect measurements. For instance, earlier in this chapter, to find the area of a figure made up of several geometric shapes, we added up the areas of all the shapes. We can get the same result indirectly, by subtracting—as in model problems 1 and 2.

MODEL PROBLEMS

1. Find the area of this figure indirectly.

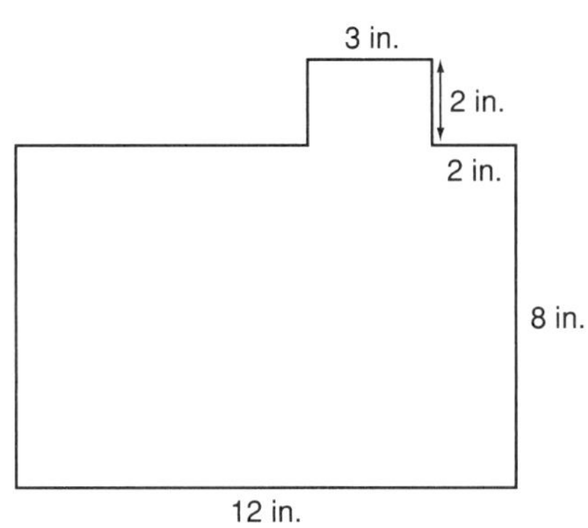

Answer: Form a rectangle by extending the sides, as shown by the dashed lines. The measurements of the new rectangle are 10 inches by 12 inches, so its area is 120 square inches.

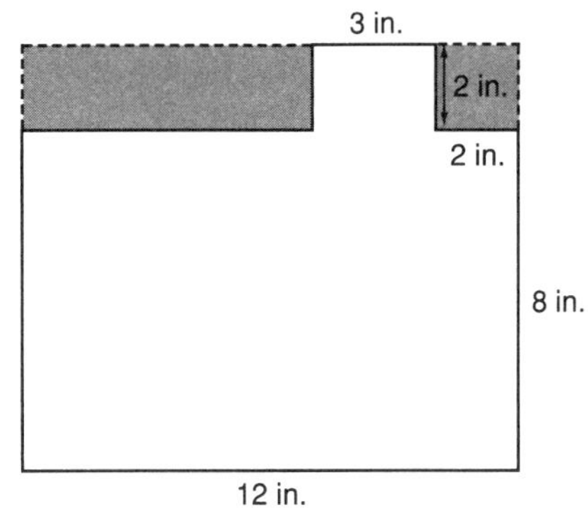

Now subtract the area of the smaller shaded region:

2 inches × 2 inches = 4 square inches

120 − 4 = 116

Then subtract the area of the larger shaded region:

2 inches × (12 inches − 3 inches − 2 inches) =
2 inches × 7 inches = 14 square inches

116 − 14 = 102

The area of the original figure is 102 square inches.

2. Indirectly find the area of the figure below.

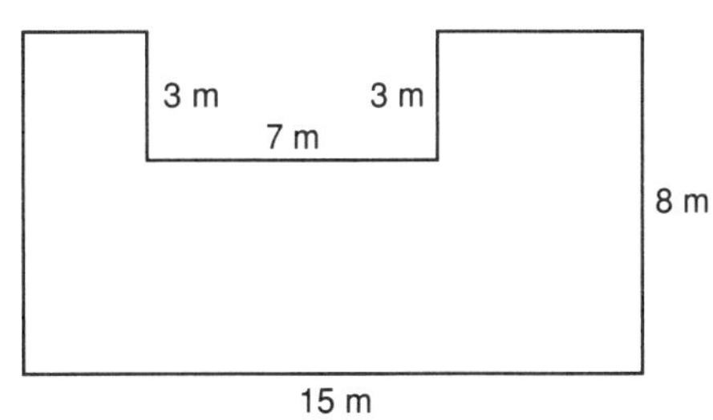

Answer: Extend one side:

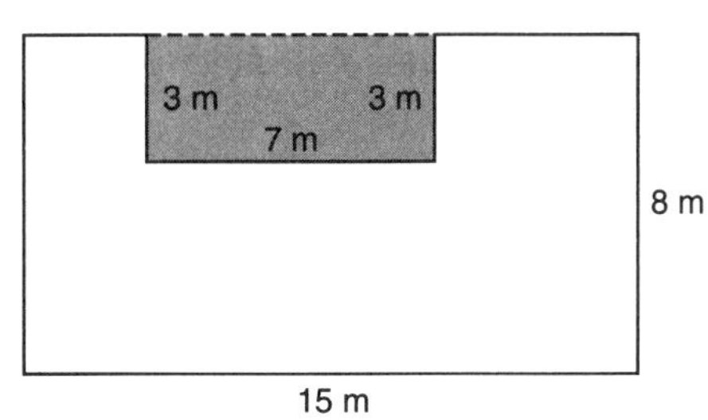

The new figure with the extended side would have an area of 15 • 8 or 120 square meters. The added area would be 7 • 3 or 21 square meters. Now subtract the added area:

120 − 21 = 99

Therefore, the area of the original figure is 99 square meters.

3. Places on the earth are located by their latitude, or angular distance from the equator. For example, Seattle, Washington, is at 47.6°N latitude. This means that the angle from the earth's radius at the equator to the point where Seattle lies is 47.6° and the direction is northward (N):

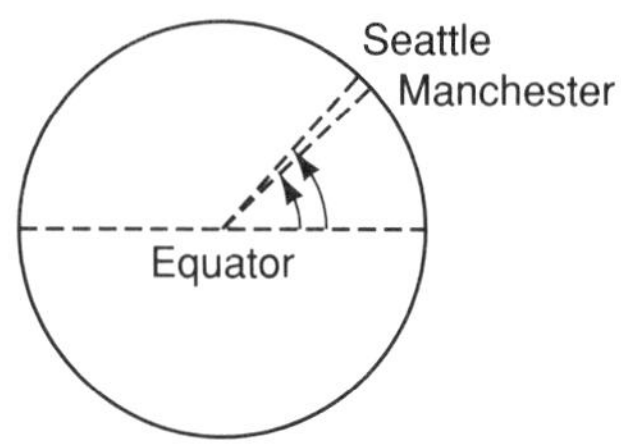

If Manchester, New Hampshire, is at 43.0°N, what is the angular measurement between Manchester and Seattle?

Answer: It is not necessary to measure the angle. Using indirect measurement, we can find that the angle between them would be 47.6° − 43°, or 4.6°.

4. Lima, Peru, is at a latitude of 12°S. Washington, D.C., is at a latitude of 38.9°N. What is the angle measurement between them?

Answer: Both locations are measured from the equator, but one is southward (S) and the other is northward (N):

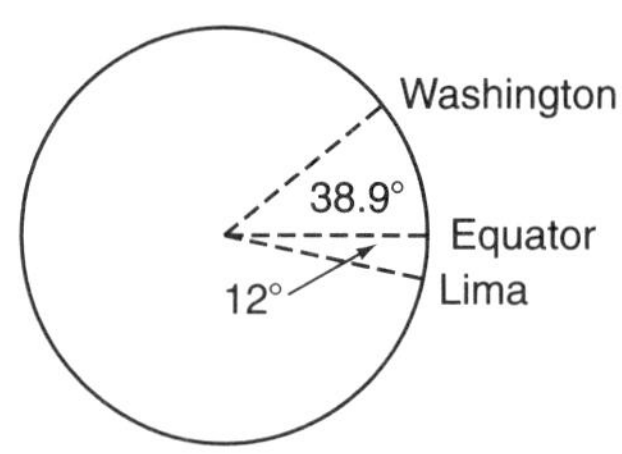

Therefore, we add the two measurements:

$12° + 38.9° = 50.9°$

5. David is flying a kite. When he is 15 feet away from his house, the kite hits the roof and falls to the ground. He knows that the house is 30 feet high. Without actually measuring the kite string, how can he calculate the length of string that was let out?

Answer: The length of the string is the hypotenuse of a right triangle whose legs are 15 feet and 30 feet. So David can use the Pythagorean theorem:

$c^2 = a^2 + b^2$

$c^2 = 15^2 + 30^2$

$c^2 = 1{,}125$

$c \approx 33.5$

The length of the string was approximately 33.5 feet.

TRY THESE ✔

1. A town park is a rectangle 80 yards wide and 100 yards long. One sidewalk goes around the park. Another sidewalk is the diagonal of the rectangle. There is a snack stand in the center of one the end of the park, as shown in the diagram. What is the **shortest** distance that Shannon will have to go if she is in-line skating on sidewalks from the snack stand to the gate?

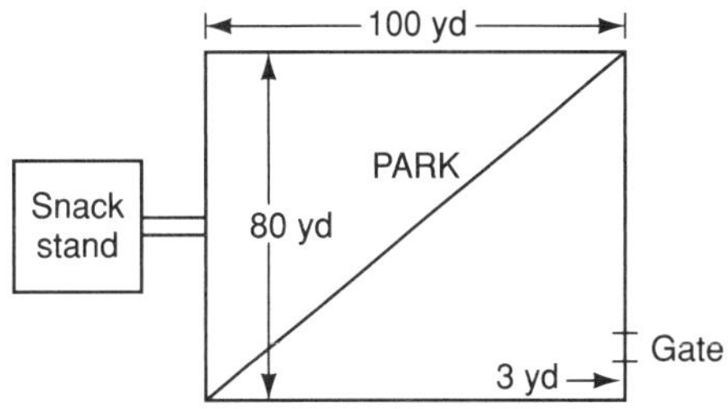

2. This design is to be cut from a piece of wood that is 4 feet by 8 feet. How many square feet of wood will be left over after the design is cut out? Use $\pi \approx 3.14$ and round to the nearest square foot.

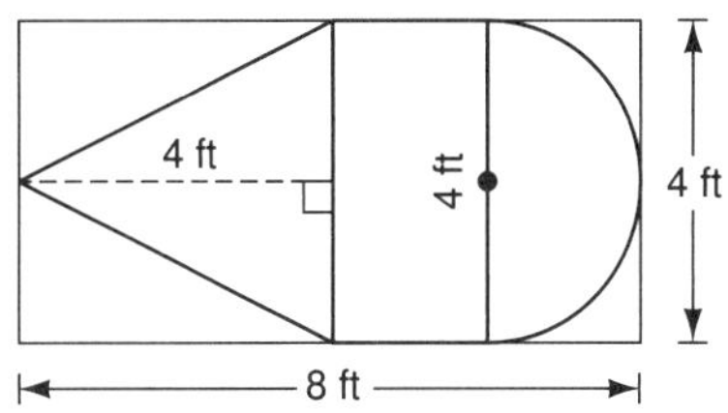

3. A cylindrical container holds exactly three balls stacked vertically. If each ball has a diameter of 2 inches, what is the volume of the container? Use $\pi \approx 3.14$ and round to the nearest cubic inch.

4. Eric wants to measure the distance across a lake from point A to point B. He lays out a right triangle on the ground with $AC = 40$ feet and $BC = 70$ feet. What is the distance from point A to point B? Round the answer to the nearest tenth.

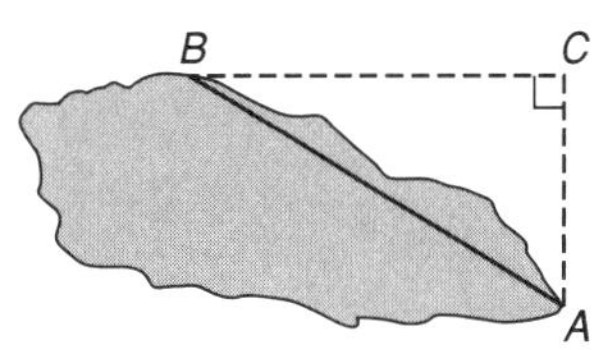

5. *MNOP* is a square with a side of 10 feet. What is the area of the unshaded region?

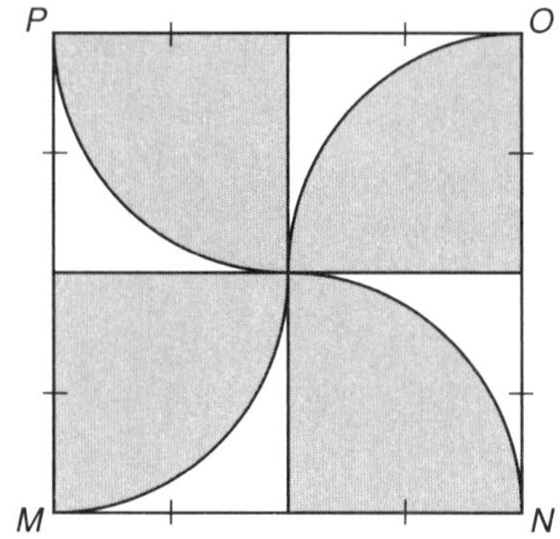

Assessment

Multiple-Choice Problems

1. The diagram below shows a picture frame: the frame is shaded, and the unshaded part is the opening for the picture. The outside is a square and the opening is a circle that covers to within 1 inch on each side of the square. What is the area of the picture frame?

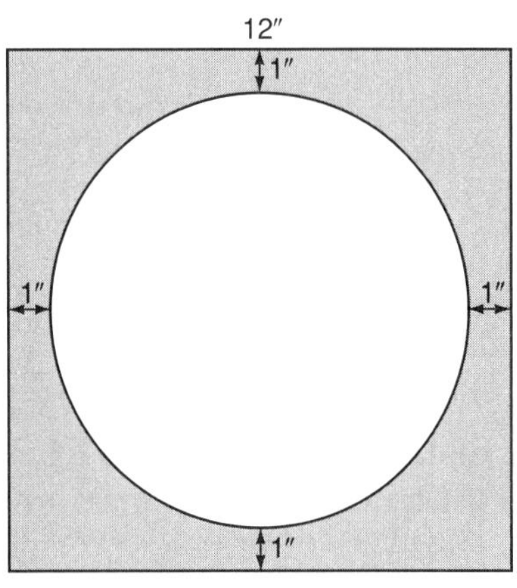

A. 30.96 square inches

B. 49.015 square inches

C. 65.5 square inches

D. 78.2 square inches

2. A 24-foot wire reaches from the top of a pole to a stake in the ground located 10 feet from the foot of the pole. How tall is the pole? Round your answer to the nearest foot.

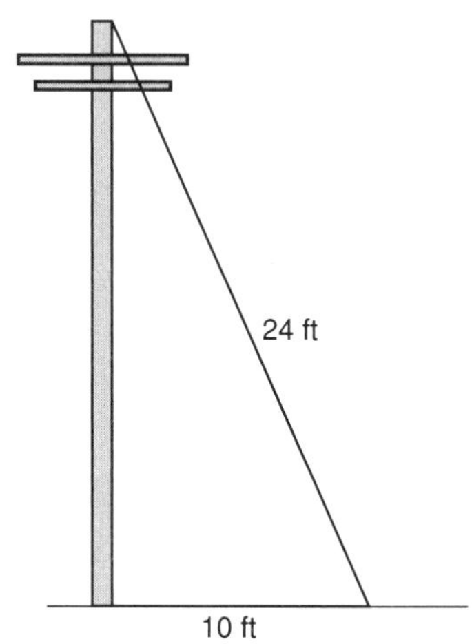

A. 26 feet

B. 22 feet

C. 20 feet

D. 18 feet

3. *ABCDEFG* is a plot of land. To find its area, surveyors mark off rectangle *MNPQ*. Using the given measurements, what is the area of the plot?

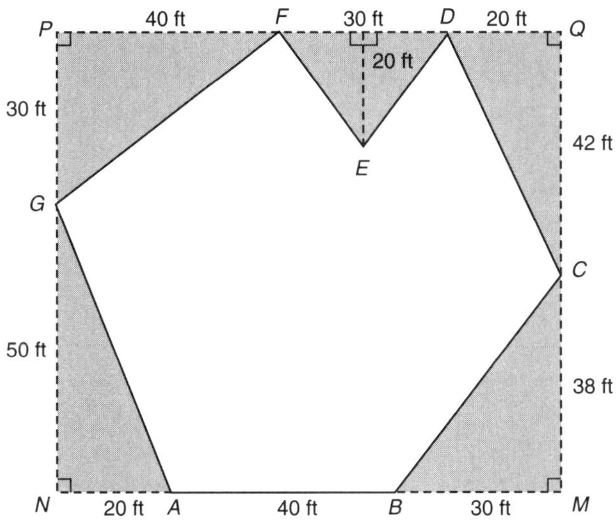

A. 2,390 square feet

B. 4,010 square feet

C. 4,810 square feet

D. 5,710 square feet

4. A piece of PVC pipe is cylindrical in shape and 200 centimeters long. Its inside diameter measures 14 centimeters and its outside diameter measures 15 centimeters. To the nearest cubic centimeter, what is the volume of PVC needed to make the pipe? (Use $\pi \approx 3.14$.)

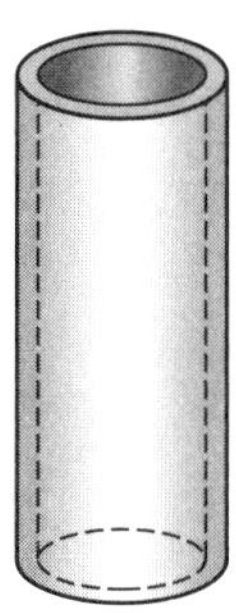

A. 18 212 cubic centimeters

B. 4553 cubic centimeters

C. 628 cubic centimeters

D. 22.765 cubic centimeters

5. Jasper walks to school using a path that cuts through two empty lots. On the map shown below, this path is *A* to *F* to *D*. When it rains, the lots flood and he has to take the path from *A* to *B* to *C* to *D*. The school (point *D*) is halfway between points *C* and *E*. What is the length of path *A-F-D* compared to the length of path *A-B-C-D*?

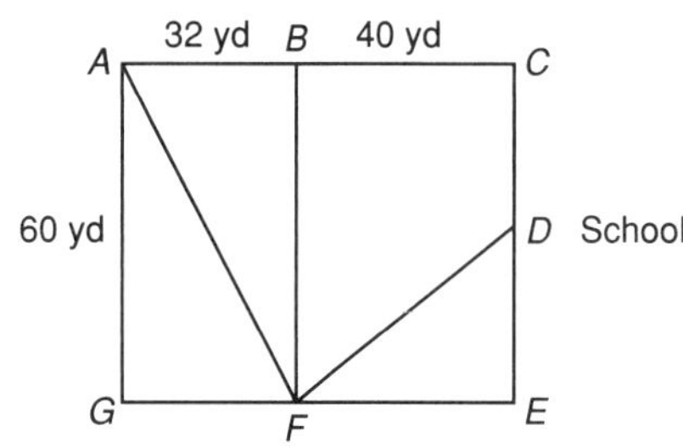

A. same length

B. 12 yards shorter

C. 16 yards shorter

D. 16 yards longer

6. A rectangular cement block contains two square holes. A side of each hole is 5 inches. The dimensions of the block are 8 inches by 16 inches by 8 inches. How many cubic inches of cement are in the block?

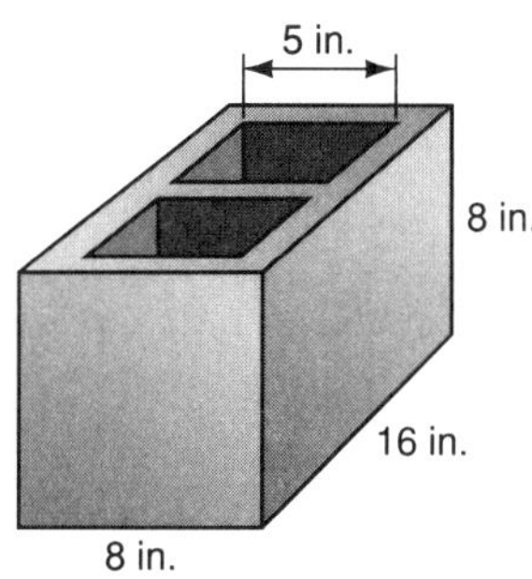

A. 400 cubic inches

B. 624 cubic inches

C. 824 cubic inches

D. 1,024 cubic inches

7. Rio de Janeiro, Brazil, is located at latitude 22.9°S. Green Bay, Wisconsin, is located at latitude 44.5°N. What is the angle measurement between the two cities?

A. 21.6°

B. 22.9°

C. 44.5°

D. 67.4°

8. To find the approximate distance between the two cities in problem 7, multiply the angle measurement by 70. What is the approximate distance between the two cities?

A. 4,718 miles

B. 3,115 miles

C. 1,603 miles

D. 1,512 miles

B SOLVING REAL-WORLD PROBLEMS INVOLVING RATES

A **rate**, also called a **rated measure**, is a comparison or ratio of two unlike quantities. Familiar rates are miles per hour and feet per second. Many ratios can be considered rated measures. For example, words typed per minute, pages read per hour, and cups per serving are all rates.

Sometimes, when we obtain a rate, we need to convert the answer to a different unit. For instance, we may get a rate expressed as cups per serving, but we want to know how many ounces per serving.

EXAMPLE

Fifteen people will be having soup for dinner. Each serving of soup contains 1.5 cups. To serve everyone, how many **ounces** of soup will be needed? There are 8 ounces in a cup.

$15 \bullet 1.5 = 22.5$ cups of soup

$22.5 \bullet 8 = 180$ ounces of soup

NOTE: The necessary conversions are listed on the FCAT Mathematics Reference Sheet.

In the example above, the conversion was from a larger unit (cups) to a smaller unit (ounces). Therefore, it was necessary to multiply the answer by the conversion factor, 8. But sometimes we must convert from a smaller to a larger unit.

EXAMPLE

A formula calls for 80 milliliters of sugar to each 1000 milliliters of water. If we are going to use 320 milliliters of sugar, how many **liters** of water should we use?

To get the answer, set up a **proportion**, a statement that two rates or ratios are equal:

$$\frac{80}{1000} = \frac{320}{x}$$

Next, solve for x: $80x = 320{,}000$, so $x = 4000$ milliliters of water.

Since there are 1000 milliliters in a liter, we divide: $\frac{4000}{1000} = 4$ liters of water.

In this problem, the conversion was from a smaller unit (milliliters) to a larger unit (liters). Therefore, it was necessary to divide.

Here are the rules for **conversion**:

- To convert from larger to smaller, multiply.
- To convert from smaller to larger, divide.

MODEL PROBLEMS

1. Jorge has been hired to move Mary to her new apartment. Mary's furniture is estimated to weigh 1,450 pounds. Martha is moving to the same area, so she hires Jorge to move her on the same day. Martha's furniture weighs about 700 pounds. Jorge is renting a van from a company that calculates van tonnage according to the amount of weight to be carried. What tonnage van should he use?

Answer:

Total weight to be carried is $1{,}450 + 700 = 2{,}150$ pounds.

Since 2,000 pounds = 1 ton, the conversion is from smaller to larger units. Divide:

$$\frac{2{,}150}{2{,}000} = 1.075 \text{ tons}$$

Therefore, Jorge should probably rent a 2-ton truck.

2. A picnic is being planned for 20 people. Each person is to get a quart of lemonade. How many gallons of lemonade should be ordered?

Answer: 20 quarts of lemonade are needed. There are 4 quarts in a gallon, so the conversion is from smaller to larger units. Divide:

$$\frac{20}{4} = 5 \text{ gallons}$$

Therefore, 5 gallons of lemonade should be ordered.

3. In a recipe for chili, it takes 2 cups of tomato sauce to make one pot of chili. If we want to make 3 pots of chili, how many ounces of tomato sauce will we need?

Answer: First calculate the amount of sauce needed for 3 pots: $2 \bullet 3 = 6$ cups of sauce. There are 8 ounces in a cup, so we are converting from larger to smaller. Multiply:

$$8 \bullet 6 = 48 \text{ ounces}$$

Therefore, we will need 48 ounces of tomato sauce.

4. A manufacturing company finds that it takes 35 minutes to assemble a certain part. How many hours will be needed to assemble 45 parts?

Answer: First calculate minutes needed: $35 \bullet 45 = 1{,}575$ minutes. There are 60 minutes in an hour, so the conversion is from smaller to larger. Divide:

$$\frac{1{,}575}{60} = 26.25 \text{ hours}$$

Therefore, it takes 26.25 hours, or 26 hours and 15 minutes, to assemble the parts.

TRY THESE

1. Shele wants to make some new curtains. For each curtain panel, she needs 45 inches of material of standard width. If she wants to make 8 panels, how many yards of material should she buy?
2. A car averages 50 miles per hour. How many minutes will it take for the car to travel 40 miles?
3. A machine part weighs 30 grams. These parts are packed 50 to a box. How many kilograms does one box weigh?
4. Five boxes are placed end to end on a shelf. Each box is 72.5 centimeters long. What is the total length of the boxes in meters?
5. The Donahue family owns a $\frac{1}{4}$-acre lot in the city. They use 5% of the property for a flower garden. How many **square feet** are in the flower garden? One acre = 43,560 sq ft.
6. The water cooler holds 10 gallons of water. Clif, the equipment manager, adds $28\frac{1}{2}$ quarts of water to fill the cooler. How many **ounces** of water were in the cooler before he filled it?
7. The dimensions of a rectangular aquarium are 48 inches by 12 inches by 18 inches. How many gallons of water will it hold if 231 cubic inches equal 1 gallon?
8. The defensive lineman picked up a fumbled ball and ran 12 yards for a touchdown. Wanting to make the distance seem much longer than it was, he expressed the length of his run in inches. How many inches did he run?

Assessment

Multiple-Choice Problems

1. Olivia can weld 7 brackets in 20 minutes. How many can she weld in an 8-hour day?

   A. 56 **C.** 168

   B. 112 **D.** 196

2. A rancher wants to fence in a field using 3 strands of wire along each side. The field is 80 yards by 100 yards. The posts are already in place. If wire cost 10 cents per foot, how much will it cost to fence the field?

   A. \$108 **C.** \$720

   B. \$324 **D.** \$7,200

3. An elephant's heart beats an average of 30 times per minute. As an average, how many times does it beat in one day?

   A. 720 times

   B. 1,800 times

   C. 21,600 times

   D. 43,200 times

4. Driving a car at a speed of 5 miles per hour is equivalent to walking at approximately what rate in feet per second?

   A. 7.3 feet per second

   B. 8.08 feet per second

   C. 73 feet per second

   D. 80.8 feet per second

5. Skye wants to measure the length of a couch at a garage sale. She uses the distance from her elbow to the tip of her longest finger and determines that the couch is 4 lengths long. She then measures the distance from her elbow to the tip of her longest finger and finds that it is 20 inches long. How long is the couch?

   A. 6 feet

   B. 6 feet 8 inches

   C. 7 feet

   D. 7 feet 4 inches

Gridded-Response Problems

6. A cylindrical storage tank is 15 feet tall with a radius of 4 feet. Knowing that 1 cubic foot = 7.5 gallons, how many gallons will this tank hold? Round your answer to the nearest tenth.

7. There are 250 milliliters of grape juice in a juice box. How many liters are there in a 12-pack of juice boxes?

8. Weather stripping is needed to go around the edges of the window shown below. How much will it cost if weather stripping is $0.05 per inch?

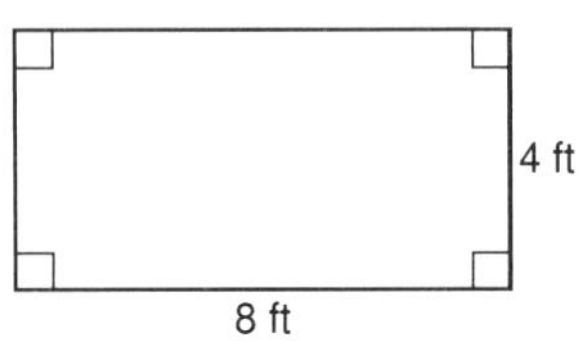

9. A butcher shop sells 4-ounce hamburger patties. The student council needs 200 patties for its fund-raiser. How many pounds of hamburger will be needed?

10. Lupe wants to put lace trim around a circular tablecloth. The tablecloth is 42 inches in diameter. How many feet of lace will she need? Use $\pi \approx \frac{22}{7}$.

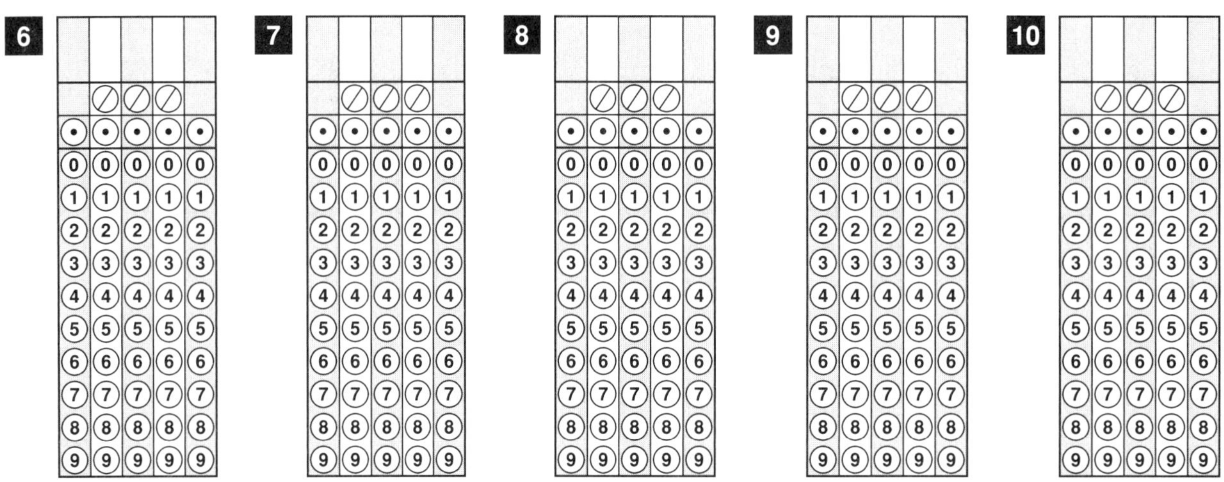

CHAPTER REVIEW

Multiple-Choice Problems

1. Anton knows that he can expect 0.2% evaporation of water in his chemistry experiment. Which of the following is closest to the amount of water he can expect to evaporate from the container shown below? Use $\pi \approx 3.14$.

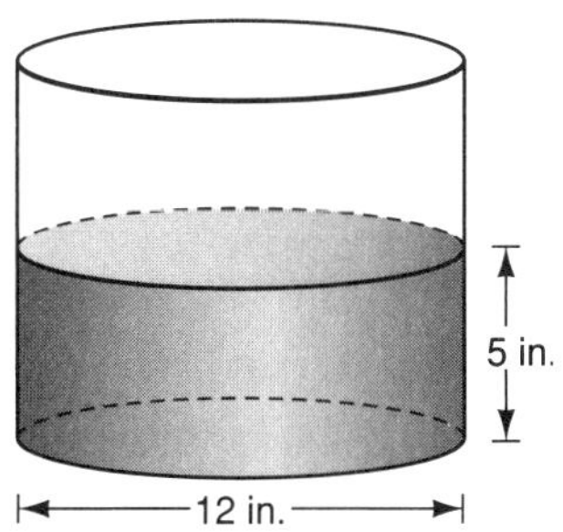

   A. 0.11 cubic inch

   B. 0.45 cubic inch

   C. 1.1 cubic inches

   D. 4.5 cubic inches

2. The 6 blades of a fan are equally spaced as shown below. What is the degree measure of $\angle a$?

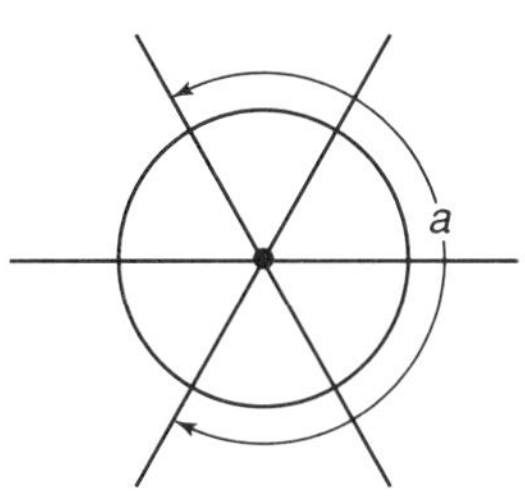

   A. 60°

   B. 120°

   C. 240°

   D. 270°

3. On a baseball field, the bases are 90 feet apart. What is the approximate distance from home plate to second base?

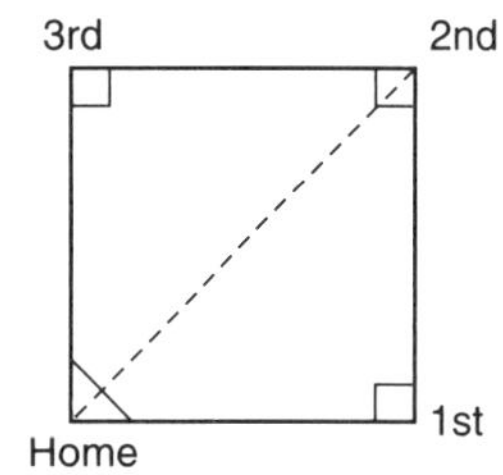

   A. 120 feet C. 153 feet

   B. 127 feet D. 180 feet

4. The surface area of a sphere is 314 square inches. What is the radius of the sphere?

   A. 25 inches

   B. 10 inches

   C. 5 inches

   D. 2.5 inches

5. A 25-foot ladder is leaning against a barn. The base of the ladder is 8 feet from the barn. How high up the side of the barn does the ladder reach? Round your answer to the nearest tenth.

   A. 18.9 feet

   B. 19.4 feet

   C. 22.6 feet

   D. 23.7 feet

6. Dr. Keller ordered a 20-gallon cylindrical container that is 2 feet tall. If 1 cubic foot = 7.5 gallons, what is the approximate radius of the container?

   A. 1.8 feet

   B. 4.9 feet

   C. 5.5 feet

   D. 6.9 feet

7. The Band Boosters sell food and soda at the football games. Last year they averaged $425 in sales per game. The table below shows the sales after the first 3 games this year. How do the sales this year compare with the average sales last year?

Game	Sales
1	$510
2	$281
3	$460

A. $8 above last year's average

B. $10 above last year's average

C. $10 below last year's average

D. $8 below last year's average

8. A round pizza with a diameter of 16 inches is cut into 9 equal pie-shaped pieces. What is the approximate length of the crust edge of one piece? Use $\pi \approx 3.14$.

A. 5.6 inches **C.** 11.2 inches

B. 9.8 inches **D.** 22.3 inches

9. A track is in the shape of a rectangle and two semicircles (half-circles). The rectangle is 4 yards by 10 yards.

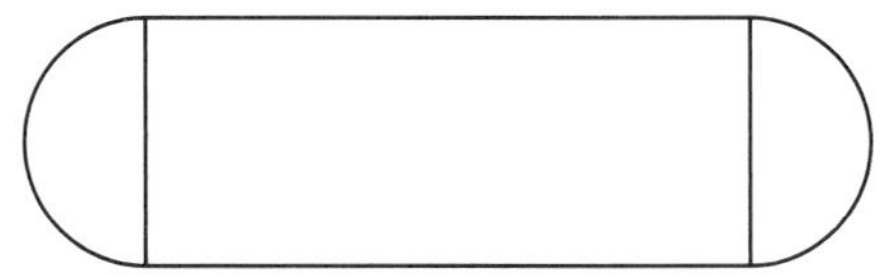

To the nearest yard, what is the perimeter of the track?

A. 28 yards **C.** 40 yards

B. 33 yards **D.** 80 yards

Gridded-Response Problems

10. Jason can run 6 feet per second. At this rate, how far can he run in 4.5 minutes?

11. Melinda wants to cut the largest possible circle out of a square sheet of paper shown below. What will be the area of the circle? Use $\pi \approx \frac{22}{7}$.

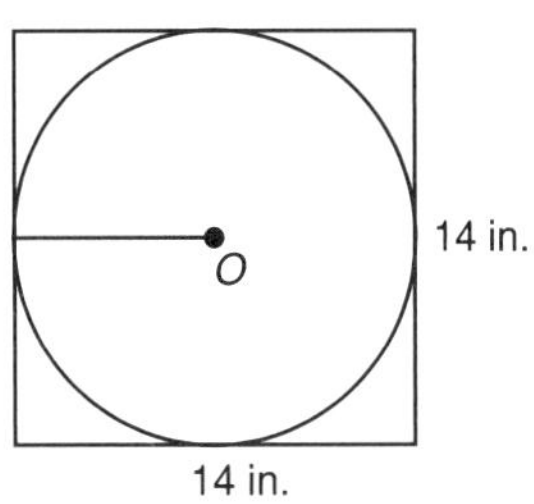

12. On their gym floor, the Rockwood Rockets have the design that is shown below. What is the area of the design?

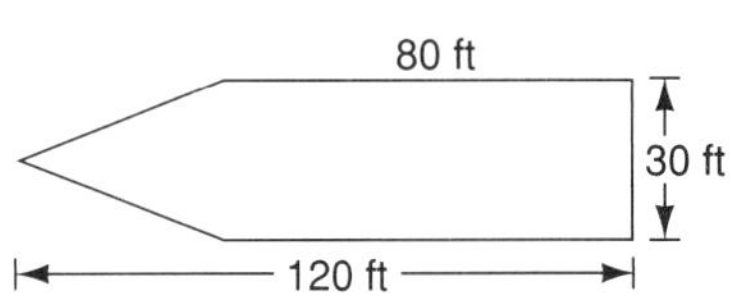

13. The Bowmans purchased a vase. The box that contains the vase is in the shape of a cube with a $2 \times 2 \times 12$-inch block in each corner, as shown in the drawing below. What is the remaining volume, in cubic inches, of the box when the blocks are inside the box?

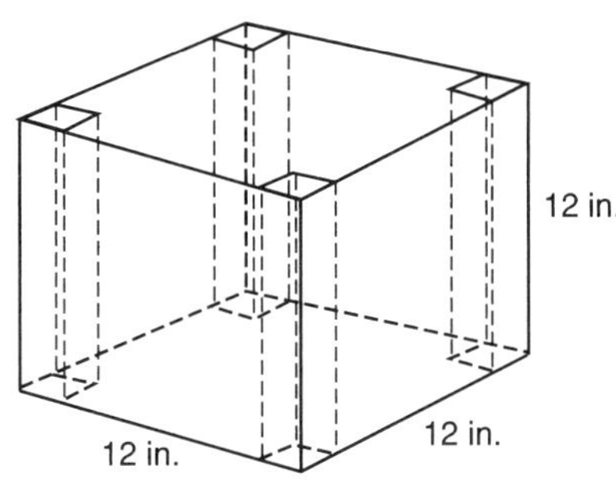

14. Sue, the navigator on a submarine, watches an approaching ship on her sonar screen. On the screen, the space between each circle represents 4 miles. If the ship is traveling at the rate of 5 miles per hour, how many hours will it take the ship to reach the stationary submarine?

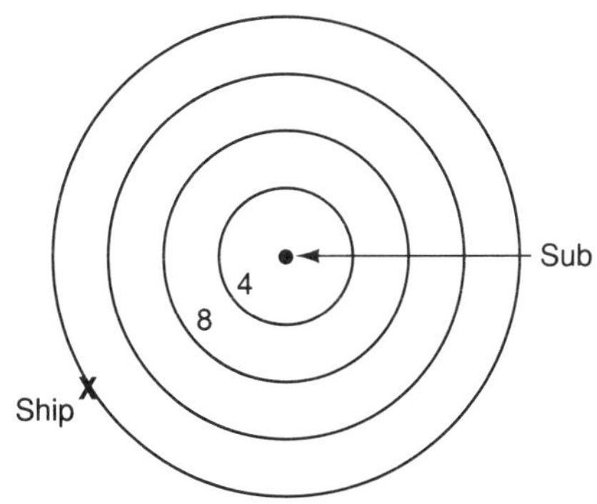

15. Two identical square pyramids have been placed with their bases touching to form a sculpture. The edge of each base is 6 feet and the total length is 20 feet. What is the volume, in cubic feet, of the sculpture?

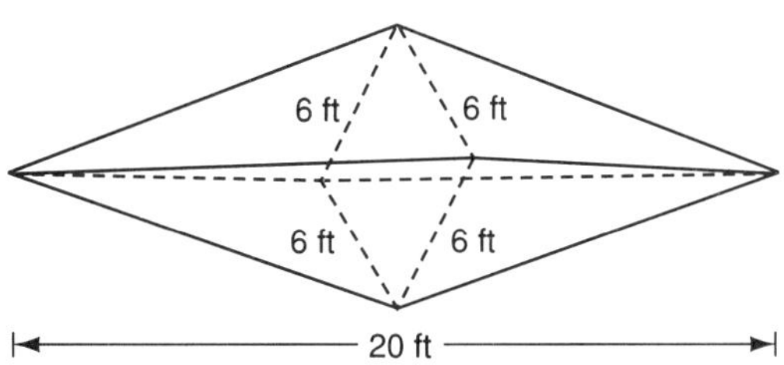

16. The label on a rectangular packing crate states that the crate will hold a maximum of 9 pounds per cubic foot. What is the total weight that the crate can hold if its measurements are 12 feet by 4 feet by 2 feet?

17. In the drafting room, a 2-foot by 3-foot rectangular grid is laid out on the top of one of the tables. Every 2 inches along each side, a line is marked forming small squares. How many 2-inch by 2-inch squares are formed on the grid?

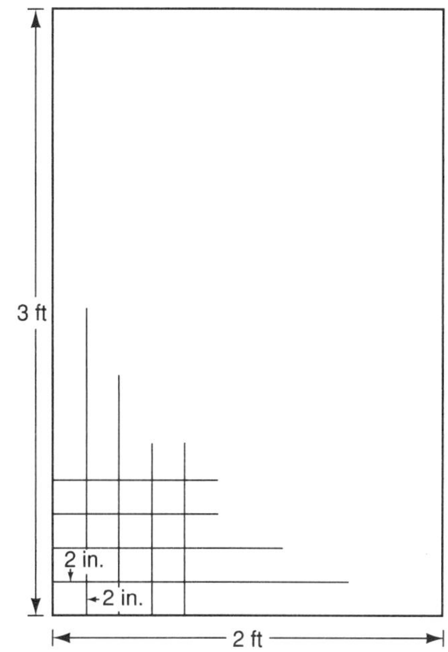

18. The area of the top of a circular can is 20.25π square centimeters. What is the diameter of the top of the can?

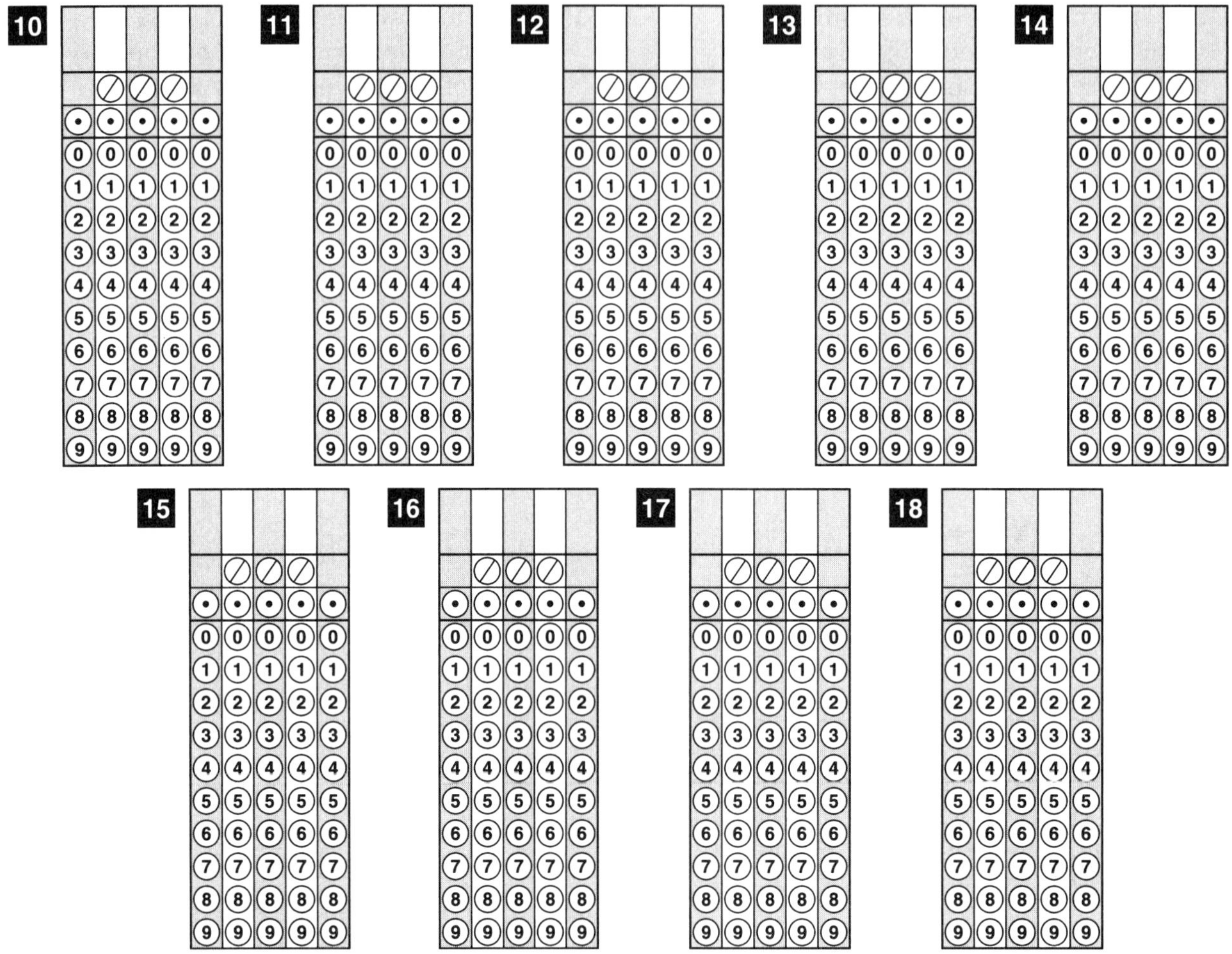

Think, Solve, Explain: Short-Response Problems

19. A baseball field is in the shape of $\frac{1}{4}$ of a circle. In spring training, the players begin practice by jogging around the field 3 times.

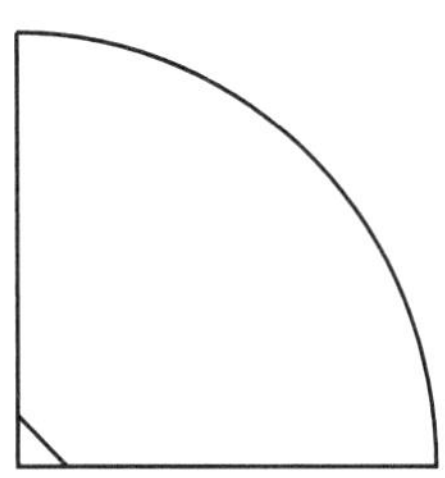

How far does each player jog? Show all work necessary to find the answer.

20. The steam engines on old trains used water to operate the engine. Along the tracks you might see a container that held the water supply for the trains. This container is in the shape of a cylinder and a cone as shown below.

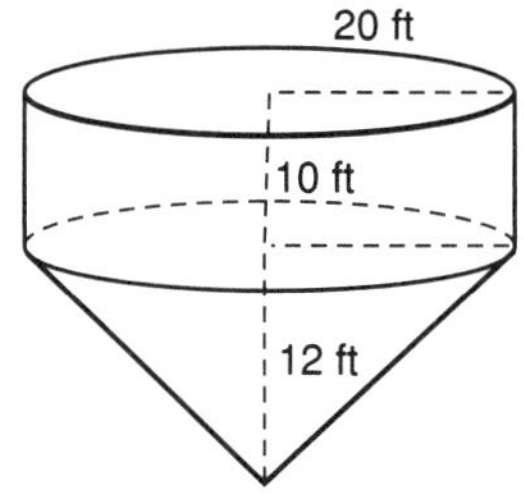

If the container is completely filled with water, what is its volume? Show all work necessary to find the answer.

21. A solid cylinder made of metal is to be melted down and recast as a solid cone. The cylinder has a radius of 6 centimeters and a height of 18 centimeters. Find the height of the recast cone if the radius of the cone is 9 centimeters. Show your work.

CUMULATIVE REVIEW

Multiple-Choice Problems

1. Between what two integers does $\sqrt{57}$ lie?

A. 6 and 7 **C.** 8 and 9

B. 7 and 8 **D.** 56 and 58

2. The distance from the earth to a star is 3 900 000 000 000 000 kilometers. How can this distance be expressed in scientific notation?

A. 3.9×10^{15}

B. 3.9×10^{14}

C. 3.9×10^{-15}

D. 3.9×10^{-14}

3. Using the Pythagorean theorem, Jacque calculated the distance between point *A* and point *B* to be $\sqrt{459}$ miles. Which of the following is a close approximation of the distance?

A. 21.4 miles

B. 25.2 miles

C. 229.5 miles

D. 459 miles

4. A drink concentrate requires 3 ounces of concentrate for every quart of water. How many pints of concentrate are needed for 5 gallons of water?

A. 6 pints **C.** 3.75 pints

B. 5.25 pints **D.** 2 pints

5. Ms. Zambito marks 15 tests in 12 minutes. What is her rate of tests marked per minute?

A. $\frac{3}{4}$ test per minute

B. 1 test per minute

C. $1\frac{1}{4}$ tests per minute

D. $2\frac{1}{2}$ tests per minute

6. What type of angle has a measurement of 115°?

A. straight **C.** right

B. obtuse **D.** acute

7. A cylindrical can with a radius of 1 inch and a height of 6 inches is filled with water. That water is then poured into a larger can that has a radius of 3 inches and a height of 16 inches. How many full small cans of water will be needed to fill up the large can?

A. 6

B. 9

C. 16

D. 24

8. Which of the following is equivalent to $3(x - 4) - 2(x + 5)$?

A. $x + 22$

B. $x - 22$

C. $x - 2$

D. $5x + 2$

9. Using aerial photography, biologists determined that there are, on average, 3.2 sharks per shaded area. Based on this average, approximately how many sharks would they expect to be in the area bounded by quadrilateral *ABCD*?

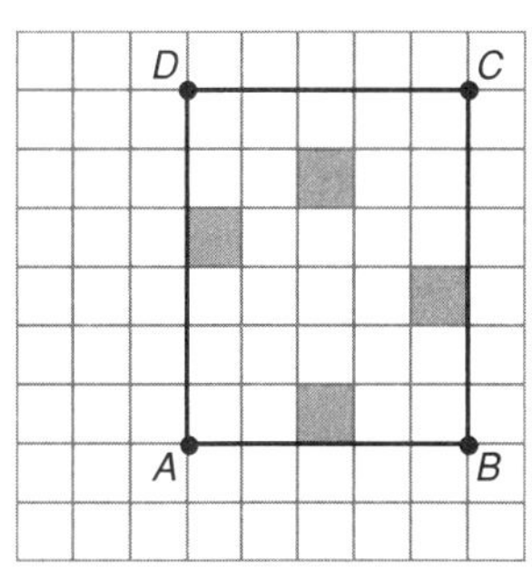

A. 86 **C.** 96

B. 90 **D.** 106

Gridded-Response Problems

10. A square, 3 inches on a side, is to be cut out of a parallelogram with sides of 7 inches and 5 inches. What is the total area of the pieces that are left?

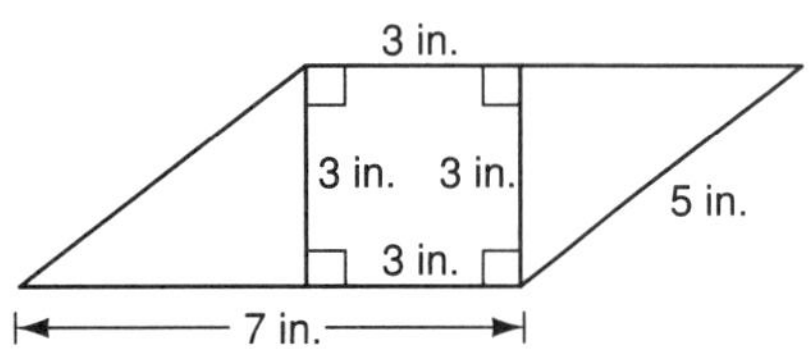

11. Elsa wants to put new tiles on her kitchen floor. The floor measures 8 feet by 10 feet. The tiles are 1 foot square and come in boxes of 30 tiles. If she cannot buy part of a box, how many boxes must she buy?

12. Jim, a member of the swim team, practiced 2 hours on Monday, 3 hours on Tuesday, $1\frac{1}{2}$ hours on Wednesday, $2\frac{1}{4}$ hours on Thursday, and 3 hours on Friday. How many hours does he have to practice on Saturday to average 3 hours per day for the 6 days?

13. Warren's job pays $150 per week. If he puts $45 into his savings account, what percent is left for him to spend?

14. There are twelve 8-ounce cans of soda in a refrigerator box. What is the equivalent amount, in quarts, of soda in the box?

15. What is the area, in square feet, of the figure shown below?

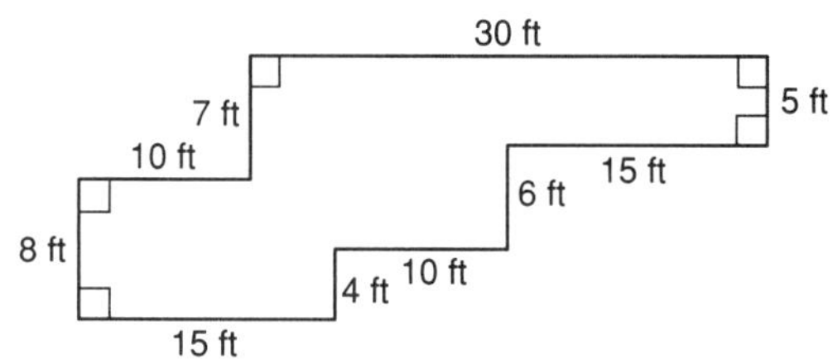

16. A sphere has a diameter of 14 centimeters. What is the surface area of the sphere? Use $\pi \approx \frac{22}{7}$.

17. Jeannie paid $199.99 for a digital camera that was originally priced at $249.99. What is the percent of decrease in the price of the camera? Express your answer to the nearest percent.

18. What is the volume of a cube that has a surface area of 294 square inches?

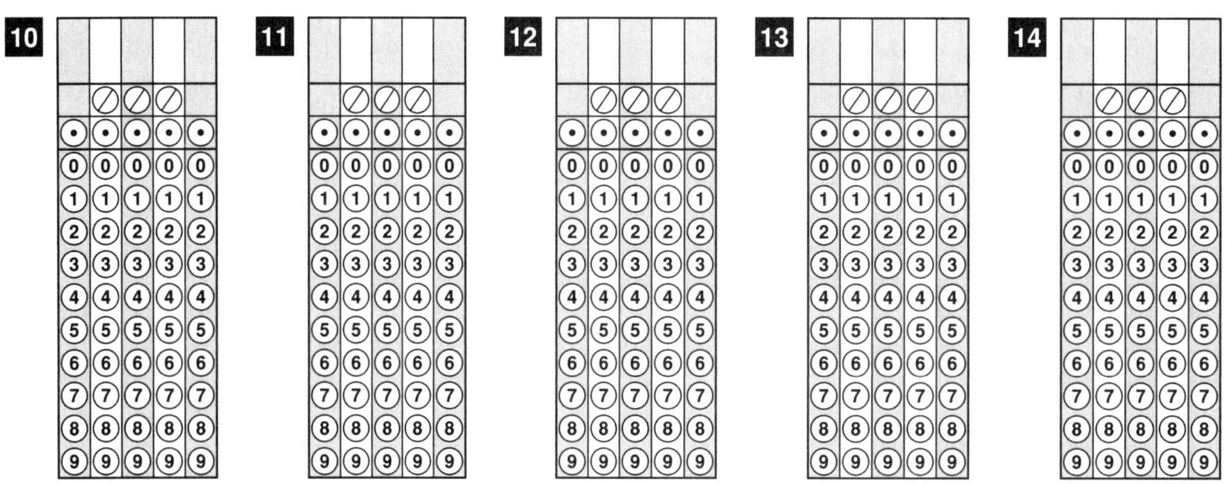

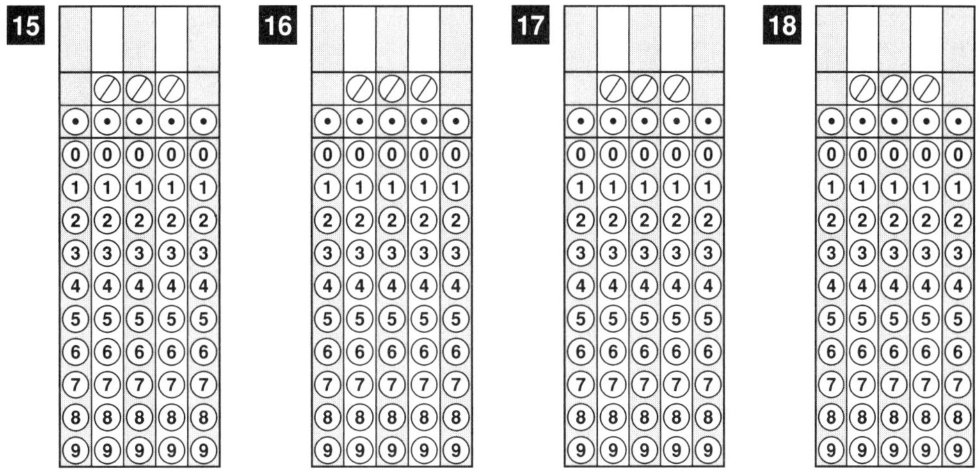

Think, Solve, Explain: Short-Response Problems

19. Aileen rode her bicycle to the bicycle repair shop and then walked home. It took her 10 minutes traveling at 9 miles per hour to get to the shop and 30 minutes to walk home. How fast was she walking?

Part A Write an equation to solve the problem.

Part B Solve the equation.

20. A large container is in the shape of a cone with a diameter of 5.2 meters and a height of 6 meters. Use $\pi \approx 3.14$

Part A Find the capacity of the container.

Part B Water is flowing into the container at the rate of 1.8 cubic meters per minute. How long, to the nearest minute, will it take to fill the container?

CHAPTER THREE

Geometry and Spatial Sense

CHAPTER CONTENTS

Section 1 Using properties and relationships of geometric shapes to construct formal and informal proofs

The study of geometry begins with three loosely defined elements: the point, the line, and the plane.

The **point** can be considered a location in space. It has no length, width, or thickness. A point is represented by a dot and named by a letter: point A.

The **line** can be defined as a set of points extending without end in two opposite directions. It has length but no width. It cannot be measured, since it is endless. A line is named by any two points on it: line AB or $\overleftrightarrow{AB}$. Or it can be named by one lowercase letter: line a.

A **plane** is a flat surface extending endlessly. It has two dimensions: length and width. A plane is named by a single capital letter: plane M.

From these three terms, we define other terms, make **postulates** (accepted as true without proof), and derive **theorems** (statements requiring proof).

This section discusses lines, angles, triangles, and quadrilaterals, describing their properties.

A LINES AND ANGLES

Lines

Here are some basic facts about the line:

- Two points **determine** or **define** a line.
- Points are **collinear** if they lie on the same line.
- A **line** contains an infinite number of points and continues indefinitely in two directions.

- A **line segment** is part of a line. A segment has two endpoints and is named by its endpoints: line segment AB or $\overline{AB}$.

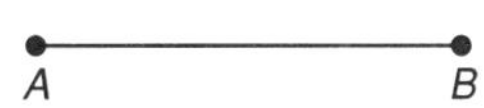

Following are more facts and definitions regarding lines.

Congruent ($\cong$) line segments have the same length. For example, if $\overline{AB} \cong \overline{CD}$, and $\overline{AB}$ is 8 inches long, then $\overline{CD}$ is 8 inches long.

Perpendicular ($\perp$) lines meet at right angles. In the diagram below, $\overleftrightarrow{AB}$ is perpendicular to $\overleftrightarrow{CD}$, or $\overleftrightarrow{AB} \perp \overleftrightarrow{CD}$.

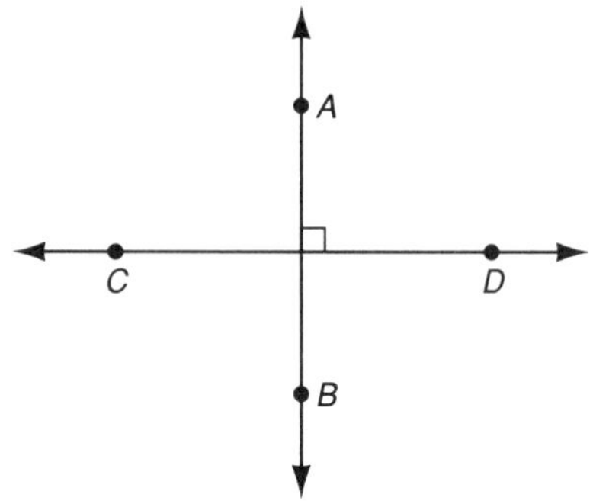

Two lines are **parallel** ($\|$) if they lie in the same plane and do not intersect no matter how far they are extended. Below, line $l \parallel$ line m.

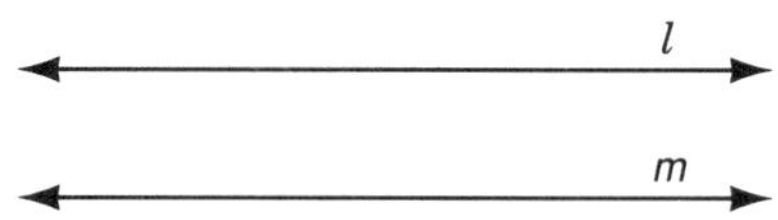

If two lines are perpendicular to the same line, then the two lines are parallel to each other. Here, line $l \perp$ line m, and line $n \perp$ line m. Therefore, line $l \parallel$ line n.

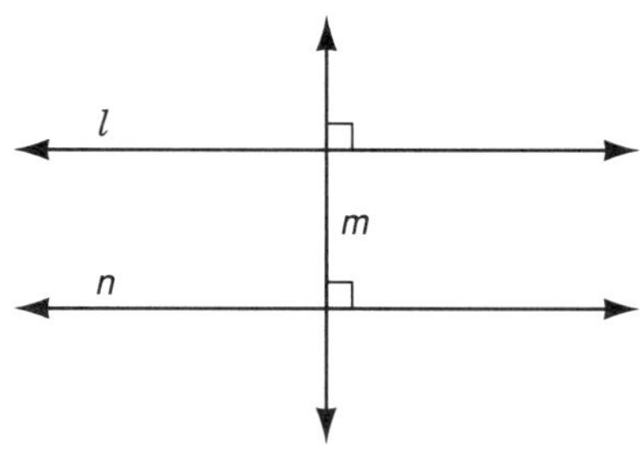

If two lines are parallel to the same line, then the two lines are parallel to each other. Line $l \parallel$ line m, and line $m \parallel$ line n. Therefore, line $l \parallel$ line n.

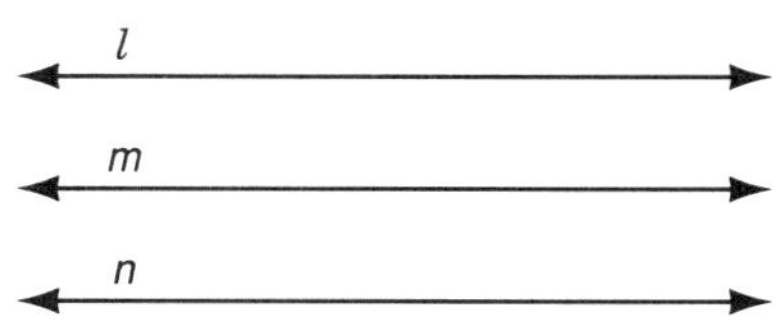

If a line is perpendicular to one of two parallel lines, it is perpendicular to the other line. Line $l \perp$ line m, and line $m \parallel$ line n. Therefore, line $l \perp$ line n.

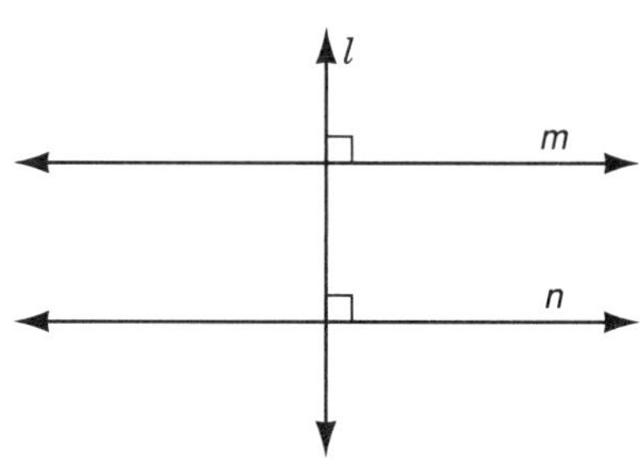

Angles

In Chapter 2, an **angle** ($\angle$) was defined as two rays with a common endpoint. The common endpoint is the **vertex**, and the rays are the **sides**. An angle can be named by its vertex: $\angle B$. It can also be named by three letters: $\angle ABC$. In this case, the vertex letter is in the middle and the other letters represent one point on each ray. Another way to name an angle is by a symbol (such as a number) within the angle: $\angle 1$.

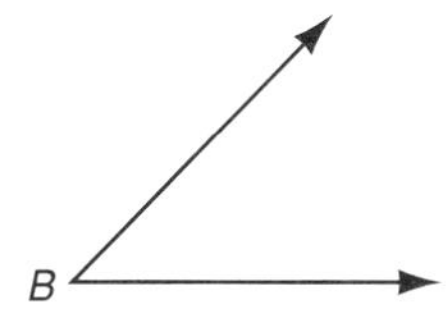

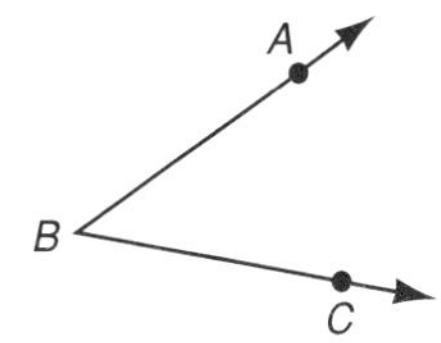

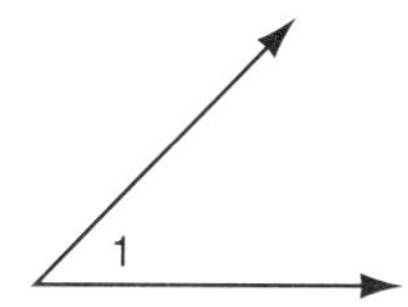

Angles are classified by number of degrees:

- An **acute angle** measures less than 90 degrees.

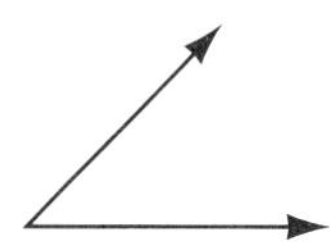

- A **right angle** measures exactly 90 degrees. Note its symbol:

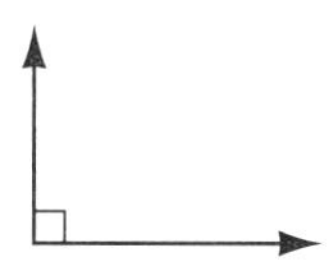

- An **obtuse angle** measures greater than 90 degrees but less than 180 degrees.

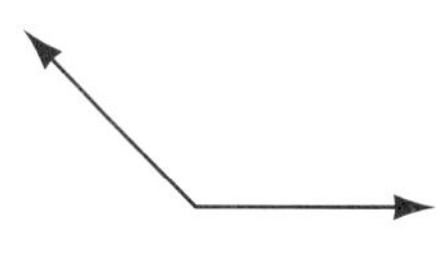

- A **straight angle** measures exactly 180 degrees.

Adjacent angles are two angles that have the same vertex and a common side but do not overlap. Angles ABD and DBC are adjacent angles because they have the same vertex B and share the side $\overrightarrow{BD}$.

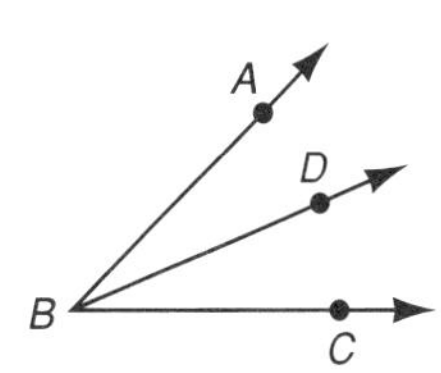

But ∠*ABD* is not adjacent to ∠*CBE*, because they overlap.

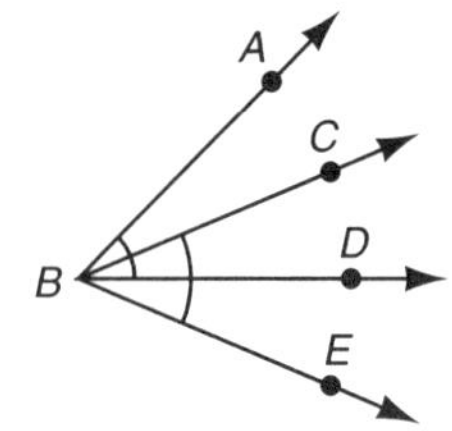

Below, ∠*ABC* is not adjacent to ∠*DCE*, because they do not have the same vertex.

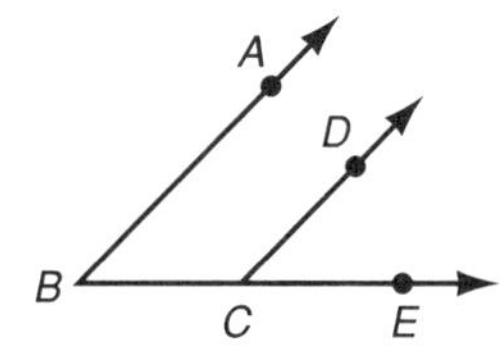

Two angles are **supplementary** when the sum of their degree measures is 180. Angles *x* and *y* are supplementary because ∠*x* + ∠*y* = a straight angle, and a straight angle measures 180 degrees. We can also say that ∠*x* is the **supplement** of ∠*y*, or that ∠*y* is the supplement of ∠*x*.

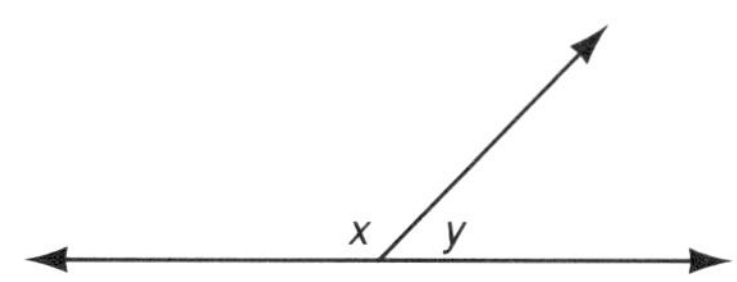

Two angles are **complementary** when the sum of their degree measures is 90. Angles *s* and *t* are complementary because ∠*s* + ∠*t* = a right angle, and a right angle measures 90 degrees. We can also say that ∠*s* is the **complement** of ∠*t*, or that ∠*t* is the complement of ∠*s*.

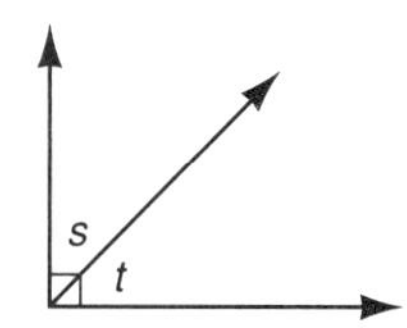

In the above examples, angles *x* and *y* are adjacent, and angles *s* and *t* are adjacent. But angles can be supplementary or complementary even if they are not adjacent. All that is required is for the sum of the measures of the angles, wherever they are located, to be 180 degrees (for supplementary angles) or 90 degrees (for complementary angles). For example, ∠*A* and ∠*B* below are supplementary and ∠*C* and ∠*D* are complementary.

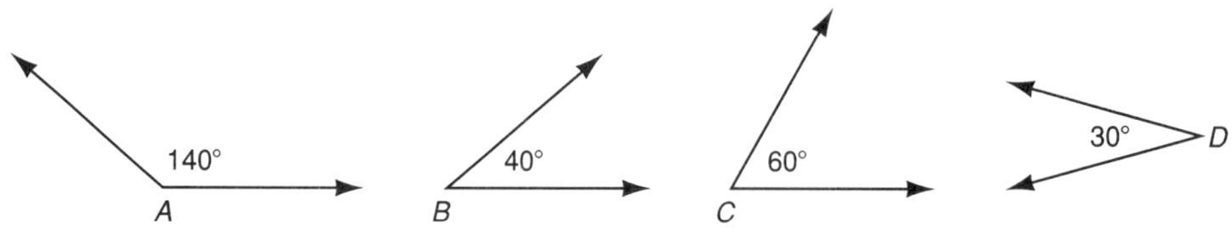

Two angles with the same measure are **congruent** angles. Here, $\angle S$ measures 40 degrees and $\angle T$ measures 40 degrees, so $\angle S \cong \angle T$.

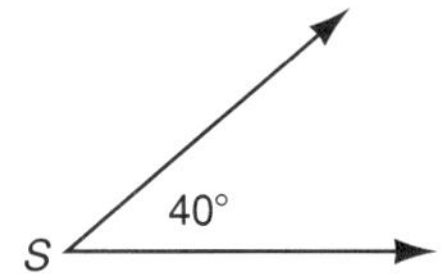

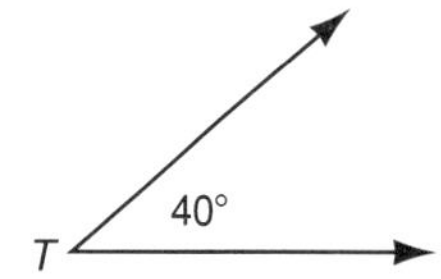

An **angle bisector** is the ray that cuts an angle into two congruent angles. Below, $\overrightarrow{OB}$ is the angle bisector of $\angle AOC$, so $\angle AOB \cong \angle BOC$. Another way to say the same thing is: measure of $\angle AOB$ = measure of $\angle BOC$, or m $\angle AOB$ = m $\angle BOC$.

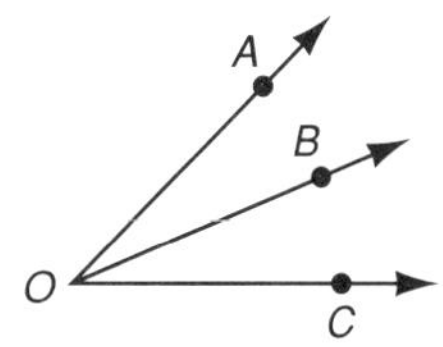

MODEL PROBLEMS Using the given diagram, answer the following questions.

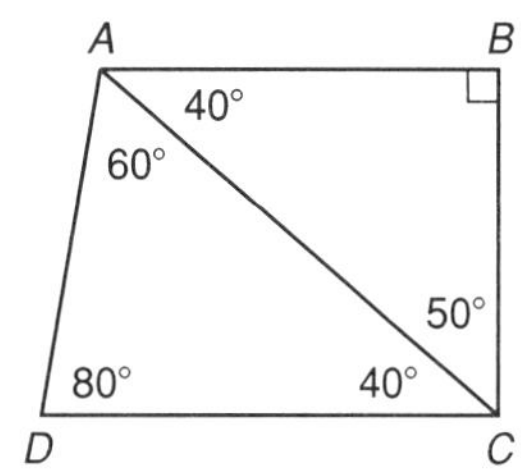

1. Name two right angles.

Answer: $\angle ABC$, $\angle BCD$

2. Name two adjacent angles that are complementary.

Answer: $\angle BCA$, $\angle ACD$

3. Name two angles that are not adjacent but are complementary.

Answer: $\angle BAC$, $\angle BCA$

4. Name an angle that is the supplement of $\angle BAD$.

Answer: $\angle ADC$

TRY THESE

Items 1–4: In the diagram, $\angle AFB$ is a right angle.

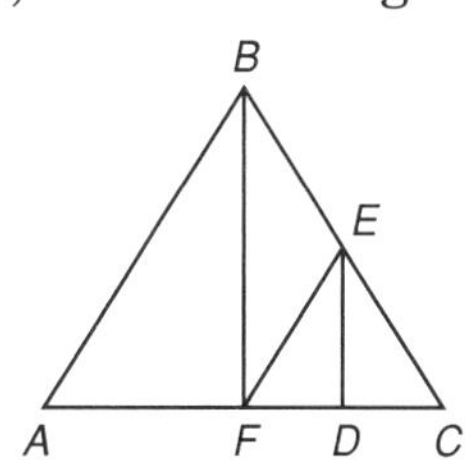

Name the following:

1. Two congruent supplementary angles.

2. Two supplementary angles that are not congruent.

3. Two complementary angles.

4. A straight angle.

Items 5–7: Use the diagram shown below.

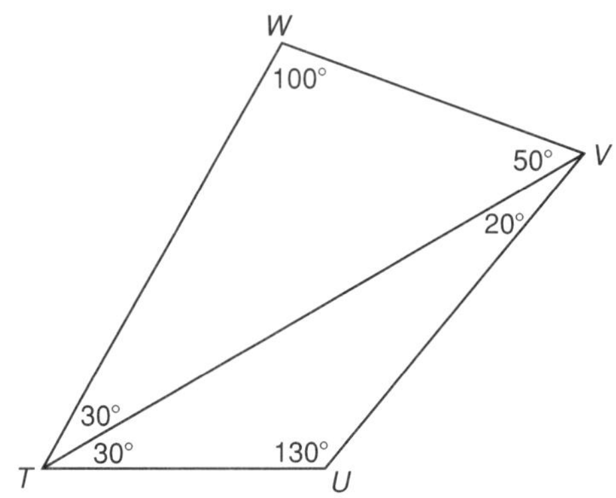

5. Which angle is bisected?

6. Name a pair of supplementary angles.

7. Name two pairs of adjacent angles.

Items 8–10: Use the diagram shown below.

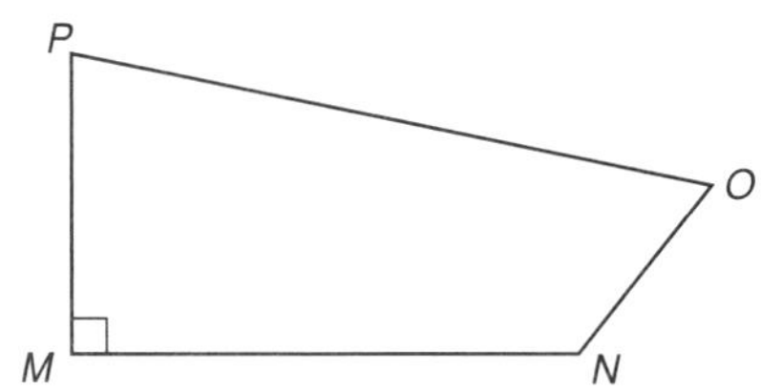

8. Name a right angle.

9. Name an obtuse angle.

10. Name the acute angles.

Intersections of lines

When two lines intersect, they form four angles. Pairs of opposite angles are called **vertical angles**. For example, $\angle 1$ and $\angle 3$ are vertical angles and $\angle 2$ and $\angle 4$ are vertical angles. Vertical angles are congruent: $\angle 1 \cong \angle 3$ and $\angle 2 \cong \angle 4$.

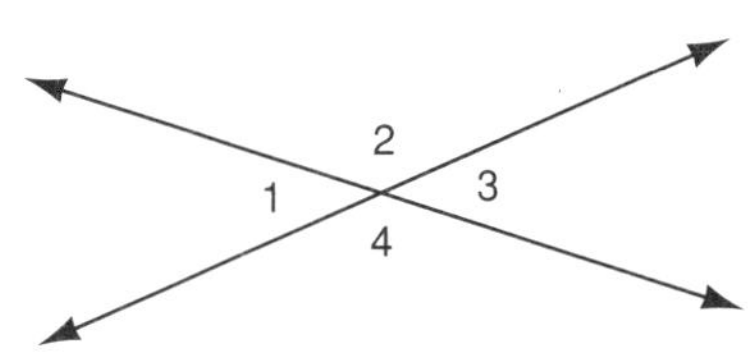

A line that intersects two or more lines is called a **transversal**. When a transversal intersects a pair of lines, four angles are formed at each intersection. Pairs of angles are named depending on their position.

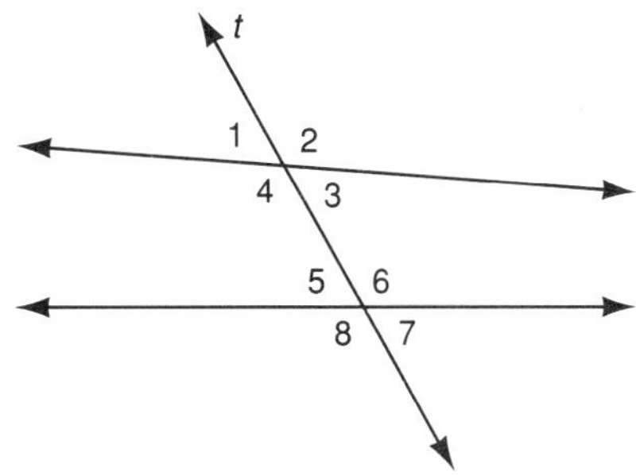

Any two angles that occupy a similar position in the two groups of four are called **corresponding angles**. In the diagram above, the following pairs are corresponding angles: 1 and 5; 2 and 6; 4 and 8; and 3 and 7.

Angles located inside the two lines intersected by the transversal are called **interior angles**. In the diagram above, angles 3, 4, 5, and 6 are interior angles. Interior angles on the same side of the transversal are angles 4 and 5 and angles 3 and 6. A pair of interior angles on opposite sides of the transversal are called **alternate interior angles**. In this case, the alternate interior angles are angles 4 and 6 and angles 3 and 5.

If two **parallel** lines are cut by a transversal, the pairs of angles have certain properties:

- Corresponding angles are congruent.
- Alternate interior angles are congruent.
- Interior angles on the same side of the transversal are supplementary.

In the diagram below, $l \parallel m$, so we have the following relationships:

Corresponding angles:

$\angle 1 \cong \angle 5$

$\angle 2 \cong \angle 6$

$\angle 4 \cong \angle 8$

$\angle 3 \cong \angle 7$

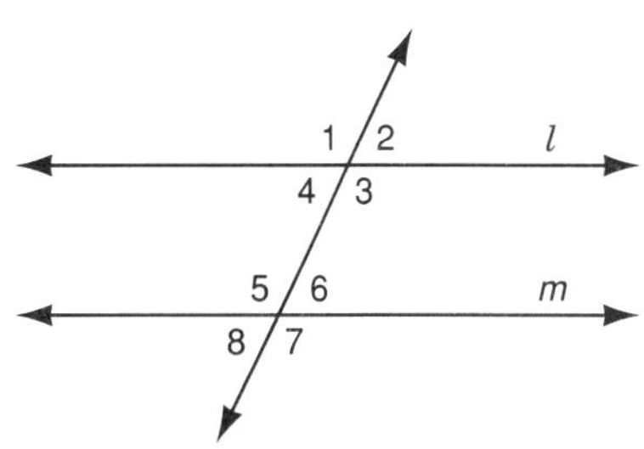

Alternate interior angles:

$\angle 4 \cong \angle 6$

$\angle 3 \cong \angle 5$

Interior angles on the same side of the transversal:

$\angle 4$ and $\angle 5$ are supplementary

$\angle 3$ and $\angle 6$ are supplementary

MODEL PROBLEMS Items 1–5: Using the given diagram, answer the following questions.

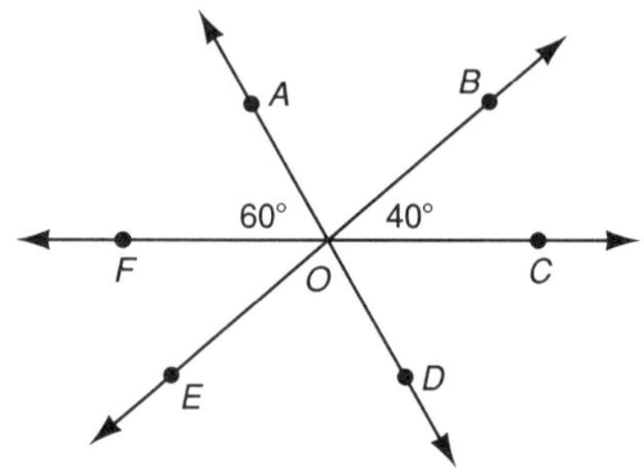

1. What is m $\angle FOE$?

Answer: 40° (because it is a vertical angle with $\angle BOC$)

2. What is m $\angle COD$?

Answer: 60° (because it is a vertical angle with $\angle AOF$)

3. What is m $\angle BOD$?

Answer: 100° (because m $\angle BOC = 40$, and m $\angle COD = 60$, and m $\angle BOC$ + m $\angle COD$ = m $\angle BOD$)

4. What is m $\angle AOB$?

Answer: 80° (because 180° − 60° − 40° = 80°)

5. What is m $\angle DOE$?

Answer: 80° (because 180° − 40° − 60° = 80°)

6. In the diagram below, line l and line m are parallel. If m $\angle a = 40$, find the measure of the other seven angles. Explain your answers.

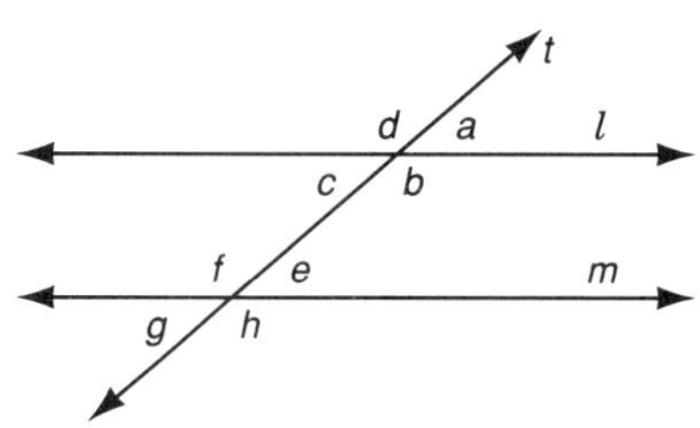

Answers:	*Explanations:*
m $\angle b = 140$	180 − 40 = 140. Angles c and b are supplementary angles and $\angle c \cong \angle a$.
m $\angle c = 40$	$\angle a$ and $\angle c$ are vertical angles.
m $\angle d = 140$	$\angle b$ and $\angle d$ are vertical angles.
m $\angle e = 40$	$\angle e$ and $\angle a$ are corresponding angles.
m $\angle f = 140$	$\angle e$ and $\angle f$ are supplementary angles.
m $\angle g = 40$	$\angle g$ and $\angle e$ are vertical angles.
m $\angle h = 140$	$\angle f$ and $\angle h$ are vertical angles. Or: $\angle g$ and $\angle h$ are supplementary angles.

TRY THESE

Items 1–6: In the diagram, OB bisects $\angle AOC$; m $\angle COD = 35$; and m $\angle GOE = 120$.

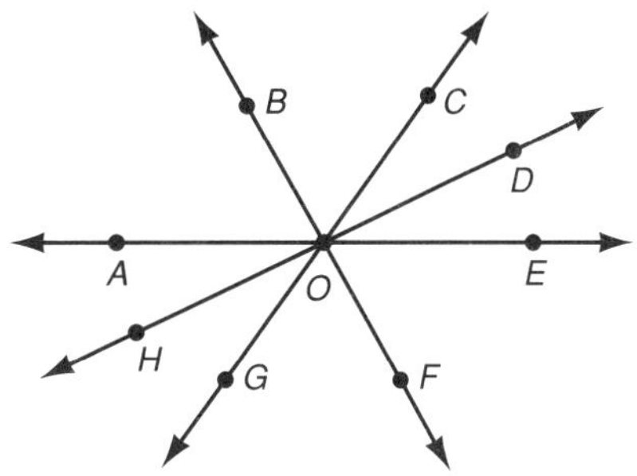

Find the measure of each of the following angles.

1. $\angle HOG$
2. $\angle HOE$
3. $\angle DOE$
4. $\angle AOC$
5. $\angle BOC$
6. $\angle HOB$

Items 7–9: In the diagram below, $L_1 \parallel L_2$ and $L_3 \parallel L_4$.

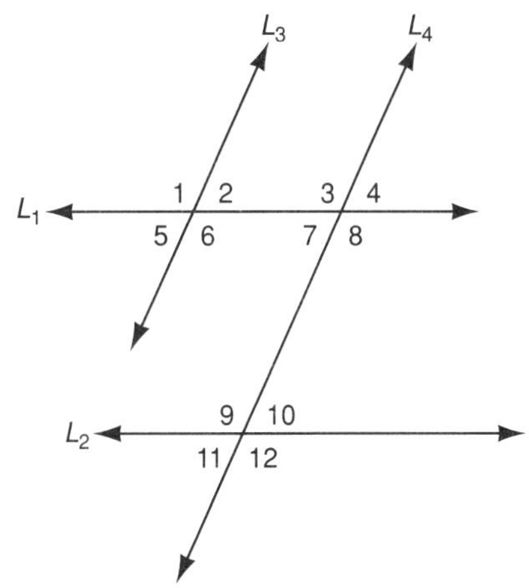

7. Name 8 pairs of corresponding angles.
8. Name 4 pairs of alternate interior angles.
9. Name 4 pairs of interior angles on the same side of the transversal.
10. In the diagram below, lines l and m are parallel and m $\angle 4 = 110$. Find the measure of the other seven angles.

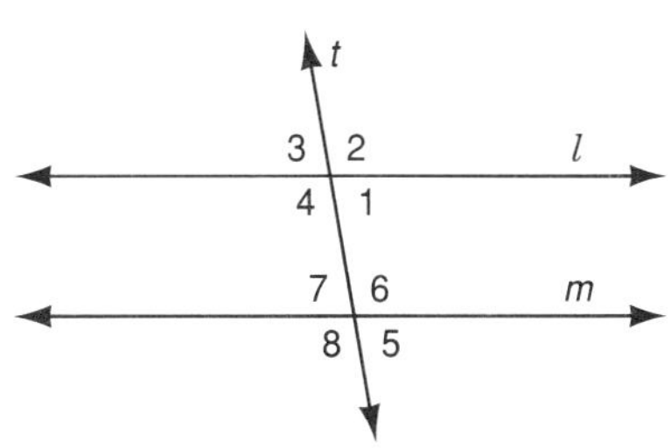

B POLYGONS

A **polygon** is a closed figure formed by line segments. Each line segment is called a **side** of the polygon.

Triangles

A three-sided polygon is called a triangle. The symbol for a triangle is $\triangle$. There are specific categories of triangles, depending on sides and angles (see also Chapter 2).

Triangle categories by sides

- **Equilateral**: All three sides are equal in length.
- **Isosceles**: At least two sides are equal in length.
- **Scalene**: No sides are equal in length.

Triangle categories by angles

- **Equiangular**: All three angles are equal in measure.
- **Right**: Has one right (90-degree) angle formed by intersecting perpendicular lines.
- **Obtuse**: Has one obtuse angle (an angle between 90 degrees and 180 degrees).
- **Acute**: Has three acute angles (angles less than 90 degrees).

Following are important facts about triangles.

The sum of the measures of the **interior angles** of a triangle is 180 degrees.

Each angle of an equiangular triangle is 60 degrees, since $180 \div 3 = 60$.

In an isosceles triangle, the angles opposite the congruent sides are congruent and the sides opposite the congruent angles are congruent. For example, $\triangle ABC$ is isosceles. If $\overline{AB} \cong \overline{BC}$, then $m\angle A = m\angle C$. If $m\angle A = m\angle C$, then $\overline{AB} \cong \overline{BC}$.

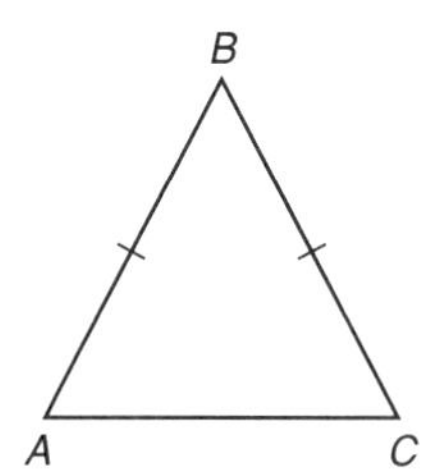

If the measures of two angles of a triangle are unequal, the measures of the sides opposite these angles are unequal in the same order. In other words, the side opposite the larger angle is the larger side.

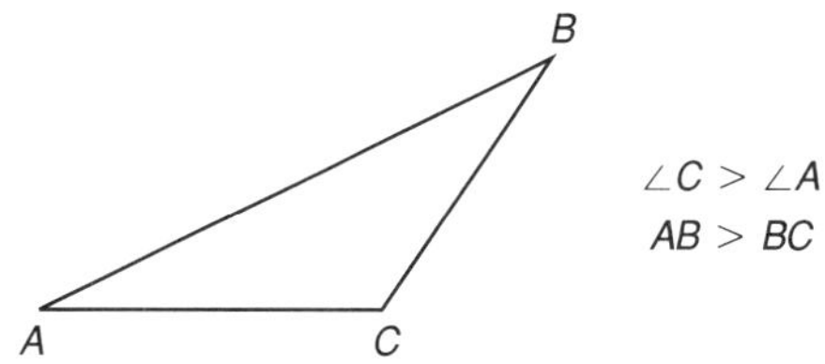

The **altitude of a triangle** is a line segment drawn from a vertex of the triangle perpendicular to the line that forms the opposite side. Altitude is often called **height** and abbreviated h.

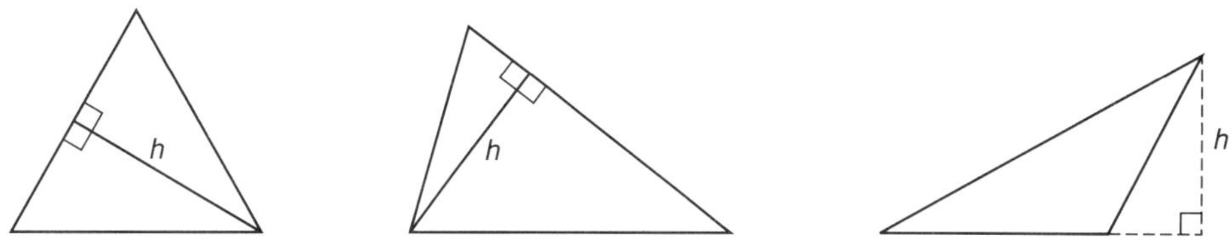

The **median of a triangle** is a line segment drawn from a vertex of the triangle to the midpoint of the opposite side. The **midpoint** is the middle point of the line segment. In other words, the midpoint divides the segment into two equal parts. Another way to say this is: The midpoint **bisects** the line segment. In the triangle below, line m is the median.

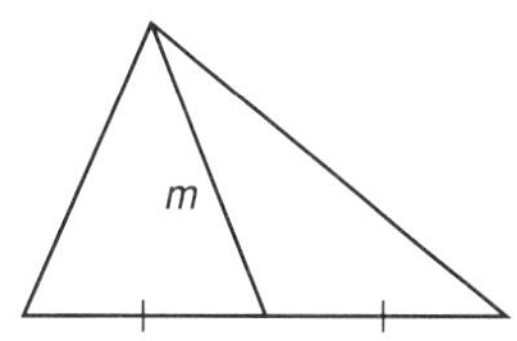

An **angle bisector** is a line segment drawn from a vertex of the triangle to the opposite side, which bisects the vertex angle. Line b below is an angle bisector.

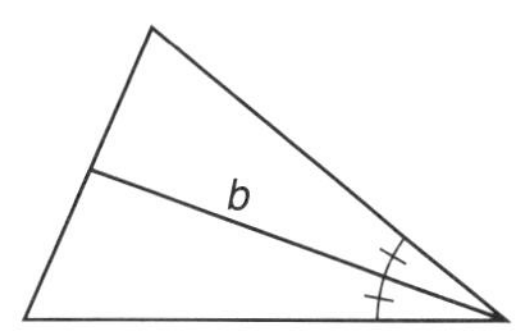

In a **right triangle**, the side opposite the right angle is called the **hypotenuse** and the other two sides are called **legs**. There is a special relationship between the hypotenuse and the legs. This relationship is stated in the **Pythagorean theorem**: The square of the hypotenuse of a right triangle is equal to the sum of the squares of the legs:

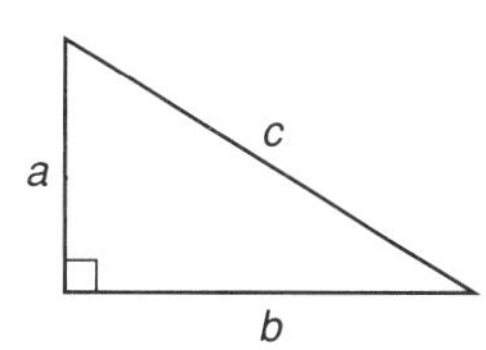

- $c^2 = a^2 + b^2$ where c is the hypotenuse and a and b are the legs.

Two special right triangles are often used. These are called the **30-60-90 triangle** and the **45-45-90 triangle**. The numbers refer to the measures of the angles in the triangle. Thus a 30-60-90 triangle has angles measuring 30, 60, and 90 degrees. A 45-45-90 triangle has angles measuring 45, 45, and 90 degrees, so it is an isosceles right triangle.

30-60-90 triangle:

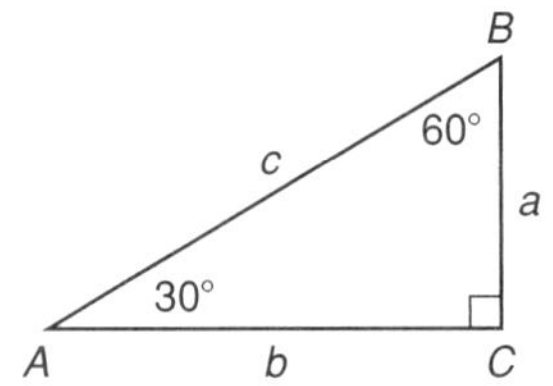

- The measure of the side opposite the 30-degree angle is $\frac{1}{2}$ the measure of the hypotenuse.
- The measure of the side opposite the 60-degree angle is $\frac{1}{2}$ the measure of the hypotenuse times $\sqrt{3}$.

For example: In triangle *ABC* above, side *a* is opposite $\angle A$, side *b* is opposite $\angle B$, and side *c* is opposite $\angle C$. The measure of $\angle A = 30$ and m $\angle B = 60$. If $c = 10$, then $a = \frac{1}{2} \bullet 10 = 5$ and $b = \frac{1}{2} \bullet 10 \bullet \sqrt{3} = 5\sqrt{3}$.

45-45-90 triangle:

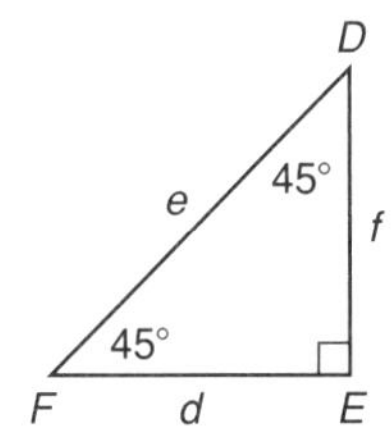

- The measure of the hypotenuse is equal to the measure of a leg times $\sqrt{2}$.

For example, in triangle *DEF* above, *d* and *f* are the measures of the equal legs and side *e* is the hypotenuse. Angles *D* and *F* are each 45 degrees, and angle *E* is a right angle (90 degrees). If the measure of leg *d* is 7, the hypotenuse would measure $7\sqrt{2}$.

Besides having three interior or inside angles, a triangle also has three exterior, or outside, angles. An **exterior angle of a triangle** is formed outside a triangle by one side of the triangle and the extension of an adjacent side. In the diagram below, $\angle CBD$ is an exterior angle of $\triangle ABC$. Angle *CBA* is adjacent to the exterior angle *CBD*. Angles *A* and *C* are called **nonadjacent (or remote) interior angles** because they are not adjacent to the exterior angle.

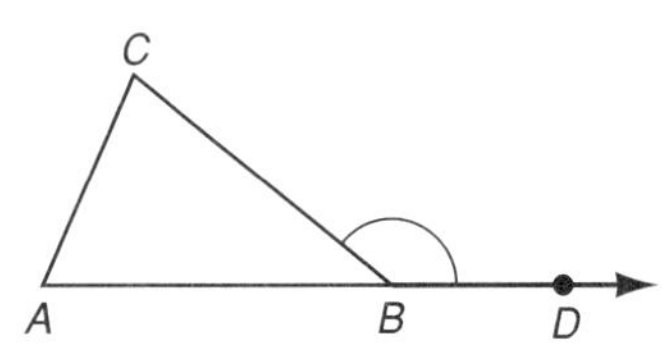

The measure of an exterior angle of a triangle is equal to the sum of the measures of the two nonadjacent (remote) interior angles. In the example above, m $\angle A$ + m $\angle C$ = m $\angle CBD$.

MODEL PROBLEMS

Items 1–3: Find the value of x.

1.

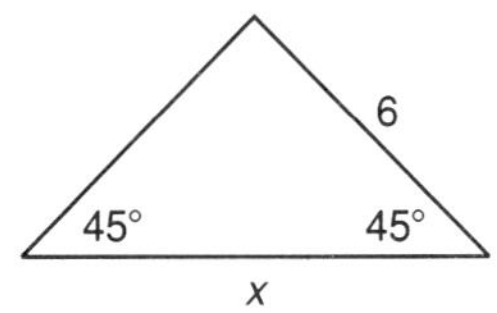

Answer: Leg times $\sqrt{2}$, in this case $6\sqrt{2}$

2.

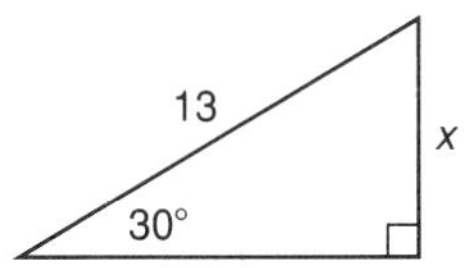

Answer: $\frac{1}{2}$ of the hypotenuse, in this case 6.5

3.

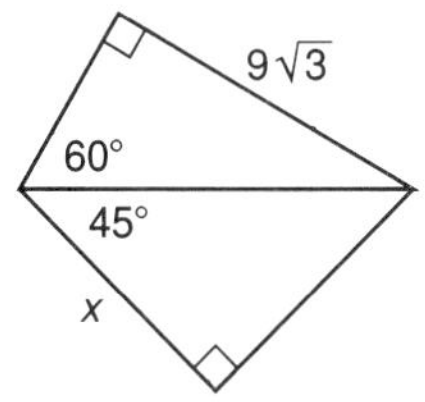

Answer: Call these triangles $\triangle$I (top) and $\triangle$II (bottom). The hypotenuse of $\triangle$I would measure 18, and this is also the hypotenuse of $\triangle$II. Therefore:

x = hypotenuse $\div \sqrt{2}$ or $\frac{18}{\sqrt{2}}$ $\left(\text{Rationalized: } \frac{18\sqrt{2}}{2} = 9\sqrt{2}\right)$

4. The vertex angle of an isosceles triangle measures 40. Find the measure of either of the exterior angles formed by extending the base.

Answer: $180° - 40° = 140°$; $140° \div 2 = 70°$; $70° + 40° = 110$ degrees

5. In $\triangle ABC$, if m $\angle B = 90$ and m $\angle C = 60$, find the measure of an exterior angle at A.

Answer: $90° + 60° = 150$ degrees

TRY THESE

Items 1–4: In the diagram below, $\overline{EB}$ is an altitude and $\overline{AD}$ is a median of $\triangle ACE$.

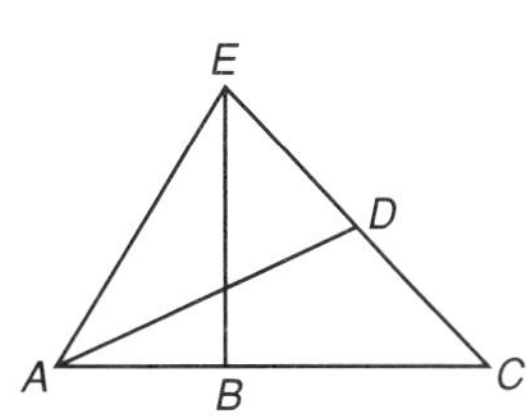

1. What two line segments are equal?

2. Name two right angles.

3. $\angle ADE$ is an exterior angle for which triangle?

4. What are the two remote interior angles of $\angle ADE$?

Items 5 and 6: Find x.

5.

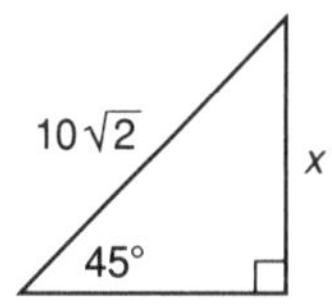

6.

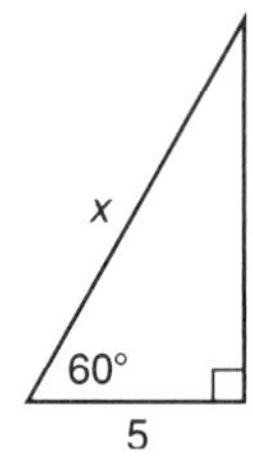

Items 7–9: Use the diagram below.

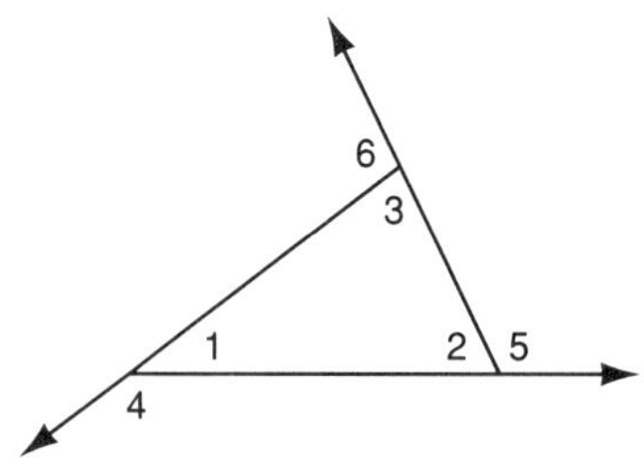

7. If m $\angle 1 = 60$ and m $\angle 2 = 40$, what is m $\angle 6$?

8. If m $\angle 2 = 65$ and m $\angle 3 = 50$, what is m $\angle 4$?

9. If m $\angle 5 = 120$ and m $\angle 1 = 40$, what is m $\angle 3$?

10. What is the area of parallelogram *WXYZ* shown below?

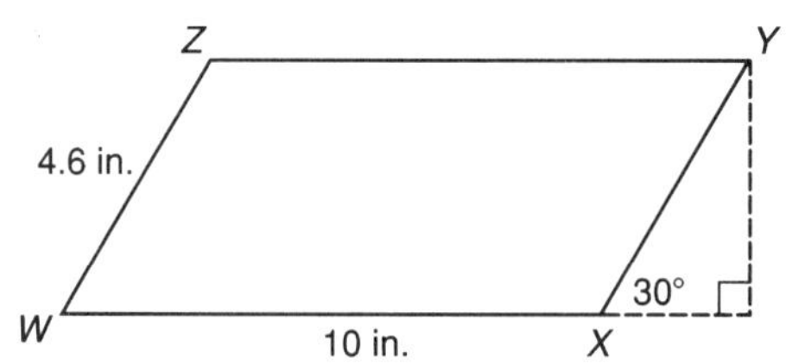

Quadrilaterals

A four-sided polygon is called a **quadrilateral**.

If both pairs of opposite sides are parallel, the quadrilateral is a **parallelogram**. A parallelogram has several special properties:

- Opposite sides of a parallelogram are congruent.
- Opposite angles of a parallelogram are congruent.
- Diagonals of a parallelogram bisect each other.

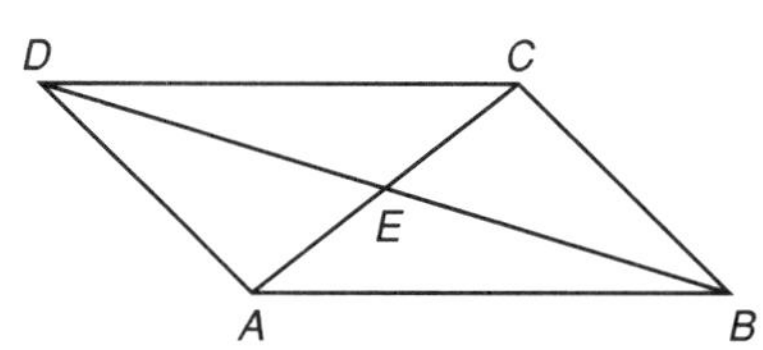

Given that $\overline{AB} \parallel \overline{CD}$ and $\overline{BC} \parallel \overline{AD}$, then:

$\overline{AB} \cong \overline{CD}$ and $\overline{AD} \cong \overline{BC}$

$\angle DAB \cong \angle DCB$ and $\angle ADC \cong \angle ABC$

$\overline{AE} \cong \overline{EC}$ and $\overline{DE} \cong \overline{EB}$

A **rectangle** is a parallelogram with four right angles. In a rectangle, the diagonals are congruent. For example, in the rectangle below, $\overline{AC} \cong \overline{BD}$.

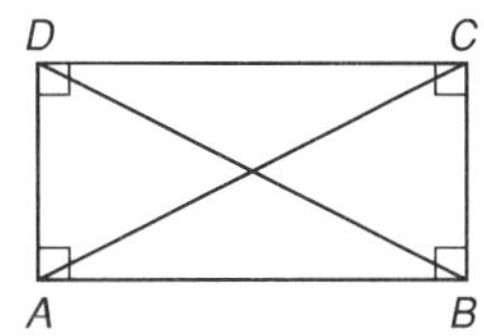

A **rhombus** is a parallelogram in which all four sides are congruent. In a rhombus, the diagonals are perpendicular and a diagonal bisects opposite angles. For example, in the rhombus below, $\overline{AC} \perp \overline{BD}$. Also, $\overline{AC}$ bisects angles DAB and DCB; and $\overline{BD}$ bisects angles ABC and ADC.

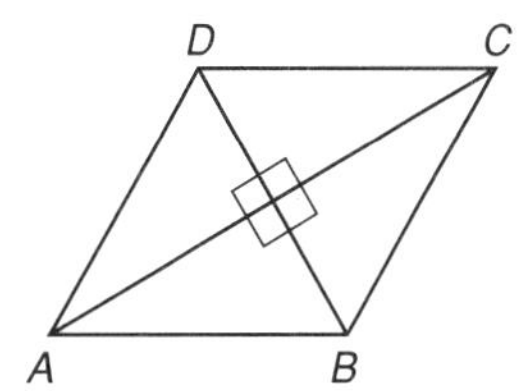

A **square** is a parallelogram with all the properties of the rectangle and the rhombus. Thus a square has four right angles and four congruent sides. $ABCD$ below is a square.

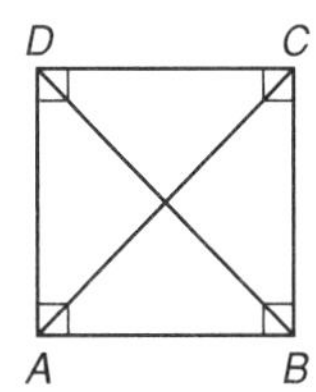

The following table shows the properties of these figures:

Property	Parallelogram	Rectangle	Rhombus	Square
Opposite sides are parallel.	✓	✓	✓	✓
Opposite sides are congruent.	✓	✓	✓	✓
Opposite angles are congruent.	✓	✓	✓	✓
Diagonals bisect each other.	✓	✓	✓	✓
Diagonals are congruent.		✓		✓
Diagonals are perpendicular.			✓	✓
Diagonals bisect opposite angles.			✓	✓
All angles are right angles.		✓		✓
All sides are congruent.			✓	✓

A **trapezoid** is a quadrilateral having exactly one pair of parallel sides. These parallel sides are called **bases**. The other two sides are called **legs**. The line segment joining the midpoints of the two legs is called the **median of the trapezoid**. The median is parallel to the bases and measures half the sum of the lengths of the bases.

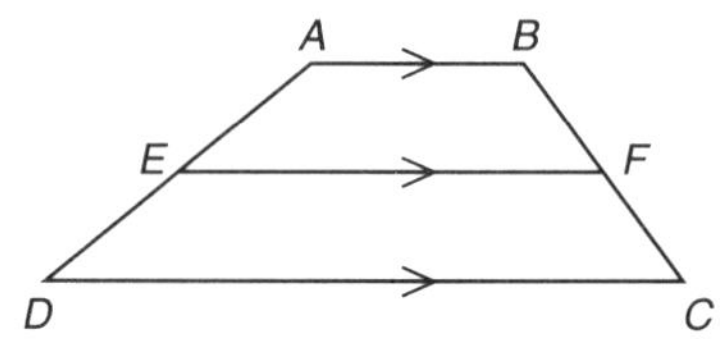

In the trapezoid above, *EF* is the median. *EF* is parallel to bases *AB* and *CD*. If $AB = 4$ inches and $CD = 12$ inches, then $EF = \frac{1}{2}(4 + 12) = 8$ inches.

Other polygons

Other polygons are listed in the table below.

Name of polygon	Number of sides
triangle	3
quadrilateral	4
pentagon	5
hexagon	6
octagon	8
decagon	10
dodecagon	12

In an **equilateral polygon**, all the sides are congruent. In an **equiangular polygon**, all the angles are congruent. A **regular polygon** is both equilateral and equiangular.

The sum of the measures of the **interior angles of a polygon** with n sides is $180(n - 2)$ degrees. For example, the sum of the interior angle of a hexagon (6 sides) is $180(6 - 2) = 720$ degrees.

In a regular polygon, the measure of each interior angle can be found by dividing this sum by the number of angles, which is the same as the number of sides. For instance, in a regular hexagon, the measure of each angle is $720 \div 6 = 120$ degrees.

The sum of the **exterior angles of a polygon** is always 360 degrees. Notice that this sum is the same regardless of the number of sides of the polygon.

In a regular polygon, the measure of each exterior angle is $\frac{360}{n}$ where n is the number of sides. This formula is found at the bottom of the first page of the Mathematics Reference Sheet.

MODEL PROBLEMS

Items 1–3: Give the best name for quadrilateral *ABCD* in each situation.

1. $\overline{AB} \cong \overline{DC}$ and $\overline{AB} \parallel \overline{CD}$

   Answer: Parallelogram

2. $\overline{AB} \parallel \overline{CD}$, $\overline{BC} \parallel \overline{AD}$, $\overline{\mathrm{AC}} \cong \overline{\mathrm{BD}}$

   Answer: Rhombus

3. $\angle A \cong \angle B \cong \angle C \cong \angle D$

   Answer: Rectangle

4. In a parallelogram, one angle measures 70 degrees. Find the measures of the other three angles.

   Answer: Opposite angles of a parallelogram are equal and consecutive angles are supplementary, so the other angles are 70, 110, and 110 degrees.

5. *ABCD* below is a trapezoid. The median *EF* measures 10 cm, and *DC* measures 12 cm. Find the length of *AB*.

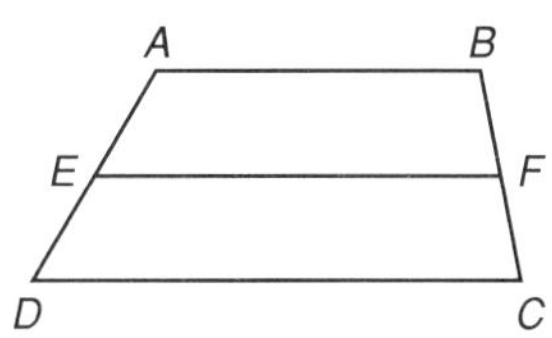

   Answer: The measure of the median $\overline{EF}$ is equal to half the sum of the lengths of the bases $\overline{AB}$ and $\overline{DC}$: $10 = \frac{1}{2}(AB + 12)$; $20 = AB + 12$; $AB = 8$ cm.

TRY THESE

1. *ABCD* is a square. $AB = 9y - 2$ and $BC = 7y + 16$. Find *y*.
2. In square *MNOP*, diagonal *MO* is drawn. Find m $\angle MON$ and m $\angle POM$.

Items 3–8: Use the diagram of parallelogram *ABCD* below.

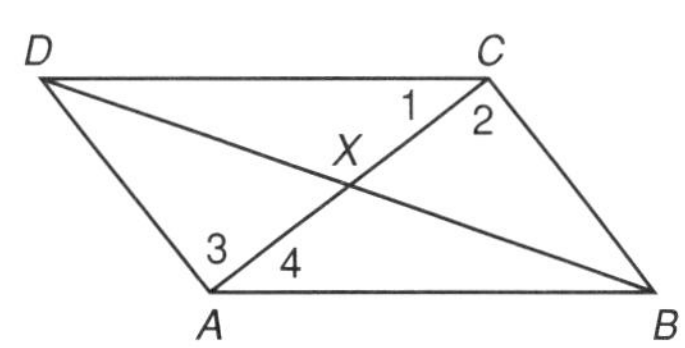

3. If $AC = 12$, then $AX = ?$
4. If $DX = 8$, then $DB = ?$
5. If m $\angle ABC = 72$, then m $\angle CDA = ?$
6. If m $\angle 1$ + m $\angle 2 = 106$, then m $\angle DAB = ?$
7. If m $\angle 3 = 88$, then m $\angle 2 = ?$
8. If m $\angle 4 = 41$, then m $\angle 1 = ?$

Items 9 and 10: *ABCD* is a rhombus.

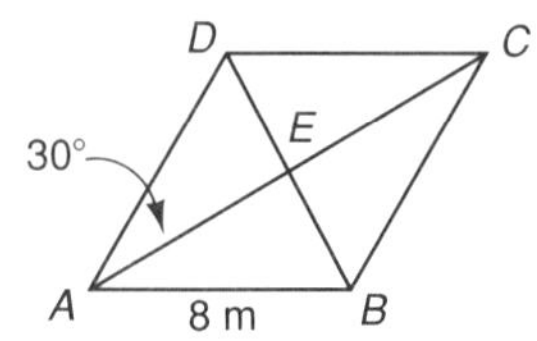

9. What is the length of *DB*?

10. What is the approximate length of *AC*?

11. What is the total number of degrees in the interior angles of an octagon?

12. In a regular octagon, how many degrees are in each interior angle?

13. What is the sum of the measures of the exterior angles of a regular decagon?

14. How many degrees are in each exterior angle of a regular decagon?

15. How many degrees are in each interior angle of a regular decagon?

16. In the diagram below, *DEFG* is a trapezoid with median $\overline{HI}$. $EF = 15$ inches and $HI = 12$ inches. How long is *DG*?

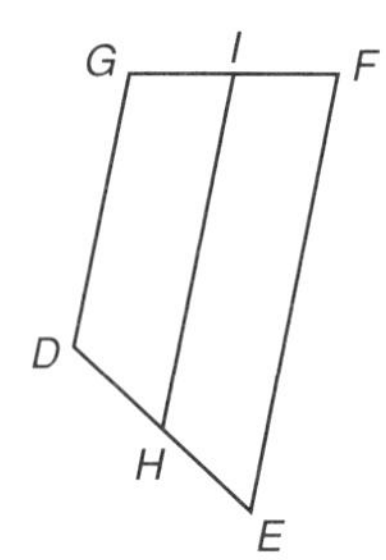

C INFORMAL AND FORMAL PROOFS

In geometry, an **informal proof** simply means giving an explanation for an answer or a conclusion: for example, "*ABCD* is a square because all four sides are congruent and all four angles are congruent."

A **formal proof** is a specific kind of step-by-step logic leading to a proof of a hypothesis. In a formal proof, each step, called a **statement**, is backed up with a mathematical **reason**. Reasons used in a formal proof include any or all of the following.

- **Given information**: Facts provided for a particular problem.
- **Definitions**: Accepted facts that have to do with meanings.

- **Postulates**: Statements that are assumed without proof.
- **Theorems**: Statements that have already been proved.

MODEL PROBLEMS

1. Given that $\angle 2 \cong \angle 3$, show by a formal proof that $\angle 1 \cong \angle 4$.

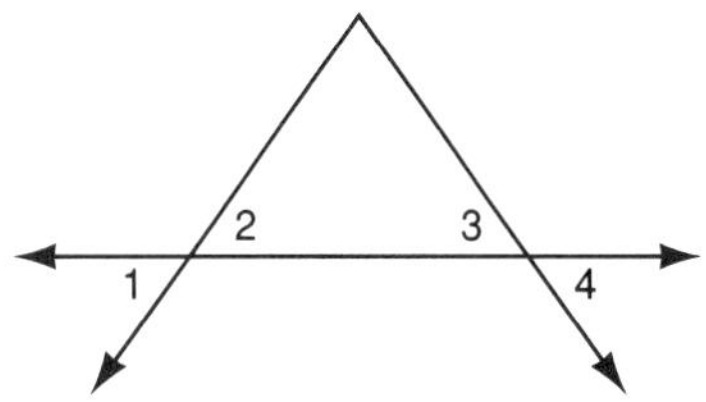

Answer:

Statements	Reasons
1. $\angle 1 \cong \angle 2$	1. Vertical angles are congruent.
2. $\angle 2 \cong \angle 3$	2. Given.
3. $\angle 1 \cong \angle 3$	3. Transitive property.
4. $\angle 3 \cong \angle 4$	4. Vertical angles are congruent.
5. $\angle 1 \cong \angle 4$	5. Transitive property.

2. Given that $m \parallel n$ and $\angle 1 \cong \angle 3$, show by a formal proof that $\angle 2 \cong \angle 3$.

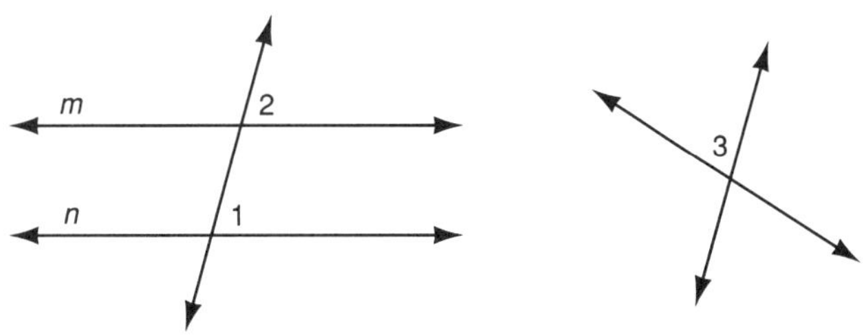

Answer:

Statements	Reasons
1. $m \parallel n$	1. Given.
2. $\angle 2 \cong \angle 1$	2. If two parallel lines are cut by a transversal, their corresponding angles are congruent.
3. $\angle 1 \cong \angle 3$	3. Given.
4. $\angle 2 \cong \angle 3$	4. Transitive property.

TRY THESE

Items 1 and 2: Given that $\angle 1 \cong \angle 3$, show by a formal proof that $\angle 2 \cong \angle 3$.

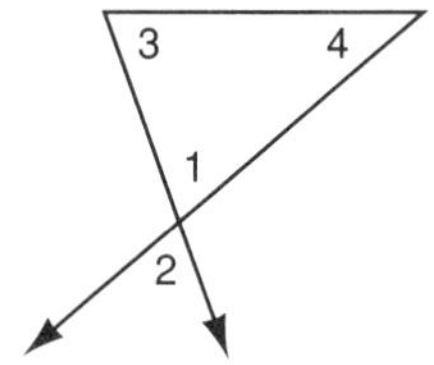

Statements	Reasons
$\angle 1 \cong \angle 3$	Given.
$\angle 1 \cong \angle 2$	**1.** ______________
$\angle 2 \cong \angle 3$	**2.** ______________

Items 3–6: Given that $\overline{HE} \perp \overline{WL}$ and $\angle HEW \cong \angle A$, supply the reasons to prove that m $\angle A = 90$.

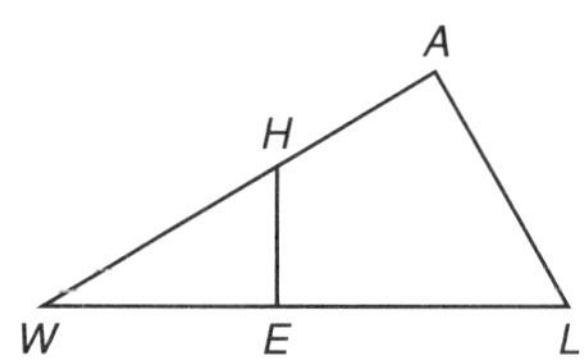

Statements	Reasons
$\overline{HE} \perp \overline{WL}$	Given.
$\angle HEW$ is a right angle	**3.** ______________
m $\angle HEW = 90$	**4.** ______________
$\angle HEW \cong \angle A$	**5.** ______________
m $\angle A = 90$	**6.** ______________

Items 7–10: $\triangle ABC$ is an equiangular triangle and $\overline{CD}$ is an angle bisector of $\angle ACB$. Supply the reasons to justify the measure of $\angle ADC$.

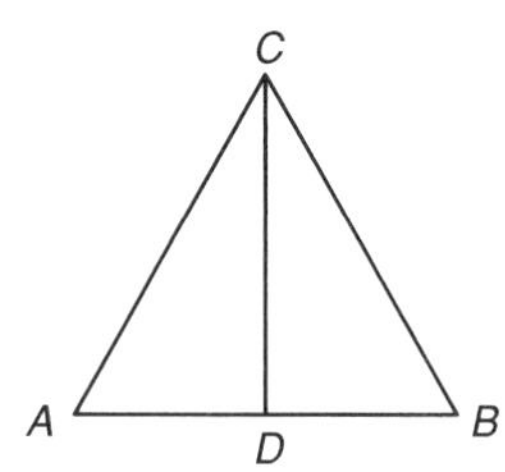

Statements	Reasons
$\triangle ABC$ is an equiangular triangle	Given.
m $\angle A$ = m $\angle B$ = m $\angle ACB = 60$	**7.** ______________
$\overline{CD}$ bisects $\angle ACB$	**8.** ______________
m $\angle ACD = 30$	**9.** ______________
m $\angle ADC = 90$	**10.** ______________

Items 11–14: $\triangle PQR$ is an isosceles triangle with base $\overline{PR}$ and m $\angle Q = 50$. Supply the reasons to show the measure of $\angle PRS$.

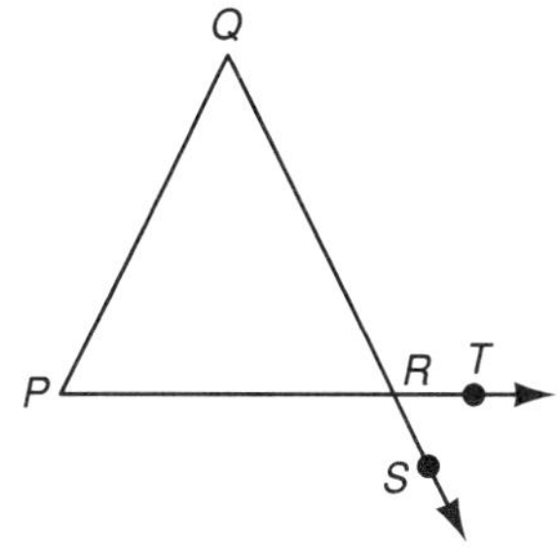

Statements	**Reasons**
$\triangle PQR$ is an isosceles triangle	Given.
$\overline{QP} \cong \overline{QR}$	**11.** ________________
$\angle P \cong \angle QRP$	**12.** ________________
m $\angle P$ = m $\angle QRP = 65$	**13.** ________________
m $\angle PRS = 115$	**14.** ________________

Assessment

Multiple-Choice Problems

Items 1–3: Refer to the diagram below.

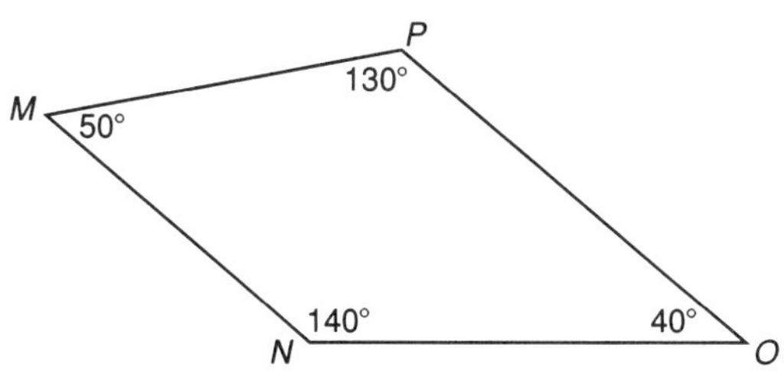

1. Which pair of angles are obtuse angles?

A. $\angle M, \angle N$ **C.** $\angle N, \angle P$

B. $\angle N, \angle O$ **D.** $\angle M, \angle O$

2. Which pair of angles are complementary angles?

A. $\angle M, \angle O$ **C.** $\angle M, \angle P$

B. $\angle N, \angle P$ **D.** $\angle N, \angle O$

3. Which pair of angles are supplementary angles?

A. $\angle M, \angle N$ **C.** $\angle N, \angle P$

B. $\angle P, \angle O$ **D.** $\angle N, \angle O$

4. Which type of angle is the supplement of an obtuse angle?

A. acute **C.** obtuse

B. right **D.** straight

5. $\overline{RS}$ is the median of $\triangle PQR$. Which of the following is a property of $\overline{RS}$?

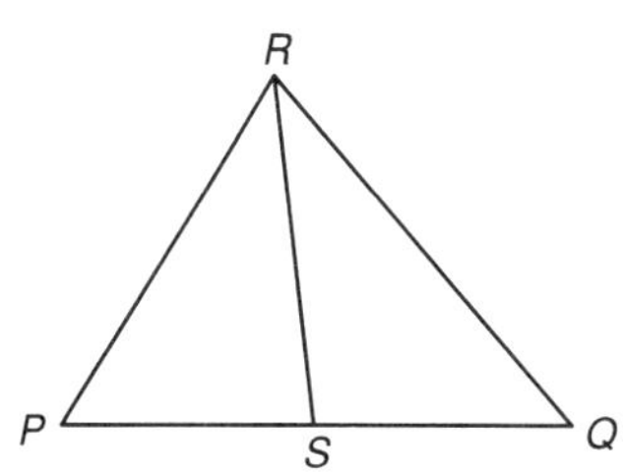

A. $\overline{RS}$ bisects $\angle PRQ$.

B. $\overline{RS}$ is perpendicular to $\overline{PQ}$.

C. $\overline{RS}$ bisects $\overline{PQ}$.

D. $\overline{RS}$ is the perpendicular bisector of $\overline{PQ}$.

6. What is the approximate length of PQ?

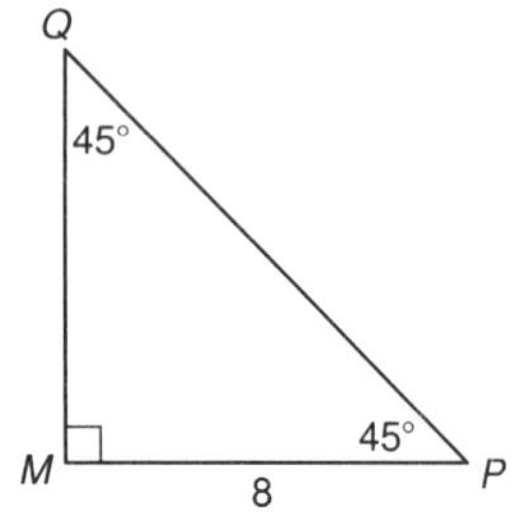

A. 4

B. 11.31

C. 13.86

D. 16

Items 7 and 8: $\triangle RST$ is equiangular with altitude $\overline{TP}$.

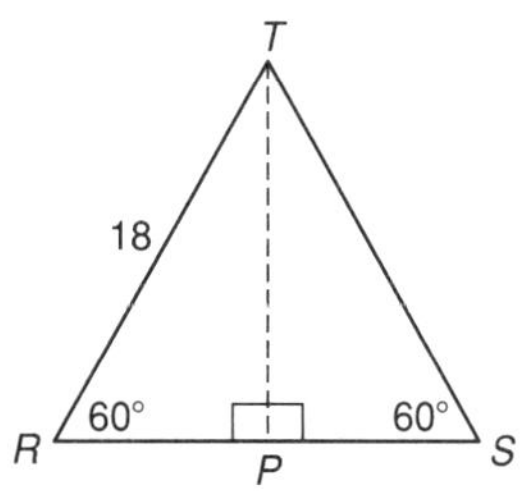

7. What is the length of $\overline{RP}$?

A. 9 **C.** 15.59

B. 12.73 **D.** 17.2

8. What is the length of altitude $\overline{TP}$?

A. 9

B. 12.73

C. 15.59

D. 17.2

Items 9 and 10: Use the diagram shown below.

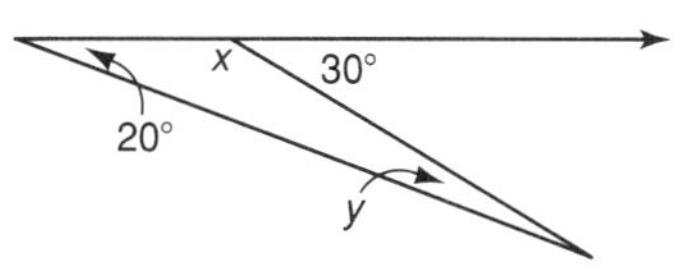

9. What is the measure of $\angle x$?

A. 10°

B. 50°

C. 150°

D. 170°

10. What is the measure of $\angle y$?

A. 10°

B. 50°

C. 150°

D. 170°

11. In the figure below, $\overline{BD}$ is an altitude of $\triangle ABC$. The proof below lists the reasons needed to justify the length of AC.

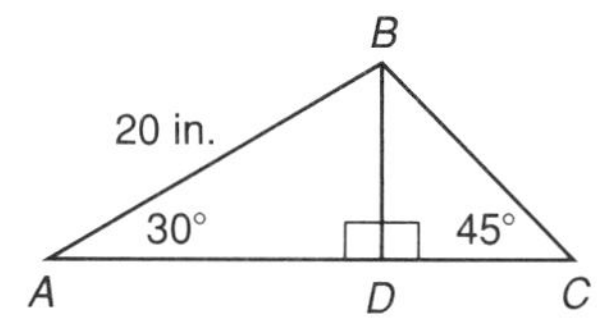

Statements	Reasons
1. $\angle ADB$ and $\angle BDC$ are right angles	An altitude is perpendicular to the opposite side.
2. $BD = 10$ in.	________________
3. m $\angle DBC = 45$	The sum of the angle measures of a triangle is 180.
4. $BD = DC$	Sides opposite equal angles are equal.
5. $DC = 10$ in.	DC and DB are equal sides.
6. m $\angle ABD = 60$	The sum of the angle measures of a triangle is 180.
7. $AD = 10\sqrt{3} \approx 17.32$	In a right triangle, the side opposite the 60° angle is $\sqrt{3}$ times the side opposite the 30° angle.
8. $AC = 17.32 + 10 = 27.32$	Addition of segments.

Which of the following phrases can be used in the reason space to justify statement number 2?

A. $\overline{DB}$ is a median to $\overline{AC}$.

B. $\overline{DB}$ is an angle bisector of $\angle ABC$.

C. $\overline{DB}$ is a bisector of $\overline{AC}$.

D. $\overline{DB}$ is the side opposite the 30° angle in a 30°-60°-90° triangle.

12. In the figure below, $\overrightarrow{EB}$ bisects $\angle CEF$. The proof on the right lists the reasons needed to justify why $\angle 1 \cong \angle 3$.

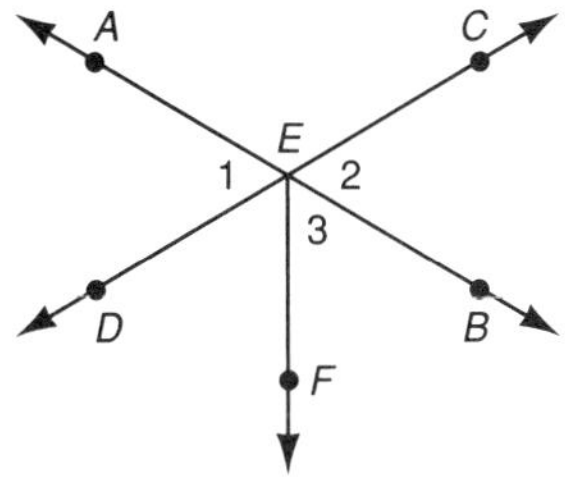

Statements	**Reasons**
1. $\overrightarrow{EB}$ bisects $\angle CEF$	Given.
2. $\angle 2 \cong \angle 3$	Definition of bisector.
3. $\angle 1 \cong \angle 2$	____________
4. $\angle 1 \cong \angle 3$	Transitive property.

Which of the following phrases can be used in the reason space to justify statement number 3?

A. $\angle 1$ and $\angle 2$ are complementary angles.

B. $\angle 1$ and $\angle 2$ are supplementary angles.

C. $\angle 1$ and $\angle 2$ are vertical angles.

D. $\angle 1$ and $\angle 2$ are adjacent angles.

Gridded-Response Problems

13. If one side of a rhombus equals 16 meters, what is the length of the adjacent side?

14. *WXYZ* is a rectangular plot of land. If $XY = 8$ kilometers and $YZ = 15$ kilometers, what is the length of *WY*?

15. What is the degree measure of an interior angle of a regular dodecagon?

16. What is the degree measure of an exterior angle of a regular pentagon?

17. The figure below is a regular octagon. How many degrees are in $\angle x$?

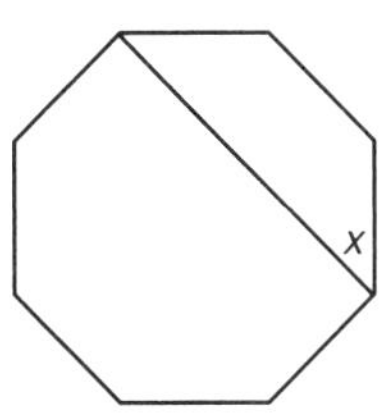

18. In the figure below, $\overline{EF}$ is a median of trapezoid *ABCD*. If $DC = 8$ and $AB = 20$, what is the measure of *EF*?

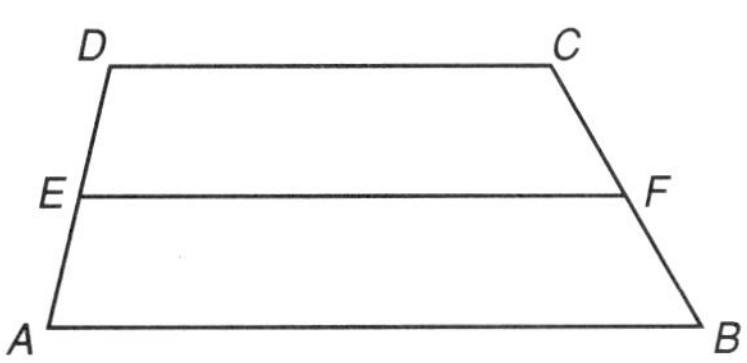

13
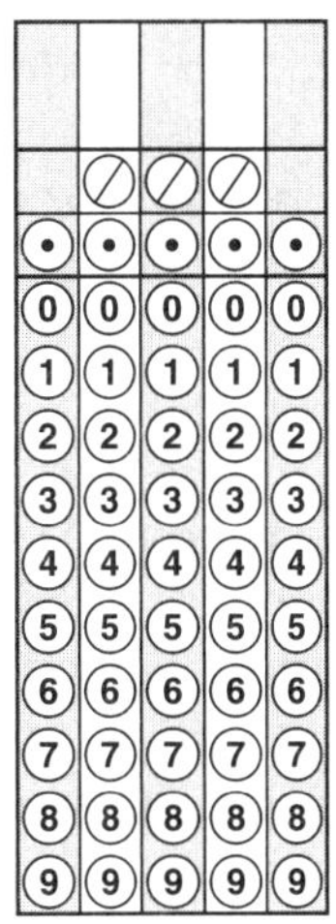

14
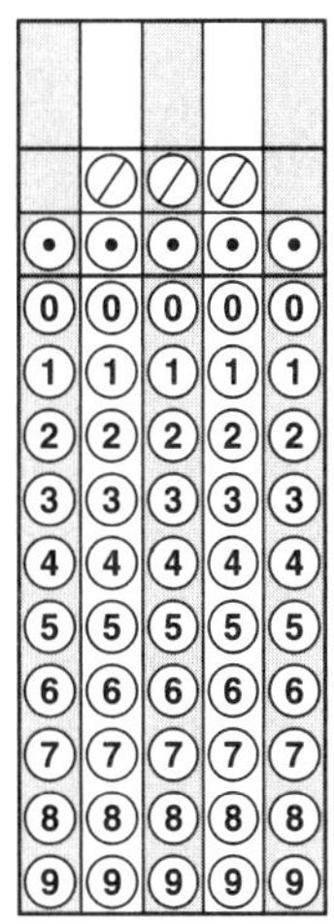

15
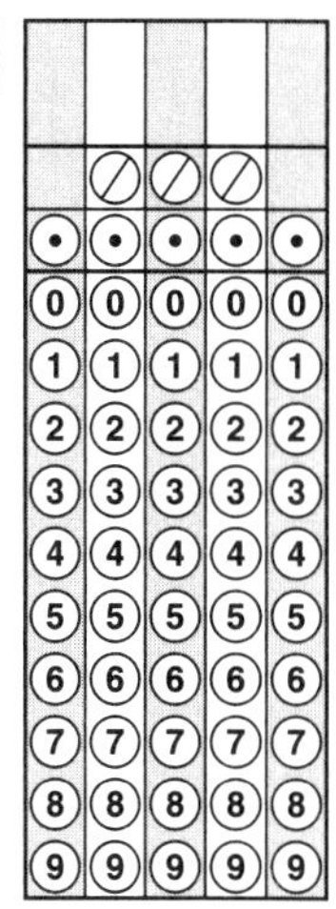

16
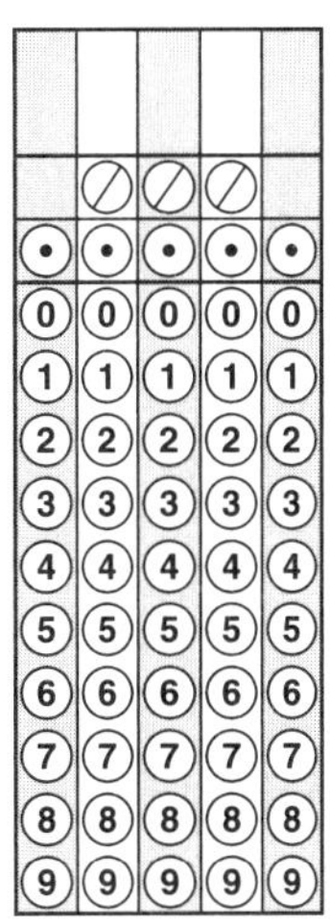

17
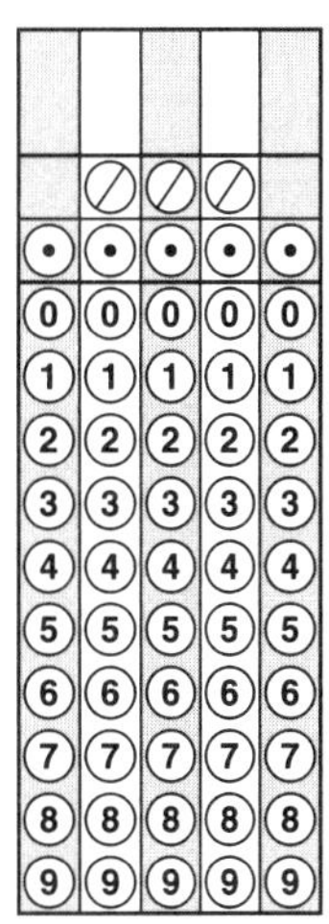

18
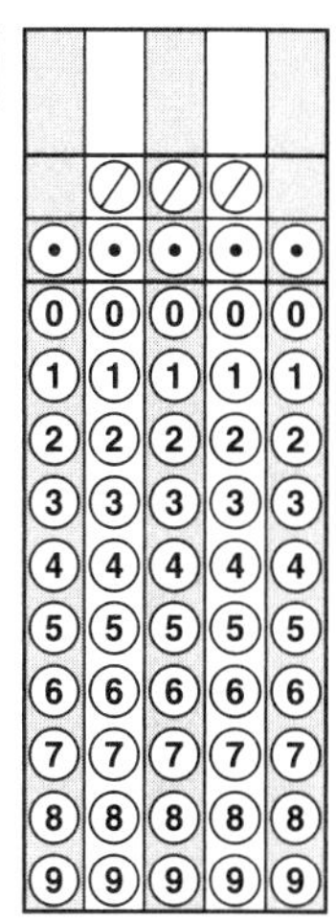

Section 2

Visualizing and illustrating ways in which shapes can be combined, subdivided, and changed

A UNDERSTANDING GEOMETRIC CONCEPTS SUCH AS CONGRUENCE, SIMILARITY, REFLECTIONS, SYMMETRY, AND TRANSFORMATIONS

Congruence

We have already used the words **congruence** and **congruent** for line segments of the same length and angles with the same measure. This section will review the **congruence of polygons**.

CONGRUENT POLYGONS

Polygons *ABCD* and *MNOP* are congruent. If we cut polygon *MNOP* out of this page and put it on top of *ABCD*, they would "match up," or **coincide**, perfectly. We can write this statement as: $ABCD \cong MNOP$.

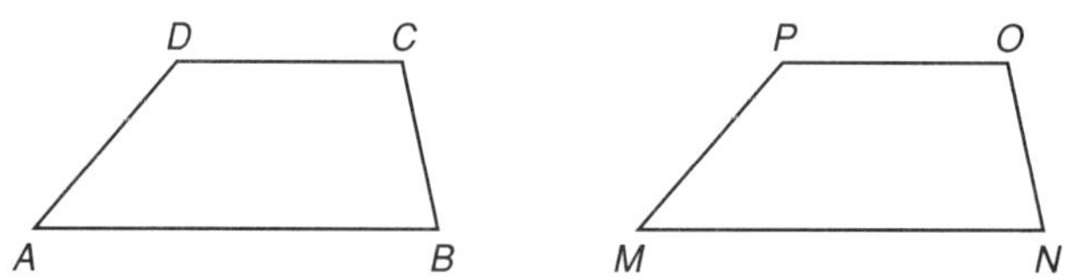

In a **congruence statement**, the letters identifying the polygons are written so that there is a one-to-one correspondence of the vertices. In our example above, vertex *A* coincides with vertex *M*, vertex *B* with vertex *N*, vertex *C* with vertex *O*, and vertex *D* with vertex *P*. Angles that "match" are called **corresponding angles**. Sides that match are called **corresponding sides**.

For polygon *ABCD* to be congruent to polygon *MNOP*, each angle in *ABCD* must be congruent to its corresponding angle in *MNOP*, and each side in *ABCD* must be congruent to its corresponding side in *MNOP*. Specifically:

Corresponding angles	Corresponding sides
$\angle A \cong \angle M$	$\overline{AB} \cong \overline{MN}$
$\angle B \cong \angle N$	$\overline{BC} \cong \overline{NO}$
$\angle C \cong \angle O$	$\overline{CD} \cong \overline{OP}$
$\angle D \cong \angle P$	$\overline{DA} \cong \overline{PM}$

Therefore, each pair of corresponding angles are congruent, and each pair of corresponding sides are congruent. So we have these facts:

- If the corresponding parts of two polygons are congruent, the polygons are congruent.
- If two polygons are congruent, their corresponding parts are congruent.

Congruent corresponding sides are usually indicated by the same slashes or hash marks. Congruent corresponding angles are indicated by the same arc marks.

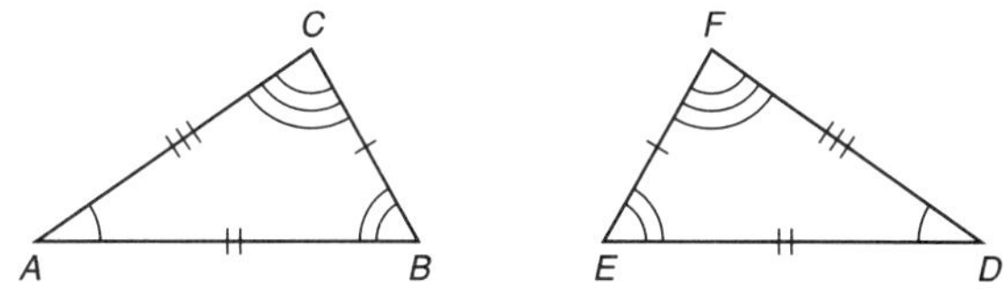

In the illustration above, six congruencies are given:

$\angle A \cong \angle D$, $\angle B \cong \angle E$, and $\angle C \cong \angle F$

$\overline{BC} \cong \overline{EF}$, $\overline{AB} \cong \overline{DE}$, and $\overline{AC} \cong \overline{DF}$

Thus we know that $\triangle ABC \cong \triangle DEF$, since their corresponding parts are congruent.

PROVING TRIANGLES CONGRUENT

To prove that two triangles are congruent, it is *not* necessary to know that all six pairs of corresponding parts are congruent. By using the properties of a triangle, congruence can be proved with less information. Following are five methods of proof.

Side-Angle-Side ≅ Side-Angle-Side (SAS) If two sides of one triangle and the **included angle** (the angle between them) are congruent to the corresponding sides and included angle of the other triangle, the triangles are congruent.

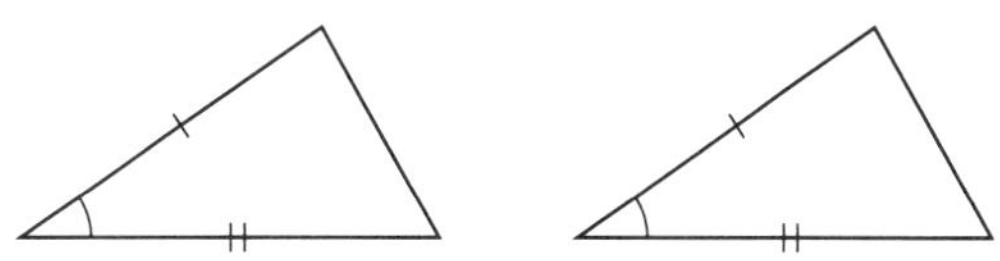

SAS ≅ SAS

Angle-Side-Angle ≅ Angle-Side-Angle (ASA) If two angles and the included side of one triangle are congruent to the corresponding angles and side of the other triangle, the triangles are congruent.

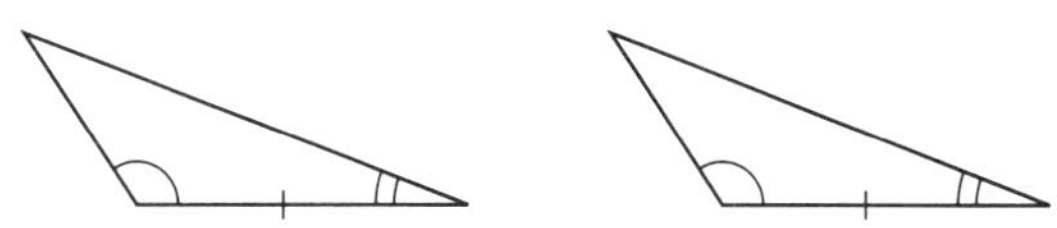

ASA ≅ ASA

Side-Side-Side ≅ Side-Side-Side (SSS) If the three sides of one triangle are congruent to the corresponding sides of the other triangle, the triangles are congruent.

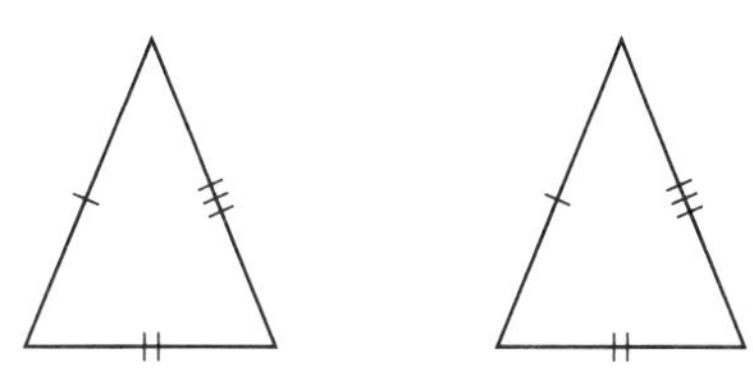

SSS ≅ SSS

Angle-Angle-Side ≅ Angle-Angle-Side (AAS) If two angles and a side opposite one of them of one triangle are congruent to two angles and the corresponding side of the other triangle, then the triangles are congruent.

AAS ≅ AAS

Right Triangles: Hypotenuse-Leg ≅ Hypotenuse-Leg (HL) If the hypotenuse and leg of one right triangle are congruent to the corresponding hypotenuse and leg of the other triangle, the triangles are congruent.

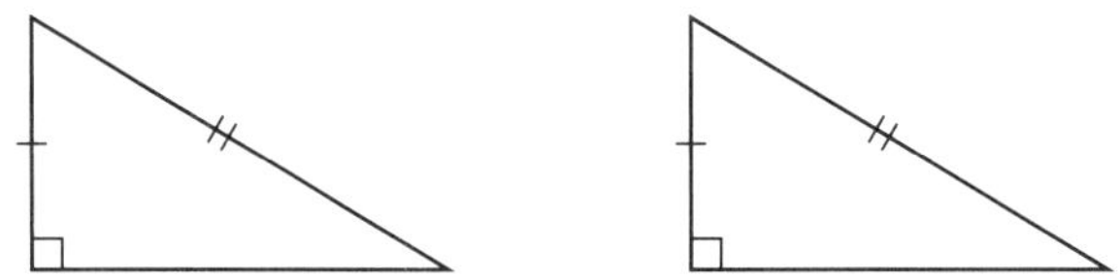

HL ≅ HL

REMEMBER: Once two triangles are proved congruent, all pairs of corresponding parts are congruent.

MODEL PROBLEMS

For each pair of triangles, determine whether △I and △II are congruent. Give your reasons.

1.

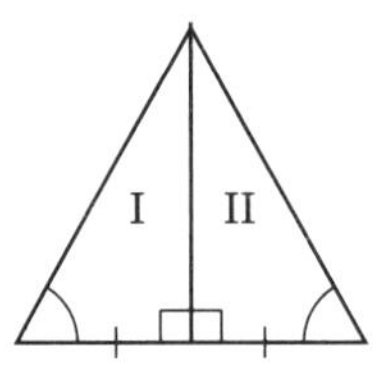

Answer: △I ≅ △II by side-angle-side (SAS), since they share a side. They are also congruent by angle-side-angle (ASA).

2.

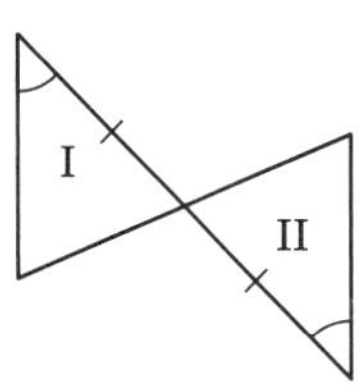

Answer: △I ≅ △II by angle-side-angle (ASA), since vertical angles are congruent.

3.

Answer: △I ≅ △II by angle-angle-side (AAS) or angle-side-angle (ASA). Since two pairs of corresponding angles are congruent, the third pair must also be congruent, because the sum of the angles in a triangle is 180 degrees.

4.

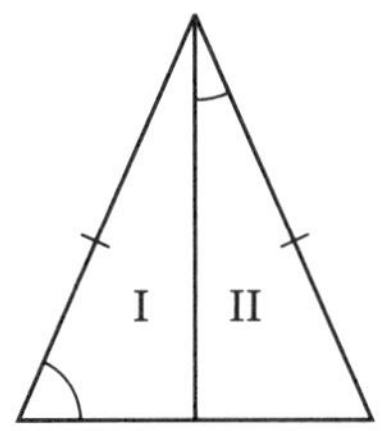

Answer: $\triangle$I and $\triangle$II are not congruent. The congruent angles are not corresponding angles.

5.

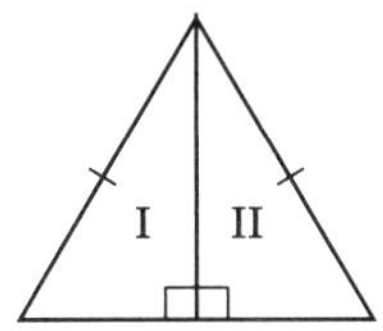

Answer: $\triangle I \cong \triangle II$ by hypotenuse-leg (HL), since these two right triangles share a leg.

TRY THESE

For each pair of triangles, determine whether $\triangle$I and $\triangle$II are congruent. Give your reasons.

1.

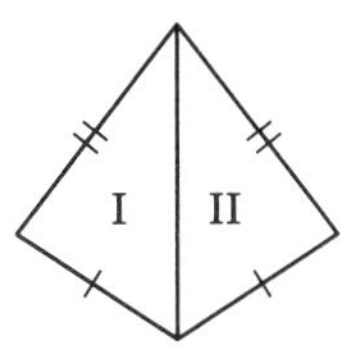

2.

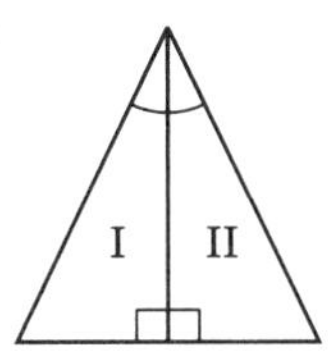

3.

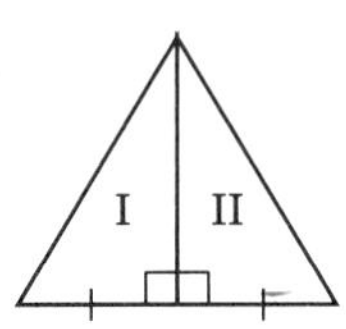

4.

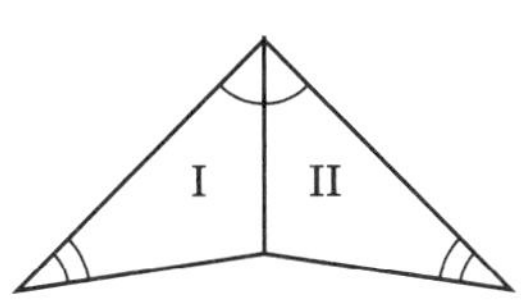

5.

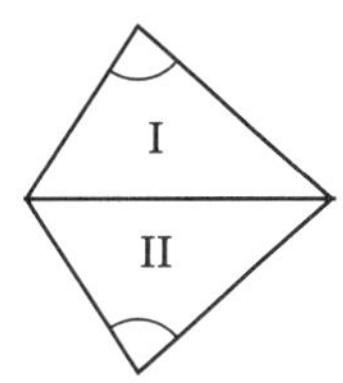

6.

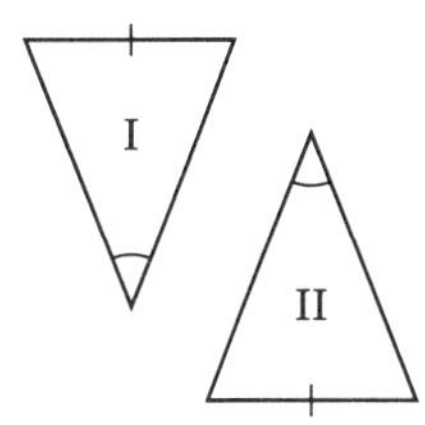

7.

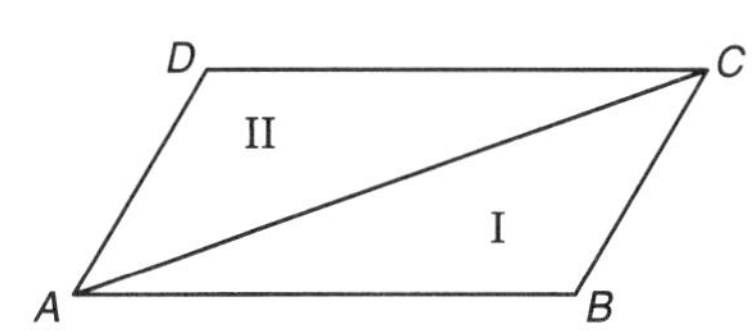

ABCD is a parallelogram

8.

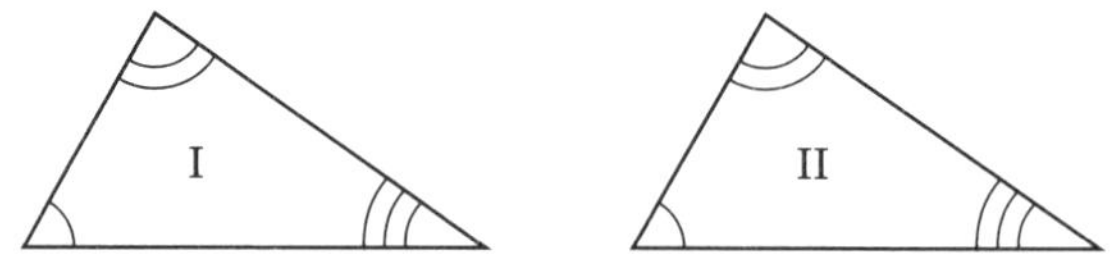

9.

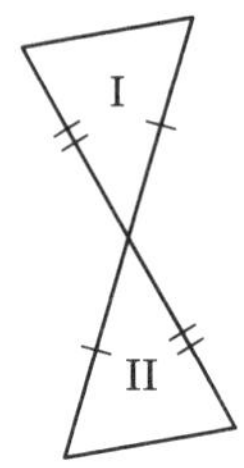

10.

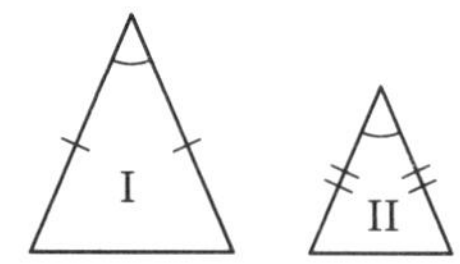

Similarity

SIMILAR POLYGONS

Congruent polygons coincide exactly with one another. That is, they have the same shape and the same size. **Similar polygons** have the same shape but not necessarily the same size.

Similar polygons (just like congruent polygons) must have a one-to-one correspondence of their vertices. In a **similarity statement**, the letters identifying the vertices are written to indicate this one-to-one correspondence. The symbol $\sim$ is used for similarity: for instance, quadrilateral $ABCD \sim$ quadrilateral $WXYZ$.

In similar polygons:

- All pairs of corresponding angles are congruent.
- All pairs of corresponding sides must have the same ratio, called the **scale factor**. In other words, the pairs of sides must be **in proportion**.

For example, $\triangle ABC$ and $\triangle DEF$ are similar, or $\triangle ABC \sim \triangle DEF$.

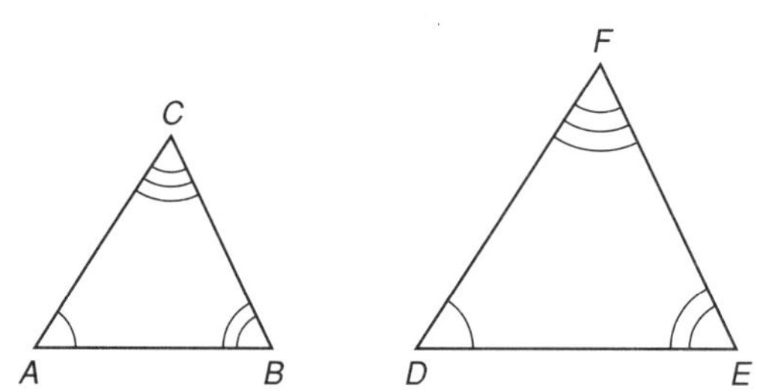

Therefore, in $\triangle ABC$ and $\triangle DEF$ we have the following relationships:

Corresponding angles	**Corresponding sides**
$\angle A \cong \angle D$, $\angle B \cong \angle E$, and $\angle C \cong \angle F$	$\frac{AC}{DF} = \frac{AB}{DE} = \frac{BC}{EF}$

PROVING TRIANGLES SIMILAR

To prove that triangles are similar, it is *not* necessary to prove all of these relationships. We can use the properties of the triangle to prove that triangles are similar with less information. Following are three methods.

Angle-Angle (AA) If two pairs of corresponding angles are congruent, the triangles are similar.

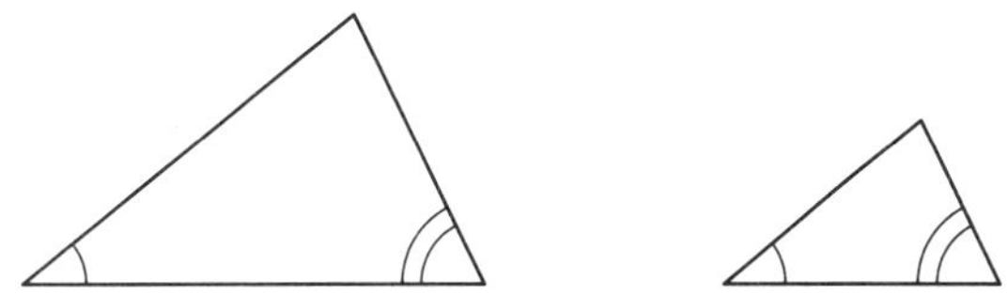

Side-Angle-Side (SAS) If two pairs of corresponding sides are in proportion and the included corresponding angles are congruent, the triangles are similar.

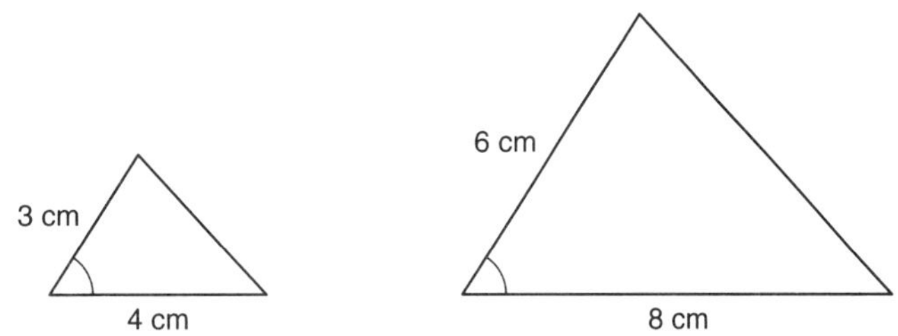

$6 = 2 \cdot 3$
$8 = 2 \cdot 4$
Scale factor is 2 : 1

Side-Side-Side (SSS) If all three pairs of corresponding sides are in proportion, the triangles are similar.

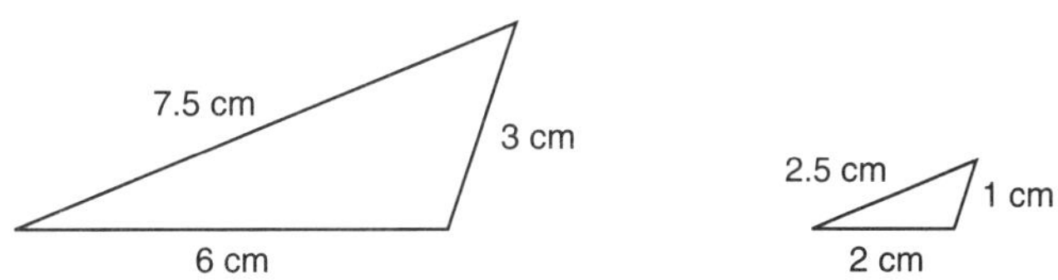

$2 = \frac{1}{3} \cdot 6$
$1 = \frac{1}{3} \cdot 3$
$2.5 = \frac{1}{3} \cdot 7.5$
Scale factor is 1 : 3

REMEMBER: If triangles are proved similar, all corresponding angles are congruent and all corresponding sides are in proportion: that is, the corresponding sides have the same scale factor.

MODEL PROBLEMS

For each pair of triangles, determine whether the triangles are similar. Give your reasons.

1.

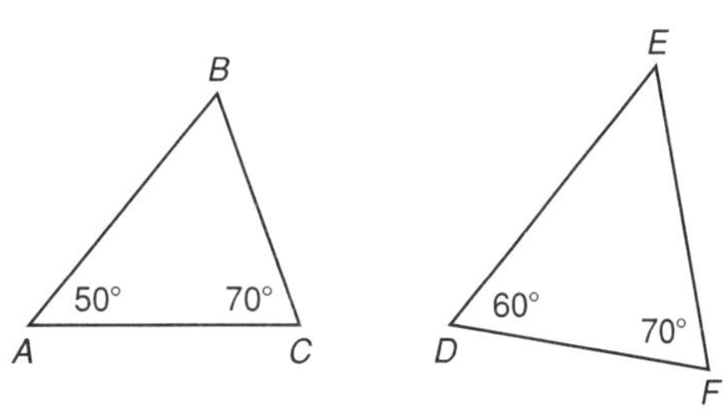

Answer: $\triangle ABC \sim \triangle DEF$. These triangles are similar by angle-angle (AA). $180 - (60 + 70) = 50$.

2.

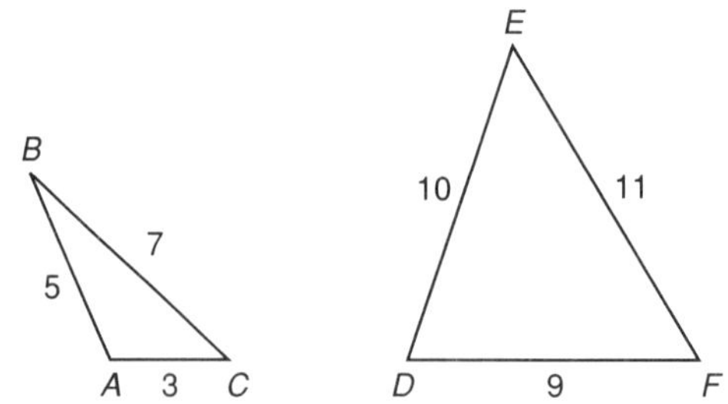

Answer: $\triangle ABC$ and $\triangle DEF$ are not similar, since their sides are not in proportion.

3.

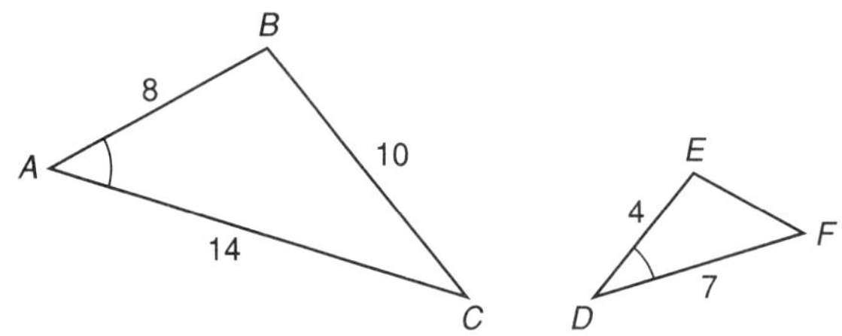

Answer: $\triangle ACB \sim \triangle DFE$. These triangles are similar by side-angle-side (SAS). The sides have a scale factor of 2 : 1.

4.

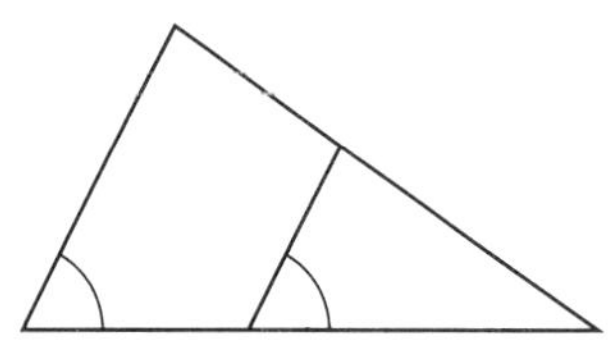

Answer: The smaller triangle is similar to the larger triangle, since they share an angle and corresponding angles are congruent (AA).

TRY THESE

Items 1–9: For each pair of triangles, determine whether the triangles are similar. Give your reasons.

1.

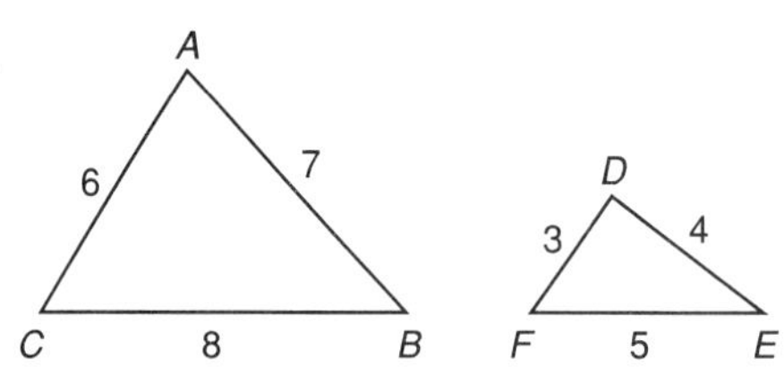

2.

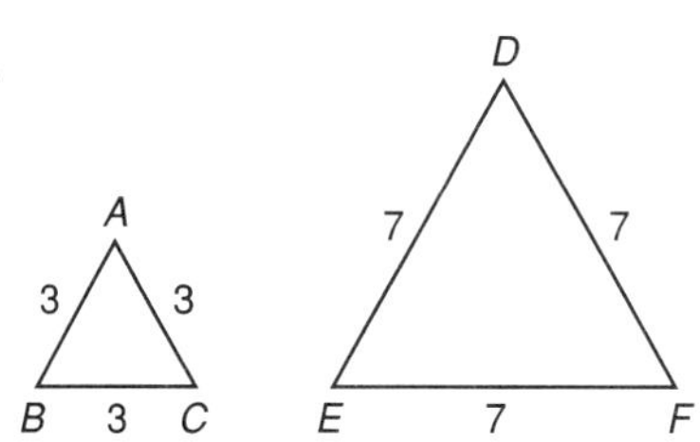

3.

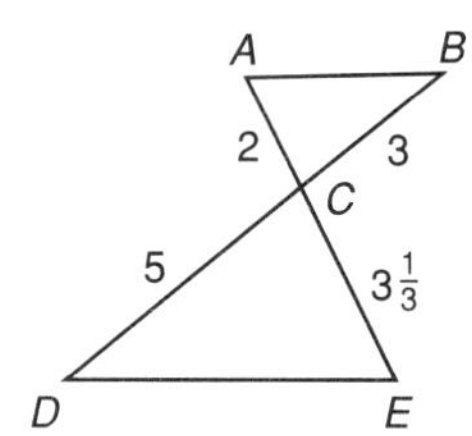

4.

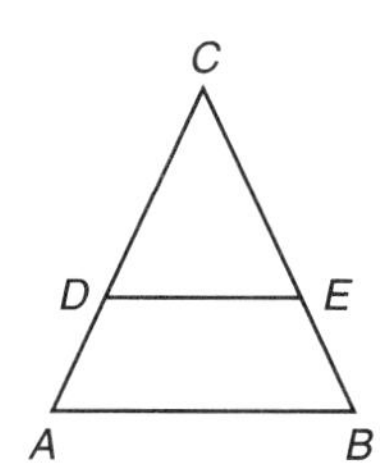

$\overline{DE} \parallel \overline{AB}$

5.

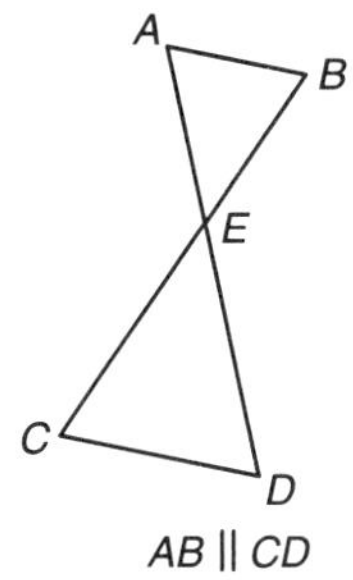

8.

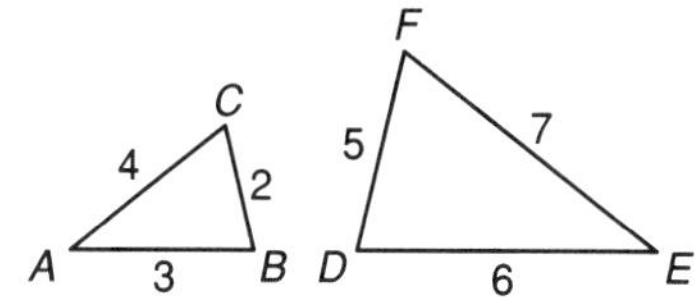

6.

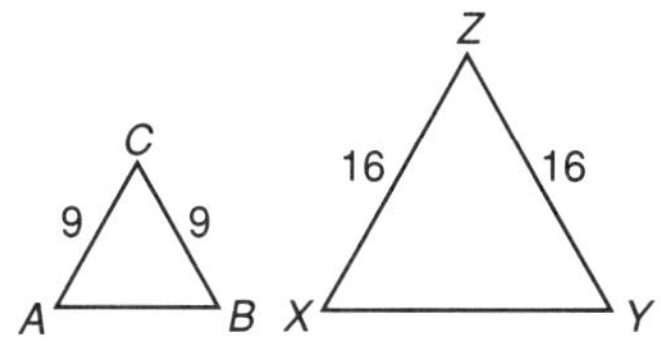

9.

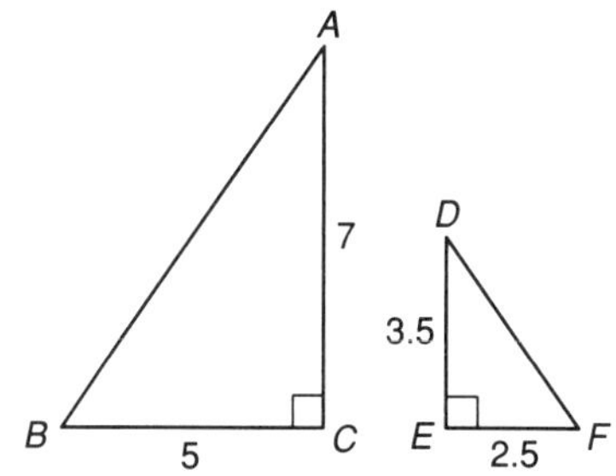

7.

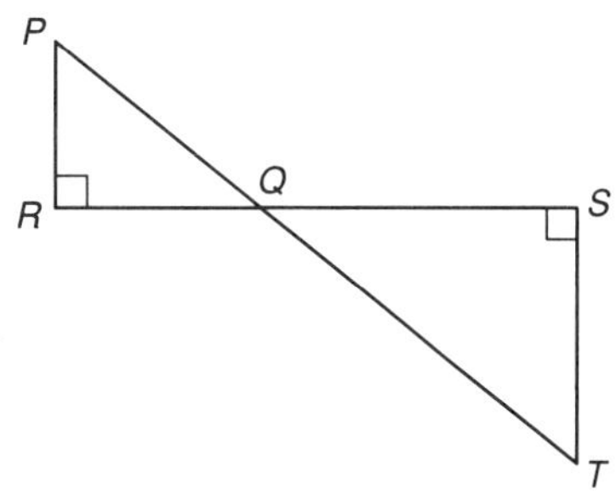

10.

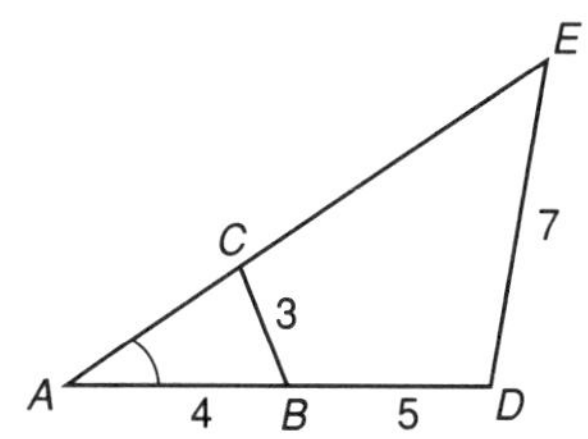

Is $\triangle ABC \sim \triangle ADE$?

Proportional lengths

Point X lies on $\overline{AB}$. Point Y lies on $\overline{CD}$. If $\frac{AX}{XB} = \frac{CY}{YD}$, we say that $\overline{AB}$ and $\overline{CD}$ are divided proportionally.

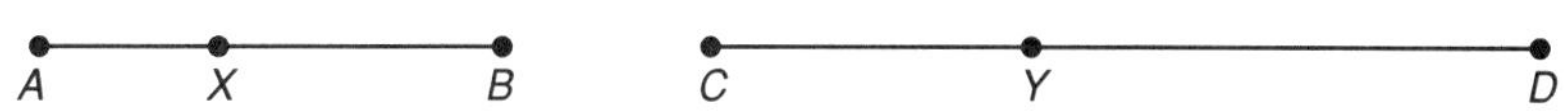

If a line that is parallel to one side of a triangle intersects the other two sides of the triangle, then it divides the intersected sides proportionally.

For example, in $\triangle ABC$ below, $\overline{PQ} \parallel \overline{AB}$. Therefore, $\frac{AP}{PC} = \frac{BQ}{QC}$.
We can also say that $\frac{AP}{AC} = \frac{BQ}{BC}$ or $\frac{PC}{AC} = \frac{QC}{BC}$.

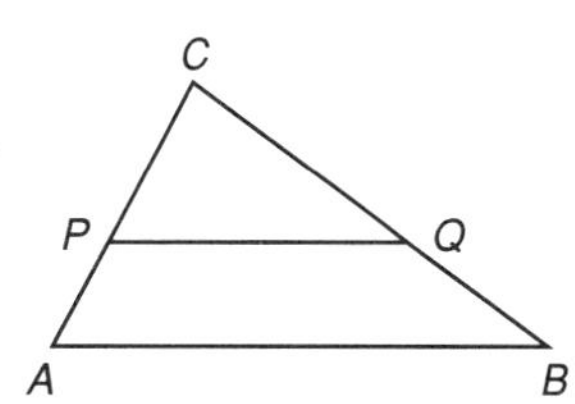

MODEL PROBLEMS

1. P and Q, respectively, divide line segments $\overline{XY}$ and $\overline{GH}$ proportionally. If $XP = 6$ units, $PY = 5$ units and $GQ = 3$ units, how long is GH?

Answer: $\dfrac{XP}{PY} = \dfrac{GQ}{QH}$. Let x = length of QH. Then:

$$\frac{6}{5} = \frac{3}{x}$$

$$6x = 15$$

$$x = \frac{15}{6} \text{ or } 2.5$$

$GQ + QH = GH$, so $GH = 3 + 2.5 = 5.5$

2. In $\triangle DEF$, $\overline{XY} \parallel \overline{DE}$. If $XF = 8$, $DX = 2$, and $EF = 12$, how long is FY?

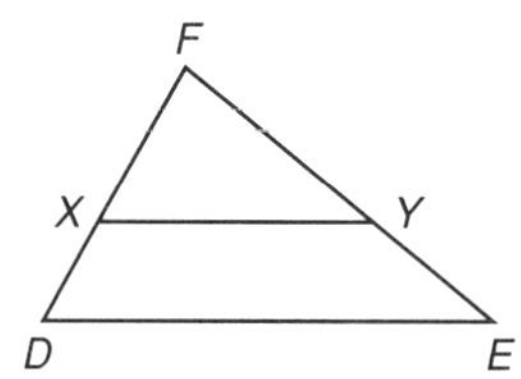

Answer: $\dfrac{XF}{DF} = \dfrac{YF}{EF}$. Let x = length of YF, which is FY. Then:

$$\frac{8}{10} = \frac{x}{12}$$

$$10x = 96$$

$$x = 9.6$$

Therefore: $FY = 9.6$

TRY THESE

Items 1–3: Respectively, B and E divide line segments $\overline{AC}$ and $\overline{DF}$ proportionally.

1. If $AB = 8$, $AC = 11$, and $DF = 6$, how long is DE?

2. If $BC = 7$, $AB = 3$, and $DF = 15$, how long is EF?

3. If $BC = 9$, $AC = 14$, and $EF = 6$, how long is DF?

Items 4–7: With $\overline{NO} \parallel \overline{PQ}$, complete the proportions.

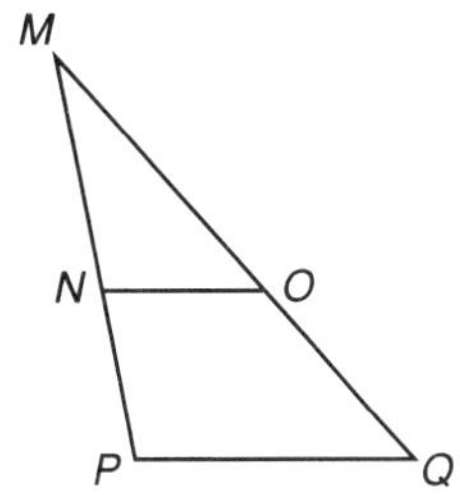

4. $\frac{MN}{NP} =$

6. $\frac{MO}{MQ} =$

5. $\frac{NP}{MP} =$

7. $\frac{MN}{NO} =$

Items 8–10: $\overline{RS} \parallel \overline{TU}$.

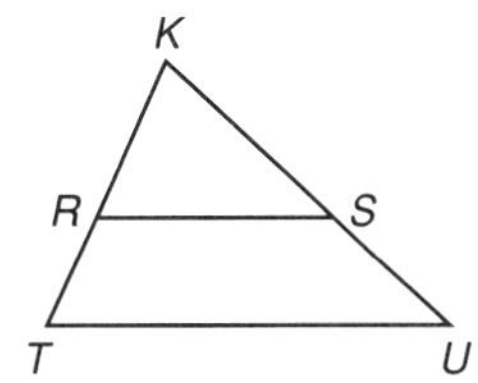

8. If $KR = 12$, $RT = 9$, and $KS = 16$, how long are KT, SU, and KU?

9. If $RT = 2$, $KS = 9$, and $KU = 12$, how long are KR, KT, and SU?

10. If $SU = 5$, $KU = 12$, and $RT = 4$, how long are KT, KR, and KS?

Reflections

The word *reflection* normally refers to objects in space, such as your reflection in a mirror or a tree's reflection in a lake or a pool. In geometry, we often consider reflections of plane figures in the same plane.

REFLECTION OF A POINT

In diagram (1) below, point A is reflected over line n to form the **image** A'. The original point, A, is called the **preimage**. Line n is called the **reflecting line** or **line of reflection**. In diagram (2), a line segment is drawn connecting A and A'. Line n is the perpendicular bisector of AA'.

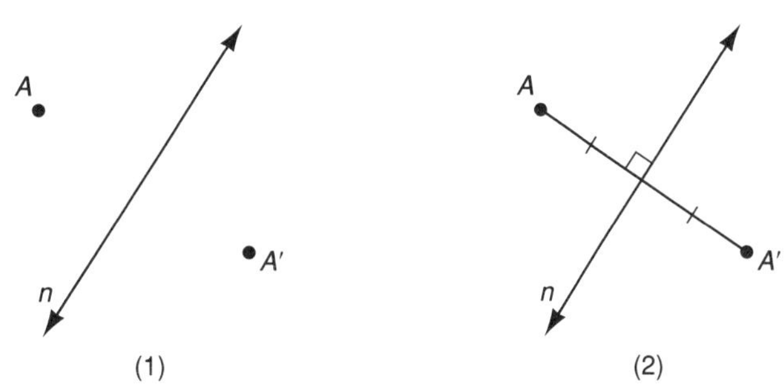

A point is a **reflection** of another point if the line segment joining the two points is perpendicular to and bisected by the line of reflection. If a point lies on the reflecting line, the point is an image of itself. In the diagram below, point A reflected over line n would be the same point.

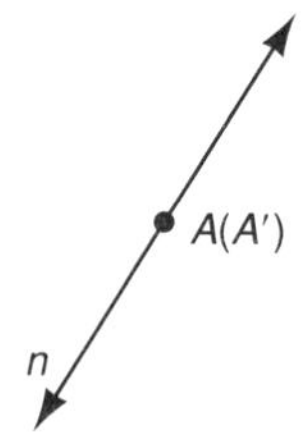

MODEL PROBLEMS

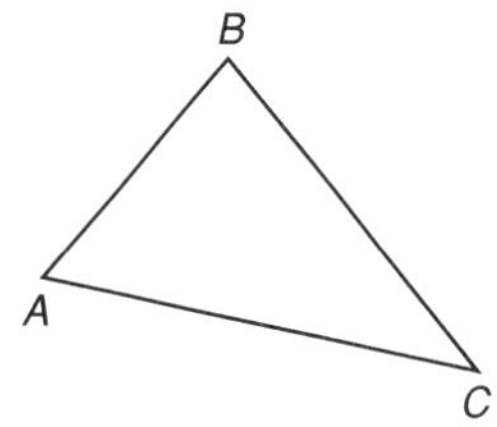

Using $\triangle ABC$, find the following:

Line m over which B is the image of A.

Line n over which C is the image of B.

Line p over which A is the image of C.

Answers:

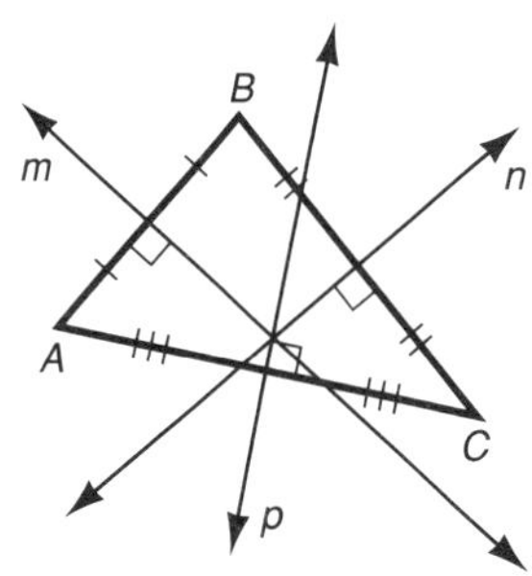

TRY THESE

1. Find the reflecting line over which the image of X is Y.

2. Find the reflecting line over which the image of Y is Z.

Items 3–5: Use the figure shown below.

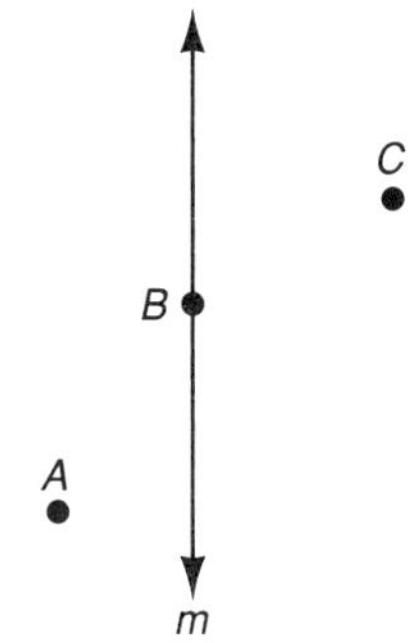

3. Show the reflection of point A over line m.

4. Show the reflection of point B over line m.

5. Show the reflection of point C over line m.

Items 6–9: Use the figure shown below.

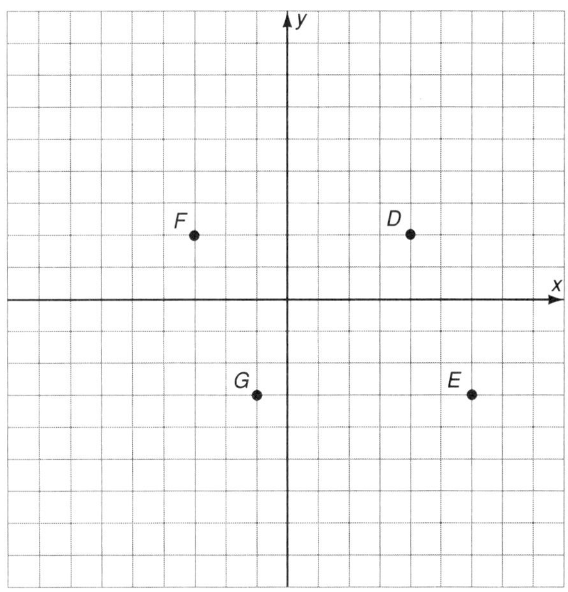

6. Show the reflection of point D over the x-axis.

7. Show the reflection of point E over the y-axis.

8. Show the reflection of point F over the x-axis.

9. Show the reflection of point G over the x-axis and then over the y-axis.

10. Find the image of point K over line m, then reflect that image over line n.

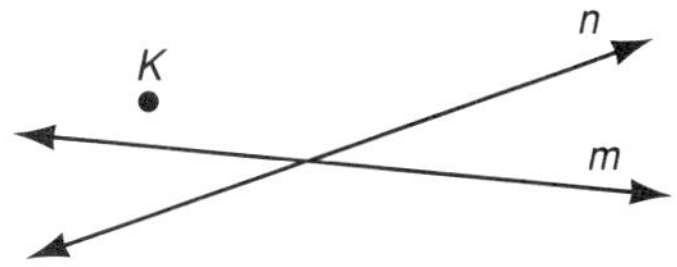

REFLECTION OF SETS OF POINTS

To reflect a set of points, reflect each point in the set. If the points in the set are collinear (lying on the same line), the images will be collinear. For example, if line segment $\overline{AB}$ is reflected over line m, the image will be line segment $\overline{A'B'}$, which is the same length as $\overline{AB}$.

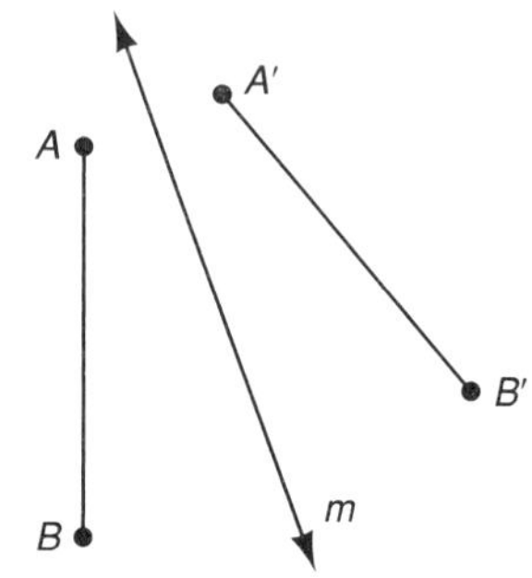

REFLECTION OF ANGLES

The reflection of an angle is an angle with the same measure. In the diagram below, the reflection of $\angle ABC$ will be $\angle A'B'C'$. The measure of $\angle ABC$ = m $\angle A'B'C'$.

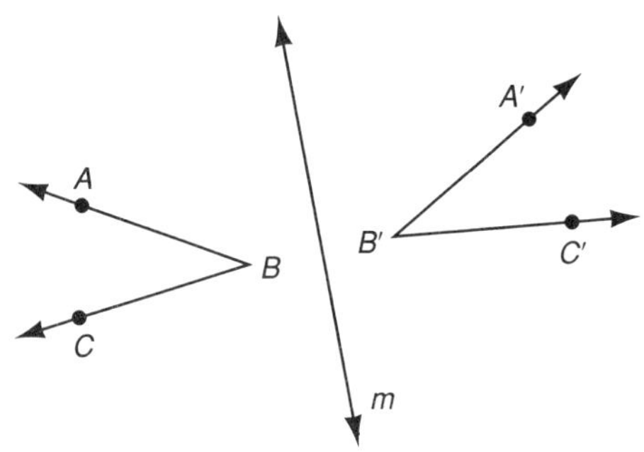

REFLECTION OF POLYGONS

When polygons are reflected, the **orientation** of the image is reversed. If the vertices of the preimage are read clockwise, the vertices of the image will be read counterclockwise. If the vertices of the preimage are read counterclockwise, the vertices of the image will be read clockwise.

Look at the example below.

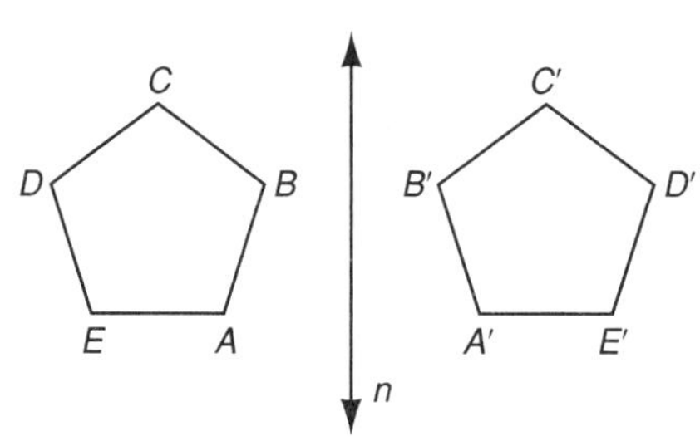

The preimage on the left would be read counterclockwise: polygon *ABCDE*. It is reflected over line *n*. Its image, on the right, would be read clockwise: polygon $A'B'C'D'E'$.

MODEL PROBLEMS

Below, *SNEB* is the reflection image of *DPFB* over line *m*.

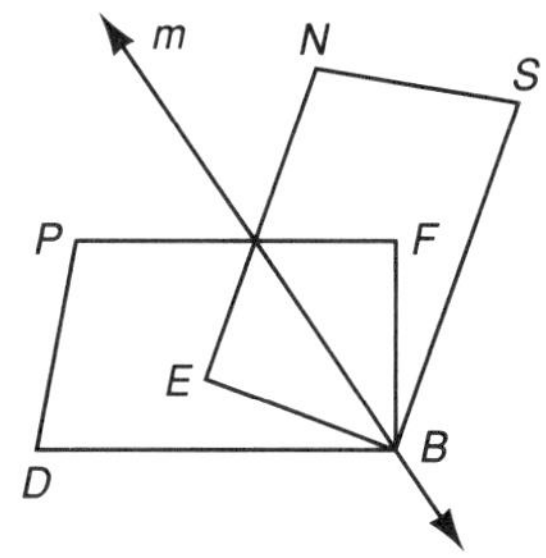

1. Which point is the reflection of point *P*? *Answer: N*
2. Which line segment is the reflection of $\overline{EN}$? *Answer:* $\overline{FP}$
3. Which angle is the reflection of $\angle DBF$? *Answer:* $\angle SBE$
4. Which line segment is equal in length to $\overline{FB}$? *Answer:* $\overline{EB}$
5. Which angle is equal in measure to $\angle D$? *Answer:* $\angle S$

TRY THESE

Items 1–5: Use the diagram from the model problem above.

1. Which point is the reflection of point *D*?
2. Which line segment is the reflection of $\overline{BS}$?
3. Which angle is the reflection of $\angle ENS$?
4. Which line segment is equal in length to $\overline{DP}$?
5. Which angle is equal in measure to $\angle S$?
6. 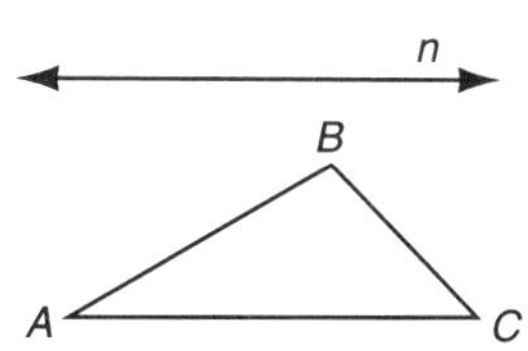

   Which triangle shows a reflection of triangle *ABC* over line *n*?

   A. 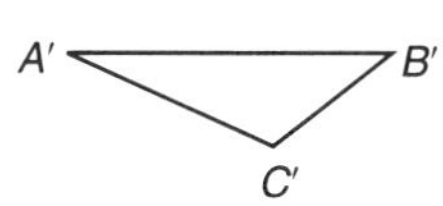

   C. 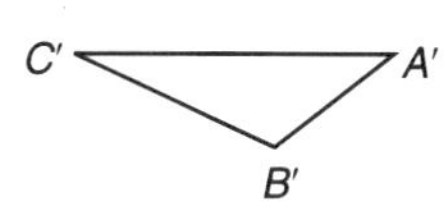

   B. 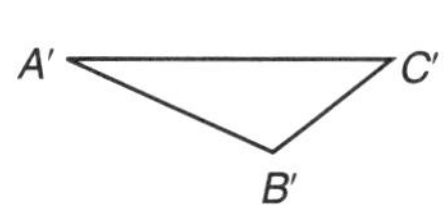

   D. 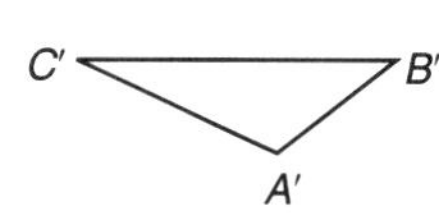

Items 7 and 8: Use the diagram below.

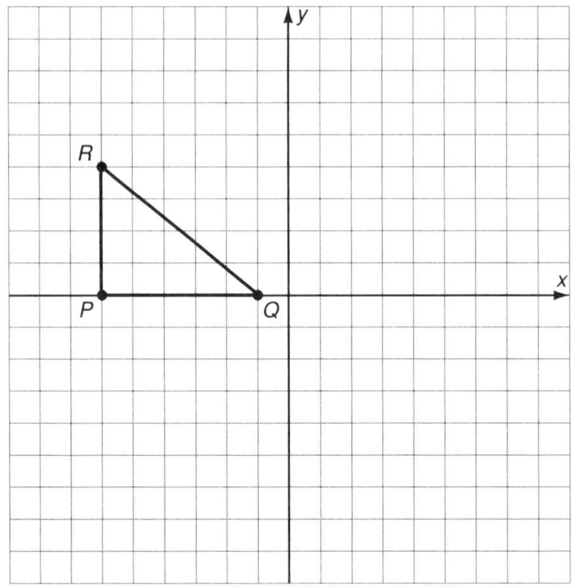

7. Show the reflection of triangle PQR over the x-axis.

8. Show the reflection of triangle PQR over the y-axis.

Items 9 and 10: Use the diagram below.

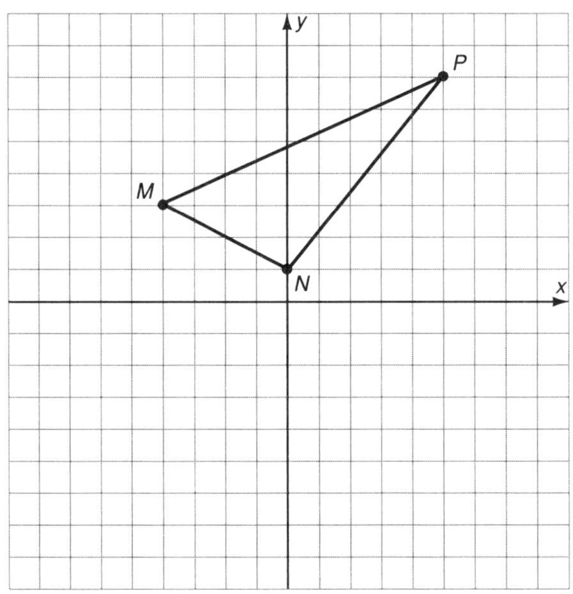

9. Show the reflection of triangle MNP over the x-axis.

10. Show the reflection of triangle MNP over the y-axis.

11. Using the diagram below, show the reflection of $RSTUVWX$ over the x-axis.

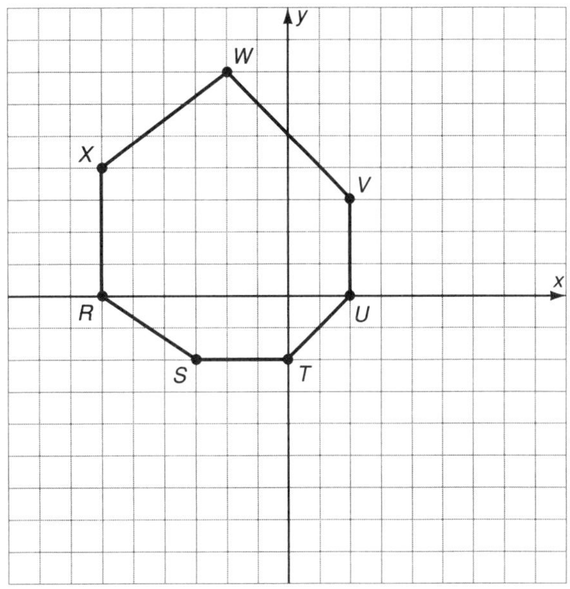

Symmetry

LINE SYMMETRY

If a figure "**maps**" or matches itself exactly over a line, we say that the figure is **symmetric over a line of symmetry**. We can also say that the figure has **line symmetry**. The figures below have line symmetry.

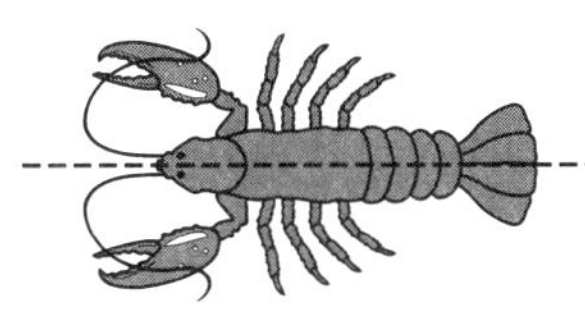

POINT SYMMETRY

A figure may have a point of symmetry such that a 180-degree turn will map the point onto itself. In that case, the figure has **point symmetry**. The figures below have point symmetry.

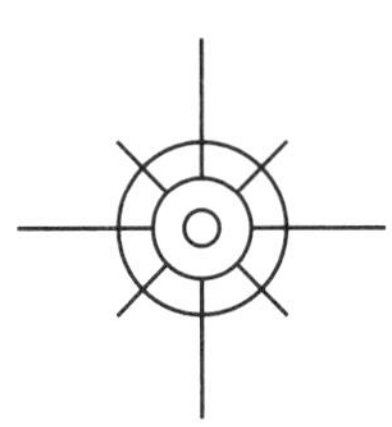

ROTATIONAL SYMMETRY

A figure may have **rotational symmetry**. This means that it will map onto itself if it is rotated on its center a certain number of degrees, less than 360 degrees. (*Less than 360 degrees* is part of this definition because *any* figure maps onto itself when rotated a full 360 degrees.) The figures below show rotations of an equilateral triangle, which has rotational symmetry (maps onto itself) at 120° and 240°.

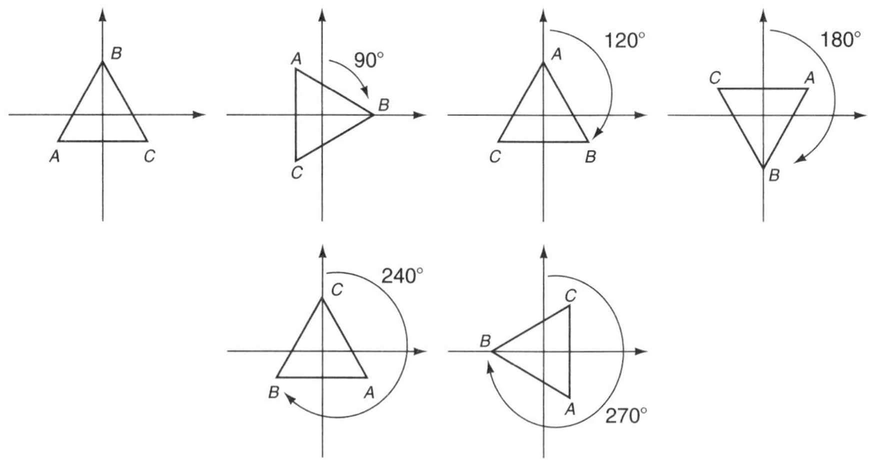

Any regular polygon has rotational symmetry. For example, when regular hexagon $ABCDEF$ is rotated $\frac{360°}{6}$, or 60°, counterclockwise about its center, the image of every point of the figure is a point of the figure. Under this rotation, $A \to B$, $B \to C$, $C \to D$, $D \to E$, $E \to F$, and $F \to A$.

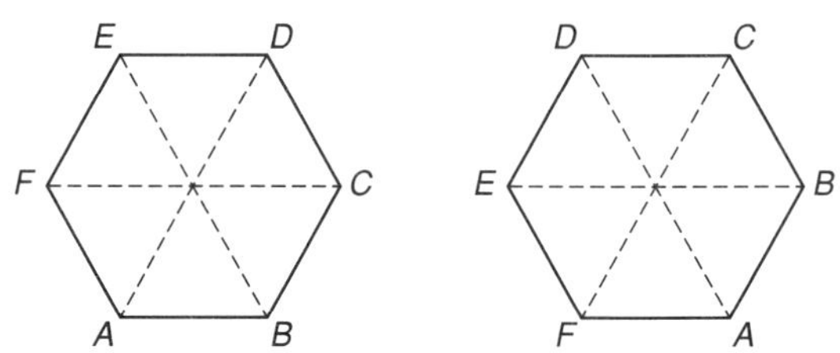

NOTE: Some figures have more than one kind of symmetry. Mapping at a 180-degree rotation is another way to describe point symmetry.

MODEL PROBLEMS

1. Which of these figures has line symmetry?

A. 

B. 

C. 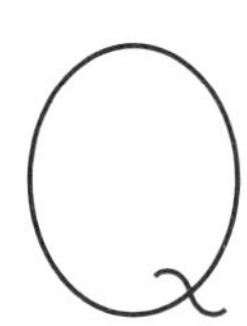

D. 

Answer: A

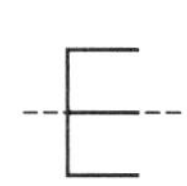

2. Which of these figures has point symmetry?

A. 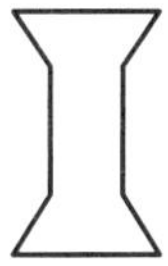 **B.**  **C.** 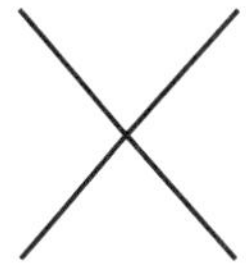 **D.** 

Answer: C

3. Which of these figures has rotational symmetry?

A.  **B.**  **C.**  **D.** 

Answer: D

TRY THESE

1. Copy the figure and draw the line of symmetry.

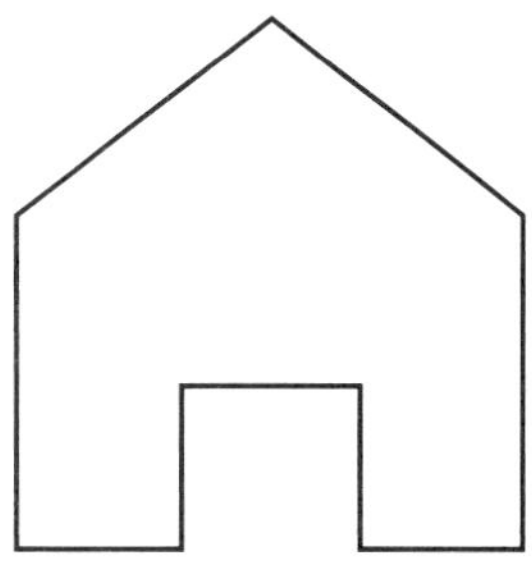

2. Copy the figure and draw and label the point of symmetry.

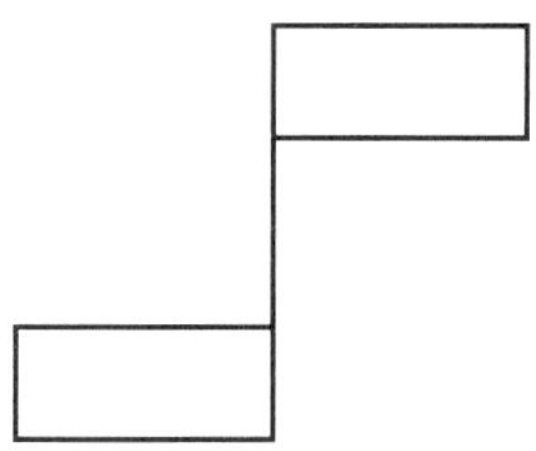

3. Copy and complete the figure to show symmetry over line m.

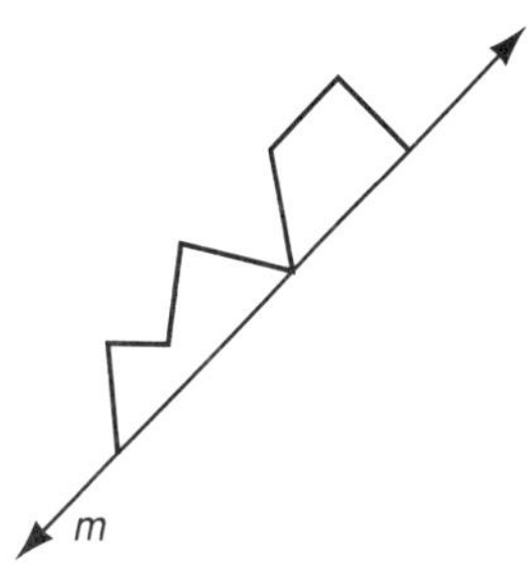

4. Draw all the lines of symmetry for the regular hexagon shown below.

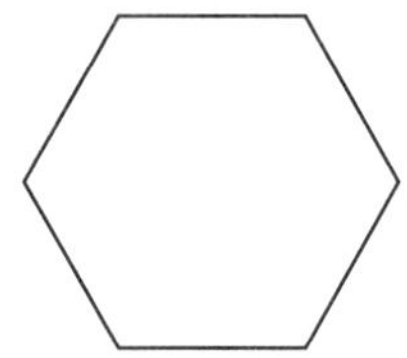

5. The parallelogram *MNOP* shown below has point symmetry. Find the location of the point and label it *R*.

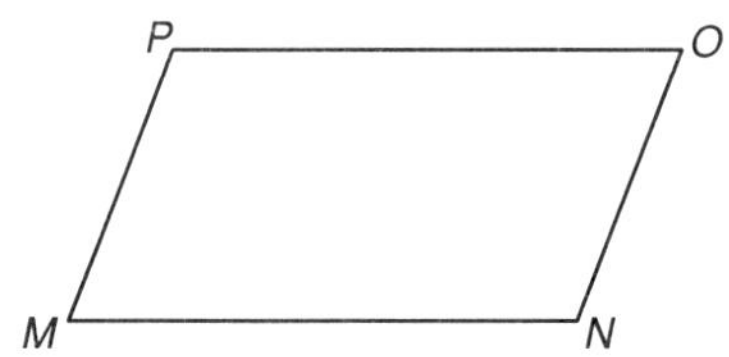

6. Find all the lines of symmetry and the point of symmetry in the figure below.

Items 7–10: Use the regular octagon shown below to answer the questions.

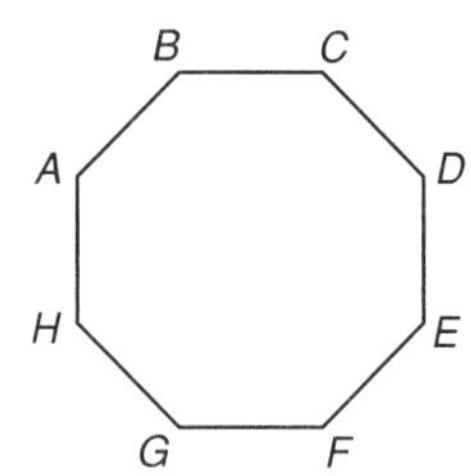

7. Does the figure have rotational symmetry?

8. Through how many degrees will the figure have to be rotated so point *A* maps onto point *H*?

9. Through what other degree rotations would the figure map onto itself?

10. Through how many degrees would the figure have to rotate in order for it to be back to the original figure?

Transformations

A **transformation** is an operation that matches a unique point with its unique image point. This is a very general idea on which many types of transformations are based. Reflection and symmetry are important in the study of transformations. Three basic concepts of transformational geometry are translation, rotation, and dilation.

TRANSLATION

The result of sliding a figure from one position to another, without turning it, is a **translation**. To get a translation, first reflect a figure over a line and then reflect it again over another line parallel to the first line. The final figure, the image, has the same "look" or orientation as the preimage. But the image is in a different place, as though it has been slid to a new position. The figure below is an example of a translation.

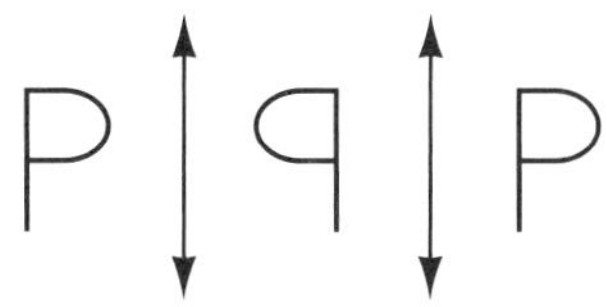

MODEL PROBLEMS

1. Which of the following shows a translation of the original?

A.  **B.** 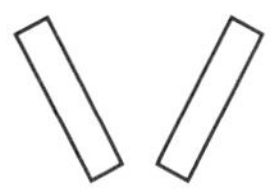 **C.**  **D.** 

Answer: C

2. Which of the following shows a translation of the original?

A. 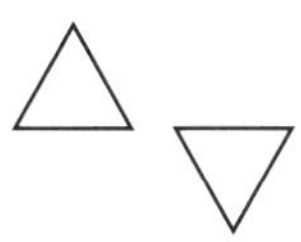 **B.**  **C.**  **D.** 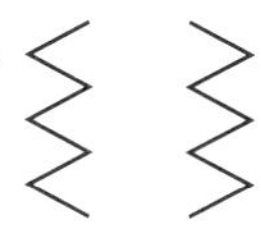

Answer: B

TRY THESE

1. State whether each of the following does or does not show a translation of the original.

a. 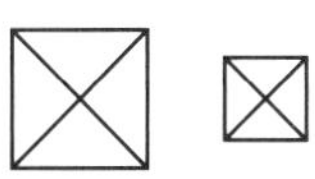 **b.**  **c.** 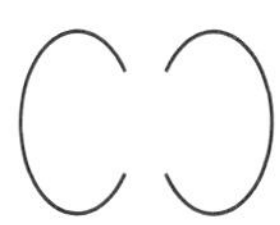

2. State whether each of the following does or does not show a translation of the original.

a.  **b.** 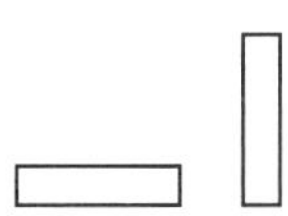 **c.** 

Items 3–6: Copy each figure and follow the directions.

3. Given that $L_1 \parallel L_2$, show the reflection of $\overline{AB}$ over L_1 and then over L_2.

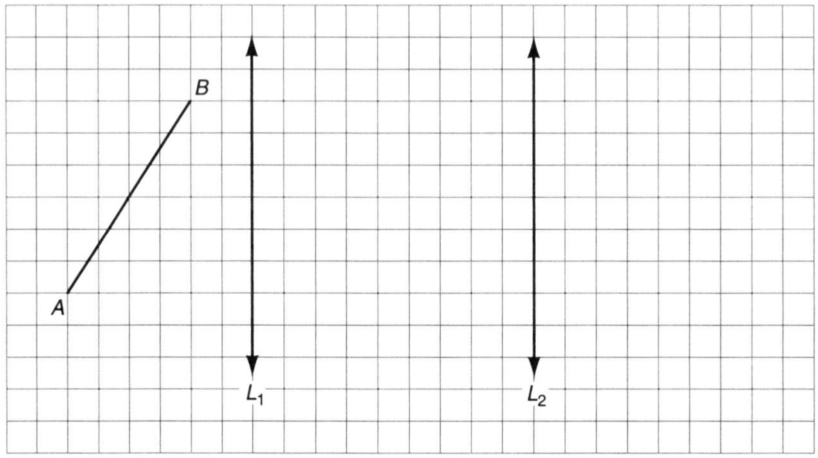

4. Given that $L_1 \parallel L_2$, show the reflection of $\overline{AB}$ over L_1 and then over L_2.

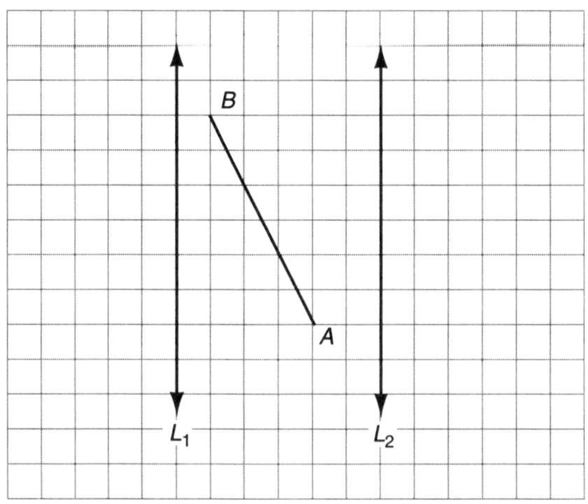

5. Given that $L_1 \parallel L_2$, show the reflection of triangle ABC over L_1 and then over L_2.

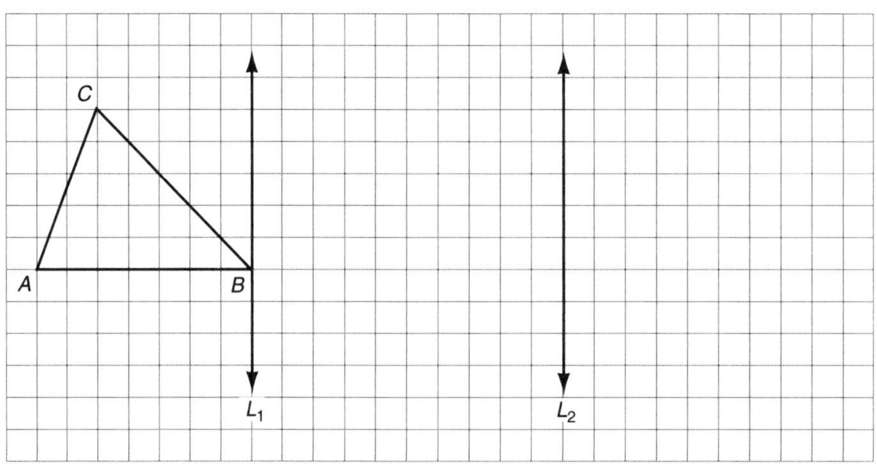

6. Given that $L_1 \parallel L_2$, show the reflection of parallelogram $ABCD$ over L_1 and then over L_2.

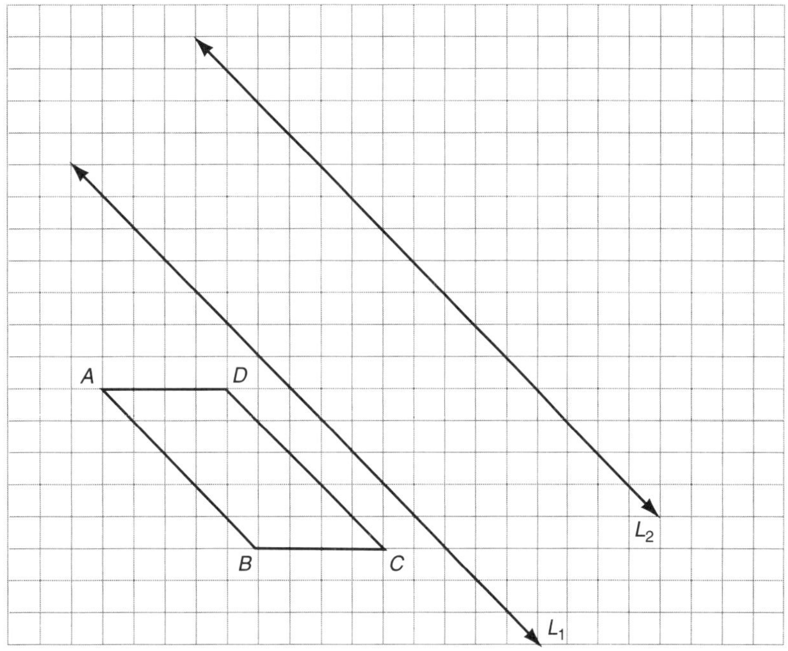

ROTATION

We have already reviewed rotational symmetry. This concept also applies when a transformation is referred to as a **rotation**. If a figure is reflected over a line and then reflected over another line that intersects the first line, the resulting image is a **rotational transformation**.

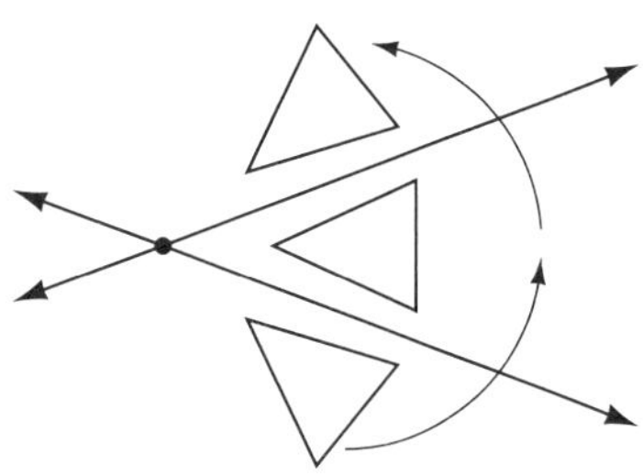

The **center** of the rotation is the point of intersection of the two lines. The measure of the angle of rotation is called the **magnitude**. The **direction** of rotation is clockwise or counterclockwise.

MODEL PROBLEMS

1. In each pair below, the original figure (preimage) is shaded. Which pair shows a 90-degree counterclockwise rotation of the original?

A. 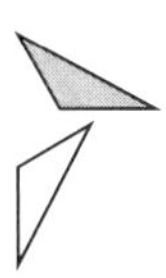 **B.** 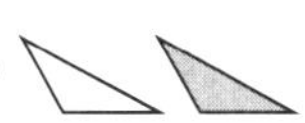 **C.** 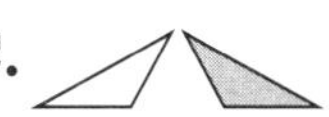 **D.** 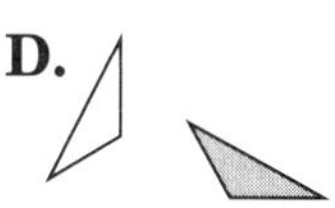

Answer: D

2. Which pair below shows a 270-degree counterclockwise rotation of the original? The preimage is shaded.

A. 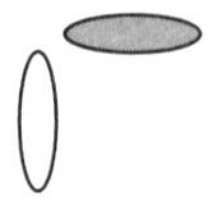 **B.** 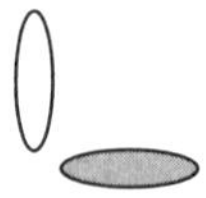 **C.**  **D.** 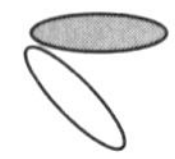

Answer: A

TRY THESE ✔

Items 1 and 2: In each case, the preimage is shaded.

1. Which pair shows a 45-degree clockwise rotation of the original figure?

A. 

C. 

B. 

D. 

2. Which pair shows a 180-degree counterclockwise rotation of the original figure?

A. 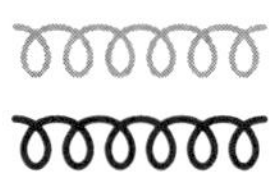 **C.** 

B.  **D.** 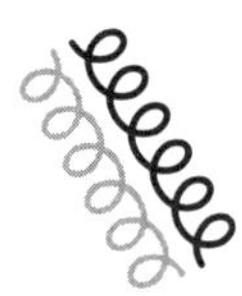

Items 3–6: Use the drawing shown below.

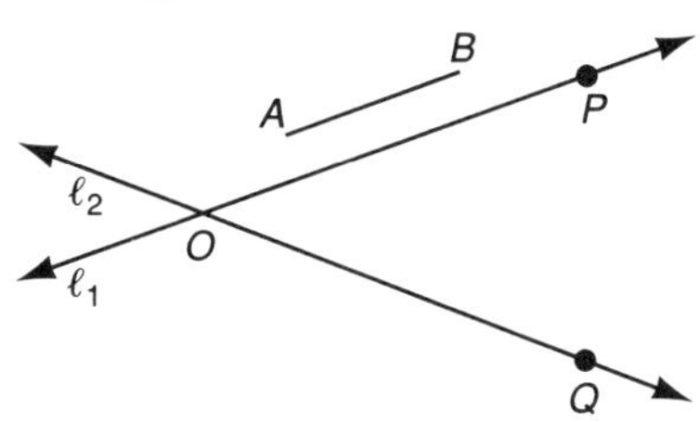

3. Show the reflection of $\overline{AB}$ over l_1 and then over l_2.

4. Use a protractor to measure the number of degrees in $\angle POQ$.

5. Draw $\overline{OA}$ and $\overline{OA''}$. What is the measure of $\angle AOA''$? This is called the magnitude of the rotation.

6. What is the relationship between the degrees in the angle formed by the intersecting lines and the magnitude of the rotation?

Items 7–10: Use the figure shown below.

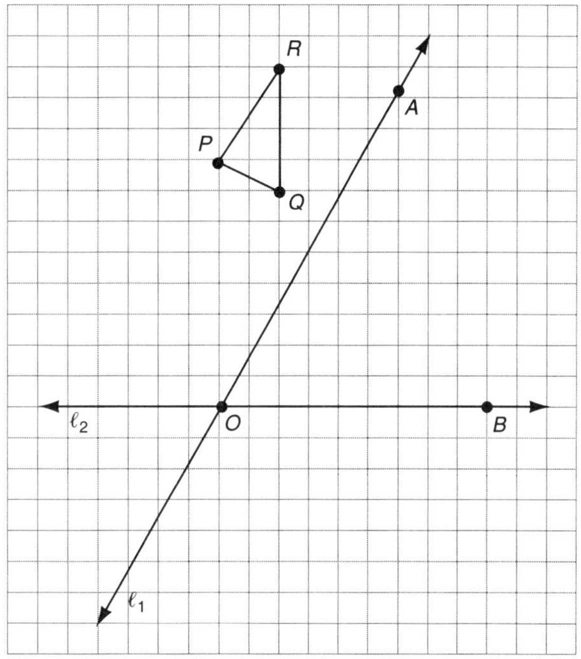

7. Reflect triangle PQR over l_1 and then over l_2.
8. Use a protractor to measure the number of degrees in $\angle AOB$.
9. Draw $\overline{OP}$ and $\overline{OP''}$. What is the measure of $\angle POP''$?
10. Draw $\overline{OQ}$ and $\overline{OQ''}$. What is the measure of $\angle QOQ''$?

DILATION

A dilation is a transformation that changes a preimage to an image that has the same shape. There is a proportional increase or decrease in size in all dimensions. For any polygon, the angles remain the same while the lengths of the sides are all increased or decreased by the same scale factor. The preimage and image are similar. For example, each side could be twice as long as in the original figure.

MODEL PROBLEMS

The diagram below shows the dilation of $\triangle ABC$ to $\triangle DEF$ with center P.

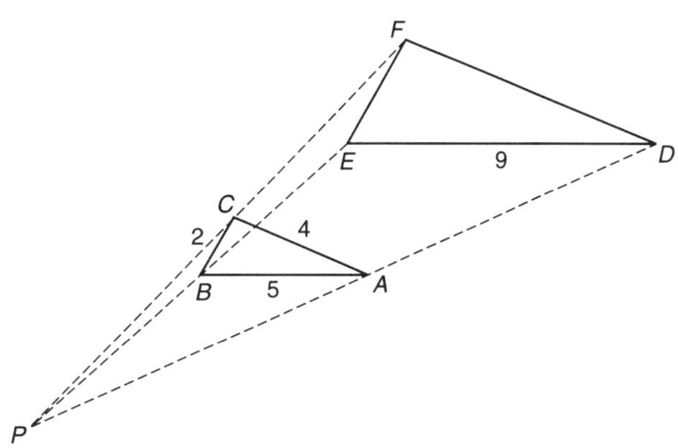

1. What is the scale factor of the dilation?

Answer: 9 to 5 or 1.8

2. What is the length of *FE*?

Answer: 2 • 1.8 = 3.6

3. What is the length of *DF*?

Answer: 4 • 1.8 = 7.2

TRY THESE

Items 1–3: Use the diagram below.

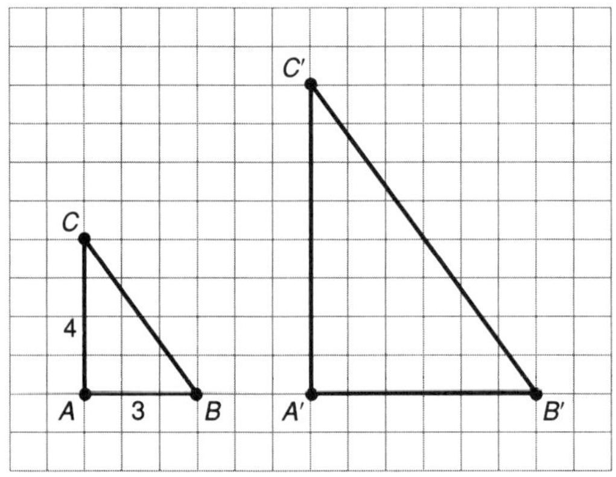

1. What are the lengths of $A'B'$, $A'C'$, CB, and $C'B'$?

2. What is the relationship between $\triangle ABC$ and $\triangle A'B'C'$?

3. How can the sizes of $\triangle ABC$ and $\triangle A'B'C'$ be compared?

Items 4–6: Use the diagram below.

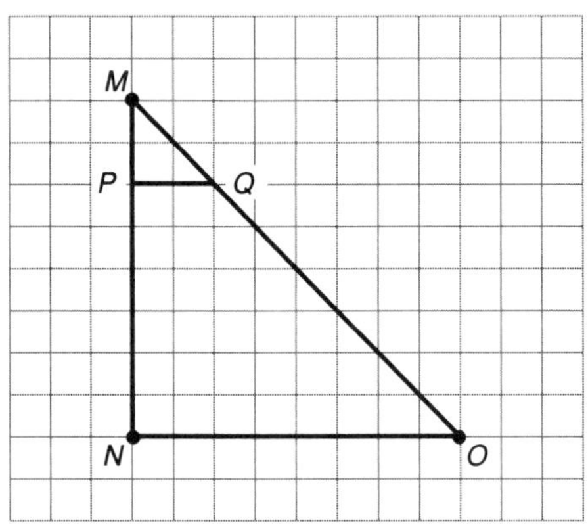

4. What are the lengths of *MP*, *PQ*, and *MQ*?

5. What are the lengths of *MN*, *NO*, and *OM*?

6. How can the sizes of $\triangle MPQ$ and $\triangle MNO$ be compared?

Combined transformations

A transformation can be a combination, or **composite**, of more than one of these operations. For example, the figure below has been rotated 90 degrees and then translated to its new position.

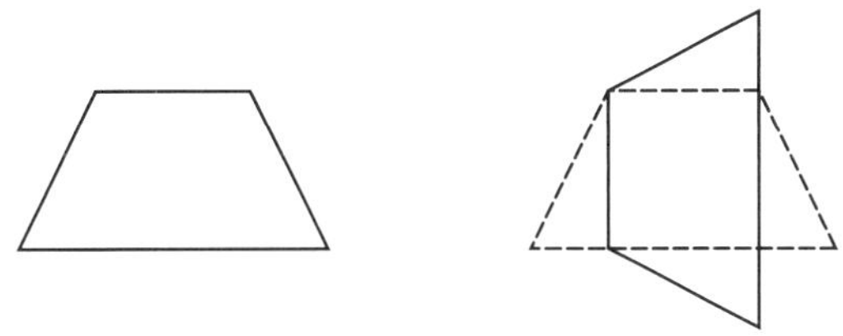

B ANALYZING AND APPLYING GEOMETRIC RELATIONSHIPS INVOLVING PLANAR CROSS SECTIONS

When a plane intersects a solid, the part of the plane bounded by the solid is called a **planar cross section**. In the three examples below, the shaded area is the planar cross section.

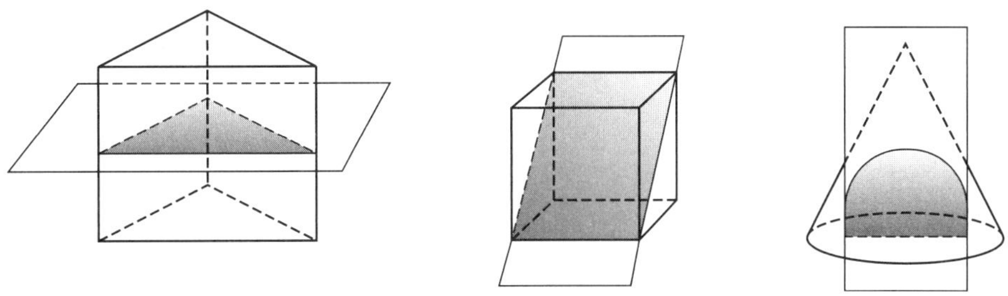

(1) (2) (3)

In diagram (1), the cross section has the same shape as the base of the solid. If the intersecting plane is parallel to the base, the cross section will be similar to the base. In diagram (2), the intersecting plane is not parallel to the base. In this case, the cross section of the cube is a rectangle. The cross section of the cone in diagram (3) is made by a plane that is perpendicular to the base but does not pass through the vertex of the cone.

Any cross section of a sphere is a circle.

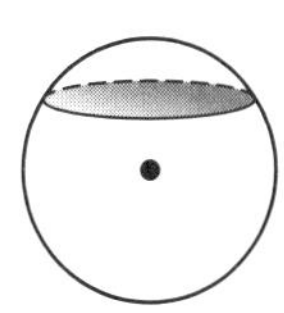

To find the area of the circle, we use the distance (d) of the cross section from the center of the sphere, the radius (r) of the sphere, and the Pythagorean theorem to find the radius (x) of the circle:

$$x^2 + d^2 = r^2$$

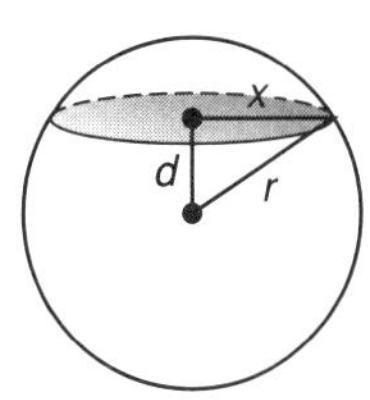

We can get information about cross sections by using the geometric properties of the solid through which the plane passes.

MODEL PROBLEMS

1. When the plane intersects the cone, which of the following will be the intersection?

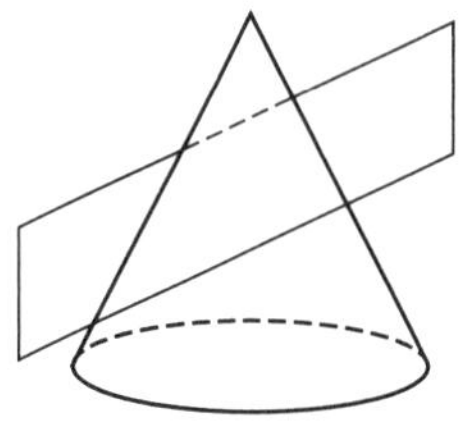

A. •

B. 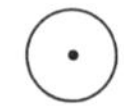

C. 

D. 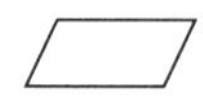

Answer: C

2. The radius of the earth is approximately 4,000 miles. What is the radius of a circle that is 2,000 miles directly above the center of the earth?

Answer: Use the Pythagorean theorem.

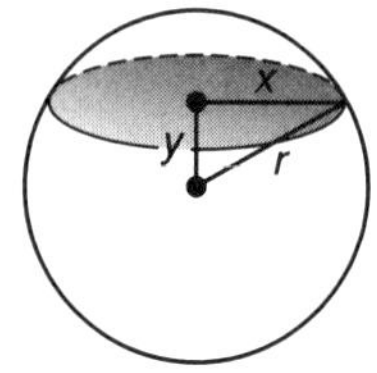

$x^2 + y^2 = r^2$

$x^2 + 2{,}000^2 = 4{,}000^2$

$x^2 + 4{,}000{,}000 = 16{,}000{,}000$

$x^2 = 12{,}000{,}000$

$x \approx 3{,}500$ miles

3. The figure below is a cube. The length of an edge is 10 units. What is the perimeter of $\triangle EGD$?

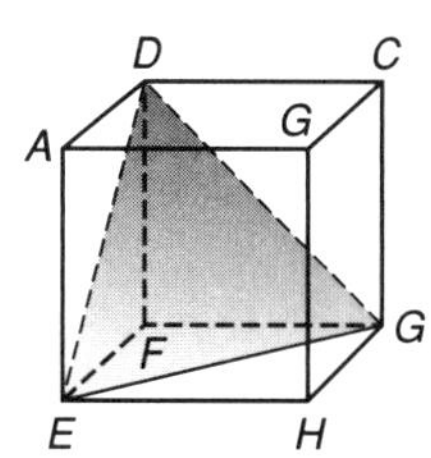

Answer: Each side of the triangle is a diagonal of a face of the square.

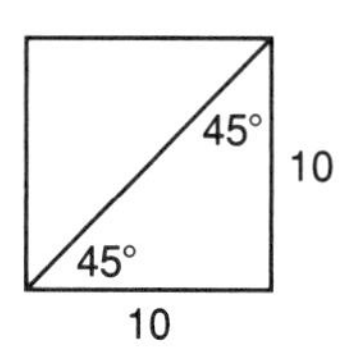

The diagonal is $10\sqrt{2}$ units.

Therefore, the perimeter of the triangle is $3 \times 10\sqrt{2} = 30\sqrt{2}$ units.

TRY THESE

1. *ABCDEF* is a cube. What type of figure is planar cross section *EDHG*?

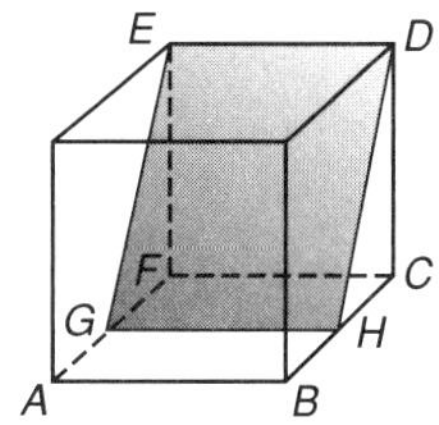

A. square

B. rectangle

C. trapezoid

D. rhombus

2. When the plane intersects the cylinder, which of the following will be the intersection?

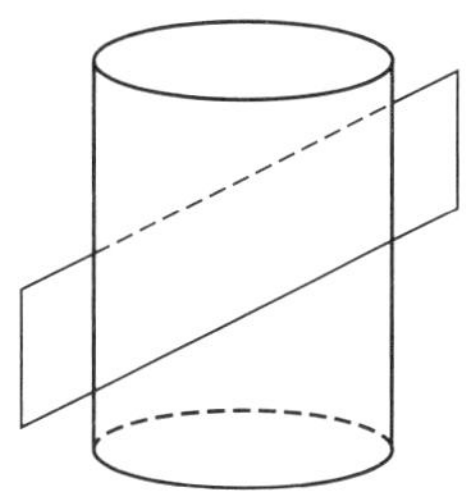

A. 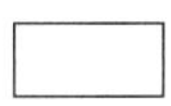

B. 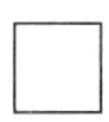

C. 

D. 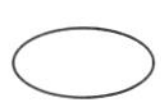

3. What is the circumference of a circle that is located 2,800 miles from the center of the earth? Give your answer in terms of π and round the coefficient to the nearest hundred. (The radius of the earth is about 4,000 miles.)

4. What is the area of the circle in problem 3? Express your answer in terms of π.

5. The base of this pyramid is a square. The shaded region is parallel to the base. What is the area of the shaded region?

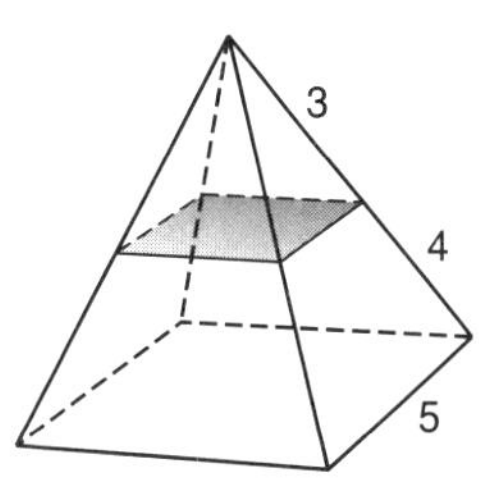

Items 6–9: In the diagram shown below, a plane parallel to the base of a right circular cone intersects the cone $\frac{1}{4}$ of the way between point C and point D. The height of the cone is 16 inches and the base radius is 7 inches.

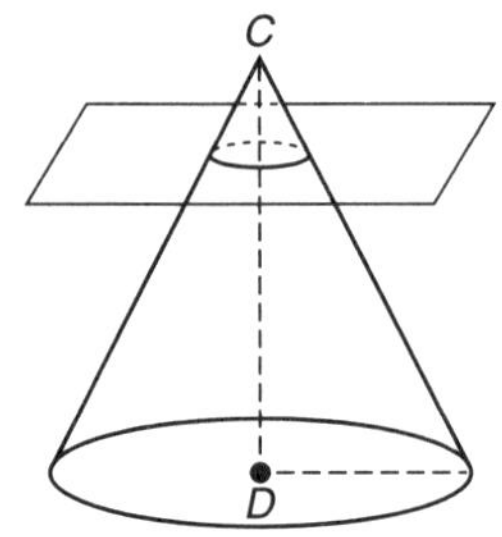

6. What is the area of the base of the smaller cone? Use $\pi \approx \frac{22}{7}$.

7. What is the volume of the smaller cone? Use $\pi \approx \frac{22}{7}$.

8. What is the slant height of the small cone?

9. What is the approximate surface area of the small cone? Use $\pi \approx \frac{22}{7}$.

Assessment

Multiple-Choice Problems

Items 1–5: Use pairs of triangles I to V.

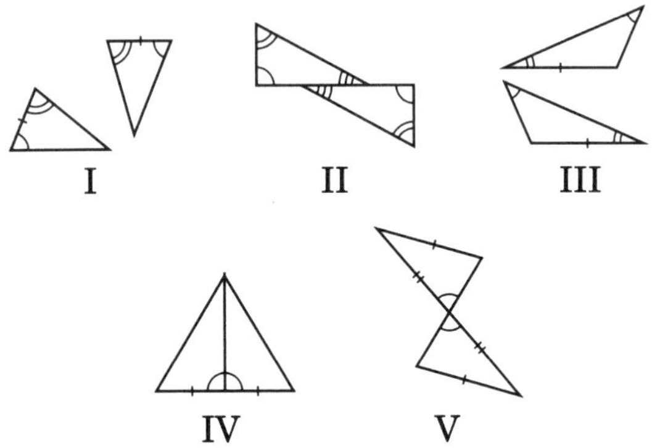

1. Which pair of triangles are congruent by side-angle-side (SAS)?

A. I

B. II

C. III

D. IV

2. Which pair of triangles are congruent by angle-side-angle (ASA)?

A. I **C.** IV

B. II **D.** V

3. Which pair of triangles are congruent by angle-angle-side (AAS)?

A. II **C.** IV

B. III **D.** V

4. Which pair of triangles can NOT be shown to be congruent?

A. I **C.** III

B. II **D.** IV

5. Which pair of triangles can be proved similar but can NOT be proved congruent?

A. I **C.** IV

B. II **D.** V

6. A light shines through a film and projects an image on a screen. If LF is 8 centimeters, LS is 2400 centimeters, and the screen image is 222 centimeters high, how high is the film image?

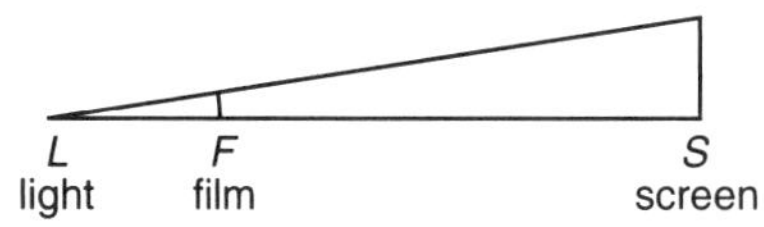

A. 0.74 cm

B. 0.86 cm

C. 74 cm

D. 86 cm

7. In the figure shown below, $\overline{TQ} \parallel \overline{SR}$. What is the length of PS?

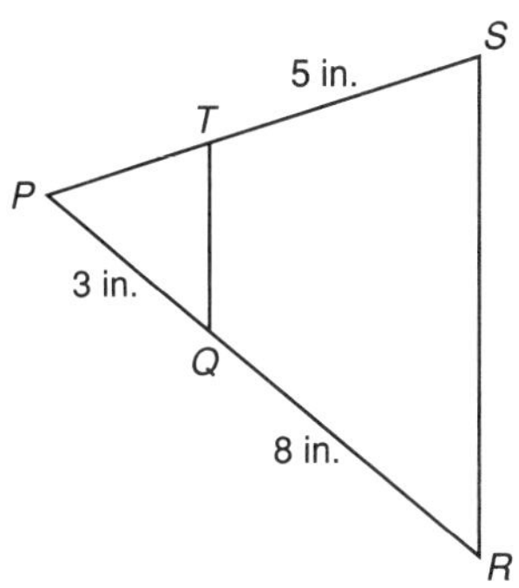

A. $1\frac{7}{8}$ in.

B. 4.8 in.

C. $6\frac{7}{8}$ in.

D. 9.8 in.

8. In the drawing below, $\triangle RST \sim \triangle NMP$. What is the length of $\overline{PN}$?

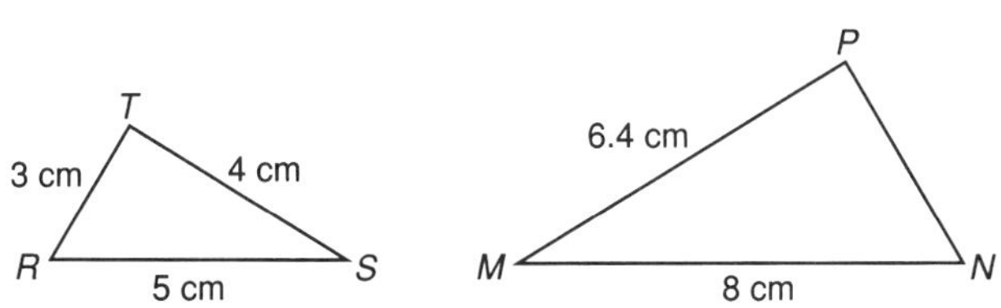

A. 4 cm

B. 4.5 cm

C. 4.8 cm

D. 5 cm

9. Polygon $ABCDE \sim$ polygon $PQRST$. Side $\overline{RS}$ in the second polygon corresponds to which side in the first polygon?

A. $\overline{AE}$

B. $\overline{DE}$

C. $\overline{CD}$

D. $\overline{BC}$

10. Pentagon $ABCDE$, shown below, has to be rotated how many degrees counterclockwise for point D to map onto point A?

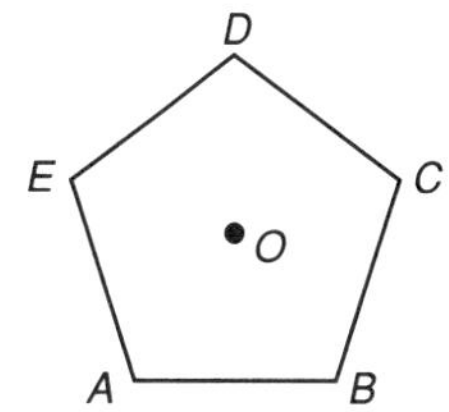

A. 45°

B. 72°

C. 90°

D. 144°

11. Which geometric figure has both horizontal and vertical line symmetry?

A. parallelogram

B. square

C. trapezoid

D. triangle

12. Arletha looked through a shop window and saw a clock, without realizing that she was really seeing the reflection of a clock in a mirror. The hands on the clock that she saw showed the time to be 4:30. What was the actual time on the clock?

A. 4:40

B. 6:30

C. 7:30

D. 8:30

13. Through which line in this figure is point C the reflection of point H?

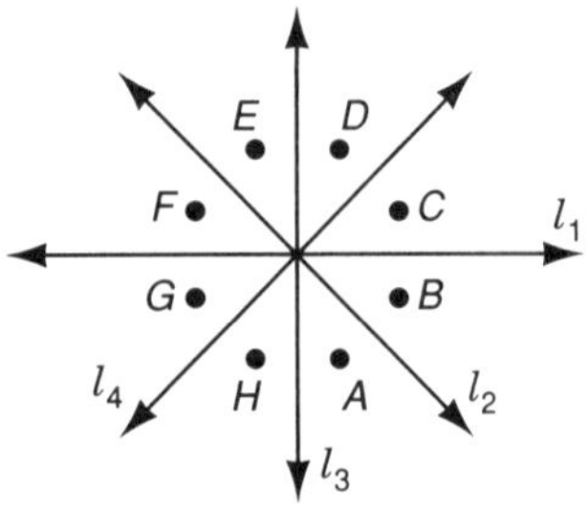

A. l_4

B. l_3

C. l_2

D. l_1

14. Triangle ABC is reflected through two parallel lines. Which drawing represents this?

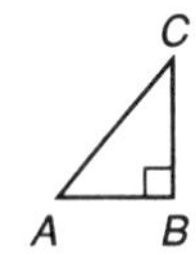

A.

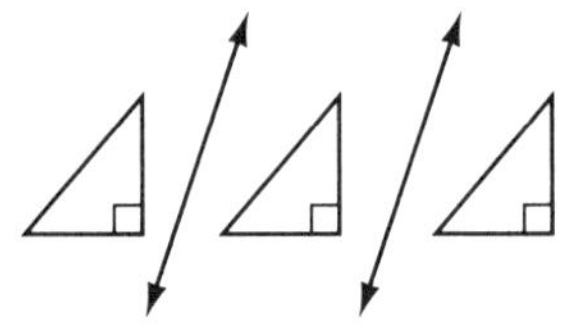

B.

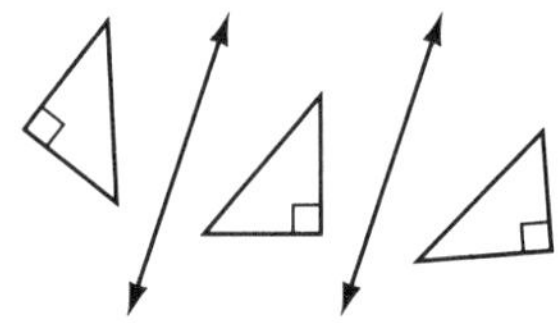

C.

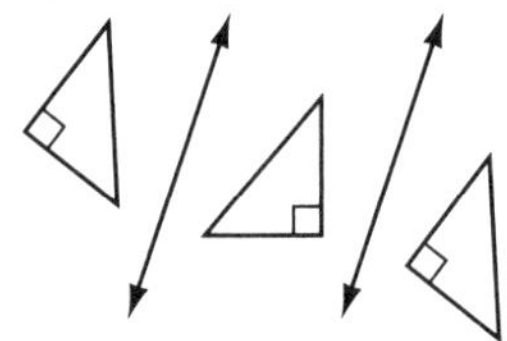

D. None of these drawings represents this reflection.

Gridded-Response Problems

15. Triangle I is dilated to form triangle II. Each side in triangle II is how many times larger than the corresponding side in triangle I?

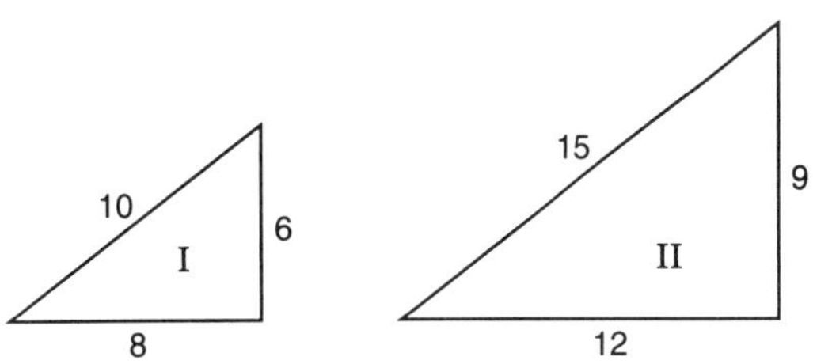

Items 16 and 17: Use the rectangular solid shown below.

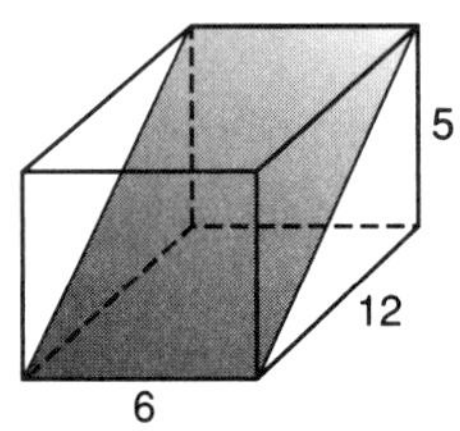

16. What is the perimeter of the shaded cross section?

17. What is the area of the shaded cross section?

Items 18 and 19: The radius of the sphere shown below is 13 meters. The shaded cross section is located 12 meters from the center of the sphere.

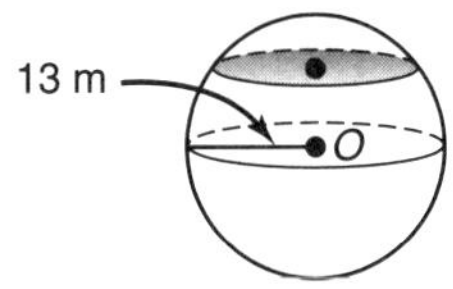

18. What is the radius of the cross section?

19. What is the area of the cross section? Use $\pi \approx 3.14$.

20. Line segments $\overline{AB}$ and $\overline{CD}$ are divided proportionally by points P and Q, respectively. What is the length of QD?

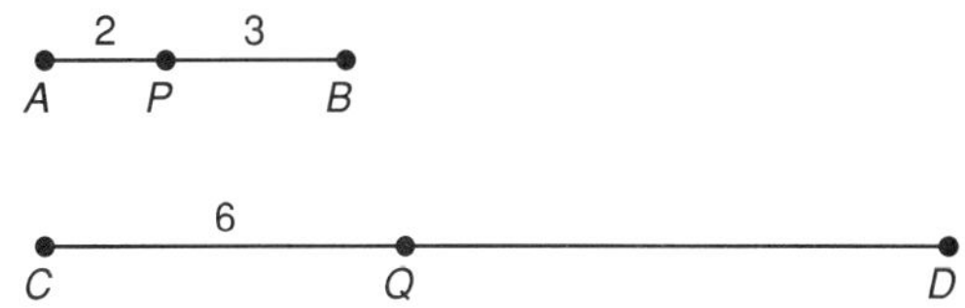

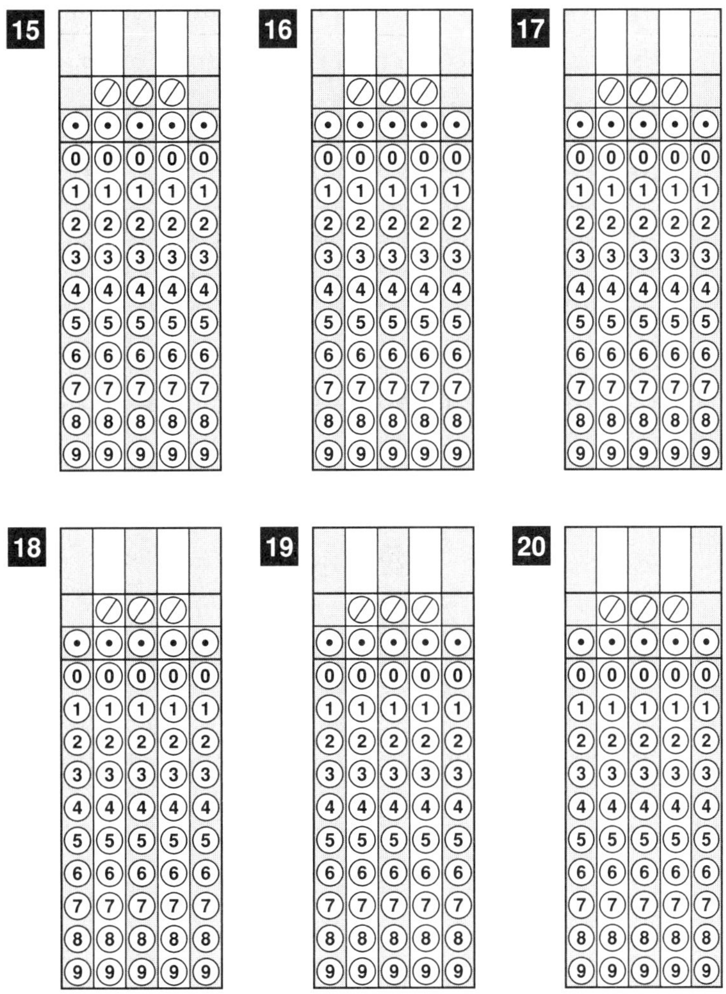

Section 3 Describing objects algebraically and using coordinate geometry to locate objects in two and three dimensions

A REPRESENTING AND APPLYING GEOMETRIC PROPERTIES AND RELATIONSHIPS TO SOLVE REAL-WORLD AND MATHEMATICAL PROBLEMS INCLUDING RATIO, PROPORTION, SIMILARITY, FORMULAS, AND ENLARGEMENTS

Ratios and proportions

A **ratio** is the quotient of two quantities, such as $\frac{a}{b}$ where b cannot be zero. A ratio can also be written as $a : b$.

A **proportion**, such as $\frac{a}{b} = \frac{c}{d}$, states that two ratios are equal. A proportion can also be written in the form $a : b = c : d$. The positions occupied by a and d are called the **extremes**, and the positions of b and c are called the **means**. In a proportion, the product of the extremes is equal to the product of the means. Algebraically, the proportion can be written as $a \bullet d = b \bullet c$ or $ad = bc$.

MODEL PROBLEMS

1. Solve the following proportion for x.

$\frac{3}{7} = \frac{5}{x}$

Answer: $3x = 35$

$x = \frac{35}{3} = 11\frac{2}{3}$

2. Solve the following proportion for x.

$\frac{9}{4} = \frac{8}{x + 2}$

Answer: $9(x + 2) = 32$ then $9x + 18 = 32$

$9x = 32 - 18$ then $9x = 14$

$x = \frac{14}{9} = 1\frac{5}{9}$

THE GEOMETRIC MEAN

If a, b, and c are positive numbers and $\frac{a}{b} = \frac{b}{c}$, then b is called the **geometric mean** between a and c. For example, to find the geometric mean between 4 and 9, you would set up the proportion $\frac{4}{b} = \frac{b}{9}$ and then multiply to get $b^2 = 36$ and $b = 6$. A geometric mean is always positive.

MODEL PROBLEMS

1. Find the geometric mean between 2 and 8.

Answer: Let b = the geometric mean. Then:

$$\frac{2}{b} = \frac{b}{8}$$

$$b^2 = 16$$

$$b = 4$$

2. Find the geometric mean between 6 and 10.

Answer: Let b = the geometric mean. Then:

$$\frac{6}{b} = \frac{b}{10}$$

$$b^2 = 60$$

$$b = \sqrt{60} = 2\sqrt{15}$$

TRY THESE

Items 1–5: Solve the following proportions for x.

1. $\frac{2}{7} = \frac{6}{x}$

2. $\frac{x}{9} = \frac{3}{5}$

3. $\frac{6}{11} = \frac{x + 1}{5}$

4. $\frac{4}{5} = \frac{3 + x}{10 + x}$

5. $\frac{1}{8} = \frac{x}{12}$

6. Find the geometric mean between 3 and 27.

7. Find the geometric mean between 13 and 25.

8. Find the geometric mean between 1 and 8.

9. 6 is the geometric mean between 5 and what number?

10. 12 is the geometric mean between 9 and what number?

RATIOS AND PROPORTIONS IN RIGHT TRIANGLES

There are some special theorems that involve ratios and proportions in right triangles.

- When an altitude is drawn to the hypotenuse of a right triangle, the two triangles formed are similar to the original triangle and to each other.

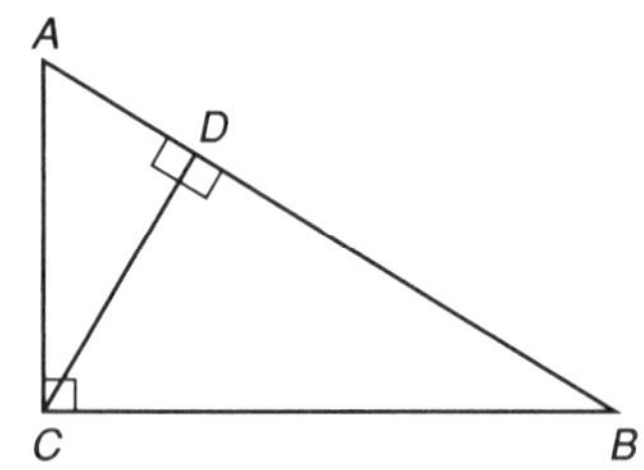

Two triangles are similar if corresponding angles are congruent and corresponding sides are in proportion. Triangle ABC is a right triangle in which the right angle is $\angle C$. An altitude is drawn from C to the opposite side $\overline{AB}$. The altitude meets $\overline{AB}$ at D. Triangle $ACD \sim \triangle CBD$ and triangles ACD and CBD are both similar to $\triangle ABC$.

- When an altitude is drawn to the hypotenuse of a right triangle, the length of the altitude is the geometric mean between the segments of the hypotenuse.

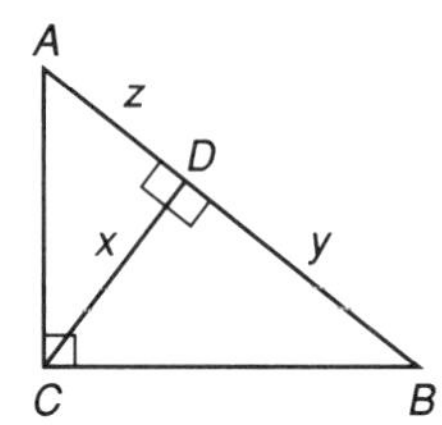

Triangle ABC is a right triangle in which the right angle is $\angle C$. An altitude is drawn from C to the opposite side $\overline{AB}$. The altitude meets $\overline{AB}$ at D. Altitude $\overline{CD}$ is the geometric mean between $\overline{BD}$ and $\overline{AD}$. Or: x is the geometric mean between y and z.

- When an altitude is drawn to the hypotenuse of a right triangle, each leg of the original triangle is the geometric mean between the hypotenuse and the segment of the hypotenuse that is adjacent to that leg.

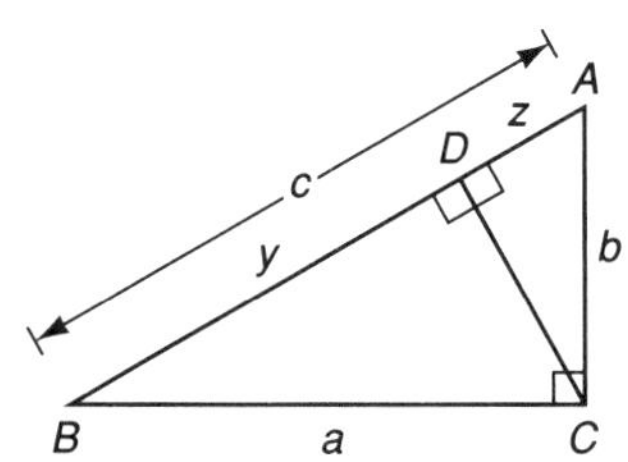

Triangle ABC is a right triangle in which the right angle is $\angle C$. An altitude is drawn from C to the opposite side $\overline{AB}$. The altitude meets $\overline{AB}$ at D. Leg $\overline{BC}$ is the geometric mean between $\overline{AB}$ and $\overline{BD}$. Leg $\overline{AC}$ is the geometric mean between $\overline{AB}$ and $\overline{AD}$. Or: a is the geometric mean between c and y; and b is the geometric mean between c and z.

$$\frac{c}{a} = \frac{a}{y} \text{ and } \frac{c}{b} = \frac{b}{z}$$

MODEL PROBLEMS

In each right triangle, an altitude is drawn to the hypotenuse. Find the values of x, y, and z.

1.

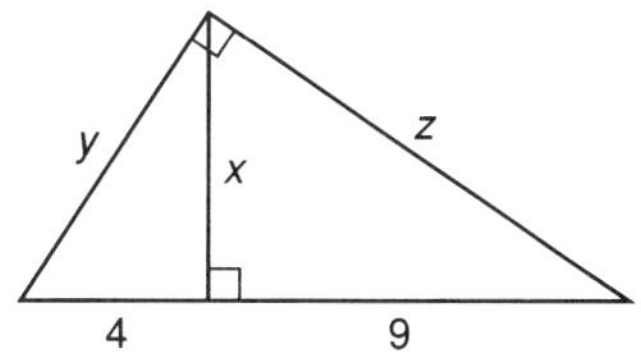

Answer:

$\frac{4}{x} = \frac{x}{9}$, so $x^2 = 36$ and $x = 6$

$4 + 9 = 13$

$\frac{13}{z} = \frac{z}{9}$ so $z^2 = 117$ and $z \approx 10.8$

$\frac{13}{y} = \frac{y}{4}$ so $y^2 = 52$ and $y \approx 7.2$

2.

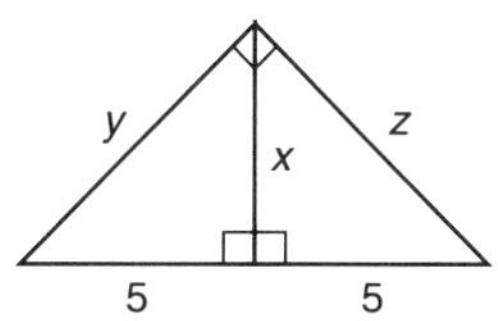

Answer:

$\frac{5}{x} = \frac{x}{5}$ so $x^2 = 25$ and $x = 5$

$5 + 5 = 10$

$\frac{10}{y} = \frac{y}{5}$ so $y^2 = 50$ and $y \approx 7.1$

$\frac{10}{z} = \frac{z}{5}$ so $z^2 = 50$ and $z \approx 7.1$

TRY THESE

Items 1 and 2: In each right triangle, an altitude is drawn to the hypotenuse. Find the values of x, y, and z. Round each answer to the nearest hundredth.

1.

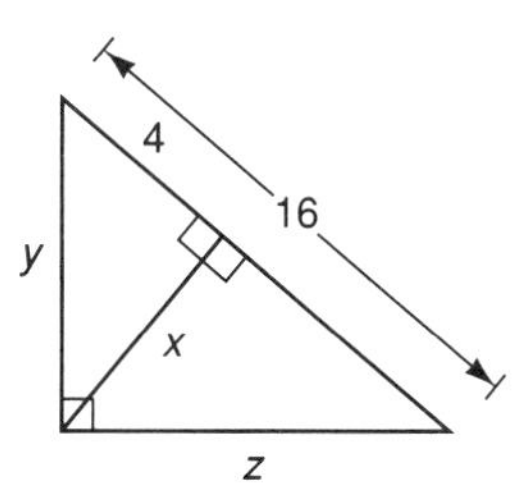

2.

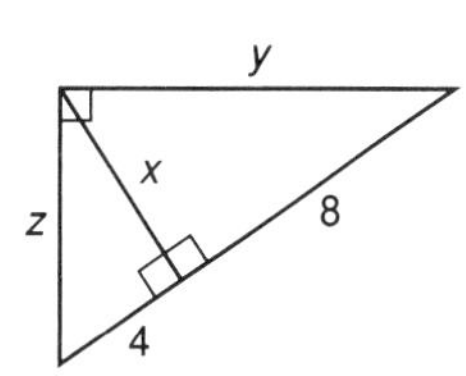

Items 3–5: Use each diagram to find a, b, c, and d. Round each answer to the nearest hundredth.

3.

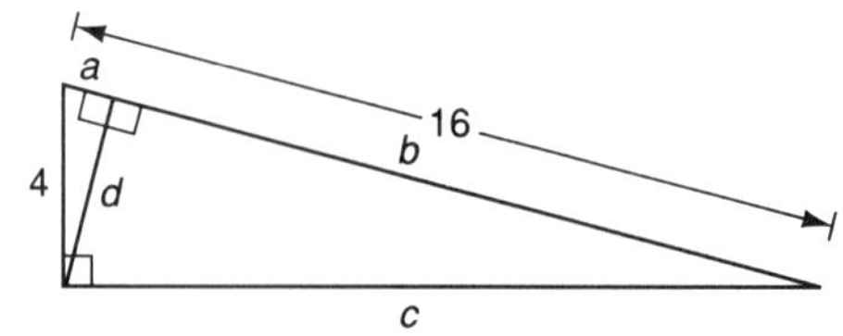

4.

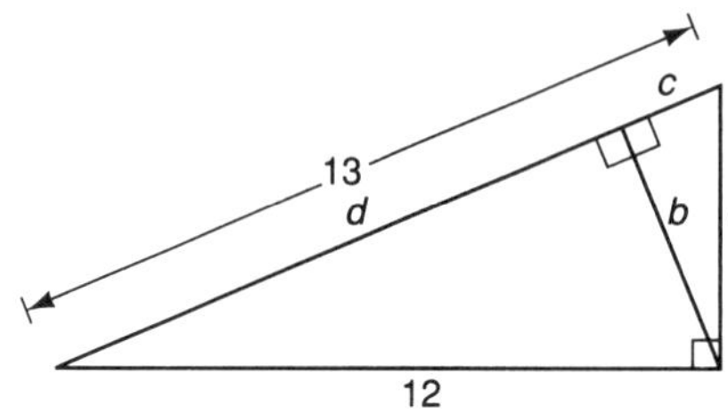

5.

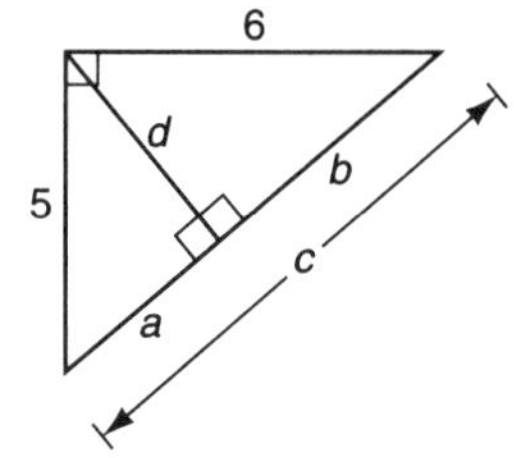

SCALE DRAWINGS

The principles of ratio and proportion are used in making **scale** models. For instance, an architect chooses a particular ratio in order to draw the plans for a building to scale. In a drawing of a room, the scale might be 1 inch = 1 foot. Then the drawing of the room would be 10 inches by 12 inches to represent an actual room that is 10 feet by 12 feet.

To determine the actual size of objects represented by a scale drawing, a proportion is set up with the **scale ratio**:

$$\text{Scale ratio} = \frac{\text{scaled measurements}}{\text{actual measurements}}$$

NOTE: Make sure the measurements in the ratios match. If you have $\frac{\text{cm}}{\text{ft}}$ on the left, you must have $\frac{\text{cm}}{\text{ft}}$ on the right.

EXAMPLE

In this drawing, the scale is 2 cm = 5 feet. What are the actual measurements of the figure?

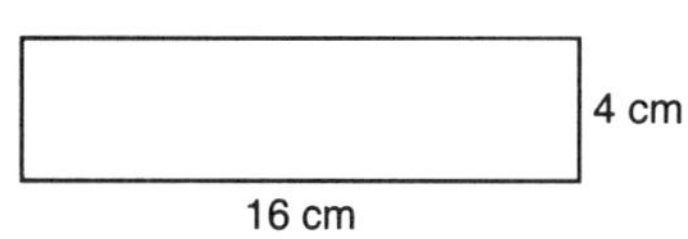

To find width, let w = width. Use: $\frac{2\text{ cm}}{5\text{ ft}} = \frac{4\text{ cm}}{w\text{ ft}}$.
Then $2w = 20$ and $w = 10$.

To find length, let l = length. Use: $\frac{2\text{ cm}}{5\text{ ft}} = \frac{16\text{ cm}}{l\text{ ft}}$.
Then $2l = 80$ and $l = 40$.

The actual dimensions are 10 feet by 40 feet.

We can use this same procedure if we know the actual measurements and we want to find the scale measurements.

MODEL PROBLEMS

1. A city map has a scale of 7 centimeters = 2 kilometers. If a distance on the map is 20 centimeters, what is the actual distance?

Answer: Let x = actual distance. Then:

$$\frac{7\text{ cm}}{2\text{ km}} = \frac{20\text{ cm}}{x}$$

$7x = 40$

$x \approx 5.7$ km

2. The distance from Fort Lauderdale to Naples is approximately 93 miles. If a map has a scale of 1 inch = 60 miles, what is the measurement between these two cities on the map?

Answer: Let x = measurement on the map. Then:

$$\frac{1\text{ inch}}{60\text{ miles}} = \frac{x}{93\text{ miles}}$$

$60x = 93$

$x \approx 1.6$ inches

TRY THESE

Items 1–5: This drawing shows the actual dimensions of a rectangular floor. Anson uses a scale of 3 inches = 10 feet to make a scale drawing. What length would he make each of the following parts?

1. XY

2. AD

3. AB

4. YC

5. PX

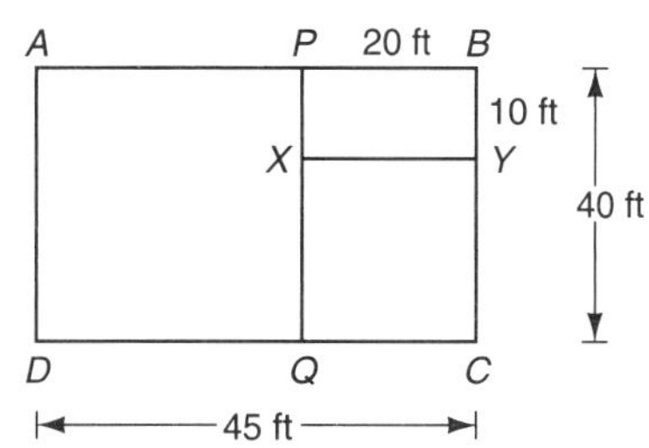

6. The actual distance between West Palm Beach and Tallahassee is 414 miles. If a map has a scale of 1 inch = 24 miles, what is the distance on the map between the two cities?

7. The distance on a map between Ocala and Naples is 17.66 inches. The scale on the map is 1 inch = 15.8 miles. What is the approximate driving distance between these two cities?

8. Tova has a canoe that is 3 feet wide and 12 feet long. She wants to construct a scale model of her canoe with a width of 2 centimeters. What will be the length of her model?

9. Jennifer has a picture of a southern bald eagle in full flight. The wingspan in the picture is 3.5 inches. The scale of the picture is 1 inch to 24 inches. What is the actual wingspan of the bird?

10. In biology, Ethan is studying a model of the human body. From his studies he knows that the small intestine is from 15 to 20 feet in length. The model is $\frac{1}{4}$ the actual size. What is the range of possible lengths of the small intestine on the model?

Similarity

INDIRECT MEASUREMENT USING SIMILAR TRIANGLES

The concepts of similar triangles, ratio, and proportion are used to measure distances indirectly. Real-world examples of indirect measurement include finding the width of a river or the height of a tree.

MODEL PROBLEMS

1. A flagpole casts a 50-foot shadow. At the same time, a man 6 feet tall, standing in front of the flagpole, casts a 7-foot shadow. What is the height of the flagpole?

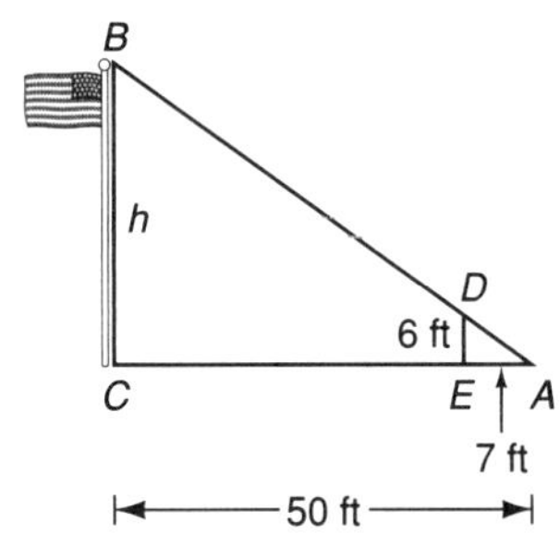

Answer: Let h = height of flagpole. Since triangles ABC and ADE are similar, we can use the ratio:

$$\frac{h}{6} = \frac{50}{7}$$

Then $7h = 300$ and $h \approx 43$ feet

2. Use the diagram below to determine the length of the pond.

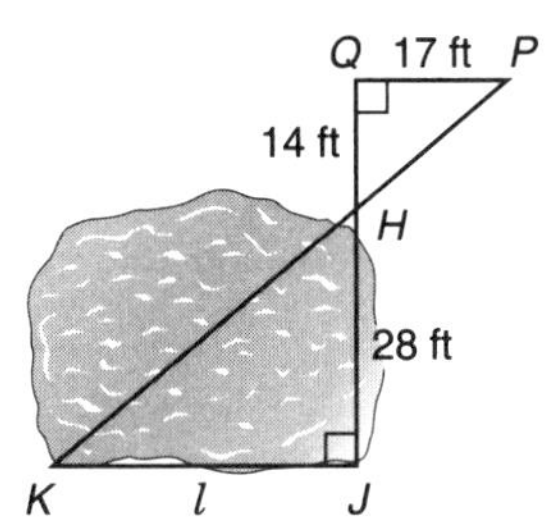

Answer: Let l = length of pond. Since triangles HJK and HQP are similar, we can use the ratio:

$$\frac{l}{17} = \frac{28}{14}$$

Then $14l = 476$ and $l = 34$ feet

TRY THESE ✔

1. A light at ground level shines on a building 20 feet away. When Ty, who is 6 feet tall, stands halfway between the light and the building, his shadow appears on the wall. How high is his shadow?

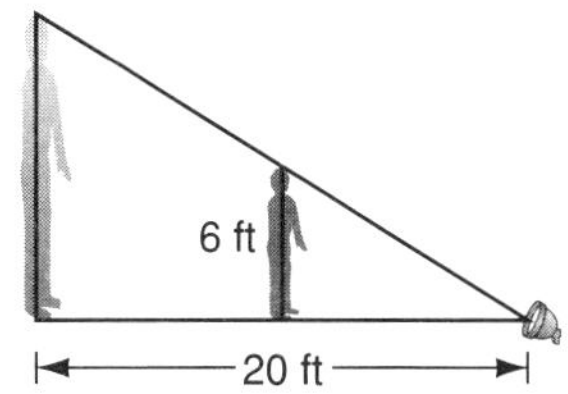

2. Use the diagram below to find AB, the width of the river.

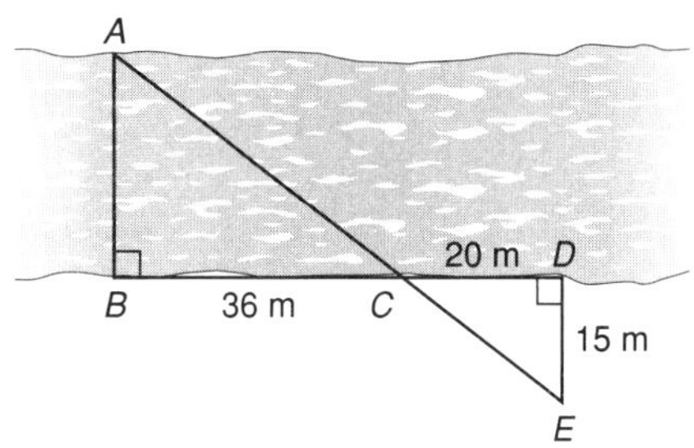

3. The shadow of an object is reflected on a wall by a light at ground level. The object is 4 feet high and 2 feet from the wall. The light is 10 feet from the wall. How high is the shadow?

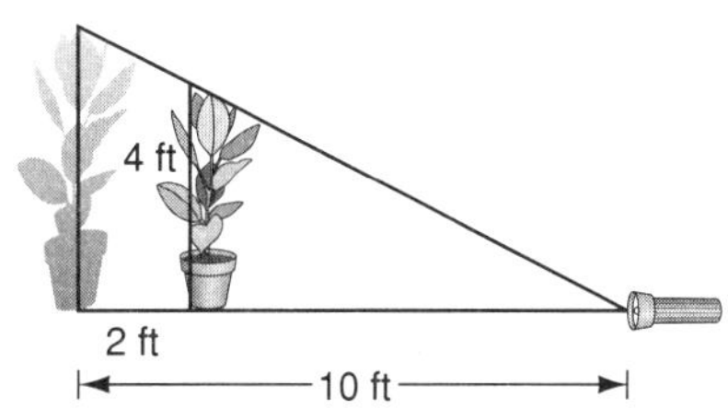

4. Lenore is 5 feet tall and casts an 8-foot shadow. At the same time a palm tree casts a 24-foot shadow. How tall is the palm tree?

5. A flagpole casts a shadow 80 feet long. At the same time, Chris, who is $6\frac{1}{2}$ feet tall, casts a 10-foot shadow. What is the height of the flagpole?

6. A 40-foot pole casts a 32-foot shadow. What is the height of a building that casts a 90-foot shadow at the same time of day?

7. It has been decided that a bridge will be erected from point A to point B. The bridge designer lays out two similar triangles to help with the calculations.

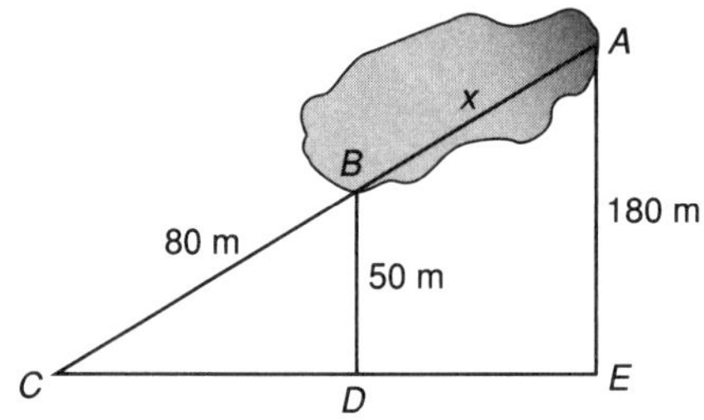

Part A Explain why $\triangle CDB \sim \triangle CEA$.

Part B Write a proportion that can be used to find x.

Part C Solve for x, the length of the bridge.

8. Mark wants to find out the size of the giant screen at the stadium. He holds an 8-inch cup 3 feet in front of his eyes. He can see both the bottom and top of the screen. The screen is h feet high and he is d feet from the screen.

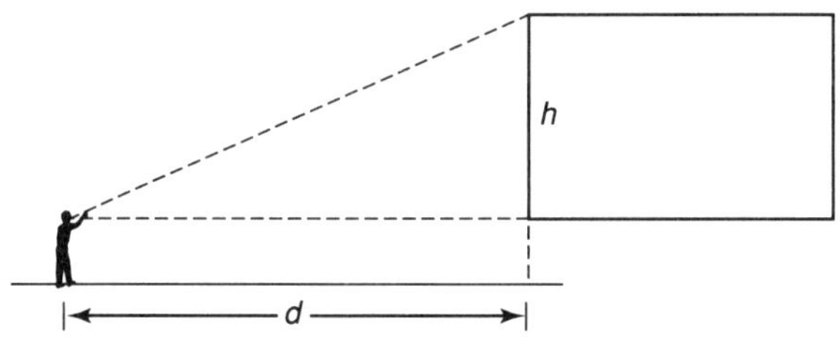

Part A Write a proportion that will represent the information that he has collected.

Part B If Mark is 180 feet from the screen, how tall is the screen?

AREAS AND VOLUMES OF SIMILAR FIGURES

There is an important relationship between the measurements of similar figures. Part of the definition of similar figures is that their corresponding sides are proportional. That is, there is a **scale factor** relating the sides of two similar figures.

This scale factor is the same for any corresponding linear (line) parts of two similar figures. For example, if two triangles are similar, their corresponding altitudes will be in the same ratio (scale factor) as a pair of corresponding sides.

EXAMPLE

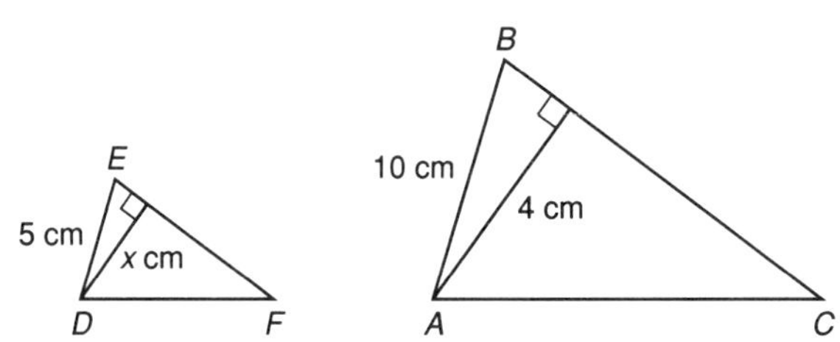

Given $\triangle ABC \sim \triangle DEF$, the scale factor would be 2:1, because $AB = (2)(DE)$.

The corresponding altitudes would have the same scale factor.

Therefore: $\frac{2}{1} = \frac{4}{x}$ so $2x = 4$ and $x = 2$

To compare the **areas of two similar figures**, we square the scale factor, since area is measured in square units.

EXAMPLE

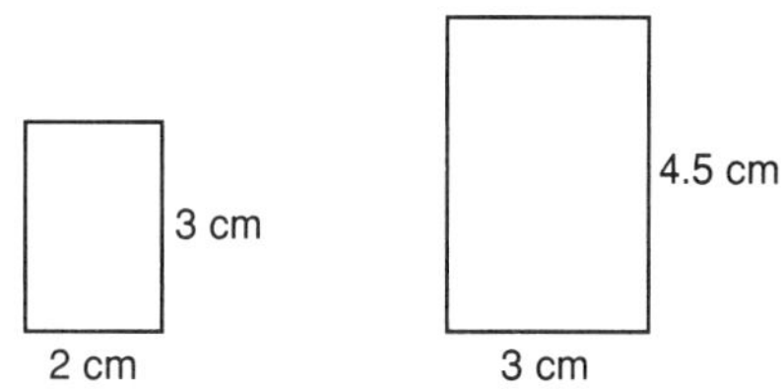

These two rectangles are similar. If the scale factor is 2:3, the areas of the two rectangles will be in the ratio of $2^2 : 3^2$ or $4 : 9$.

To demonstrate this, use the formula for area of a rectangle: $A = lw$. For the smaller rectangle $A = 2$ cm • 3 cm = 6 square cm. For the larger rectangle $A = 3$ cm • 4.5 cm = 13.5 square cm. Set up a proportion relating these areas to the scale factor for the areas, 4 : 9.

$$\frac{4}{9} = \frac{6}{13.5}$$

$$13.5 \bullet 4 = 9 \bullet 6$$

$$54 = 54$$

We can also compare the **volumes of two similar figures** by using the scale factor. Volume is measured in cubic units, so to compare the volumes of two similar figures we use the cube of the scale factor. If the scale factor of two solids is $a : b$, the ratio of their volumes will be $a^3 : b^3$.

EXAMPLE

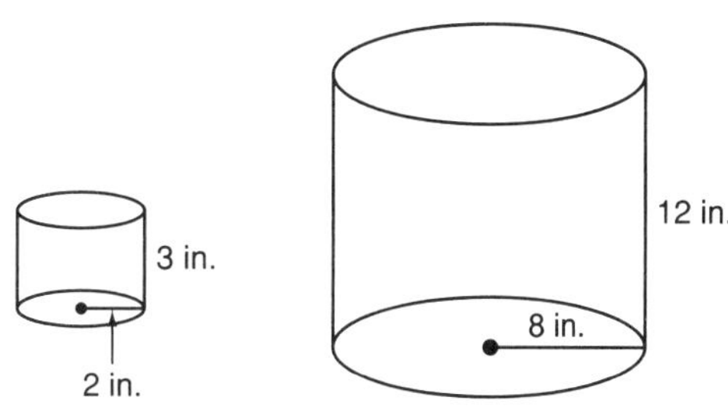

These two cylinders are similar. The scale factor is 1:4. Therefore, the ratio of their volumes is $1^3 : 4^3$ or 1:64. In other words, the volume of the larger cylinder is 64 times the volume of the smaller one.

Check by using the formula for the volume of a cylinder $V = \pi r^2 h$.

Small cylinder: $V = \pi \bullet 2^2 \bullet 3 = 12\pi$

Large cylinder: $V = \pi \bullet 8^2 \bullet 12 = 768\pi$

Ratio: $\frac{12\pi}{768\pi} = \frac{1}{64}$

MODEL PROBLEMS

1. The ratio of the altitudes of two similar triangles is 3:5. What is the ratio of their areas?

Answer: $\frac{3^2}{5^2} = \frac{9}{25}$

2. The sides of two similar regular pentagons are in the ratio of 2:7. If the area of the smaller pentagon is 8 square centimeters, what is the area of the larger pentagon?

Answer: Let A = area of the larger pentagon. Set up a proportion:

$$\frac{2^2}{7^2} = \frac{4}{49} = \frac{8}{A}$$

$4A = 392$

$A = 98$ sq cm

3. The ratio of the areas of the faces of two cubes is 9:16. What is the ratio of their volumes?

Answer: The linear ratio would be $\frac{3}{4}$. Therefore:

Volume ratio $= \frac{3^3}{4^3} = \frac{27}{64}$

TRY THESE

Items 1–4: Use the two similar triangles shown below.

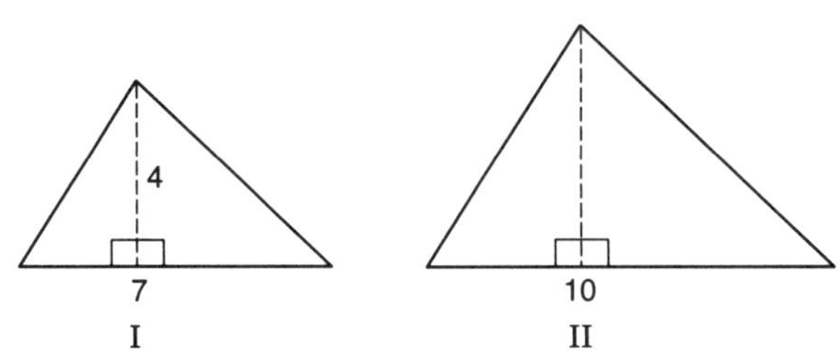

1. What is the ratio of the corresponding sides of △I to △II?

2. What is the ratio of the altitudes of △I to △II?

3. What is the ratio of the area of △I to the area of △II?

4. What is the area of △II? Round your answer to the nearest hundredth.

Items 5 and 6: The area of △III, which is simlar to △I above is 56 square units.

5. What is the ratio of the area of △I to the area of △III?

6. Find the measures of the altitude and the base of △III?

Items 7–12: The dimensions of rectangular box I are one-third the size of the dimensions of rectangular box II. Box II has dimensions of 6 feet by 4 feet by 2 feet.

7. What is the length, width, and height of box I?

8. What is the ratio of a corresponding pair of sides of box I to box II?

9. Find the surface area of each box.

10. What is the ratio of the surface area of box I to the surface area of box II?

11. Find the volume of each box.

12. What is the ratio of the volume of box I to the volume of box II?

13. Cubes P and Q are similar. The volume of P is 27 times the volume of Q. What is the length of y?

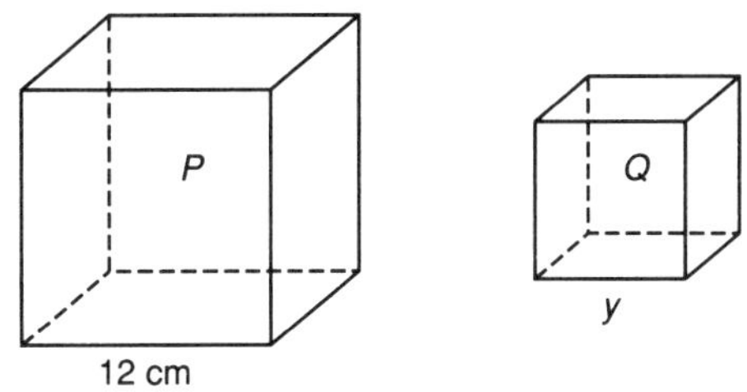

14. If the area of $\triangle X$ is 20 sq mm, what is the area of $\triangle Y$?

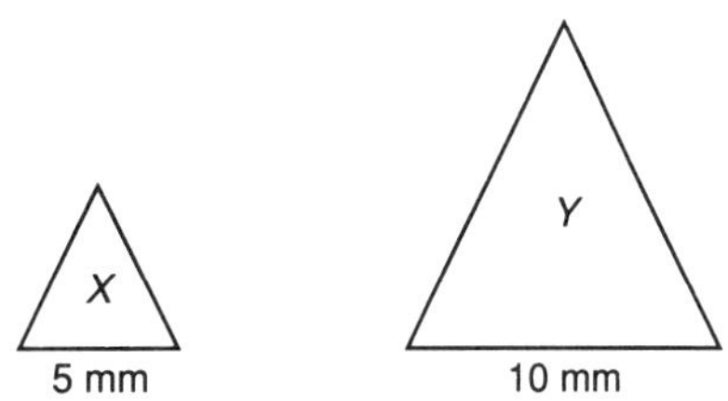

15. What is the measure of the edge of a cube that has a volume 8 times the volume of the cube shown below?

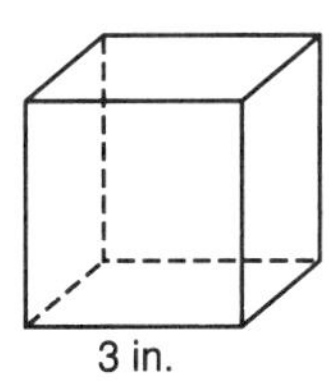

Assessment

Multiple-Choice Problems

1. In the drawing below, $\overline{CD}$ is an altitude of right triangle ABC. If $AD = 5$ and $DB = 7$, what is the length of $\overline{DC}$?

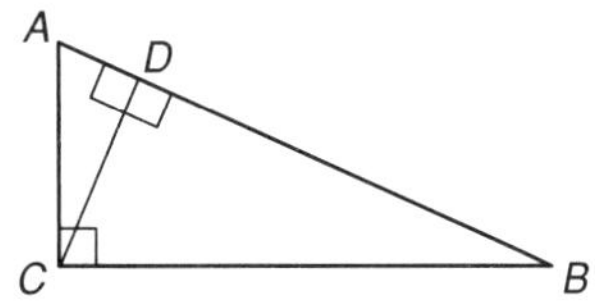

A. 1.4 **C.** 12

B. 5.9 **D.** 35

2. This diagram is part of a map.

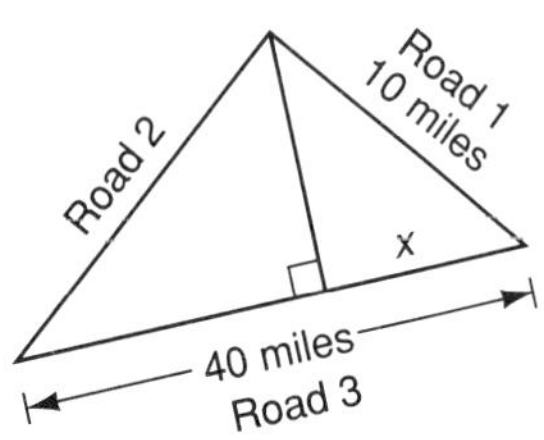

The angle formed by Road 1 and Road 2 is a right angle. Road 3 is 40 miles long and connects Roads 1 and 2. A new road is to be constructed directly from the intersection of Roads 1 and 2 to Road 3. How far will it be from the intersection of Roads 1 and 3 to the new road?

A. 2.5 miles **C.** 10.5 miles

B. 5 miles **D.** 20 miles

3. The cones below are similar. If the radii of the cones are 5 cm and 2 cm, what is the ratio of their volumes?

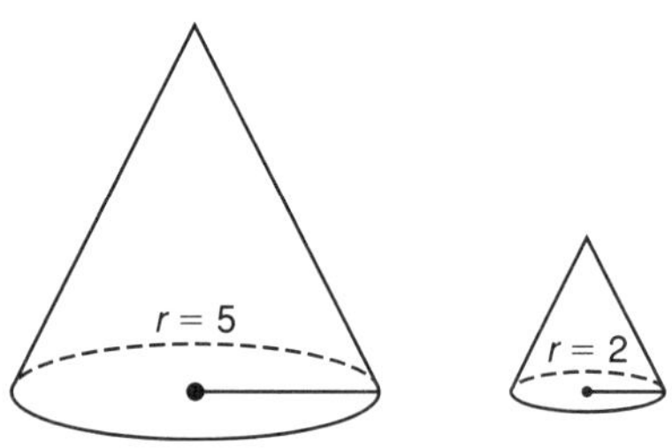

A. 5:2 **C.** 25:4

B. 15:6 **D.** 125:8

4. What is the ratio of the surface areas of the cones in problem 3?

A. $\frac{5}{2}$

B. $\frac{15}{6}$

C. $\frac{25}{4}$

D. $\frac{125}{8}$

5. The edge of one cube is one-third the size of the edge of another cube. If the volume of the smaller cube is 5 cu in., what is the volume of the larger cube?

A. 405 cu in.

B. 135 cu in.

C. 45 cu in.

D. 15 cu in.

6. Shown below, quadrilateral $RSTU \sim$ quadrilateral $YXWZ$. What is the length of TU?

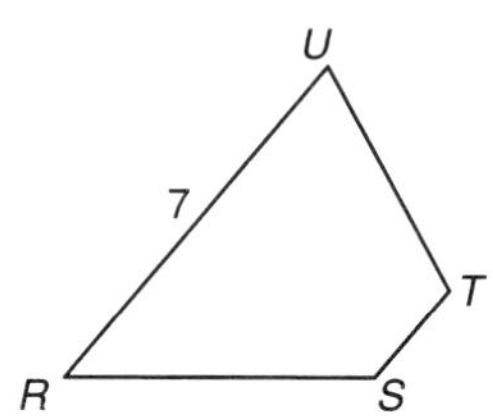

A. 1.75

B. 5.25

C. 7

D. 8.75

Gridded-Response Problems

7. An architect is making a scale drawing of a floor plan where 1 inch = 18 feet. If an opening between two rooms measures $\frac{5}{8}$ inch on the scale drawing, what is the measurement of the actual opening?

8. Dwayne leans a 20-foot ladder against his house. The ladder makes an angle of 60 degrees with the ground. To support the ladder, Dwayne uses a brace (BD) from the edge of the house perpendicular to the ladder. How long is the brace (BD), to the nearest hundredth?

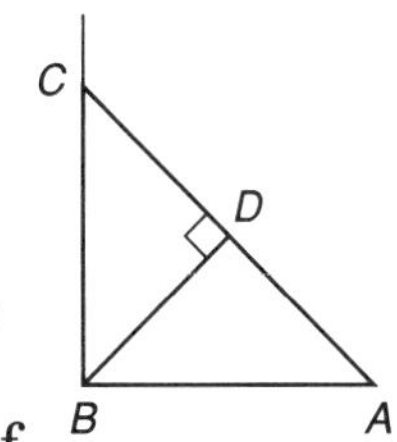

9. Gutzon Borglum, the sculptor of Mount Rushmore, created a scale model of the four presidents' faces that were to be carved into the mountain. He used a scale of 1 inch to 12 feet. (That is, 1 inch on the model equaled 12 feet on the mountain.) If the presidents' noses are each 20 feet long on Mount Rushmore, how long is each nose on the scale model?

10. Use the information in problem 9. If the heads on the scale model were each 5 inches tall, how tall is each head on Mount Rushmore?

11. Shown below, pentagon $ABCDE \sim$ pentagon $FGHIJ$. The area of the larger pentagon is 376 square units. What is the area of the smaller pentagon?

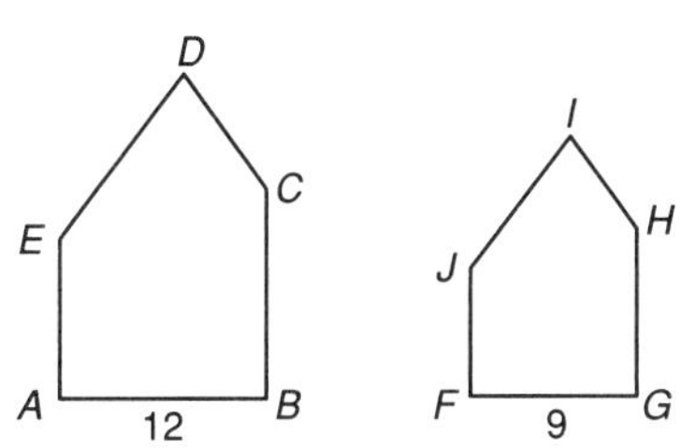

12. The two cylinders shown below are similar. If the volume of the smaller one is 16 cubic units, what is the volume of the larger one?

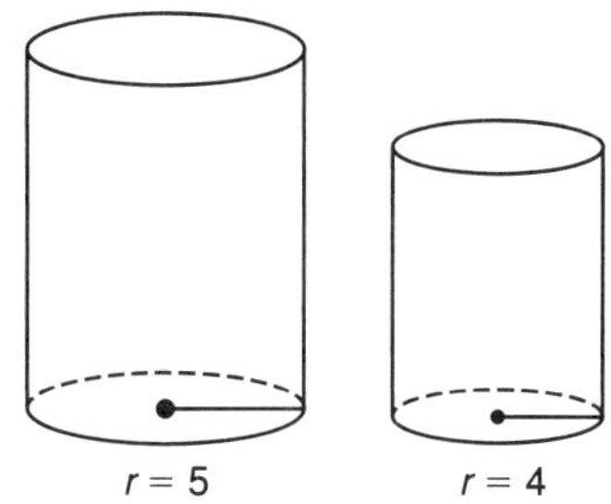

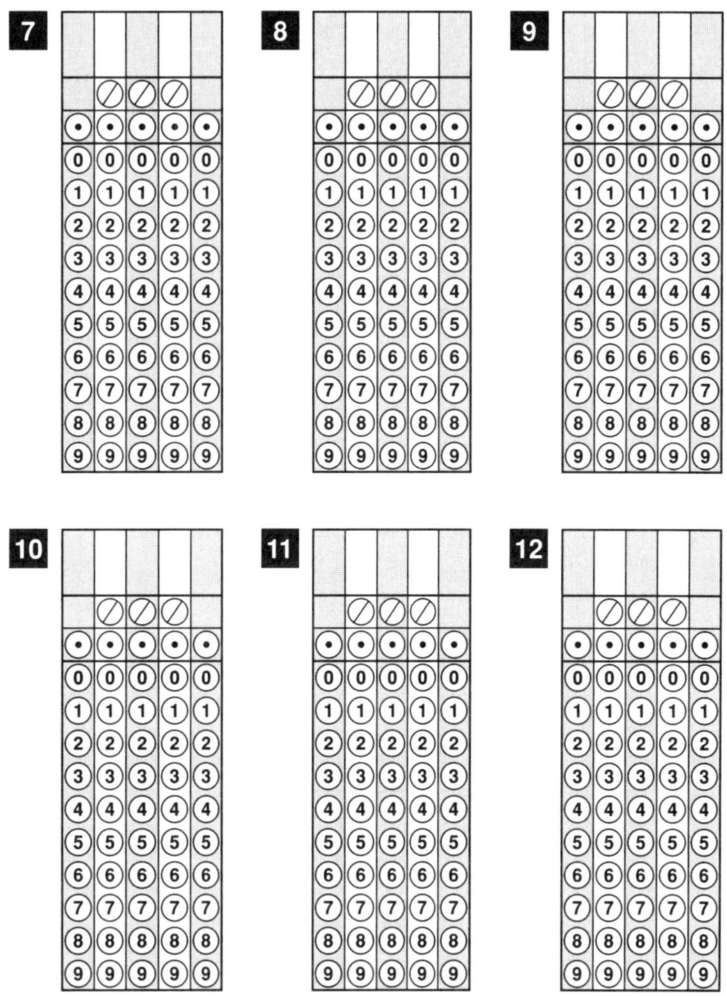

Think, Solve, Explain: Extended-Response Problems

13. According to an ordinance in Central City, no sign can be erected that is more than 22 feet high. Burger Palace has just put up a new sign, and the city manger has come to inspect it. To measure the height of the sign, the manager, who is 6 feet tall, stands near the sign pole. He measures the shadow of the pole on the ground and then measures his own shadow. The drawing below shows his measurements.

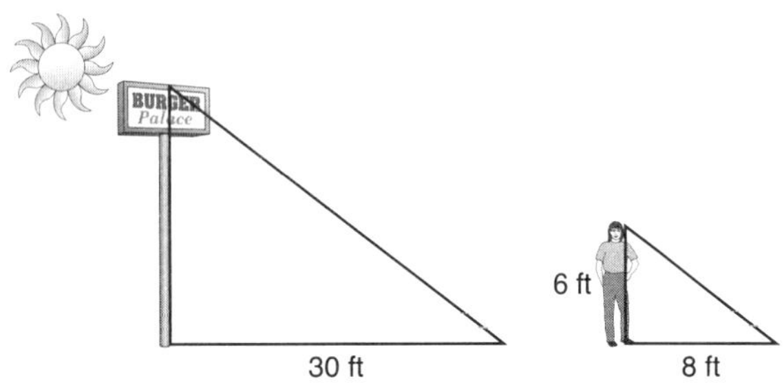

Part A Using geometric concepts, explain why these two triangles are similar.

Part B Set up a proportion that can be used to find the height of the sign.

Part C Solve the proportion to find the height of the sign. Can the sign stay, or will it have to be taken down?

14. $\overline{BE}$ is parallel to $\overline{CD}$. The area of trapezoid $BEDC$ is 45 square units. $AE = 4$ units. $ED = 2$ units.

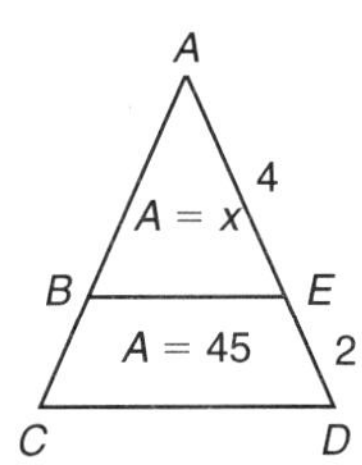

Part A Using geometric concepts, explain why $\triangle ABE \sim \triangle ACD$.

Part B What is the ratio of corresponding sides?

Part C If x is the area of $\triangle ABE$, express the area of $\triangle ACD$ in terms of x.

Part D Set up a proportion comparing the areas.

Part E Solve the proportion for x, the area of the smaller triangle.

15. Marsha is trying to find the distance across a lake from point A to point B. She has drawn two triangles with $\overline{CD}$ and $\overline{AB}$ both perpendicular to $\overline{EA}$.

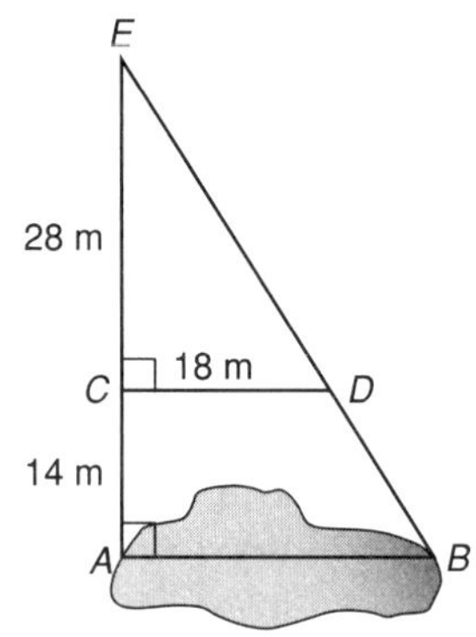

Part A Explain why $\triangle ECD \sim \triangle EAB$.

Part B Write a proportion that can be used to find the distance between point A and point B.

Part C Solve to find AB.

B APPLYING AND VERIFYING PROPERTIES OF TWO-DIMENSIONAL FIGURES INCLUDING DISTANCE, MIDPOINT, SLOPE, PARALLELISM, AND PERPENDICULARITY USING RECTANGULAR COORDINATE SYSTEMS AND FINITE GRAPHS

Coordinate (Cartesian) plane

To locate a point on a plane, a reference point is needed. This requires **coordinate geometry**.

On a plane, draw a horizontal line, referred to as the ***x*-axis**. Perpendicular to the x-axis, draw a vertical line, known as the ***y*-axis**. The point at which these axes intersect is called the **origin**.

The x-axis and y-axis are number lines, with the origin taking the position of zero. On the x-axis, the numbers to the right of zero are positive, and the numbers to the left of zero are negative. On the y-axis, the numbers above the origin are positive, and the numbers below the origin are negative.

The intersection of the two axes divides the plane into four **quadrants**, numbered I, II, III, and IV.

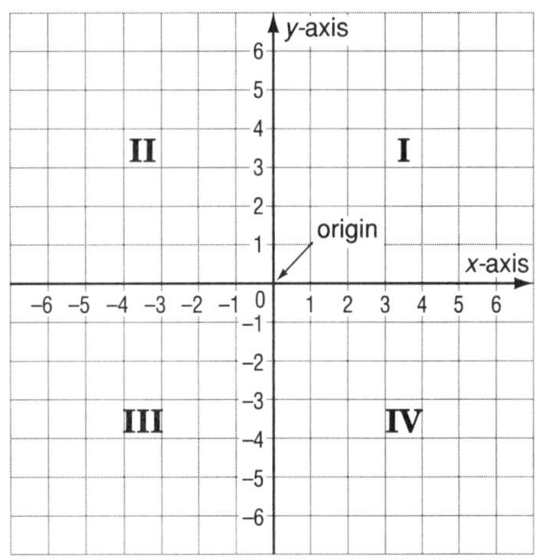

To locate a point on the plane, we use an **ordered pair** of the form (x, y). The first number refers to the position in the x direction. The second number refers to the position in the y direction.

EXAMPLE

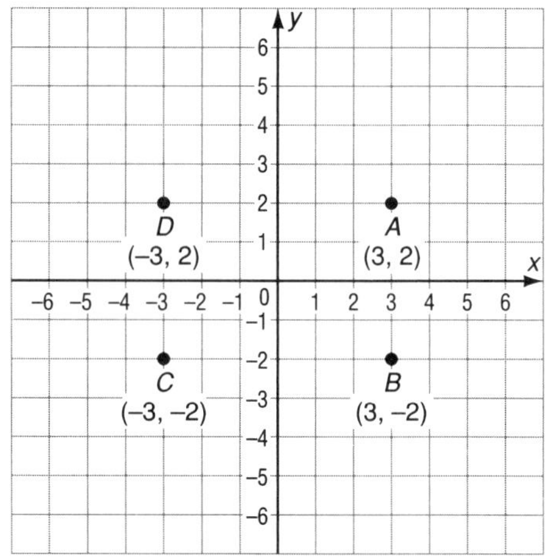

On the graph above, point A is indicated by the ordered pair (3, 2).

Point B is indicated by the ordered pair (3, −2).

Point C is indicated by the ordered pair (−3, −2).

Point D is indicated by the ordered pair (−3, 2).

MODEL PROBLEMS

1. Plot the following ordered pairs: E (−5, 2); F (0, −2); G (2, −5); H (−3, 0)

Answer:

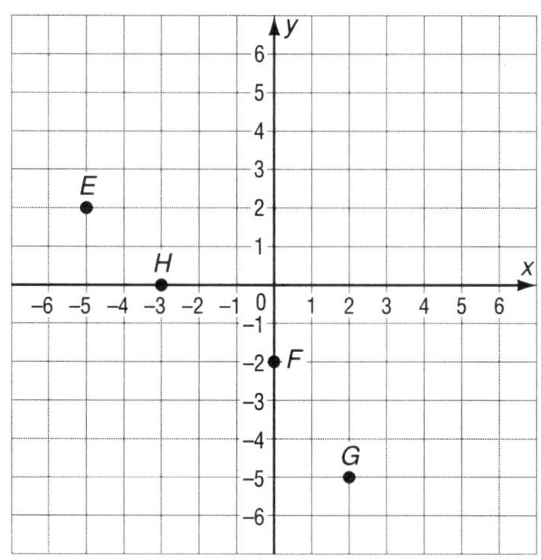

2. Using the graph below, give the ordered pairs of points M, N, and P.

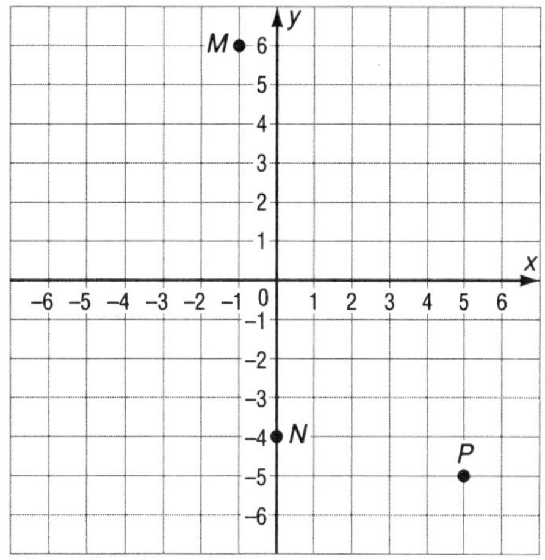

Answer: $M\ (-1, 6)$; $N\ (0, -4)$; $P\ (5, -5)$

TRY THESE

1. Plot the following ordered pairs: $E\ (-4, -2)$; $F\ (3, 0)$; $G\ (5, -2)$; $H\ (0, -2)$.
2. Using the graph below, give the ordered pairs of points M, N, and P.

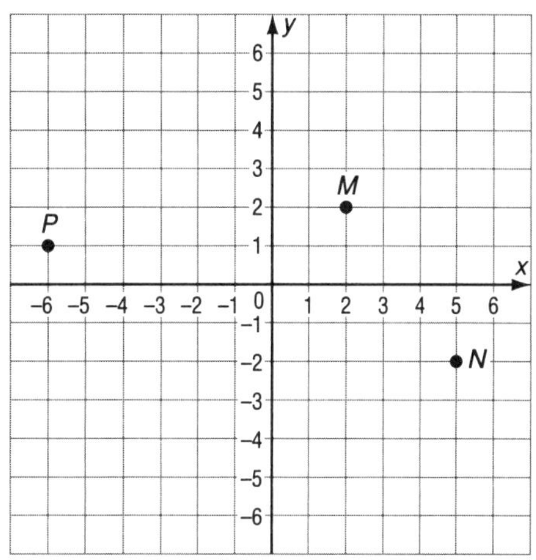

3. Using the graph below, give the ordered pairs of points A, B, C, D, E, and F.

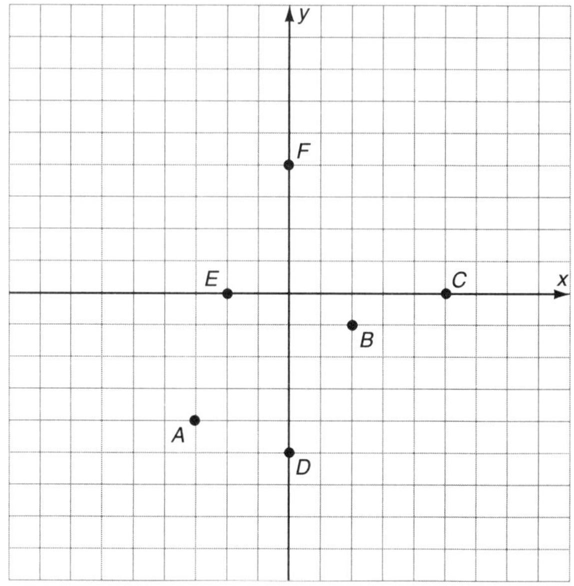

Distance between two points

To find the distance between two points on a coordinate graph, first determine if the points are on the same horizontal or vertical line.

If the points are on the same horizontal line, simply find the difference of the x-coordinates. If the points are on the same vertical line, find the difference of the y-coordinates. In each case, the difference is the distance between the two points. Remember that this distance must be a positive number. If the difference of the coordinates comes out negative, use the absolute value.

If the points are not on a vertical or horizontal line but rather on a diagonal line, you need to use a **distance formula**. This formula is based on the Pythagorean theorem. Given points (x_1, y_1) and (x_2, y_2), the distance d between them is:

$$d = \sqrt{(x_2 - x_1)^2 + (y_2 - y_1)^2}$$

MODEL PROBLEMS

1. Find the distance d between A (3, 5) and B (−2, 5).

Answer:

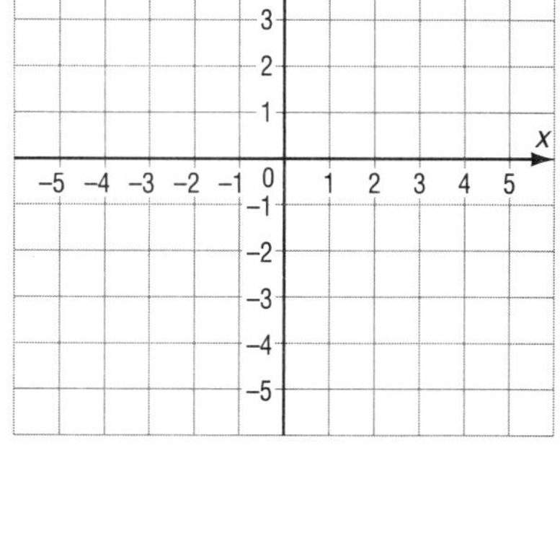

Points A and B have the same y-coordinate. Therefore, they lie on the same horizontal line.

Subtract: $d = 3 - (-2) = 5$ units

2. Find the distance d between $C\ (2, -3)$ and $D\ (2, 6)$.

Answer:

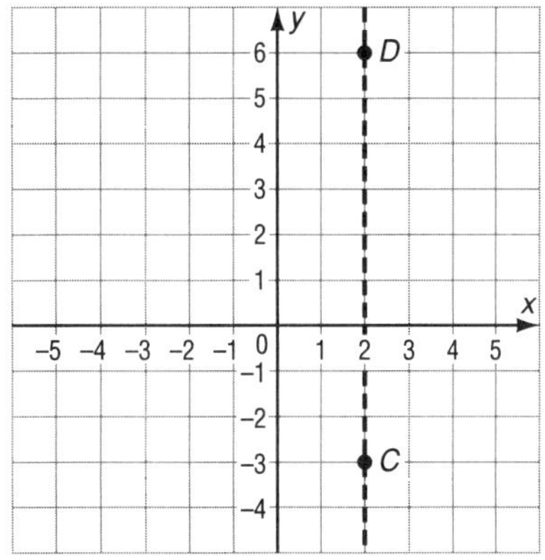

Points C and D have the same x-coordinate. Therefore, they lie on the same vertical line.

Subtract: $d = -3 - 6 = -9$ and $|-9| = 9$ units

3. Find the distance between $G\ (-4, 2)$ and $H\ (2, -1)$.

Answer:

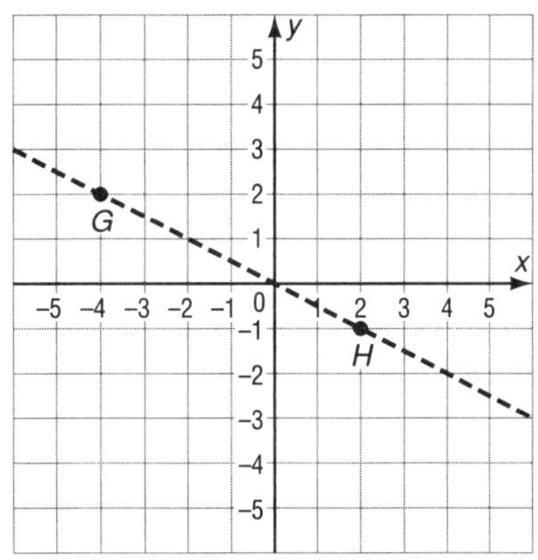

Points G and H do not lie on a horizontal or vertical line. Therefore, use the distance formula:

$$d = \sqrt{(x_2 - x_1)^2 + (y_2 - y_1)^2}$$

$$d = \sqrt{(-4 - 2)^2 + [2 - (-1)]^2}$$

$$d = \sqrt{(-6)^2 + 3^2} = \sqrt{45} = 3\sqrt{5} \approx 6.71$$

TRY THESE

Find the distance between the points in each pair.

1. (2, 1) and (2, 6)

2. (4, −3) and (0, −3)

3. (2, −3) and (−2, 0)
4. (3, 2) and (−1, −2)
5. (5, 0) and (−4, −1)
6. (−1, 3) and (4, −2)
7. (−7, 2) and (0, −5)
8. (−5, −4) and (8, −3)
9. (0, 6) and (1, −9)
10. (−3, −7) and (−8, −4)

Midpoint formula

The **midpoint** between two points is the point located halfway between them. Finding the midpoint can be thought of as finding the average of the x-coordinates and the average of the y-coordinates. For points (x_1, y_1) and (x_2, y_2), the midpoint is:

$$\left(\frac{x_1 + x_2}{2}, \frac{y_1 + y_2}{2}\right)$$

EXAMPLE

The midpoint of A (4, −6) and B (−2, 4) is:

$$\left(\frac{4 + (-2)}{2}, \frac{-6 + 4}{2}\right) = \left(\frac{2}{2}, \frac{-2}{2}\right) \text{ or } (1, -1)$$

MODEL PROBLEMS

1. Find the midpoint of the line segment with the endpoints (2, −7) and (0, −4).

   Answer: $\frac{2 + 0}{2} = 1$ and $\frac{-7 + (-4)}{2} = -\frac{11}{2}$

   Midpoint: (1, −5.5)

2. Find the midpoint of the line segment with the endpoints (0, −6) and (5, 8).

   Answer: $\frac{0 + 5}{2} = 2.5$ and $\frac{-6 + 8}{2} = 1$

   Midpoint: (2.5, 1)

3. Find the endpoint B of line segment AB if the coordinates of A are (3, 5) and the midpoint of $\overline{AB}$ is (6, 8).

   Answer:

   $2(6) = 12$ and $12 - 3 = 9$ $\left(\text{Algebraically, solve } \frac{x + 3}{2} = 6\right)$

   $2(8) = 16$ and $16 - 5 = 11$ $\left(\text{Algebraically, solve } \frac{y + 5}{2} = 8\right)$

   B (9, 11)

TRY THESE

1. Find the midpoint of a line segment with the endpoints (−2, 2) and (6, −4).
2. Find the midpoint of a line segment with the endpoints (−3, 7) and (−7, −5).
3. Find the other endpoint of the line segment *CD* if the coordinates of *D* are (−3, 8) and the midpoint of $\overline{CD}$ is (3, 9).
4. Find the midpoint of a line segment with the endpoints (−4, 6) and (3, 4).
5. Find the midpoint of a line segment with the endpoints (−2, −5) and (−4, 5).
6. Find the other endpoint of the line segment *CD* if the coordinates of *D* are (2, 9) and the midpoint of $\overline{CD}$ is (5, 1).
7. Find the midpoint of a line segment with the endpoints (6, −2) and (−7, −9).
8. Find the midpoint of a line segment with the endpoints (4, 7) and (−2, −1).
9. Find the other endpoint of the line segment *CD* if the coordinates of *D* are (6, −2) and the midpoint of $\overline{CD}$ is (1, 8).
10. Find the other endpoint of the line segment *CD* if the coordinates of *D* are (−4, 3) and the midpoint of $\overline{CD}$ is (2, −5).

Slope of a line

The steepness of a line on a coordinate plane is called its **slope**. Slope is measured as a ratio: the vertical (y) change between any two points on a line divided by the horizontal (x) change between the points. Vertical change is called **rise** and horizontal change is called **run**, so we can say that slope $= \frac{\text{rise}}{\text{run}}$ or "rise over run."

Rise (vertical change) is the difference of the y-coordinates, and run (horizontal change) is the difference of the x-coordinates. So the formula for slope is:

$$m = \frac{y_2 - y_1}{x_2 - x_1}$$

The m is the symbol for slope, and the formula is read as "slope equals the change in y divided by the change in x."

If the vertical change (rise) is a greater amount than the horizontal change (run), the line is steeper. If the horizontal change (run) is greater than the vertical change (rise), the line is less steep.

If the line goes up from left to right, the slope is **positive**. If the line goes down from left to right, the slope is **negative**.

A **horizontal line** has a slope of 0, since there is no vertical change.

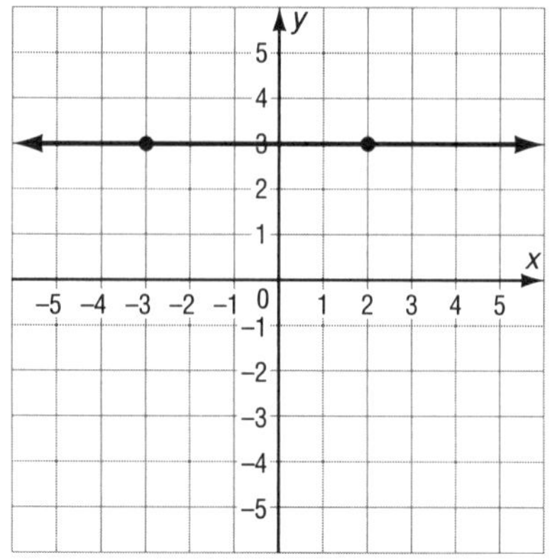

In the graph above, points (2, 3) and (−3, 3) lie on a horizontal line: $m = \frac{3 - 3}{-3 - 2} = \frac{0}{-5} = 0.$

A **vertical line** is said to have an **undefined slope**. This is because there is no change in x and division by 0 is undefined.

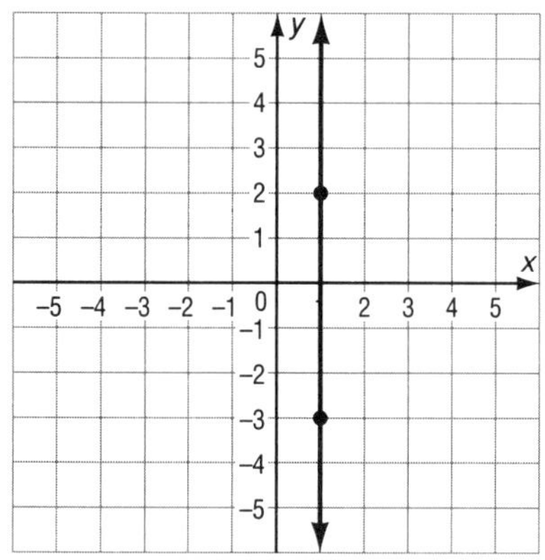

In the graph above, points (1, 2) and (1, −3) lie on a horizontal line: $m = \frac{-3 - 2}{1 - 1} = \frac{-5}{0}$ (undefined).

MODEL PROBLEMS

1. Find the slope m of the line containing points (3, −2) and (−5, 4).

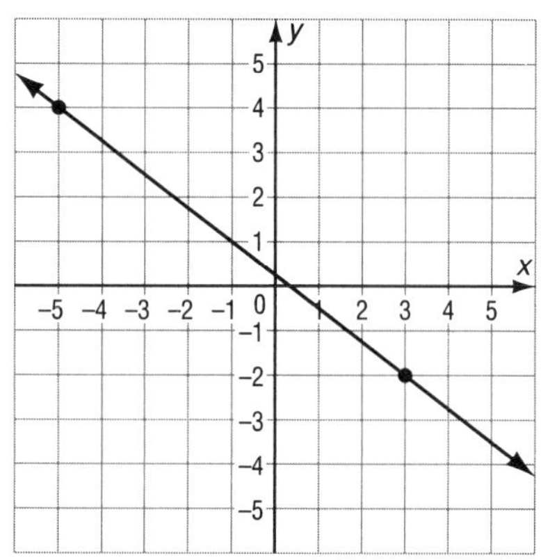

Answer: $m = \frac{4 - (-2)}{-5 - 3} = \frac{6}{-8} = -\frac{3}{4}$

Items 2 and 3: Find the slope m of the line connecting each pair of points.

2. (2, 5) and (−1, 5)

Answer: $m = \frac{5 - 5}{-1 - 2} = \frac{0}{-3} = 0$

3. (6, 3) and (6, −2)

Answer: $m = \frac{-2 - 3}{6 - 6} = \frac{-5}{0}$ (undefined)

TRY THESE

Find the slope m of the line containing the following pairs of points. State whether the slope is positive, negative, 0, or undefined.

1. (2, 1) and (5, 4)

2. (−1, 4) and (−4, 1)

3. (3, −2) and (−1, −5)

4. (0, −2) and (−7, 4)

5. (6, 3) and (−4, 1)

6. (−9, 2) and (−1, −2)

7. (−2, 5) and (3, 5)

8. (4, −3) and (−6, −2)

9. (−3, 7) and (1, 3)

10. (−4, −2) and (−4, 8)

Slopes of parallel and perpendicular lines

Two or more **parallel lines** have the same slope.

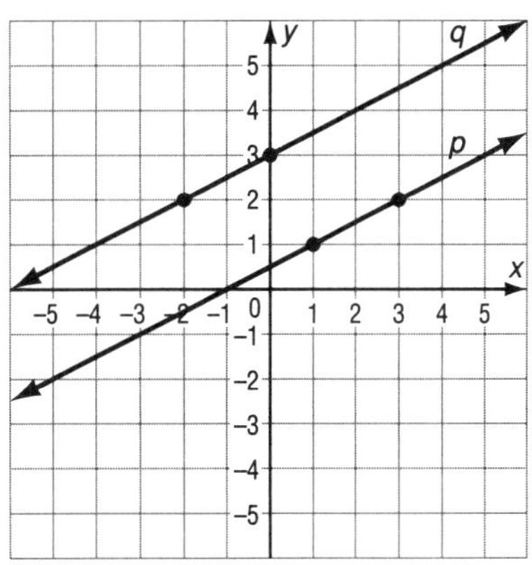

In the graph above, line p contains points (1, 1) and (3, 2) and $m = \frac{2 - 1}{3 - 1} = \frac{1}{2}$. Line q contains points (0, 3) and (−2, 2) and $m = \frac{2 - 3}{-2 - 0} = \frac{-1}{-2} = \frac{1}{2}$. Therefore, $p \parallel q$.

For two **perpendicular lines**, one slope is the negative reciprocal of the other slope.

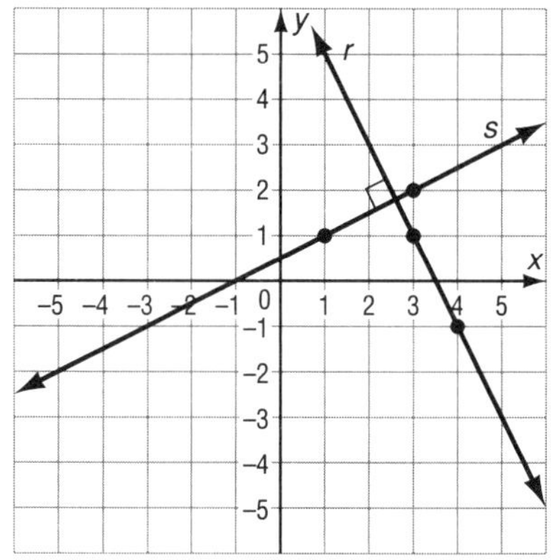

In the graph above, line s contains points (1, 1) and (3, 2) and $m = \frac{2-1}{3-1} = \frac{1}{2}$. Line r contains points (4, −1) and (3, 1) and $m = \frac{1-(-1)}{3-4} = \frac{2}{-1}$ or $-\frac{2}{1}$. These two slopes are negative reciprocals of each other. Therefore $r \perp s$.

MODEL PROBLEM

Determine the slopes of lines AB, CD, GH, and JK. Which lines are parallel? Which lines are perpendicular?

A (3, 8) and B (1, 4)

C (0, −6) and D (4, −5)

G (5, 1) and H (1, 0)

J (3, 5) and K (5, 4)

Answer: Slope of $\overleftrightarrow{AB} = \frac{2}{1}$. Slope of $\overleftrightarrow{CD} = \frac{1}{4}$. Slope of $\overleftrightarrow{GH} = \frac{1}{4}$.

Slope of $\overleftrightarrow{JK} = \frac{-1}{2}$ or $-\frac{1}{2}$.

Therefore:

$\overleftrightarrow{AB} \perp \overleftrightarrow{JK}$ (negative reciprocal slopes)

$\overleftrightarrow{CD} \parallel \overleftrightarrow{GH}$ (equal slopes)

TRY THESE

Items 1–4: Find the slope m of the line connecting each pair of points.

1. A (4, 8) and B (1, 4)
2. C (1, −6) and D (4, −2)
3. G (3, 3) and H (1, 0)

4. $J\,(6, -1)$ and $K\,(3, 1)$
5. State which lines in problems 1–4 are parallel and which are perpendicular.

Items 6 and 7: Plot the points $A\,(-3, -2)$, $B\,(2, -1)$, $C\,(5, 2)$, and $D\,(0, 1)$. Connect the points to form quadrilateral *ABCD*.

6. Find the slope of each side of the quadrilateral.
7. Find the length of each side and determine what type of figure is quadrilateral *ABCD*.

Items 8–10: Plot the points $E\,(-3, 4)$, $F\,(1, 1)$, $G\,(-2, -3)$, and $H\,(-6, 0)$. Connect the points to form quadrilateral *EFGH*.

8. Find the slope of each side.
9. Find the length of each side.
10. What type of quadrilateral is *EFGH*? Explain your answer.

NONLINEAR GRAPHS

When the graph of a set of points forms a straight line, it is called **linear**. But not all relations are linear. Some graphs have **nonlinear** shapes.

A relation that is **quadratic** will form a **parabola** when graphed, and so it is said to be **parabolic**. For the equation $y = x^2$, the coordinates would include (0, 0), (1, 1), (2, 4), (−1, 1), (−2, 4) and the graph would be the one shown below.

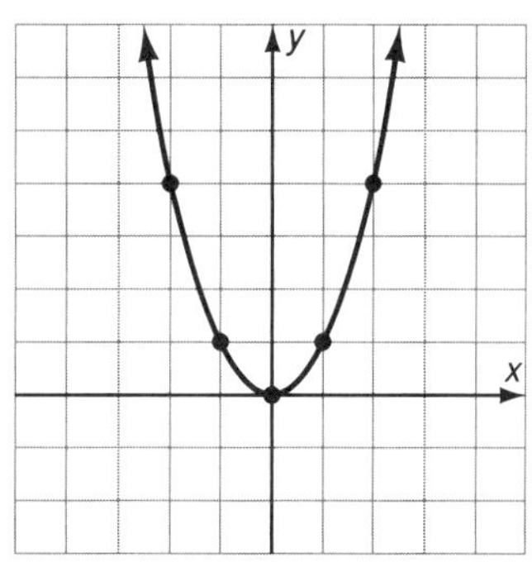

The lowest or turning point of this graph is (0, 0) and is called the **vertex**. Notice that this graph is symmetric to the y-axis. In this case the y-axis is called the **axis of symmetry** and can be written as the vertical line $x = 0$.

NOTE: The vertex of a parabolic graph may be a high point instead of a low point when the graph opens downward.

The graph of $y = x^2 + 2$ would have the same shape as $y = x^2$, but translated (or "slid") up two units. The vertex of this parabola is (0, 2) and the axis of symmetry is $x = 0$.

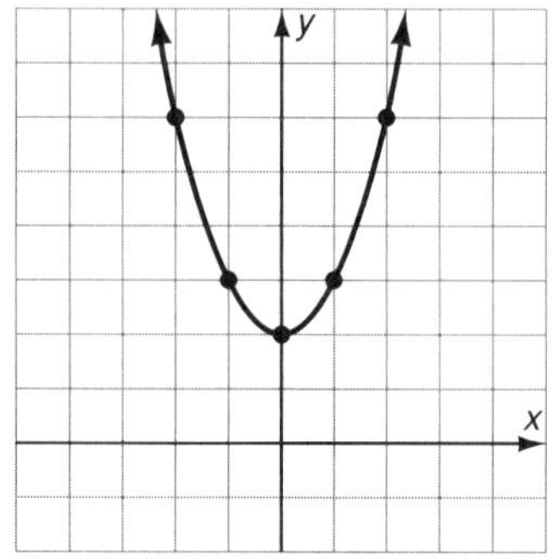

The graph of $y = (x - 2)^2$ would be the same as $y = x^2$, but translated (slid) two units to the right. The vertex of this parabola is (2, 0) and the axis of symmetry is $x = 2$.

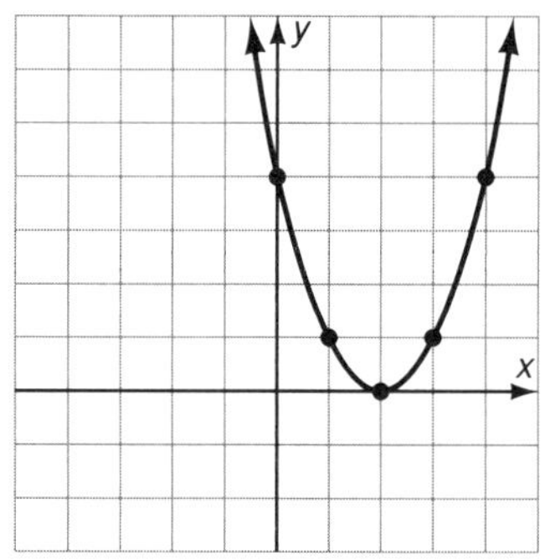

MODEL PROBLEMS

1. What is the vertex (the lowest point) on this graph?

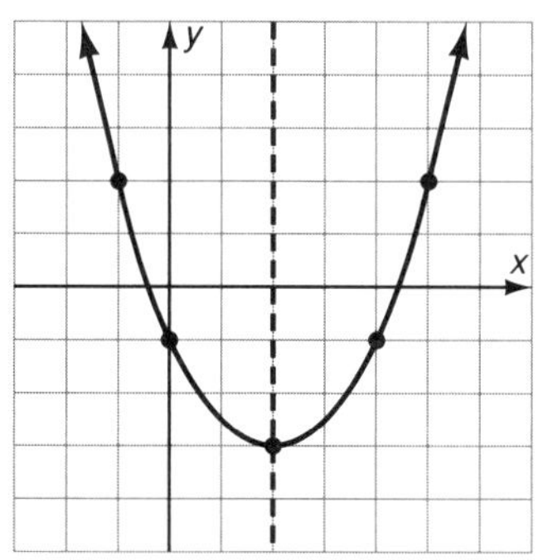

Answer: (2, −3)

2. What is the line of symmetry on the graph?

Answer: $x = 2$

TRY THESE

Items 1–3: Use the graph shown below.

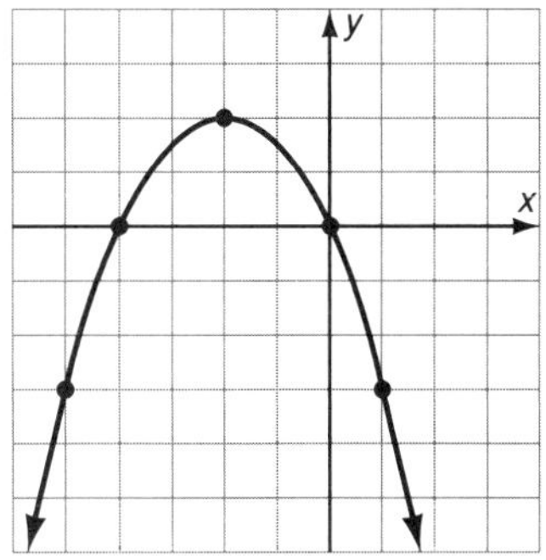

1. What is the vertex (the highest point) on this graph?
2. What is the line of symmetry on the graph?
3. For what values of x are the y values greater than 0?
4. Using the graph in the model problem, for what values of x are the y values negative?

Items 5 and 6: The graph below represents the height of a completed pass in football and the number of seconds the football was in the air.

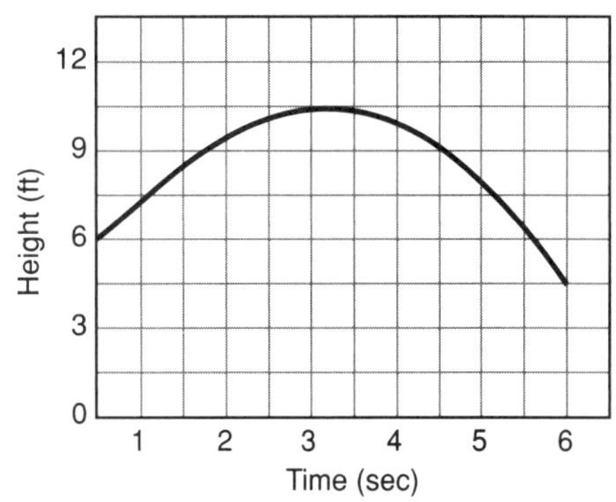

5. During what time interval was the pass over 9 feet above the ground?
6. What was the highest point of the pass and when did this occur?

Items 7–10: The circle shown below has its center at the origin and passes through the point (3, 1).

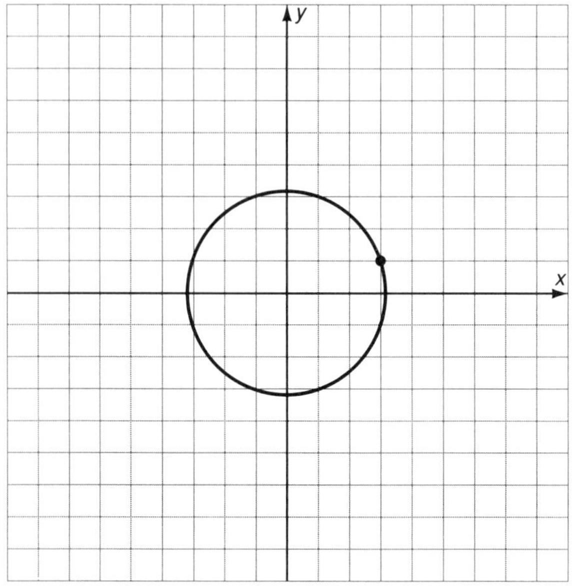

7. What is the length of the radius of the circle? Round your answer to the nearest hundredth.
8. What is the length of the diameter of the circle?
9. For what values of x does y have values either greater than 1 or less than -1?
10. For what values of x does y have values either greater than 2 or less than -2?

Assessment

Multiple-Choice Problems

1. Point M has the coordinates $(-4, 2)$. In what quadrant is M located?
   A. I
   B. II
   C. III
   D. IV

2. Line segment XY has endpoints of $(-5, -6)$ and $(3, -2)$. What is the approximate length of segment XY?
   A. 5
   B. 6
   C. 8
   D. 9

3. What is the slope of the line containing M and N in the graph below?

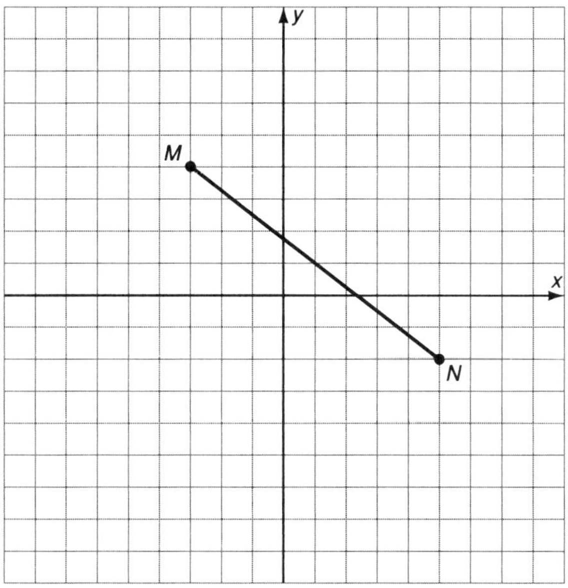

A. 1
B. $\frac{3}{4}$
C. $-\frac{3}{4}$
D. -1

Items 4 and 5: Slopes of six lines, L_1, L_2, L_3, L_4, L_5, and L_6, are respectively: $\frac{2}{3}$, $\frac{3}{4}$, $\frac{-4}{3}$, $\frac{3}{2}$, $\frac{-3}{2}$, and $\frac{6}{8}$.

4. Which line is parallel to L_2?

A. L_1
B. L_3
C. L_5
D. L_6

5. Which line is perpendicular to L_1?

A. L_3
B. L_4
C. L_5
D. L_6

6. What is the midpoint of the line segment connecting points $X(-3, 6)$ and $Y(5, 2)$?

A. (1, 4)
B. (4, 4)
C. (2, 8)
D. (8, 8)

Gridded-Response Problems

7. George is giving Walter directions on how to get to his house from school. He tells Walter to go 12 blocks south and then 5 blocks east. What is the straight-line distance from school to George's house?

8. What is the slope of the line passing through points R and S?

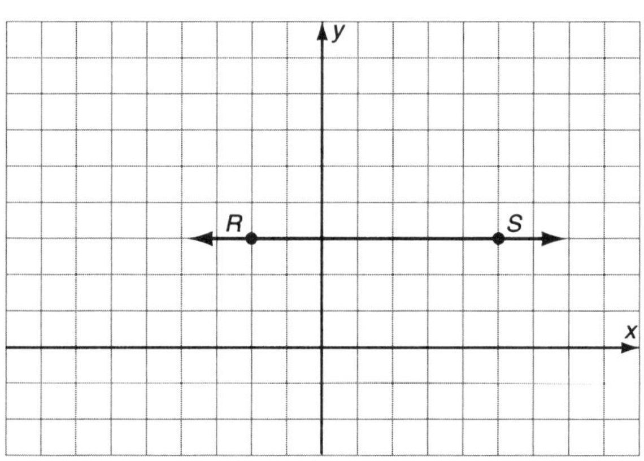

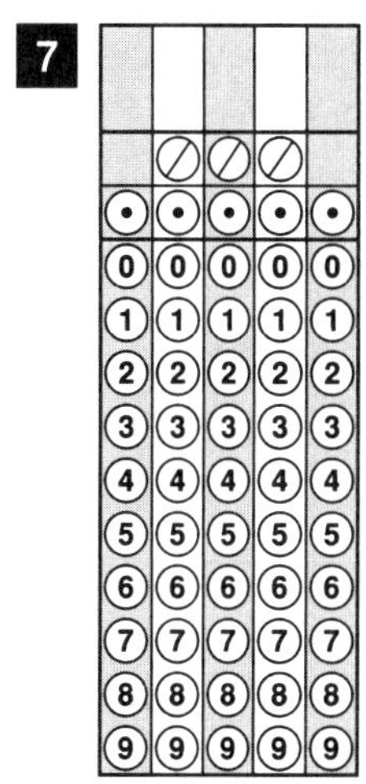

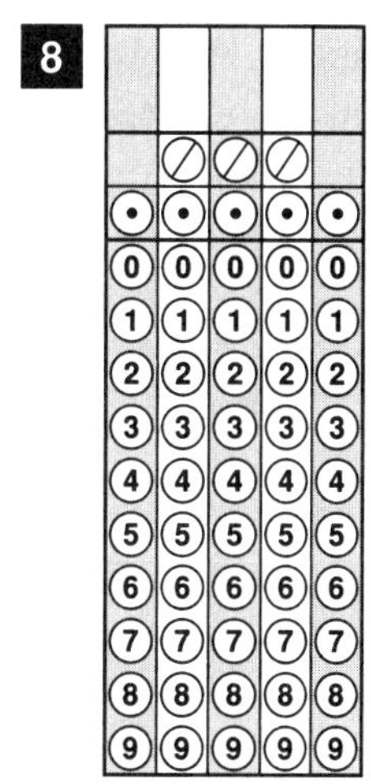

9. Three students realize that their houses lie in a straight line. Vic's house can be considered to be at the coordinates (2, 3). Lil's house, which is halfway between Vic's house and Roger's house, is located at (5, 5). What is the y-coordinate of Roger's house?

10. What is the slope m of the line parallel to the line containing points $(2, -5)$ and $(-3, -7)$?

11. What is the slope m of the line perpendicular to the line containing points $(-6, -2)$ and $(4, -5)$?

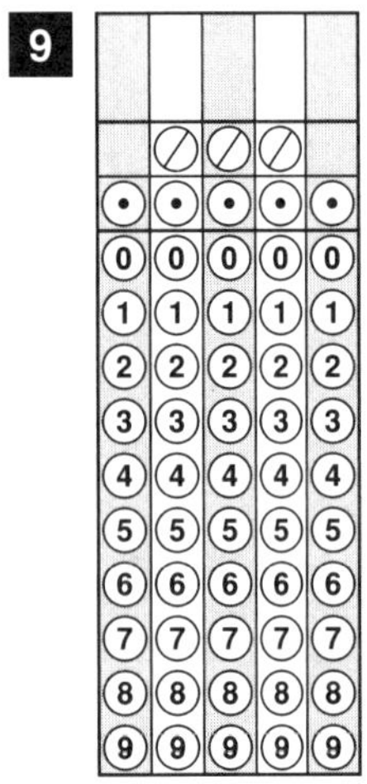

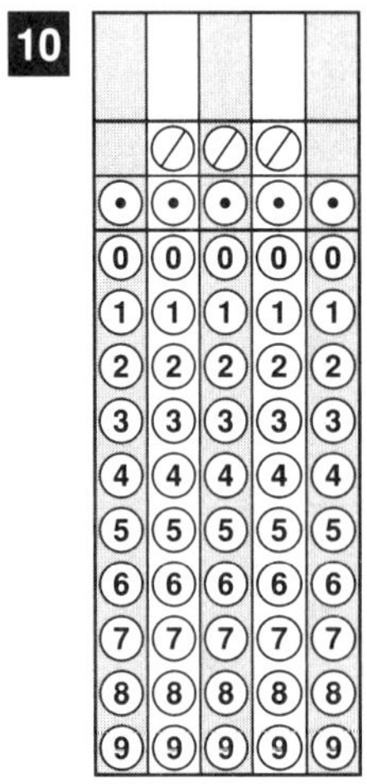

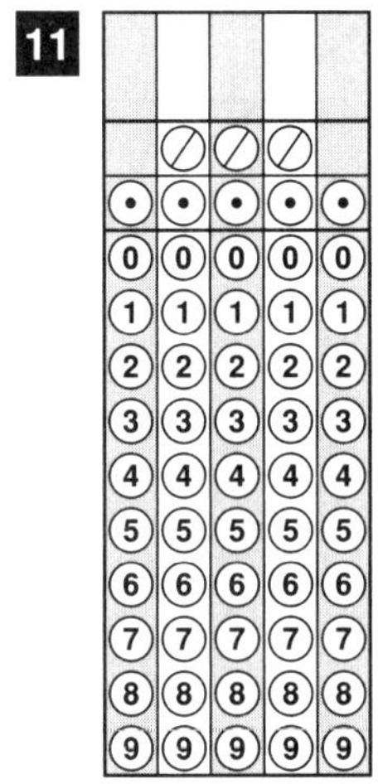

Think, Solve, Explain: Short-Response Problems

12. Elena is standing on the foul line, ready to attempt to throw a basketball into the basket. A diagram with measurements is shown below. Elena is at B.

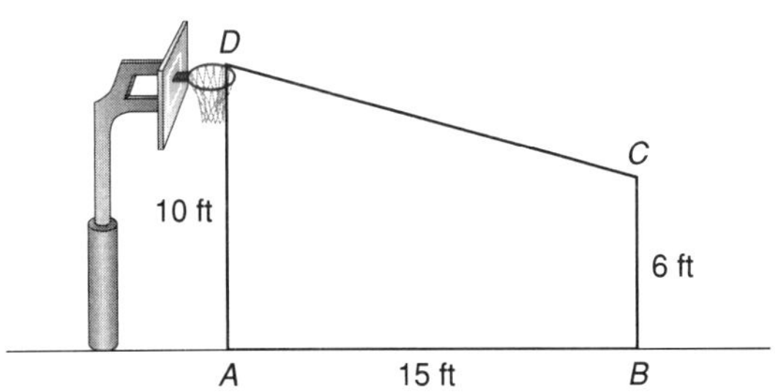

Part A If point A is considered (0, 0), what are the coordinates of points B, C, and D?

Part B What is the distance d from the point of release (C) to the basket (D)? Round your answer to the nearest tenth.

13. The graph below shows the path of a baseball after it was hit.

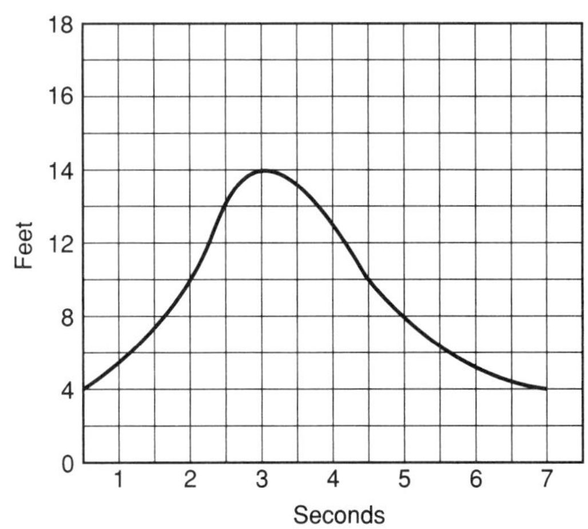

Part A When does it reach the highest point, and how high is that point?

Part B During what time interval was the ball above 10 feet?

Part C According to the graph, does the ball ever hit the ground? Explain.

14.

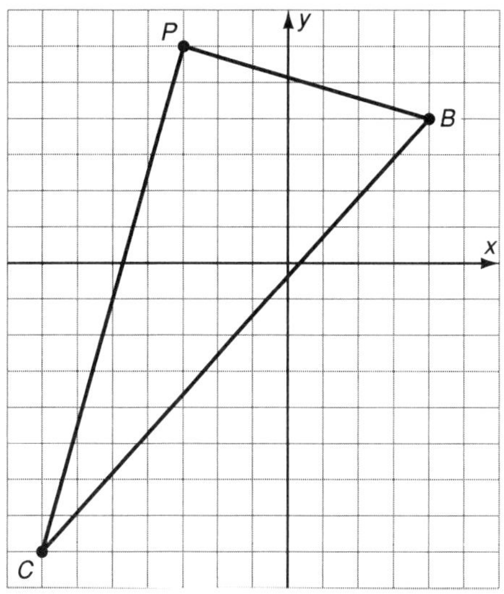

Part A Find the slope of each side of this triangle.

Part B Use these slopes to determine what kind of triangle is shown. Explain how you arrived at your answer.

15.

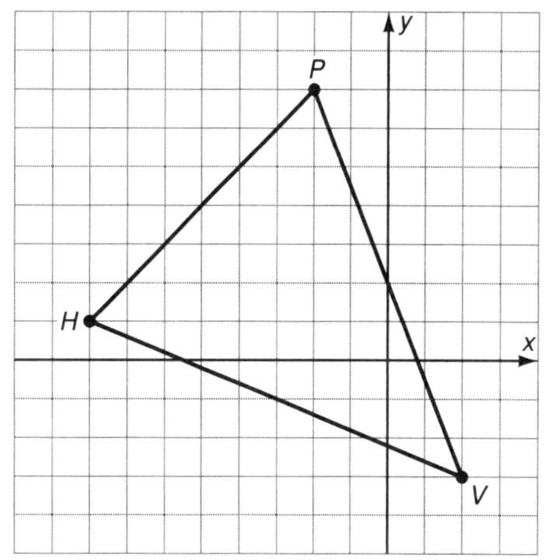

Part A Find the length of each side of this triangle.

Part B Using these lengths, explain what kind of triangle it is.

CHAPTER REVIEW

Multiple-Choice Problems

1. What type of angle is the supplement of an acute angle?

A. acute **C.** obtuse

B. right **D.** straight

2. Two angles are congruent and complementary. What is the measure of each angle?

A. 30°

B. 45°

C. 60°

D. 90°

3. Triangle *ABC* is congruent to triangle *DEF*.

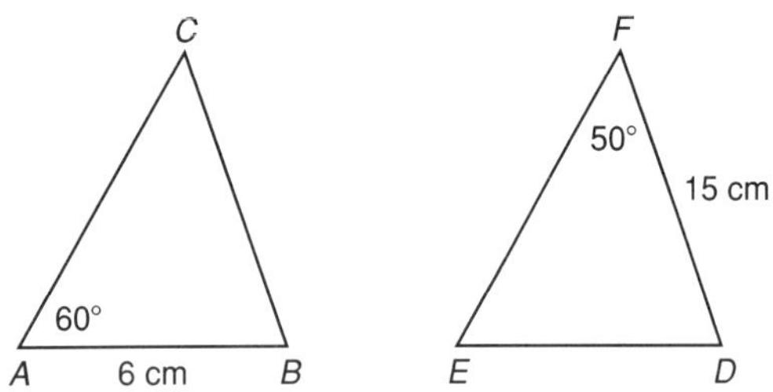

Which statement is FALSE?

A. $\angle E$ = 60 degrees

B. AC = 15 cm

C. $\angle C$ = 50 degrees

D. DE = 6 cm

4. The radius of a sphere is 25 centimeters. A cross section is located 7 cm below the center of the sphere. What is the circumference of the cross section?

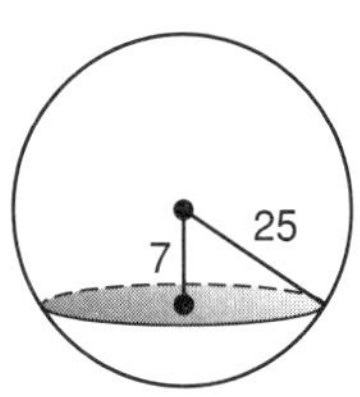

A. 7π **C.** 25π

B. 24π **D.** 48π

5. The drawing below illustrates a roof truss at Jacee's house. How long is the support $\overline{AC}$?

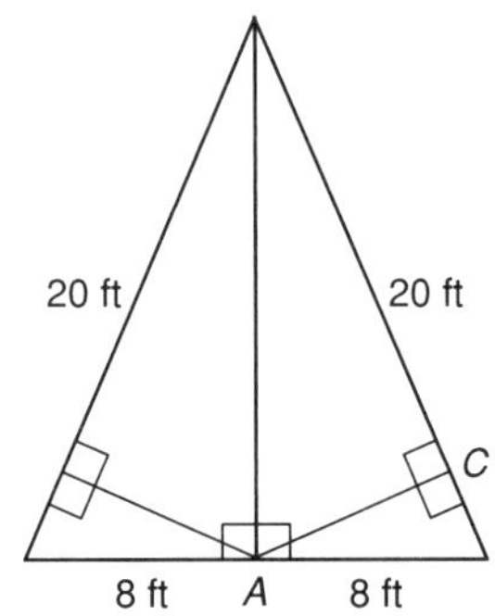

A. 2.5 ft **C.** 4 ft

B. 3.2 ft **D.** 4.5 ft

6. Two similar cylinders are being made as props for a play. The volume of the larger cylinder must be 8 times larger than volume of the smaller one. If the smaller cylinder has a radius of 10 inches, what will be the radius of the larger cylinder?

A. 20 inches **C.** 100 inches

B. 50 inches **D.** 1,000 inches

7. Which of the following is NOT a planar cross section from the right square pyramid?

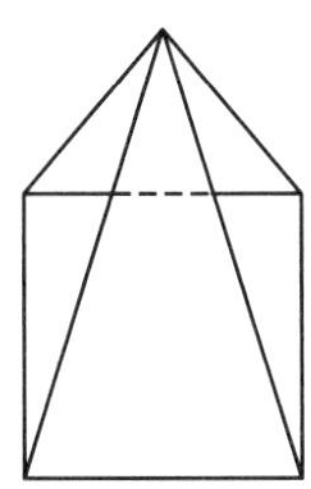

A. 

B. 

C. 

D. 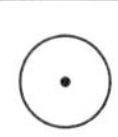

8. What is the slope of a line perpendicular to a line passing through point A (1, 7) and point B (9, 3)?

   A. -2

   B. $-\frac{1}{2}$

   C. $\frac{1}{2}$

   D. 2

9. What is the length of a line segment drawn from point A (5, −2) to B (−3, −7)?

   A. 5

   B. $\sqrt{29}$

   C. $\sqrt{39}$

   D. $\sqrt{89}$

10. Jolanda's father built an ultralight airplane. The diagram below shows the heights it reached over the time it flew. During what time interval did it fly above 150 feet?

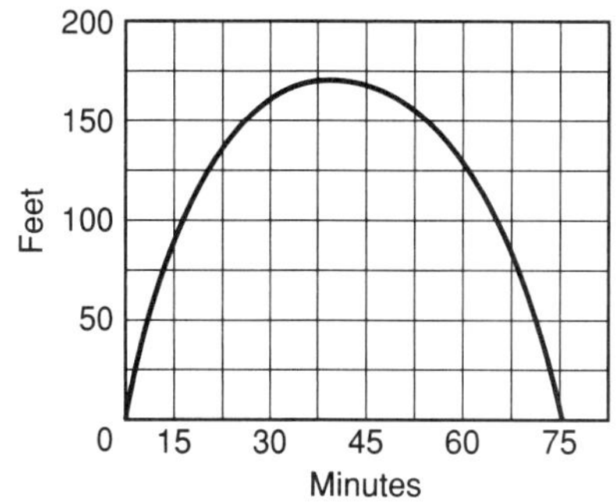

   A. between 0 and 25 minutes

   B. between 25 and 55 minutes

   C. between 55 and 75 minutes

   D. It never flew above 150 feet.

Gridded-Response Problems

11. Point A is reflected over the x-axis. What is the y-coordinate of point A'?

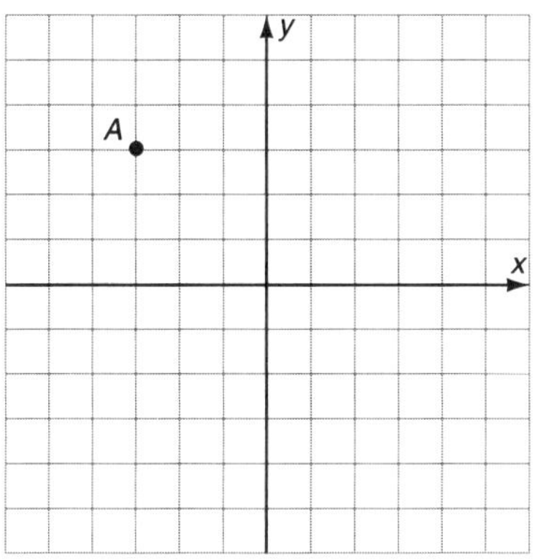

12. Two braces are needed to support a large ladder. The diagram shows the measurements for placing them. If the shorter brace is 14 feet long, what is the length of the longer brace?

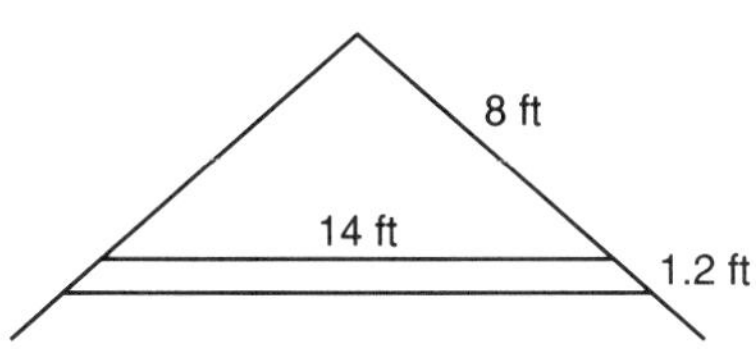

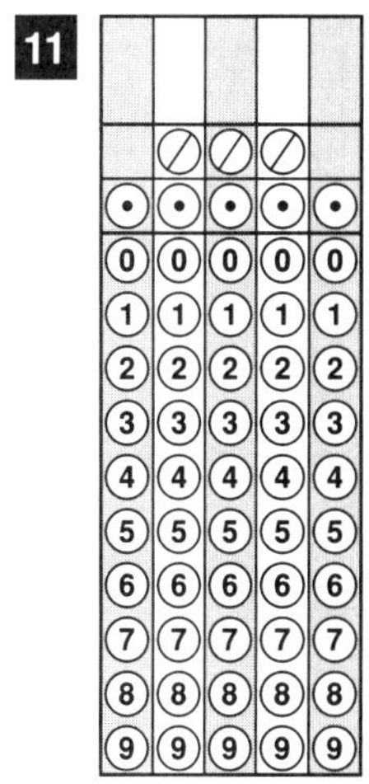

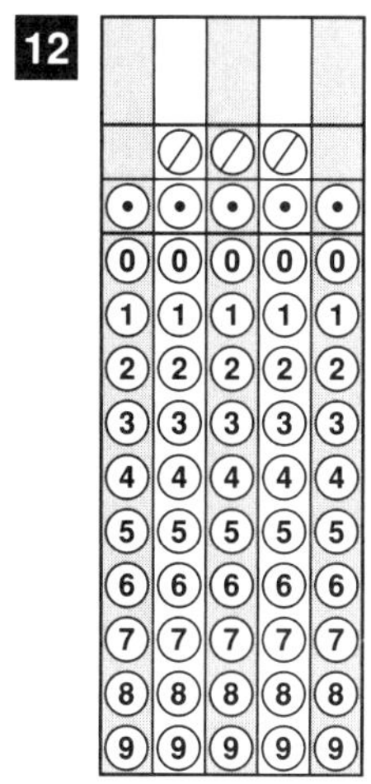

13. Jennifer has a dollhouse that is based on a 1 to 9 scale. A model cradle, 4 inches long, represents an actual cradle that is how many feet long?

14. In this rectangular solid, a triangular cross section is shaded. What is the area of the cross section?

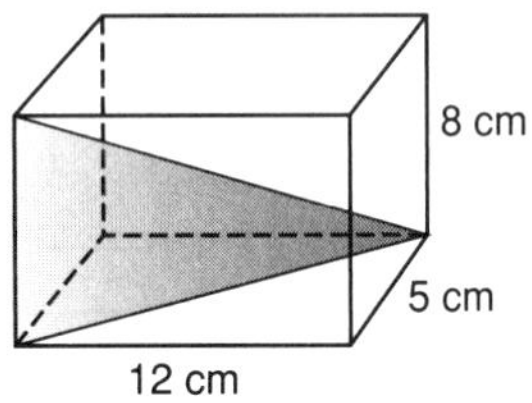

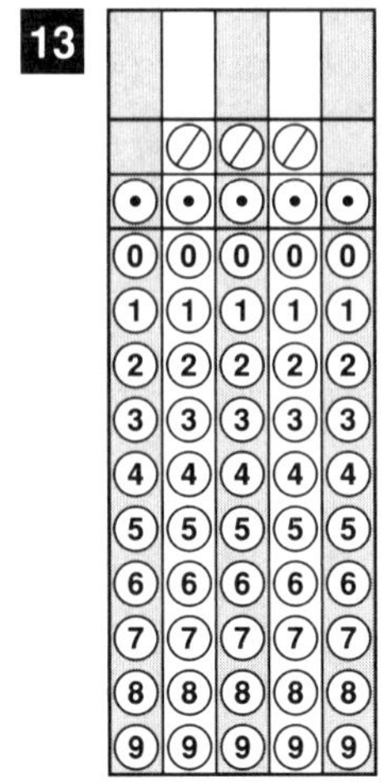

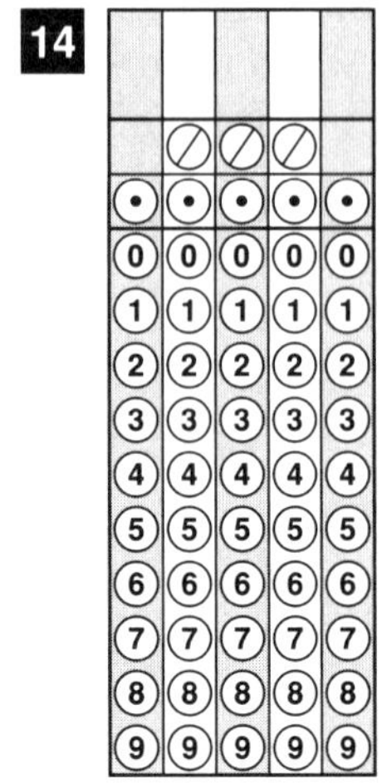

Think, Solve, Explain: Short-Response Problems

15. Triangle ABC has vertices of A $(-2, -2)$, B $(2, 4)$, and C $(4, -4)$.

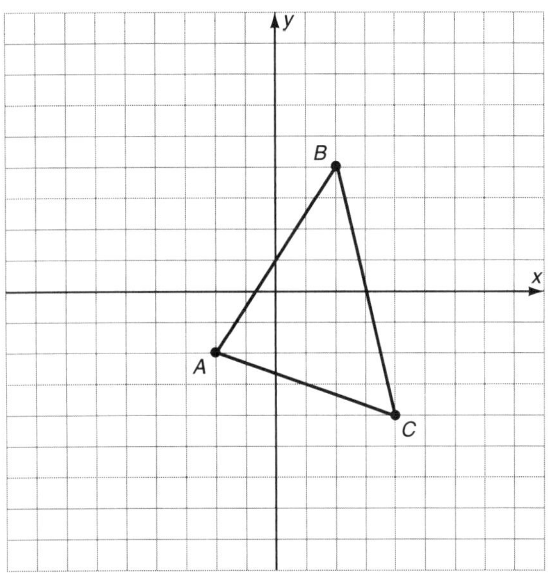

Part A Find the coordinates of D, the midpoint of $\overline{AB}$, and E, the midpoint of $\overline{BC}$.

Part B Find the slopes of $\overline{DE}$ and $\overline{AC}$.

Part C What is the relationship between $\overline{DE}$ and $\overline{AC}$?

16. Points A $(5, 7)$ and B $(-3, -1)$ are endpoints of the diameter of a circle.

Part A What are the coordinates of the center of the circle?

Part B What is the length of the radius of the circle?

Part C Is point $(7, 2)$ on the circle? Show reasons to justify your answer.

Think, Solve, Explain: Extended-Response Problems

17. The surface area of the larger of two cones is 9 times greater than the surface area of the smaller cone.

Part A If the volume of the larger cone is 270π cubic inches, what is the volume of the smaller cone?

Part B The volume of a cylinder that is 30 inches tall is the same as the volume of the larger cone. What is the radius r of the cylinder?

18. Danielle wants to find the distance across the lake from point A to point B. She lays out $\overline{AC} = 30$ meters and $\overline{CB} = 80$ meters to form a triangle. She marks off $\overline{ED}$ parallel to $\overline{AB}$.

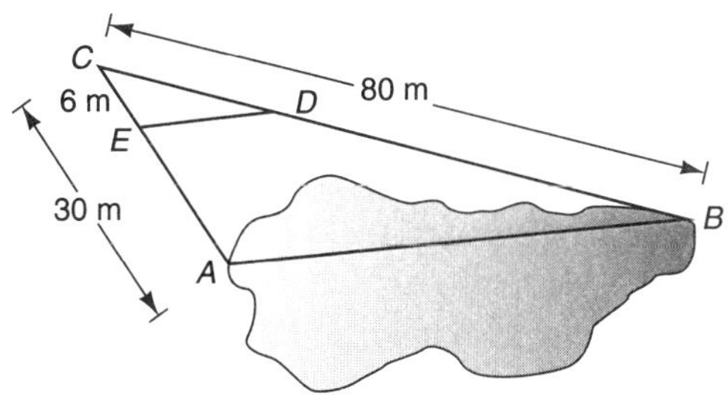

Part A Explain why $\triangle CED \sim \triangle CAB$.

Part B If $CE = 6$ m and $ED = 9$ m, what is the length of CD?

Part C What is the length of AB?

CUMULATIVE REVIEW

Multiple-Choice Problems

1. Which one of the following expressions is equivalent to $3(x - 6) - 2(x - 5)$?

A. $5x - 28$ **C.** $x - 28$

B. $5x - 8$ **D.** $x - 8$

2. How much money can Ruthie save if she buys a \$65 dress on sale at 20% off?

A. \$1.30 **C.** \$30

B. \$13.00 **D.** \$52

3. What transformations must the letter **E** go through in the following example?

A. reflection, rotation

B. translation, rotation

C. dilation, reflection

D. dilation, rotation

4. In the diagram below, $\overline{DE} \parallel \overline{AB}$. What is the length of $\overline{AB}$?

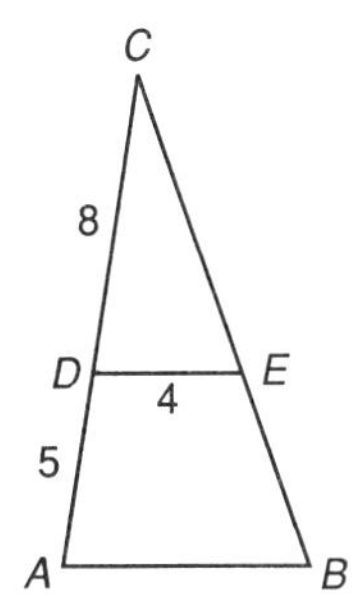

A. 2.5 **C.** 6.5

B. 6.4 **D.** 8.9

5. Mike bought a soda cube (24 twelve-ounce cans) at the store. How many gallons of soda are in the cube?

A. 2 gallons **C.** $2\frac{1}{3}$ gallons

B. $2\frac{1}{4}$ gallons **D.** $2\frac{1}{2}$ gallons

6. What is the volume of a sphere that has a surface area of π square centimeters?

A. $\frac{\pi}{6}$ cu cm **C.** $\frac{\pi}{48}$ cu cm

B. $\frac{\pi}{18}$ cu cm **D.** $\frac{\pi}{64}$ cu cm

7. Which letter has **no** lines of symmetry?

A. A **C.** R

B. C **D.** X

8. Tina's coach has her skating two circles that form the shape of the number 8. She starts where the two circles meet and goes counterclockwise around the bottom circle and then crosses over and goes clockwise around the top circle ending up where she started. If the radius of each circle is 7 feet, how many feet does Tina skate if she makes 4 trips around the figure? Use $\pi \approx \frac{22}{7}$.

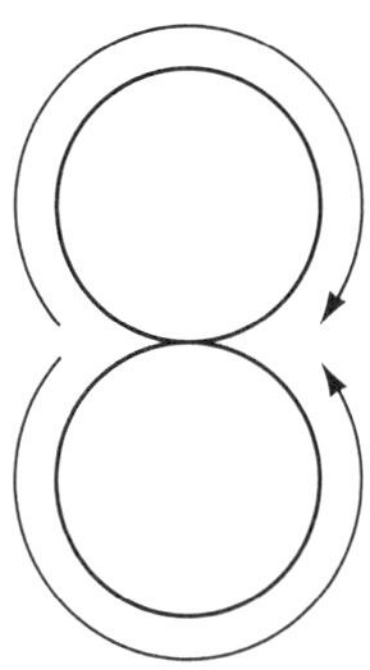

A. 88 feet **C.** 264 feet

B. 176 feet **D.** 352 feet

9. In the diagram below, points B and E divide line segments AC and DF proportionally. If $AB = 9$, $AC = 15$, and $DF = 25$, how long is EF?

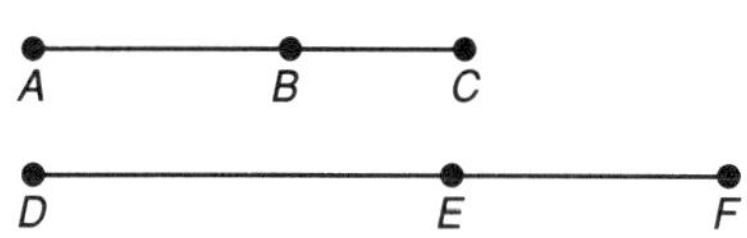

A. 5 **C.** 15

B. 10 **D.** 20

10. The vertex angle of an isosceles triangle is 20°. How many degrees are in the angle labeled x?

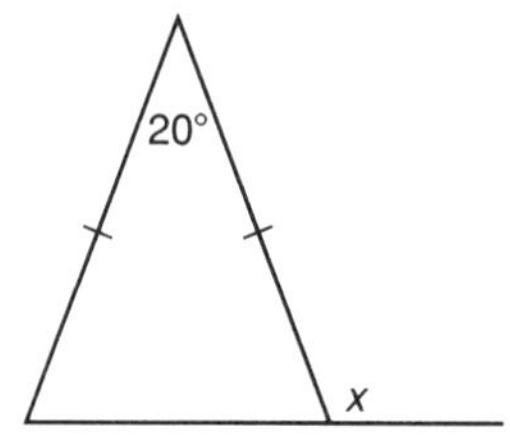

A. 80°

B. 90°

C. 100°

D. 120°

11. On a map, 1 centimeter = 32 miles. If two cities are 6.2 centimeters apart on the map, what is the approximate distance between the two cities?

A. 20 miles

B. 50 miles

C. 200 miles

D. 240 miles

12. As shown in the diagram below, the shaded section of the bleachers at the concert can seat approximately 450 people. If the concert is sold out, what is the approximate number of people in the bleachers?

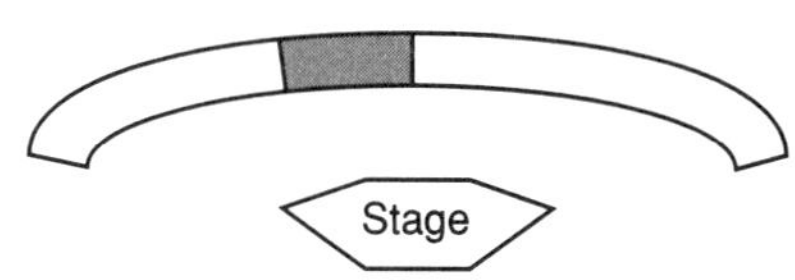

A. 1,926 people

B. 2,153 people

C. 2,687 people

D. 3,294 people

Gridded-Response Problems

13. What is the geometric mean between 12 and 3?

14. What is the slope of the line that passes through the points $(-3, -2)$ and $(2, 7)$?

15. What is the distance between the points $(5, 9)$ and $(-1, 1)$?

16. What is the surface area of a rectangular box that is 3 feet long, 2 feet wide, and 1 foot high?

17. What is the value of five cubed?

18. In parallelogram $ABCD$, $\angle D = 50°$. How many degrees are in $\angle C$?

19. Quadrilateral $ABCD \cong$ quadrilateral $MNOP$. $AB = 3$, $BC = 4$, $CD = 5$, and $DA = 7$. How long is ON?

20. A 5-foot woman casts a 9-foot shadow. At the same time, what will be the length of a shadow cast by a 30-foot tree?

21. Half of the band members will be in 5 equal groups so they can spell out the word LIONS. What fractional part of the band will be in each group?

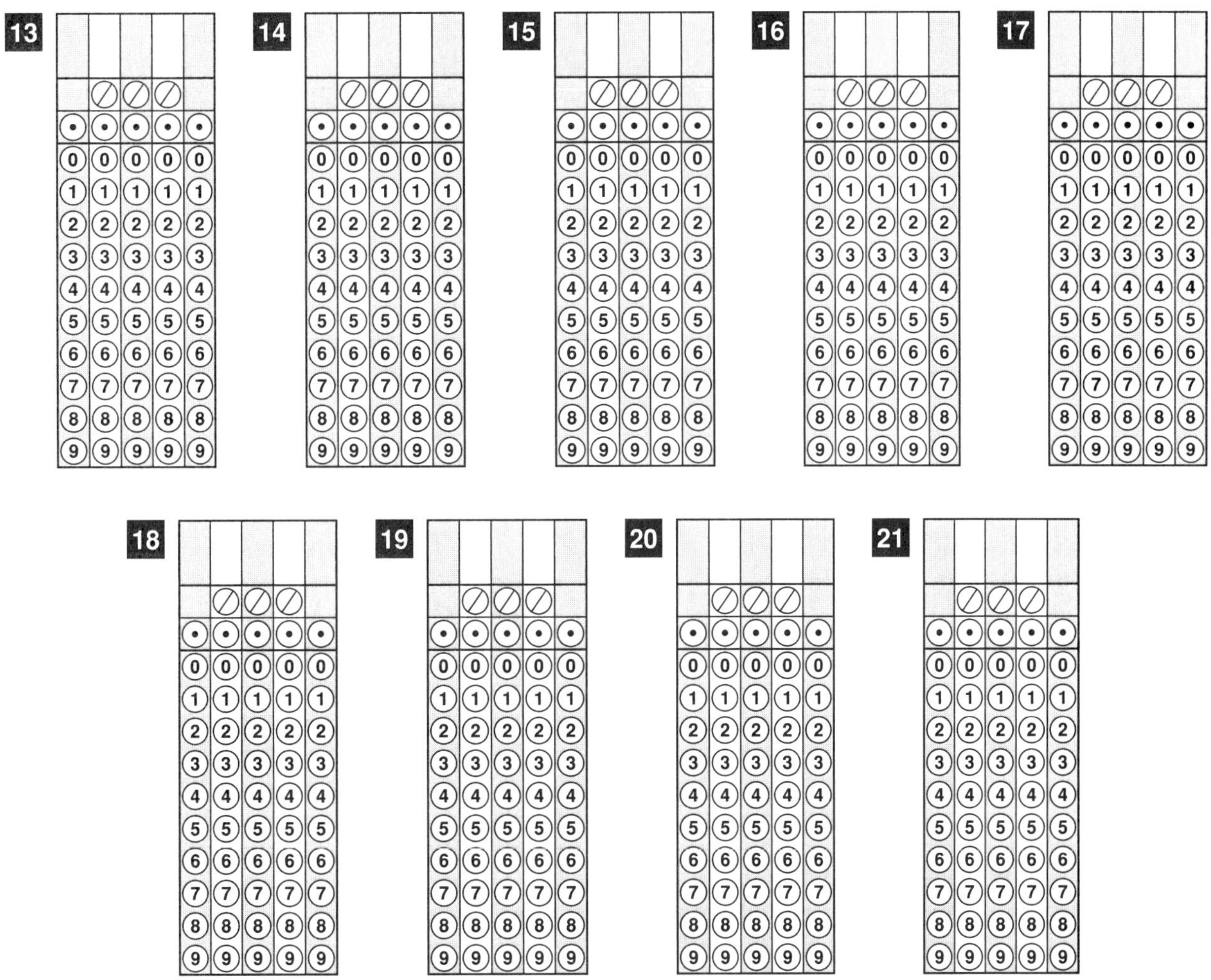

Think, Solve, Explain: Short-Response Problems

22. The Norman window shown below is made up of a semicircle and a rectangle.

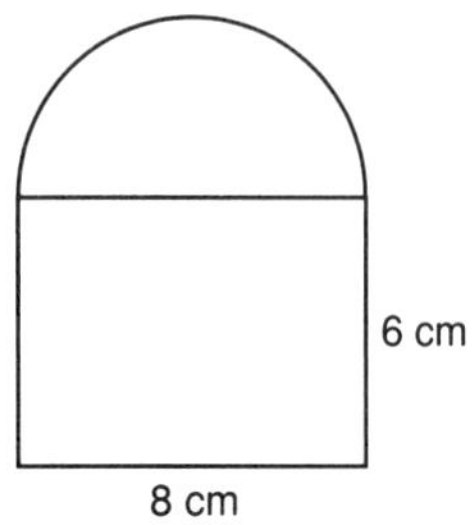

Part A Write a formula for the area A of the entire window.

Part B Find the area of the window. Use $\pi \approx 3.14$.

23. Roger plotted the vertices of triangle ABC on his paper as shown below.

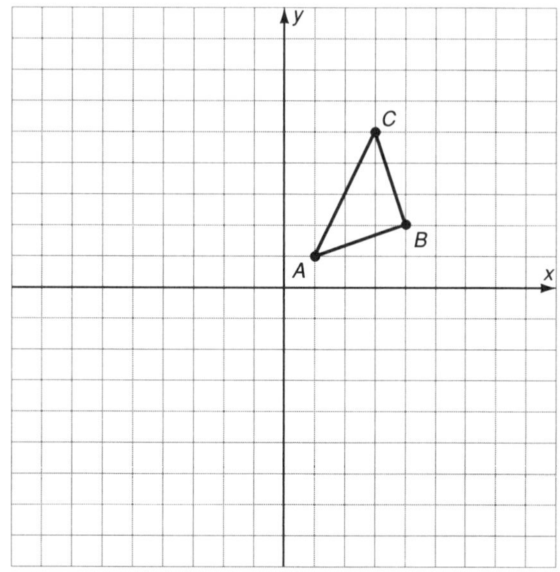

Part A Find the slope of each side.

Part B Based on the slopes, what type of triangle is $\triangle ABC$? Explain your answer.

Think, Solve, Explain: Extended-Response Problem

24. Two loading ramps meet the ground at the same angle. The heights of the docks for these ramps are 6 feet and 8 feet. The ramp to the higher dock meets the ground 20 feet from the building.

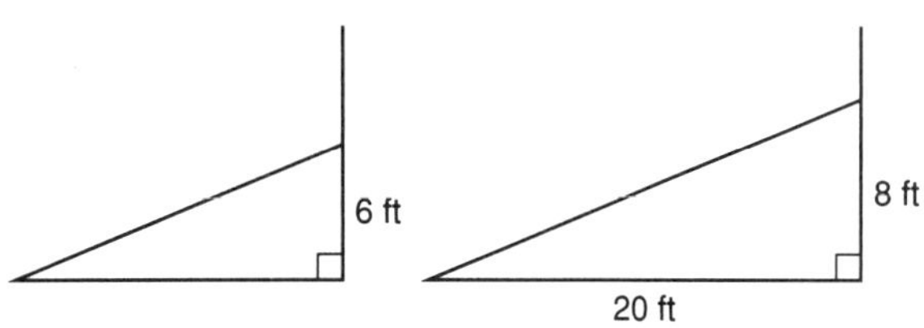

Part A The triangles formed by the ramps, the ground, and the wall of the building are similar. Why? Explain using geometric concepts or in a formal proof.

Part B What is the distance from the building to the bottom of the lower ramp? Set up a proportion to solve for the distance. Solve the proportion.

Part C You now replace these ramps with two new ramps. For each new ramp you increase the angle it makes with the ground. You do not change the height at which it meets the wall. What will happen to the distance from the bottom of the ramp to the wall? Explain.

CHAPTER FOUR

Algebraic Thinking

CHAPTER CONTENTS

Section 1

Describing, analyzing, and generalizing patterns, relations, and functions

A USING WORDS, SYMBOLS, VARIABLES, TABLES, AND GRAPHS TO DESCRIBE, ANALYZE, AND GENERALIZE PATTERNS, RELATIONSHIPS, AND FUNCTIONS

Patterns

In the real world and in mathematics, some problems can be solved by discovering a pattern in the given information and using that pattern to draw a conclusion. The pattern might be numerical, visual, or a combination of both.

NUMERICAL PATTERNS

Arithmetic Sequences Three or more terms with a common difference are called an **arithmetic sequence**.

EXAMPLE

3, 5, 7, 9, . . . is an arithmetic sequence.
The common difference is 2:

$5 - 3 = 2$; $7 - 5 = 2$; and $9 - 7 = 2$

Notice that the fourth term of this sequence is 9. To obtain 9, we add 2 three times to the first term, 3.

If we want to know the 30th term, we need to add 2 to 3 (the first term) 29 times. This addition will give us 61:

$$3 + (29 \times 2) = 61$$

If we want to determine the 75th term:

$$3 + (74 \times 2) = 151$$

MODEL PROBLEMS

1. Given the following arithmetic sequence: 8, 14, 20, 26, . . .

a. Find the common difference.

Solution:

$14 - 8 = 6$

$20 - 14 = 6$

$26 - 20 = 6$

Answer: The common difference is 6.

b. Find the 42nd term.

Solution: $8 + (41 \bullet 6) = 254$

Answer: The 42nd term is 254.

2. Given the following arithmetic sequence: 28, 14, 0, −14, . . .

a. Find the common difference.

Solution:

$14 - 28 = -14$

$0 - 14 = -14$

$-14 - 0 = -14$

Answer: The common difference is −14.

b. Find the 16th term.

Solution: $28 + [15 \bullet (-14)] = -182$

Answer: The 16th term is −182.

TRY THESE

1. Given the following arithmetic sequence: 6, 13, 20, 27, . . .

a. Find the common difference.

b. Find the 32nd term.

2. Given the following arithmetic sequence: 18, 14, 10, 6, . . .

a. Find the common difference.

b. Find the 20th term.

3. Given the following arithmetic sequence: 3, 7, 11, 15, . . .

a. Find the common difference.

b. Find the 8th term.

c. Find the sum of the 8 terms.

4. Given the following arithmetic sequence: 12, 7, 2, −3, . . .

a. Find the common difference.

b. Find the 18th term.

5. Given the following arithmetic sequence: 2.3, 3.4, 4.5, 5.6, . . .

a. Find the common difference.

b. Find the 25th term.

6. Given the following arithmetic sequence: $\frac{3}{4}, \frac{5}{8}, \frac{1}{2}, \frac{3}{8}, \ldots$

a. Find the common difference.

b. Find the 21st term.

7. Given the following arithmetic sequence: 6.3, 8.5, 10.7, 12.9, . . .

a. Find the common difference.

b. Find the 12th term.

8. Given the following arithmetic sequence: 102, 86, 70, 54, . . .

a. Find the common difference.

b. Find the 15th term.

9. When Ceila was 6 years old her dad gave her an allowance of \$5 a week. Each year he increased the allowance by \$2. What will be Ceila's allowance when she is 18?

10. At Lou's new job, he was given a starting salary of \$450 a week. If his work is satisfactory he is guaranteed a \$20 per week increase every three weeks. Lou's work has been great. What is Lou's salary after 36 weeks?

Geometric Sequences Three or more terms with a pattern showing a common ratio are called a **geometric sequence**.

EXAMPLES

5, 10, 20, 40, . . . is a geometric sequence. The common ratio is 2:

$$\frac{10}{5} = 2; \frac{20}{10} = 2; \text{ and } \frac{40}{20} = 2$$

12, 6, 3, $\frac{3}{2}, \frac{3}{4}$, . . . is a geometric sequence. The common ratio is $\frac{1}{2}$:

$$\frac{6}{12} = \frac{1}{2}; \frac{3}{6} = \frac{1}{2}; \frac{\frac{3}{2}}{3} = \frac{1}{2}; \text{ and } \frac{\frac{3}{4}}{\frac{3}{2}} = \frac{1}{2}$$

To find a particular term in a geometric sequence, use repeated multiplications on a calculator. For example, suppose you want the 10th term of the sequence 5, 10, 20, 40, ... First determine the common ratio. In this case the ratio is 2. Continue with the following steps:

Keystroke	Term	Result
2		
×		
5	1st term	
=	2nd term	10
=	3rd term	20
=	4th term	40
=	5th term	80
=	6th term	160
=	7th term	320
=	8th term	640
=	9th term	1280
=	10th term	2560

MODEL PROBLEMS

1. Given the following geometric sequence: 8, 40, 200, 1,000, . . .

   a. Find the common ratio.

   Solution: $\frac{40}{8} = 5$ and $\frac{200}{40} = 5$ and $\frac{1{,}000}{200} = 5$

   Answer: The common ratio is 5.

   b. Find the 7th term.

   Solution: Use a calculator.

   Press [5] [×] [8] =

   Then press the [=] key five more times.

   Answer: The 7th term is 125,000.

2. Given the following geometric sequence: 8, 4, 2, 1, . . .

   a. Find the common ratio.

   Solution: $\frac{4}{8} = \frac{1}{2}; \frac{2}{4} = \frac{1}{2}; \frac{1}{2} = \frac{1}{2}$

   Answer: The common ratio is $\frac{1}{2}$.

   b. Find the 8th term.

   Solution: Use a calculator.

   Press [.] [5] [×] [8] [=]

   Then press the [=] sign six more times.

   Answer: The 8th term is 0.0625.

TRY THESE

1. Given the following geometric sequence: 9, 3, 1, $\frac{1}{3}$, . . .

   a. Find the common ratio.

   b. Find the 7th term.

2. Given the following geometric sequence: −5, 10, −20, 40, . . .

   a. Find the common ratio.

   b. Find the 8th term.

3. Given the following geometric sequence: −3, −6, −12, −24, . . .

   a. Find the common ratio.

   b. Find the 8th term.

4. Given the following geometric sequence: $\frac{1}{2}, \frac{1}{6}, \frac{1}{18}, \frac{1}{54}, \ldots$

   a. Find the common ratio.

   b. Find the 6th term.

5. Given the following geometric sequence: $1, -3, 9, -27, \ldots$
   - **a.** Find the common ratio.
   - **b.** Find the 7th term.
   - **c.** Find the sum of the first 7 terms.
6. Given the following geometric sequence: $1.4, 2.8, 5.6, 11.2, \ldots$
   - **a.** Find the common ratio.
   - **b.** Find the 9th term.
7. Given the following geometric sequence: $\frac{1}{8}, -\frac{1}{4}, \frac{1}{2}, -1, \ldots$
   - **a.** Find the common ratio.
   - **b.** Find the 8th term.
8. Given the following geometric sequence: $100, 20, 4, \frac{4}{5}, \ldots$
   - **a.** Find the common ratio.
   - **b.** Find the 7th term.

Items 9 and 10: A red shirt looses 1% of its color (fades) with each washing.

9. What percent of the color is lost after 9 washings?
10. What percent of the color still remains after 22 washings?

VISUAL PATTERNS

When we look at wallpaper, a carpet, or a fabric, a definite visual pattern may appear. It may be a simple pattern such as this:

□ □□ □□□

Or it may be a more complicated pattern such as this:

❑ ❒ ❑ ❒

To analyze a visual pattern, look at the items in the pattern in successive order. Then change the visual pattern to a numerical pattern.

MODEL PROBLEMS

1. How many rectangles are needed to represent the 6th term of this pattern?

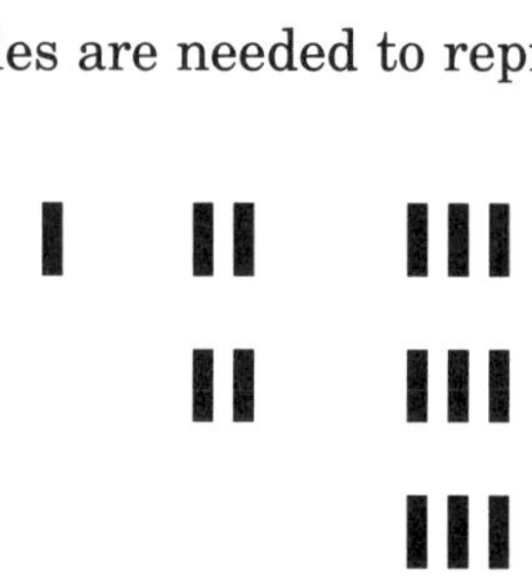

Solutions: Use the *visual pattern*. The first item, or term, is one rectangle. The second term is a set of two by two rectangles. The third term is a set of three by three rectangles.

Answer: The sixth term would therefore be a set of six by six rectangles.

Use the *numerical pattern*. The number of rectangles is 1, 4, 9, . . . These are perfect squares, giving the pattern 1, 4, 9, 16, 25, 36, 49, . . .

Answer: The sixth term would be 36 or 6×6, a set of six by six rectangles.

2. Look at the pattern MATHMATHMATH . . . . What letter would be in the 85th position?

Solution:

The visual pattern is a repetition of the letters in the word MATH. Every four letters, the pattern repeats. Group the letters by fours and number each letter:

M A T H | M A T H | M A T H

1 2 3 4 | 5 6 7 8 | 9 10 11 12

To find the number of repetitions, divide:

$$\frac{85}{4} = 21\frac{1}{4}$$

Thus the number of repetitions is 21, with 1 left over.

Answer: The 85th position is the letter M.

TRY THESE

1. Look at the pattern below. What is the position of the $\times$ in the last group?

☐☐☐☐☐ ☐☒☐☐☐ ☐☐☐☐☐ ☐☐☐☐☐

☒☐☐☐☐ ☐☐☐☐☐ ☐☐☒☐☐ ☐☐☐☐☐

2. Given the following pattern, describe the 10th figure.

○ △ ○ □ ○ . . .

3. Find the next four letters to complete the following pattern: A M B B L L C K D D . . . .

4. Given the arithmetic sequence 36, 49, 62, 75, . . . , find the 23rd term.

5. Given the geometric sequence 3, 6, 12, 24, . . . , find the 10th term.

6. Given the pattern 1, -1, 3, 1, 5, ... What must be added to the first term in order to get the second term? What must be added to the second term to get the third term, and so on? Find the 11th term.

7. What is the missing term in the following pattern?
5 7 ? 15 23 36 57

8. Given the pattern BALLADBALLAD . . . . What letter is in the 125th position?

9. In the following pattern, where do the letters X, Y, Z go?

```
A      E F   H I   K L M N           T   V W
_____________________________________________
  B C D    G     J         O P Q R S   U
```

10. A grocery store clerk is making a display with cans. There is to be 1 can on the top row, 2 cans on the next row down, 3 cans on the third row down, and so forth. What is the total number of cans in the display if it is 10 rows high?

Relationships

Often, we need to relate two sets of information. For example, we might want to compare the weather in Tampa with the weather in Miami during five days in December. Such a comparison can be called a **relationship**. Relationships can be described in several ways.

WORDS

We can use words to describe a relationship. For example:

> On Monday, Tuesday, and Wednesday, the average daytime temperature in Miami was eight degrees higher than the average daytime temperature in Tampa. On Thursday and Friday the average daytime temperature in Miami was five degrees higher than the average daytime temperature in Tampa.

SYMBOLS AND VARIABLES

We can describe a relationship using symbols or variables. For example, let m = average daytime temperature in Miami and t = average daytime temperature in Tampa. Then the temperatures on Monday, Tuesday, and Wednesday can be expressed as:

$m = t + 8°$ or $t = m - 8°$

The temperatures on Thursday and Friday can be expressed as:

$m = t + 5°$ or $t = m - 5°$

TABLES

We can use tables to describe relationships. For example, here is a table expressing these temperatures:

Average Daily Temperature (°F) for a 5-Day Period in December

Day	Miami	Tampa
Monday	82°	74°
Tuesday	80°	72°
Wednesday	75°	67°
Thursday	74°	69°
Friday	76°	71°

GRAPHS

We can also express a relationship by using a graph, such as this line graph of the temperatures.

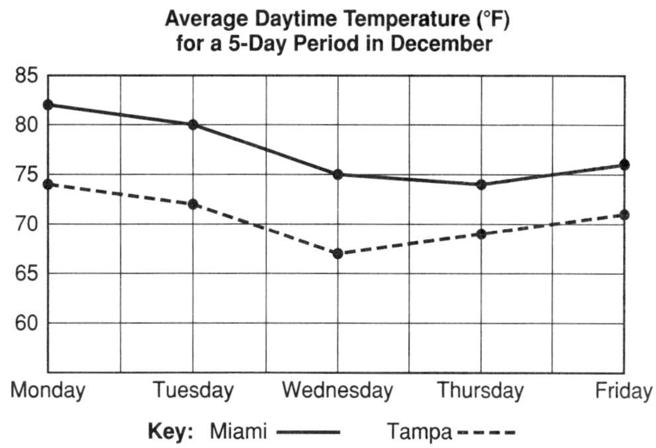

Another kind of graph can be used to show the relationship between speed or distance and time.

EXAMPLES

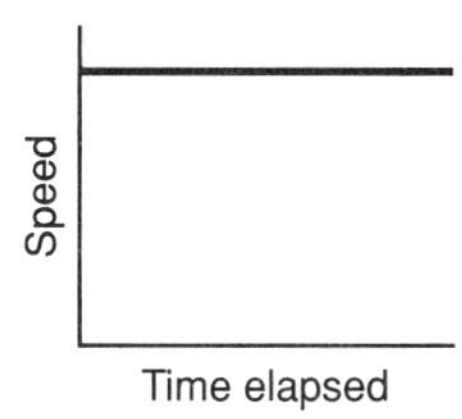

a person driving at a constant rate of speed.

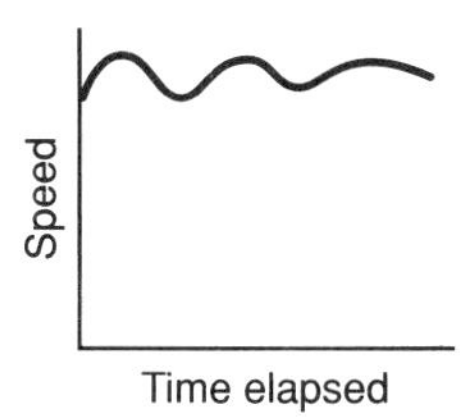

a person speeding up, slowing down, speeding up, slowing down approaching a constant speed

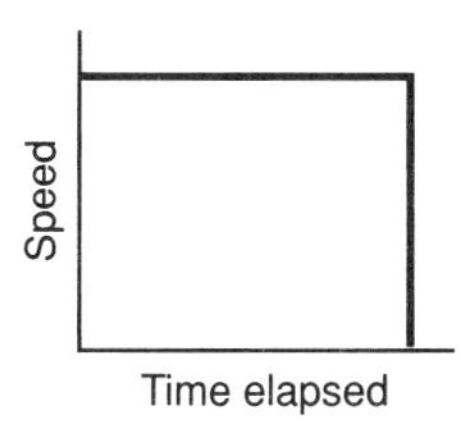

a person moving at a constant rate of speed and instantly stopping

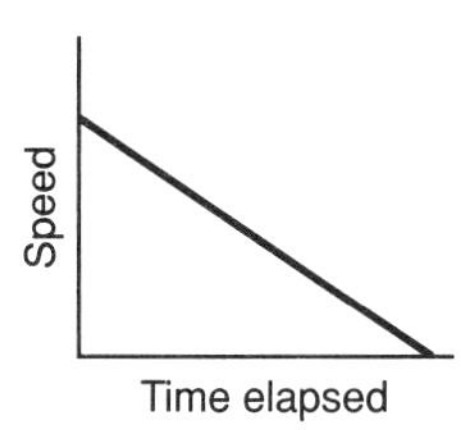

a person coming to a gradual stop to let people out of the car

ORDERED PAIRS: DOMAIN AND RANGE

The following table is another example of a relationship.

Career Touchdowns by the Top 5 Miami Dolphns

Player	TDs
Clayton	82
Moore	75
Duper	59
Csonka	57
Chambers	39

This relationship of player to career touchdowns can be written as a set of **ordered pairs**:

{(Clayton, 82), (Moore, 75), (Duper, 59), (Csonka, 57), (Chambers, 39)}

For each ordered pair, the first element is the player's name and the second element is the number of touchdowns he made. If we listed every Miami Dolphin, past and present, the set would contain many more ordered pairs. A **relation** can be defined as a set of ordered pairs.

Within a set of ordered pairs, the first elements of the pairs form a set, or **subset**, that is called the **domain**. In the example of the Dolphins, the domain is the top five players:

{Clayton, Moore, Duper, Csonka, Chambers}

The second elements within a set of ordered pair form a subset that is called the **range**. In the example above, the range is the number of touchdowns:

{82, 75, 59, 57, 39}

MODEL PROBLEMS

Given the following relation: {(2, 5), (3, 6), (4, 5), (5, 6)}

1. What is the domain?

Answer: {2, 3, 4, 5}

2. What is the range?

Answer: {5, 6}

ARROW DIAGRAMS

Another way to show a relation is with **arrow diagrams**. The example below shows the relation is *less than*, using the set of numbers {2, 3, 4}. Let a and b stand for elements of the set where $a < b$. Then the relation contains the ordered pairs {(2, 3), (2, 4), (3, 4)}.

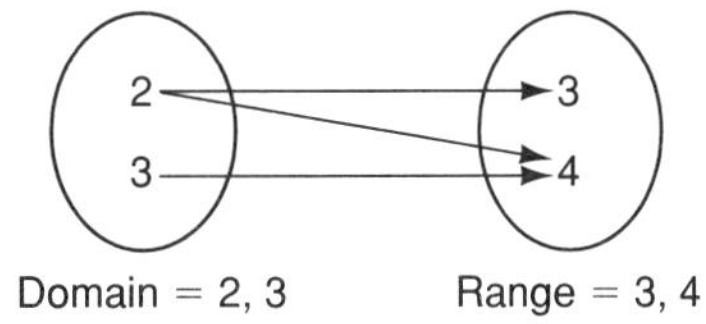

MODEL PROBLEMS

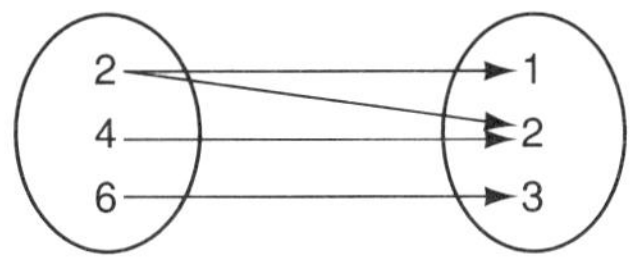

1. What are the ordered pairs of the relation in this arrow diagram?

   Answer: {(2, 1), (2, 2), (4, 2), (6, 3)}

2. Give the domain of the relation.

   Answer: {2, 4, 6}

3. Give the range of the relation.

   Answer: {1, 2, 3}

TRY THESE

1. The drawing represents which of the following?

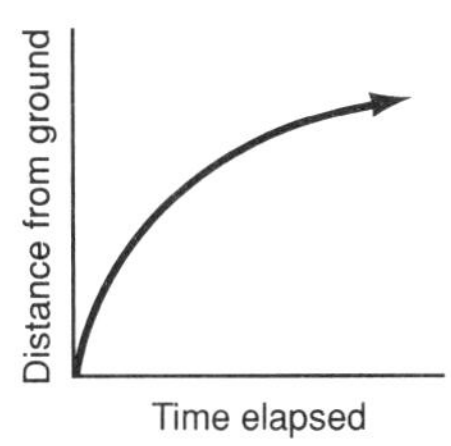

   A. someone swinging on a swing

   B. launching of a rocket

   C. someone climbing a hill

   D. someone dropping an object out of a third-story window

2. Use the following relation: {(a, z), (b, y), (c, x), (d, w)}.

   a. Give the domain of the relation.

   b. Give the range of the relation.

3. {(1, 4), (3, 5), (5, 4), (7, 5)}

   a. State the domain of this relation.

   b. State the range of the relation.

   c. Draw an arrow diagram for this relationship.

4. 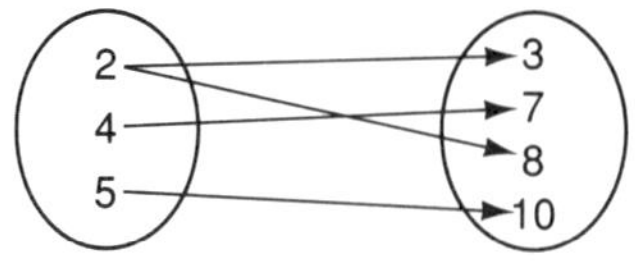

a. Write the ordered pairs of the relation in this arrow diagram.

b. Give the domain of the relation.

c. Give the range of the relation.

Items 5–8: For each relationship, find the domain and range and then draw an arrow diagram.

5. {(1, 1), (2, 4), (3, 9), (4, 16), (5, 25)}

6. {(3, 6), (4, 6) (5, 6)}

7. {(3, 8), (5, 8), (9, 8)}

8. {(2, 1), (2, 2), (2, 3), (2, 4)}

Items 9–11: For each arrow diagram, list the ordered pairs of the relationship and state the domain and range.

9. 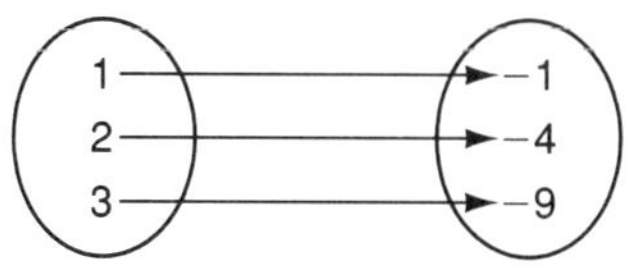

10. 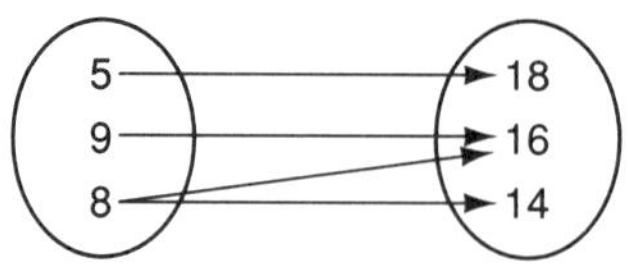

11. 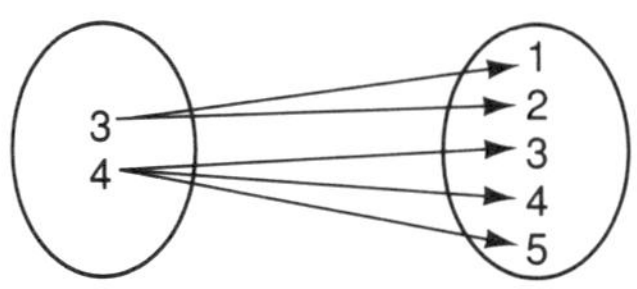

Functions

From relationships, we can form something that, in mathematics, is called a "function." A **function** is defined as a relation in which each member of the domain *corresponds to one and only one* member of the range.

For example, look at relations A and B below.

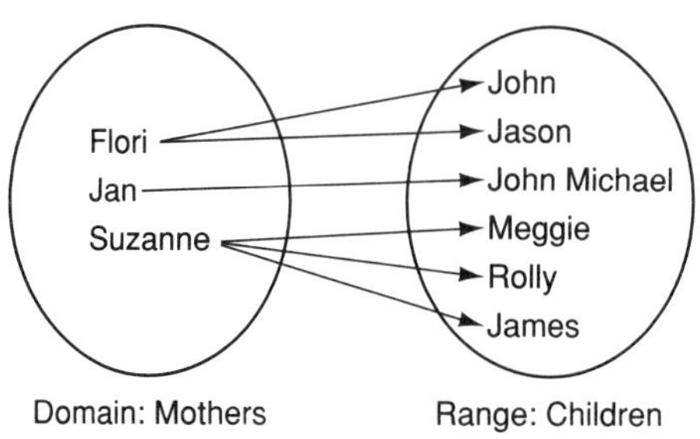

Relation A

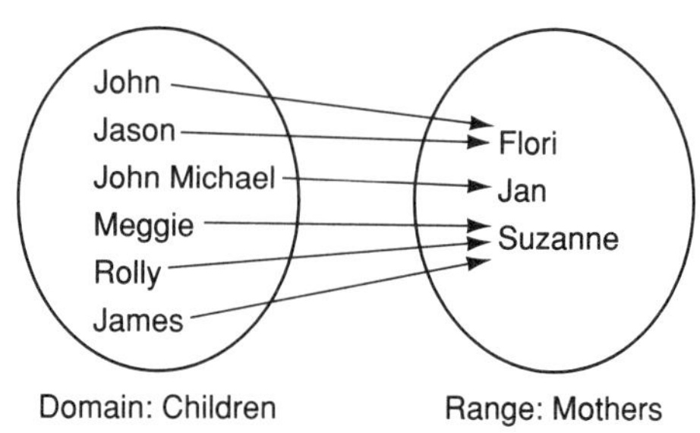

Relation B

- In relation A, each mother can have more than one child, but each child can have one and only one mother. Thus in relation A, Flori has two children, Jan has one child, and Suzanne has three children. Every mother in the domain corresponds to 1, 2, or 3 children in the range.
- In relation B, every child in the domain corresponds with one and only one mother.
- In this example, A and B are both relations, but only B is a function. This is because in B, each member of the domain corresponds to one and only one member of the range.

Another example of a function is the relationship between a number and its double. Look at the numbers 0, 1, 2, 3, and 4. Their relationship with their doubles can be written as ordered pairs: {(0, 0), (1, 2), (2, 4), (3, 6), (4, 8)}. This is a function, since no two ordered pairs have the same first element. That is, no member of the domain corresponds with more than one member of the range.

MODEL PROBLEMS

1. For each arrow diagram, determine if the relation is or is not a function.

a.

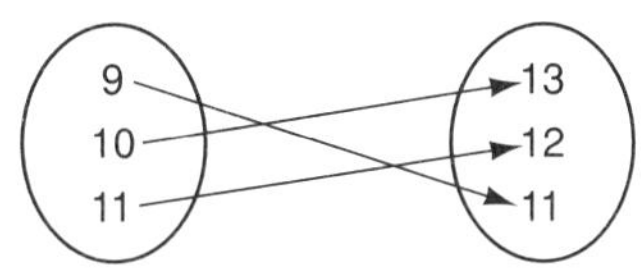

Answer: This is a function. Its range is {11, 12, 13}.

b.

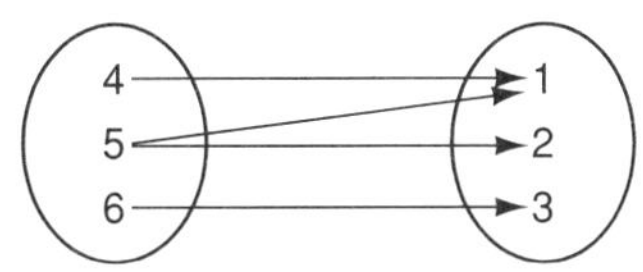

Answer: This is not a function because 5, in the domain, corresponds with both 1 and 2 in the range.

c.

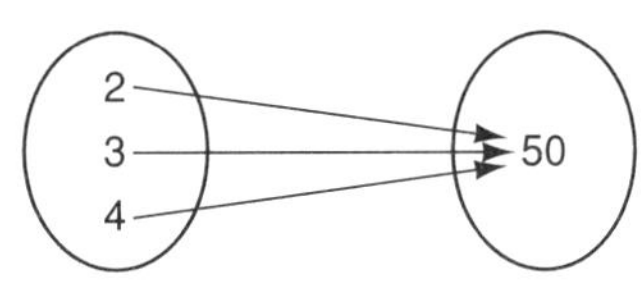

Answer: This is a function. Its range is {50}.

2. Which relations are functions?

a. {(2, 3), (5, 6), (7, 3)}

Answer: Function

b. {(5, 7), (8, 4), (5, 1)}

Answer: Not a function, since 5 corresponds to 7 and 1.

c. {(1, 2), (4, 8)}

Answer: Function

TRY THESE

Items 1 and 2: Identify each relation as a function or not a function.

1. a. {(4, 7), (2, 8), (5, 2), (1, 0), (9, 10)}

b. {(5, 5), (6, 6), (7, 7), (8, 8)}

2. a. {(2, 3), (2, 4), (2, 5), (2, 6), (2, 7)}

b. {(1, 2), (2, 2), (3, 2), (4, 2), (5, 2)}

3. a.

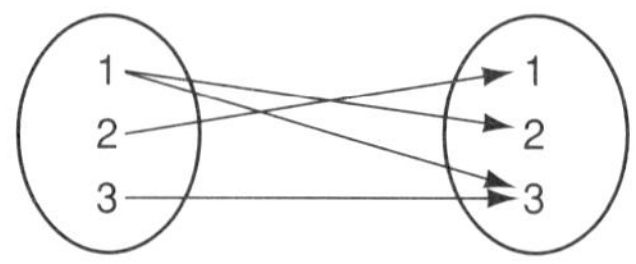

b.

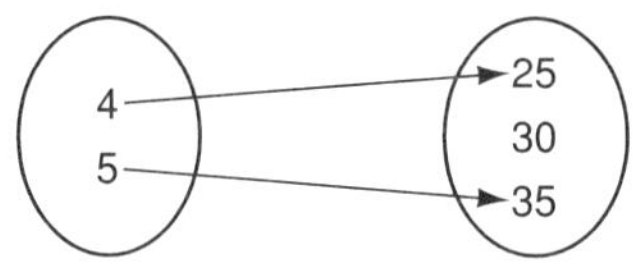

4. a.

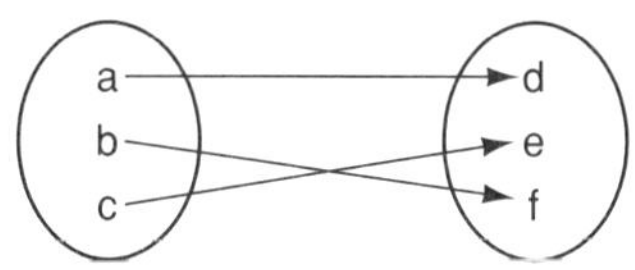

b.

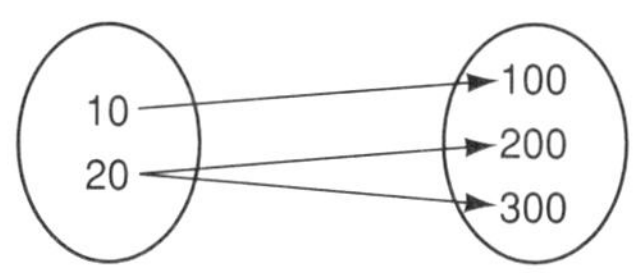

5. The table below records the height of Robert Wadlow, the world's tallest man, at several ages.

Age	Height (feet)
8	6.0
12	6.9
16	7.8
20	8.6
22	8.9

a. Write the ordered pairs of this relation (age, height).

b. Write the domain.

c. Write the range.

d. Is this relation a function?

FUNCTION NOTATION

If we look at a set of ordered pairs, we can often describe the set by stating a rule in symbols.

For example, consider this set of numbers and their doubles: {(0, 0), (1, 2), (2, 4), (3, 6), (4, 8)}.

If we expand the domain to include all real numbers, the number of ordered pairs would be infinite. If we let a variable, such as x, stand for any member of the domain, then $2x$ is the corresponding member of the range. The ordered pair can be written as $(x, 2x)$. This relation can be stated in words: *The function of* x *is* 2x. A familiar notation for this function is $f(x) = 2x$. Another example of a set of ordered pairs is: {(0, 0), (1, 1), (2, 4), (3, 9), (4, 16)}. If the domain of this set is expanded to include all real numbers, the rule for the set in **function notation** or **function form** is $f(x) = x^2$.

MODEL PROBLEM

Given the following set of ordered pairs, state in function notation a rule that describes the set:

{(0, 1), (1, 2), (2, 3), (3, 4), (4, 5)}

Answer: $f(x) = x + 1$

Ordered pair: $(x, x + 1)$

TRY THESE

For each given set of ordered pairs, state in function form a rule that describes it.

1. {(1, 0), (3, 2), (5, 4), (7, 6), (9, 8)}
2. {(0, 1), (1, 3), (2, 5), (3, 7), (4, 9)}
3. {(0, 0), (1, 1), (2, 8), (3, 27), (4, 64)}
4. {(0, 0), (1, 0), (2, 0), (3, 0)}
5. {(0, 0), (1, 0.25), (4, 1), (8, 2)}
6. {(0, 0), (1, 1), (2, 2), (3, 3), (4, 4)}
7. {(0, 1), (1, 4), (2, 7), (3, 10), (4, 13)}
8. {(0, −1), (1, 1), (2, 3), (3, 5), (4, 7)}
9. {(0, 0), (1, 1), (2, 4), (3, 9), (4, 16)}
10. {(0, 6), (2, 6), (4, 6), (8, 6), (10, 6)}

GRAPHING FUNCTIONS

Ordered pairs can be placed, or **plotted**, on a **coordinate graph** (see Chapter 3). This can be useful in our study of functions.

Each member of an ordered pair is called a **coordinate**. The first coordinate of the ordered pair is the x-value and is measured on the horizontal axis, the ***x*-axis**. In function vocabulary, this is the member of the domain.

The second coordinate of the ordered pair is the y-value and is measured on the vertical axis, the **y-axis**. In function vocabulary, this is said to be the corresponding member of the range.

The ordered pairs A (2, 3), B (−2, 3), C (−2, −3), and D (2, −3) are graphed below.

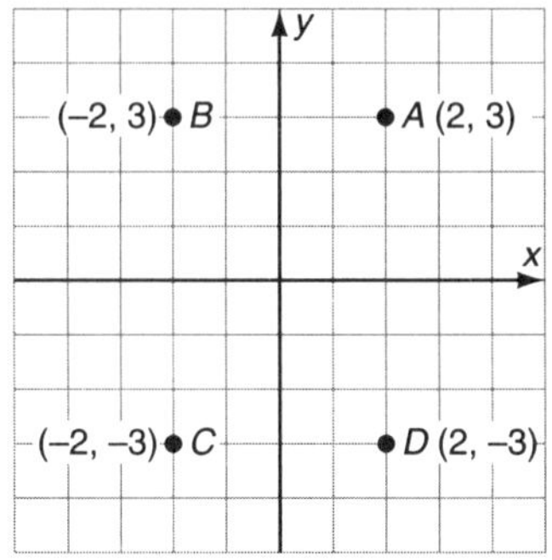

If a domain is increased so that it includes all real numbers, the number of ordered pairs will be infinite. To get an idea of the possible ordered pairs, we can graph some ordered pairs of the function and then connect them in sequence from left to right. Below are graphs of the functions $f(x) = 2x$, $f(x) = x^2$, and $f(x) = x + 1$ that we found earlier. These rules can also be written as $y = 2x$, $y = x^2$, $y = x + 1$. In this case we say that y *is a function of* x.

{(−1, −2), (1, 2), (2, 4), (3, 6), (4, 8)}
$f(x) = 2x$

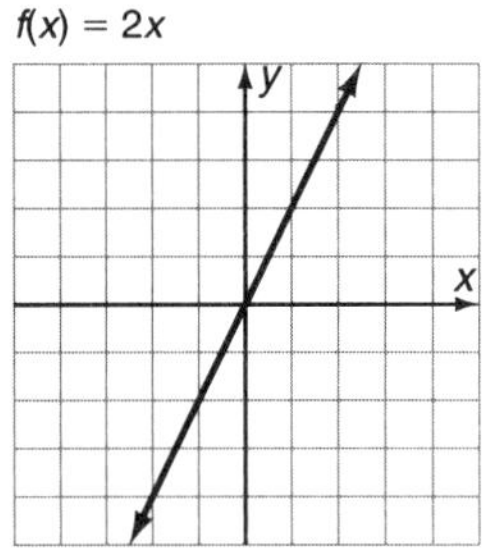

{(−1, 1), (1, 1), (2, 4), (3, 9), (4, 16)}
$f(x) = x^2$

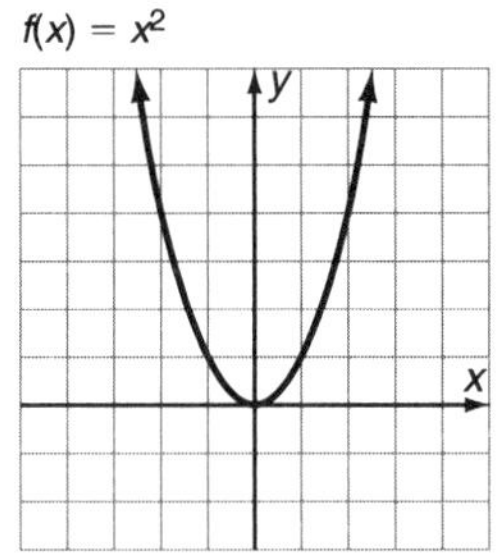

{(−1, 0), (1, 2), (2, 3), (3, 4), (4, 5)}
$f(x) = x + 1$

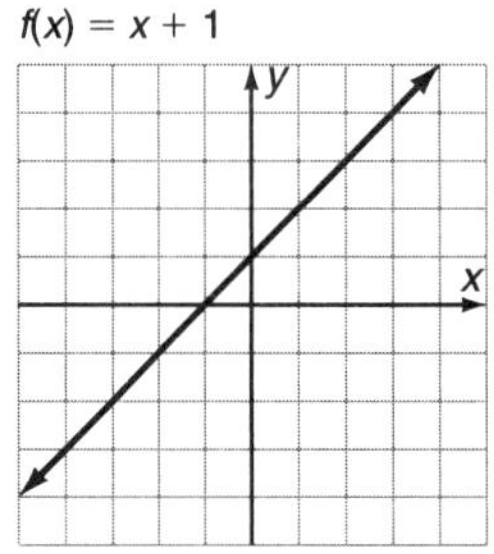

We can determine whether a graph is or is not a function by using the **vertical line test**. A function is defined as a relation in which one and only one value of the range (y) corresponds to each value of the domain (x). If a graph is a function and we draw a vertical line anywhere on the coordinate grid, that line will intersect the graph at no more than one point. All points on the vertical line will have the same x value. Therefore, only one y-value is part of the graph of a function. If there is more than one y-value, the graph is not a function. Look at the following examples.

EXAMPLE 1

$y = x^2$

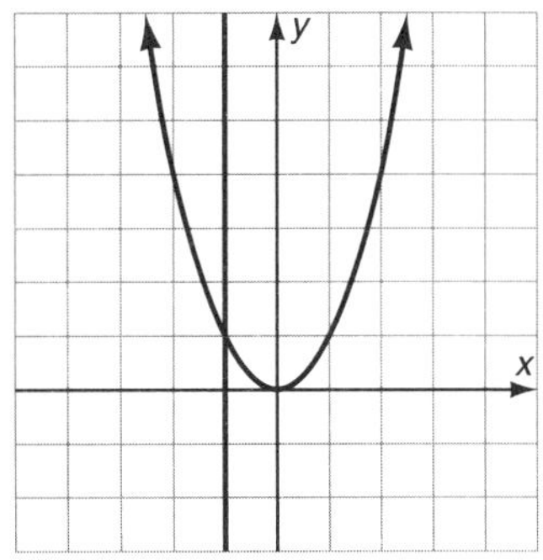

The graph above passes the vertical line test, because any vertical line will intersect the graph at only one point. Therefore, $y = x^2$ is a function.

EXAMPLE 2

$y^2 = x$

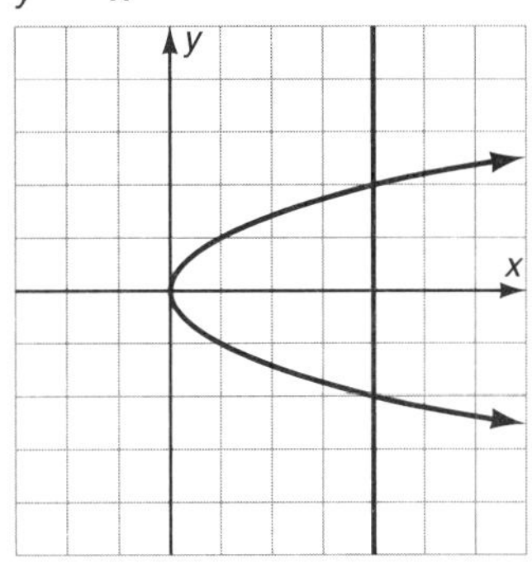

This graph does *not* pass the vertical line test, because the vertical line intersects the graph at more than one point. Therefore, $y^2 = x$ is *not* a function.

MODEL PROBLEMS

Using the vertical line test, determine which of the following graphs are functions:

1.

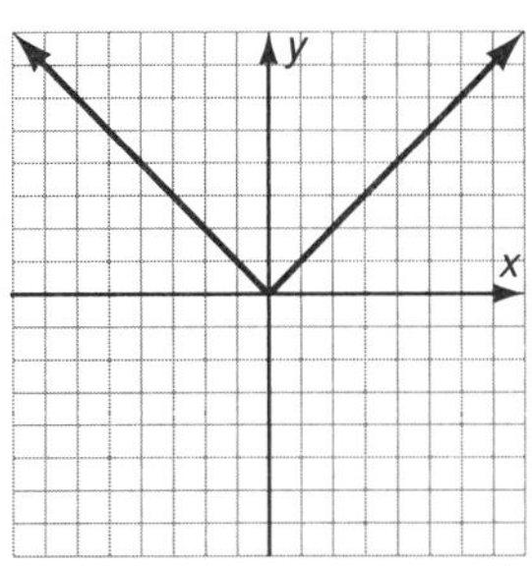

Answer:

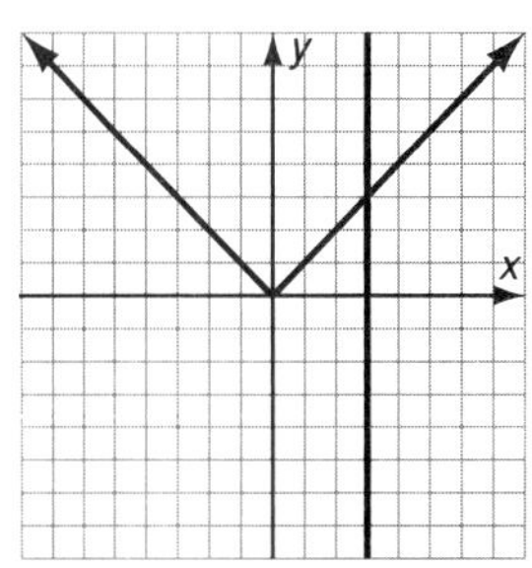

Function

2.

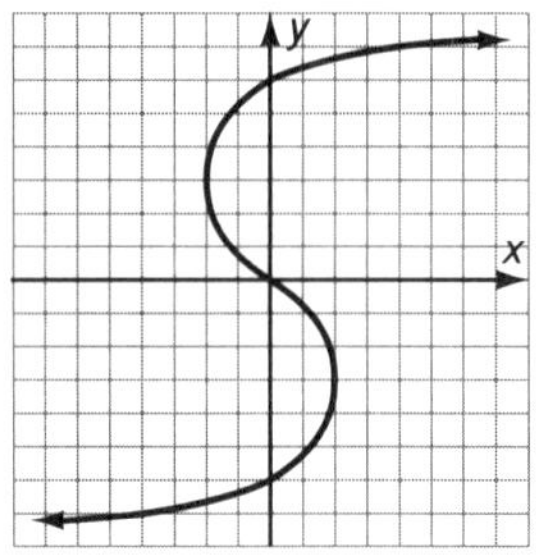

Answer:

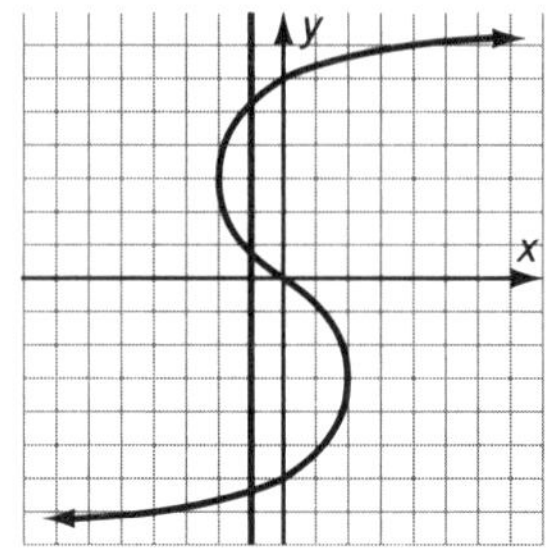

Not a function

3.

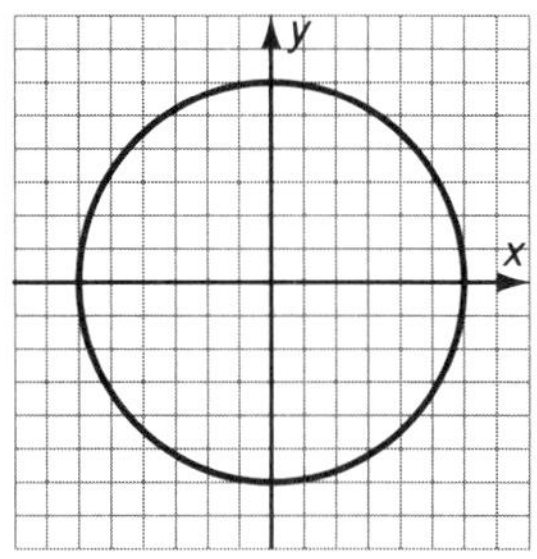

Answer:

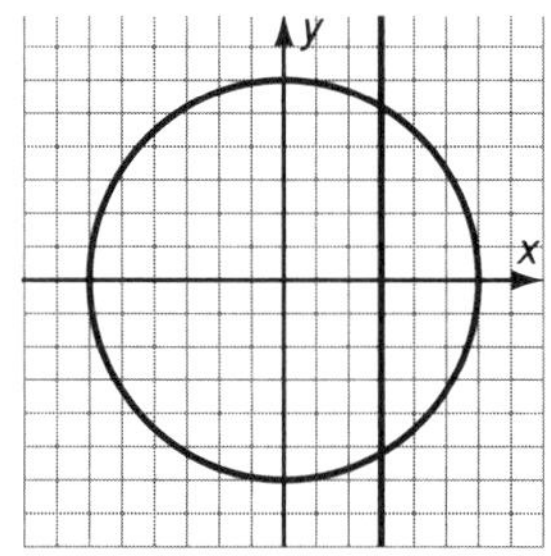

Not a function

4.

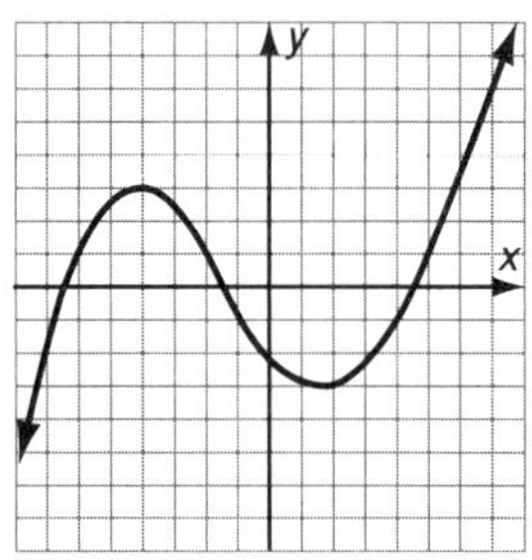

Answer:

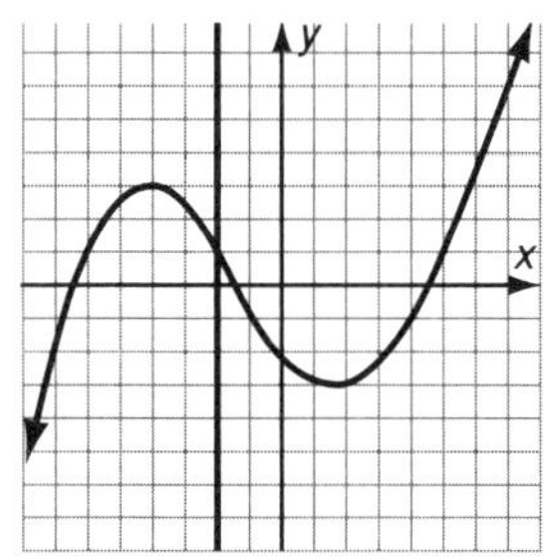

Function

TRY THESE

Using the vertical line test, determine which of the following graphs are functions.

1.

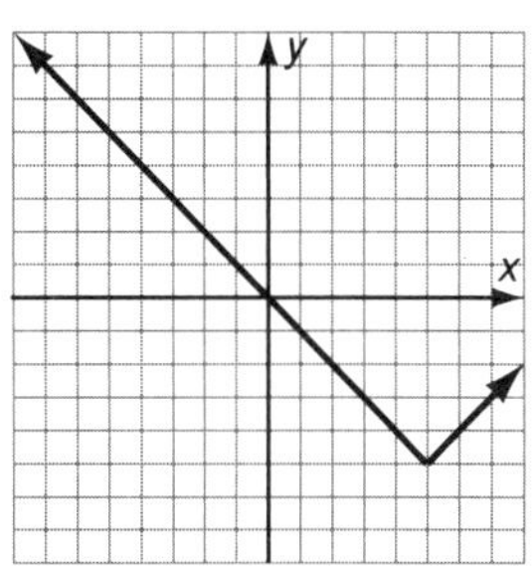

2.

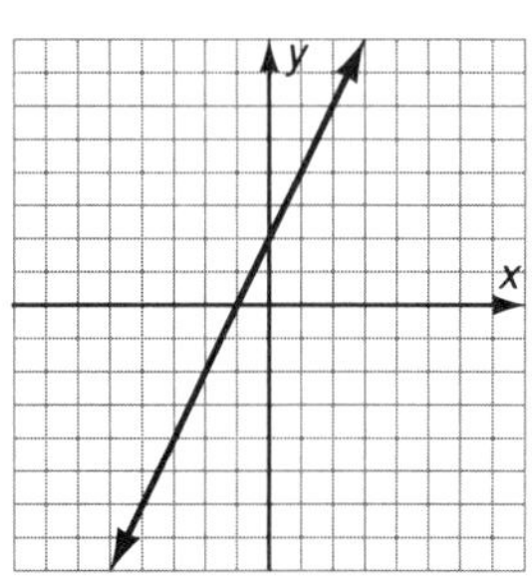

3.

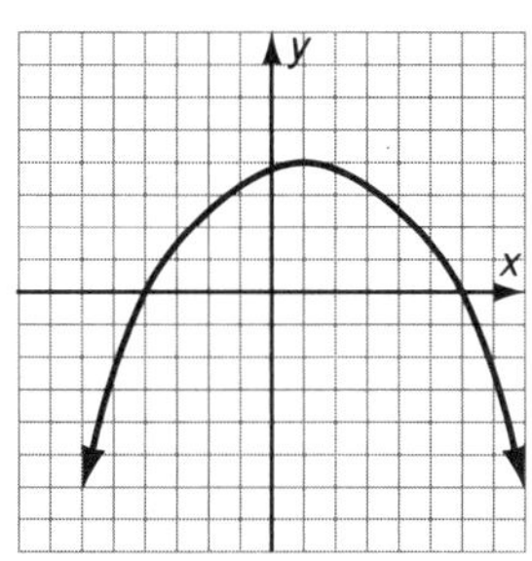

4.

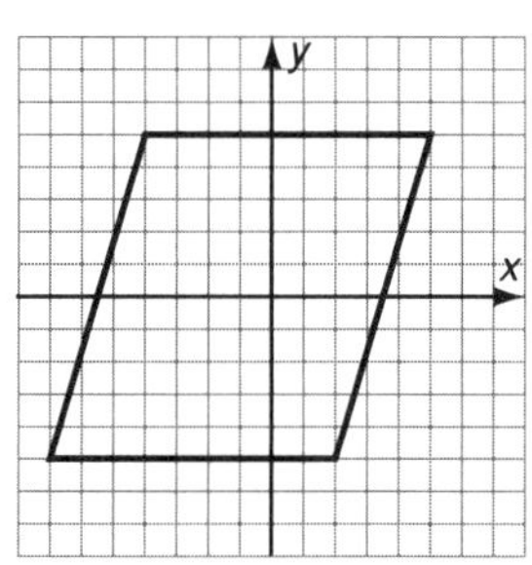

5. 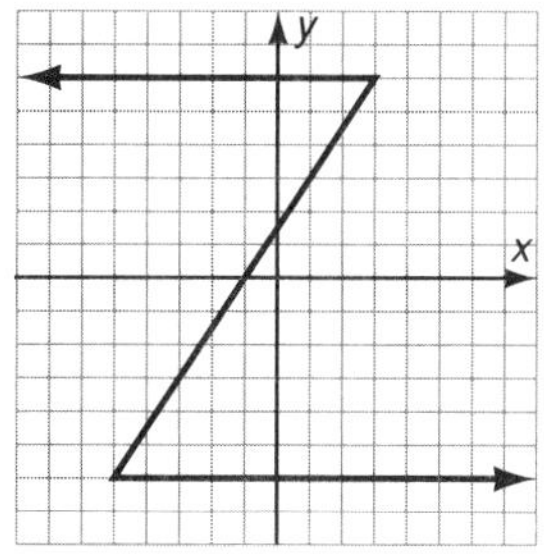

8. 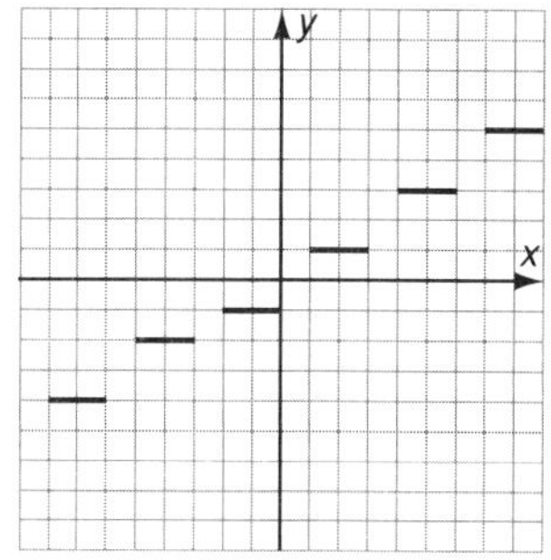

6. 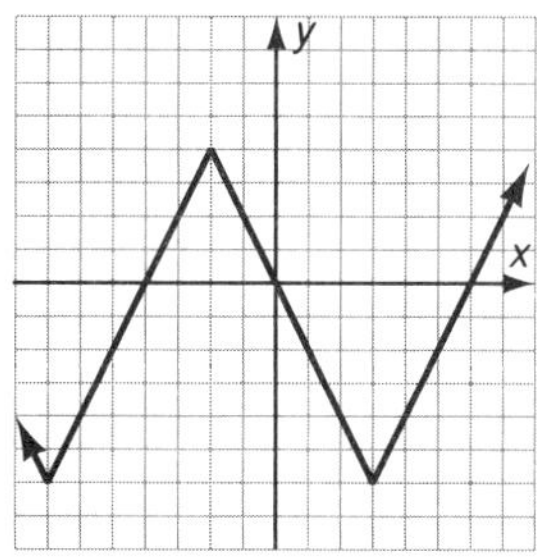

9. 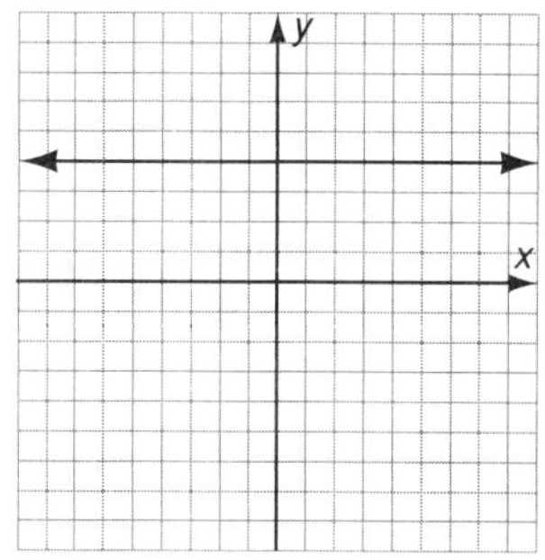

7. 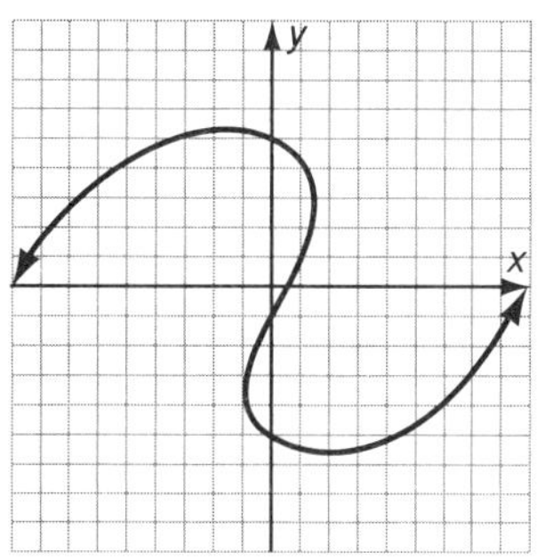

10. 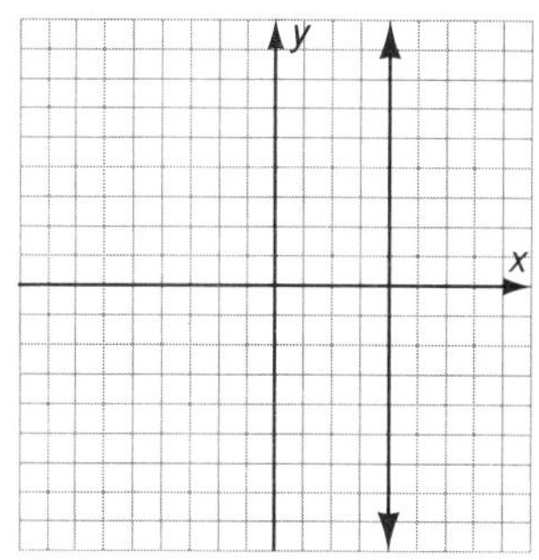

APPLICATIONS OF FUNCTIONS

Functions are used in real-world problems. For example, suppose that your school club is selling candy bars to raise money. You might use the following table to find the cost of more than 1 candy bar.

Number of candy bars	Cost
1	\$1.25
2	\$2.50
3	\$3.75
4	\$5.00
5	\$6.25

You can also write a rule or **formula**, letting a variable such as x represent the number of candy bars. Then \$1.25x would represent the total cost of x candy bars:

$f(x) = \$1.25x$ or $y = \$1.25x$

Another kind of function is a repair person's rate of pay. Suppose this person charges an initial fee of \$50 to come to your home and then \$35 per hour until the job is completed:

Let h = numbers of hours worked

$f(h)$ = total cost of the job

$f(h) = \$50 + \$35h$ or $y = \$50 + \$35h$

Any formula is really a form of a function. For example, the area A of a circle depends on the radius of the circle: $A = \pi r^2$. We can say that the area is a function of the radius or A is a function of r.

MODEL PROBLEMS

1. The rental rate for a car is \$40 per day and \$0.35 per mile. Write an equation representing the cost of renting this car for 1 day.

   Answer: Let x = number of miles driven and y = cost of renting the car for 1 day. Then:

   $y = \$40 + \$0.35x$

2. The pressure on the bottom of a water tank is 62.4 times the depth of the water in the tank. Write an equation to represent this.

   Answer: Let d = depth of water and p = pressure. Then:

   $p = 62.4d$

TRY THESE

Items 1–3: Write an equation for each situation.

1. Cost C of printing a booklet at \$1.25 plus \$0.15 per page (p).
2. Perimeter P of an equilateral triangle with side s.
3. Cost C of renting a rug cleaner at \$10 plus \$0.75 per hour (h).
4. Perimeter P of a rectangle with length l and width w.
5. Salary S of \$5.50 per hour for the first 40 hours and \$8.25 for each hour (h) over 40 hours.
6. Cost C for a plumber who charges \$50 for a service call plus \$25 per hour (h) for time worked.
7. Area A of a square with a side s.
8. Cost C for a party when the hotel charges \$150 plus \$5.80 per person (p) who attends the party.
9. Amount A that a salesperson earns from an 8% commission on sales s.
10. Refer back to the example of the candy bars on page 265.
    a. What is the cost of 8 bars?
    b. How many bars would cost \$12.50?

Assessment

Multiple-Choice Problems

1. Which of the following relations is a function?

   A. $\{(3, 5), (2, 6), (1, 7), (0, 6), (1, 8)\}$

   B. $\{(4, 5), (4, 6), (4, 7), (4, 8)\}$

   C. $\{(1, 6), (2, 6), (3, 6), (4, 6), (5, 6)\}$

   D. $\{(2, 6), (6, 2), (1, 2), (2, 1)\}$

2. Which is NOT a function?

   A. $\{(0, -1), (-1, -2), (-2, -3), (-3, -4)\}$

   B. $\{(7, -3), (8, -3), (-3, 8), (3, 7)\}$

   C. $\{(4, 5), (5, 4), (4, 6), (6, 4)\}$

   D. $\{(3, 9), (2, 8), (1, 7), (0, 6)\}$

3. A function is described by the equation $3n - 1 = m$. What value of m completes the table?

n	0	2	5	6	11
m	−1	5	14	?	32

   A. 15 **C.** 17

   B. 16 **D.** 24

4. Find the number that completes this table.

x	1	2	3	4	5	6	7	8
y	1	1	2	3	5	8	?	21

   A. 12 **C.** 16

   B. 13 **D.** 20

5. Which decimal represents this visual pattern?

   △ ▽ ◁ ◁ . . .

   A. 0.3333333 . . .

   B. 0.12331233 . . .

   C. 0.32132132 . . .

   D. 0.31313131 . . .

6. Which graph represents a child climbing up a slide and then sliding down?

A.

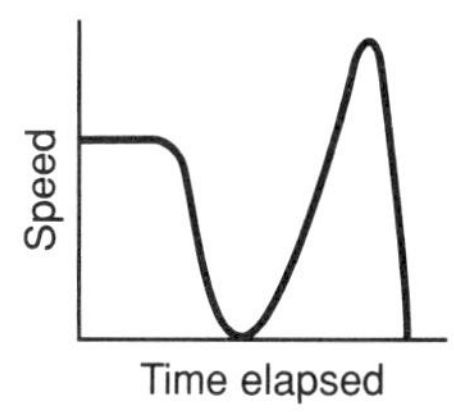

C.

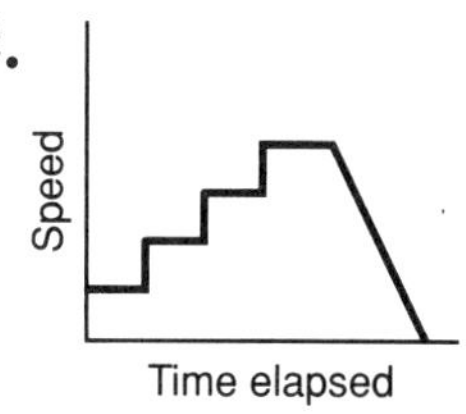

B.

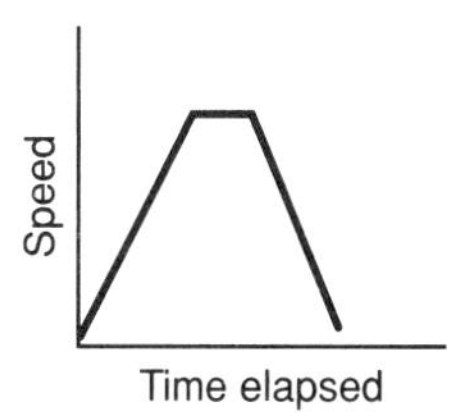

D.

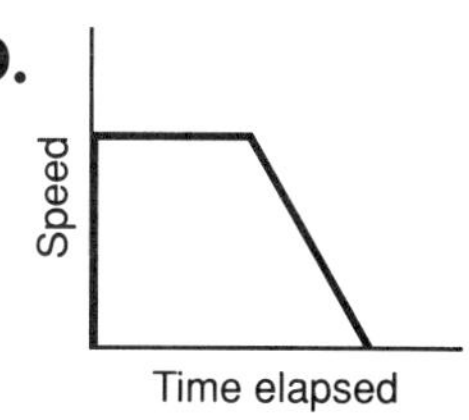

Gridded-Response Problems

Items 7 and 8: The school's mini-theater is shaped like a trapezoid. There are 46 seats in the first row. Each row has 2 fewer seats. There are 14 rows in the theater.

7. How many seats are in the 12th row?

8. How many seats are in the mini-theater?

9. Find the missing number.

f	−2	−1	0	1	2	3	6
h	−8	−1	0	1	8	27	?

10. The expression $4(x + 1)$ defines y. Find the missing number.

x	0	3	7	8	11	20	25
y	4	16	32	36	?	84	104

Items 11 and 12: Robert is stacking logs for his father. The log pile is in the shape of a trapezoid with 20 logs in the bottom row and 5 logs in the top row. Each row has one less log than the row below it.

11. How many rows of logs are in the log pile?

12. What is the total number of logs in the pile?

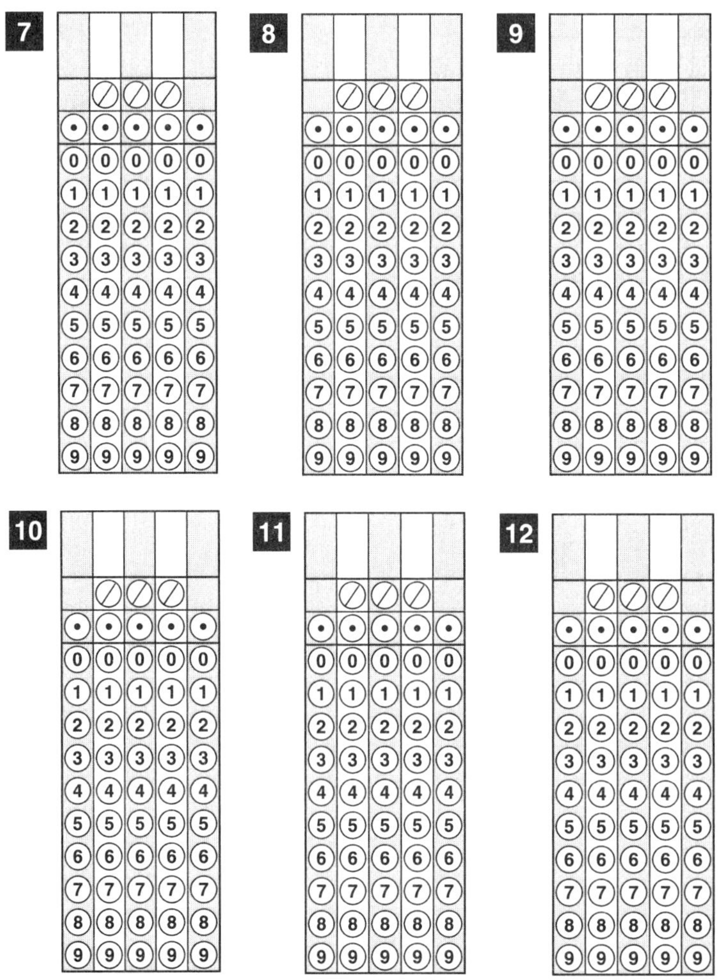

B CHANGING PARAMETERS TO DETERMINE THE IMPACT ON A GIVEN FUNCTION

In a relation, a function, or an equation, an unknown quantity is called a **variable**. A known quantity—a given number—is called a **constant** or a **parameter**. Here are two examples of these concepts.

EXAMPLE 1

If Dyane earns \$7.25 per hour and works for x hours in one day, her earnings for that day can be expressed as \$7.25x. The variable is x and the parameter is \$7.25.

If she gets a raise, the parameter will change. For example, if she gets a raise of \$0.50 per hour, the amount she earns in one day will be expressed as \$7.75x.

EXAMPLE 2

Jeff wants to enlarge a fenced-in portion of his backyard. This portion is rectangular, and its present dimensions are 15 feet long by 10 feet wide. Because of the way it is placed, its width cannot be changed. Jeff wants an area of 200 square feet. To get this new area, how much will he have to increase the length?

One parameter, width (10 feet), must stay the same. The other parameter, length (15 feet), will change. The area of a rectangle is length times width. To get the new area, 10 times some number must equal 200:

$$w \times l = A$$
$$10 \times l = 200$$
$$l = \frac{200}{10}$$
$$l = 20$$

Therefore, the parameter of length must change (increase) from 15 feet to 20 feet.

MODEL PROBLEMS

1. Mr. Coe takes automobile trips to see clients. He is reimbursed at a rate of \$0.45 per mile. His average weekly mileage is 350 miles. How much more will he receive per week if the rate is increased to \$0.50 per mile?

Solution:

Present reimbursement = 350 × \$0.45 = \$157.50

Future reimbursement = 350 × \$0.50 = \$175.00

Increase in reimbursement = \$175.00 − \$157.50 = \$17.50

Alternative solution:

Future rate = \$0.50

Present rate = \$0.45

Increase in rate = future rate minus present rate =

\$0.50 − \$0.45 = \$0.05

Increase in reimbursement = \$0.05 × 350 = \$17.50

Answer: He will receive \$17.50 more per week.

2. A manufacturer wants to reduce the volume of a cylindrical can by reducing its height. The present radius of the can is 2 inches, and the present height is 21 inches. The reduced volume of the new can must be 176 cubic inches. By how many inches must the height of the old can be decreased? Use the formula for volume of a cylinder $V = \pi r^2 h$. Use $\pi \approx \frac{22}{7}$.

Solution:

$V = \pi r^2 h$

$176 \approx \frac{22}{7} \times (2 \times 2) \times h$

$176 = \frac{88}{7} \times h$

$\frac{7}{88} \times 176 = \frac{88}{7} \times \frac{7}{88} \times h$

$14 = h$, so the height of the new can is 14 inches

Subtract:
(height of old can) − (height of new can) = amount of decrease
21 − 14 = 7

Answer: The height of the old can must be decreased by 7 inches.

TRY THESE

1. The cost of 7 pizzas is \$77. You have collected \$100 from your class to buy pizzas. How many pizzas can you buy? Show your work.

2. On an automobile trip last summer, the Harris family averaged 50 miles per hour. On the first day they drove for 7 hours and traveled 350 miles. On the second day they drove for 5 hours, stopped for lunch, and then drove for 4 more hours. How many miles did they drive on the second day? Show your work.

Items 3–5: Mike records the number of gallons of gasoline that he buys each month. Last month he bought 89 gallons of gas. His car gets 23 miles per gallon. Mike is considering buying a new car that gets 34 miles per gallon.

3. How much farther, in one month, will Mike be able to drive with the new car if he uses the same number of gallons of gasoline?

4. If Mike drives the same distance with his new car as he did in a month with his old car, how many gallons of gasoline will he use?

5. If the price of gasoline remains constant at \$2.94 per gallon, how much will Mike save this month by driving the same distance with his new car?

6. A company ships most of its items in boxes that are shaped like cubes with edges of 12 inches. How much more volume will the boxes have if the base remains the same but the height is increased by 3 inches?

7. The company wants to keep the height of its newest package at 15 inches and the base a square. The volume has to be 2,940 cubic inches. What will be the necessary dimensions of the base?

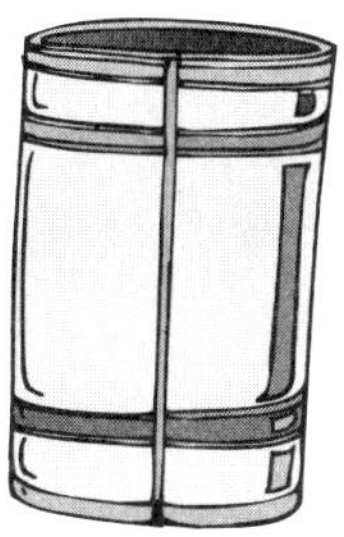

8. The company decided that the parts could be shipped in a cylindrical container with a height of 15 inches. If the volume remains at 2,940 cubic inches, what is the radius of the container? Use $\pi \approx 3.14$.

9. Frank has a garden that is 9 feet by 15 feet that he wants to enlarge. In order to triple the area of the garden, he plans to double the length. What change must he make in the width?

10. Wallpaper that is 34 inches wide has four 3-inch blue stripes alternating with 5 white stripes. Another pattern of the same overall width has five 2-inch blue stripes alternating with 6 white stripes. What is the difference in the size of the white stripes in the two types of wallpaper?

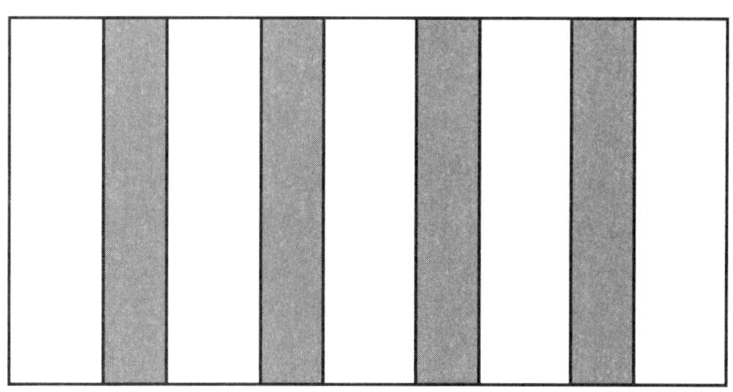

Assessment

Multiple-Choice Problems

1. A newspaper carrier earns $0.11 for each paper delivered. Brandon's route consists of 76 customers. If Brandon could increase the number of customers by 15, how much more would he earn each day?

   A. $1.65 **C.** $8.36

   B. $2.65 **D.** $10.01

2. Angelique made arrangements to rent a bus for a class trip. She received a quote of $630. Each student would have to pay an equal share of the cost. Originally, 45 students signed up for the trip, but only 40 students actually rode the bus. Only those riding the bus had to pay. How much less would it have been per student if all 45 who signed up had actually ridden the bus?

   A. $1.00 **C.** $3.50

   B. $1.75 **D.** $4.00

3. Secondmaid Orange Juice Company has been selling concentrated orange juice in 6-centimeter-tall cans that have a radius of 4 centimeters. As a marketing tool, they want to increase the radius of the can to 5 centimeters so that it will hold 2.5 times more concentrate. What will be the height of the new container?

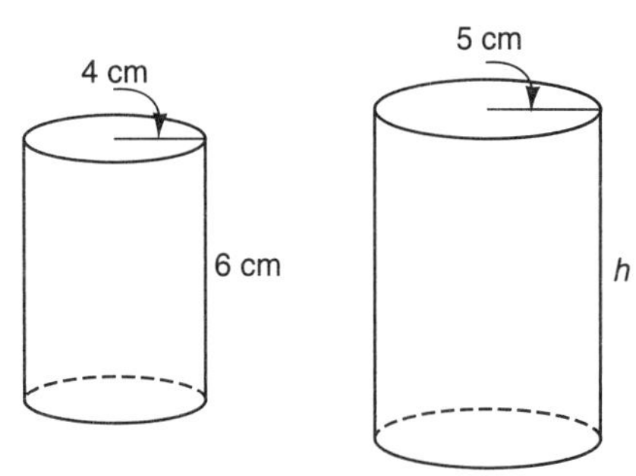

   A. 3.84 centimeters

   B. 9.6 centimeters

   C. 12.3 centimeters

   D. 15 centimeters

4. On her paper, Megan drew a square with a side of 3 inches. Her teacher then asked her to draw two more squares, one with the original side doubled and one with the original side squared. What is the difference in the areas of the two new squares?

   A. 9 square inches

   B. 27 square inches

   C. 45 square inches

   D. No difference in area

Gridded-Response Problems

5. Lloyd needs to decrease the area of his garden by one-third. At the present time his garden is 20 feet by 12 feet. If he reduces the width from 12 feet to 10 feet, what will be the new length?

6. At Suncoast High School, the ratio of boys to girls participating in sports is 7 to 4. Last year the total number of students participating in sports was 154. This year the total number increased to 220. How many more girls participated this year than last year?

7. A cylindrical container has a height of 12 inches and a diameter of 6 inches. How many times more liquid is contained in this container than a cylindrical container that is 12 inches high and has a diameter of 5 inches?

8. Delroy has just finished tiling the floor in his bedroom. The dimensions of the floor are 12 feet by 15 feet. He used tiles that measured 12 inches by 12 inches and cost $1.65 each. Now, the same tiles are on sale at $1.05 each. How much would Delroy have saved if he had bought the tiles on sale?

5
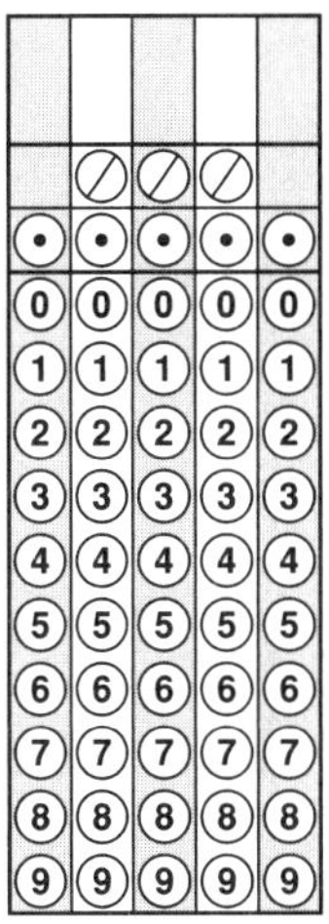
6
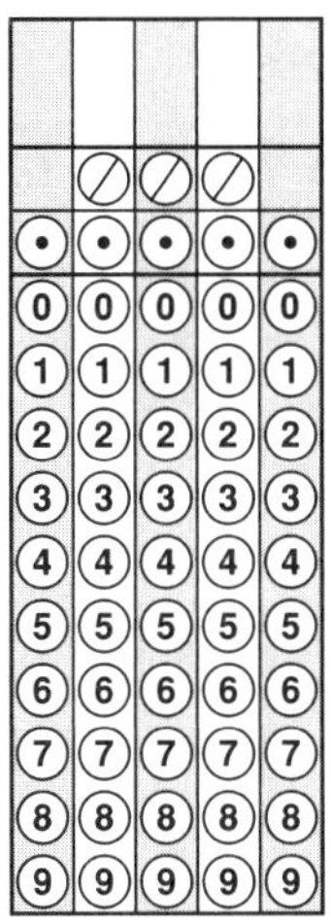
7
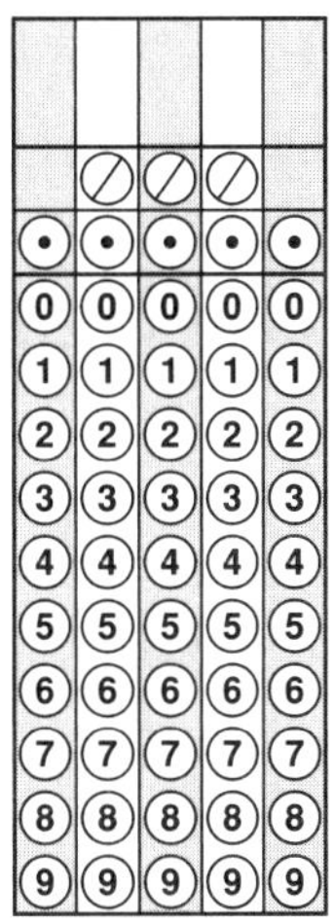
8
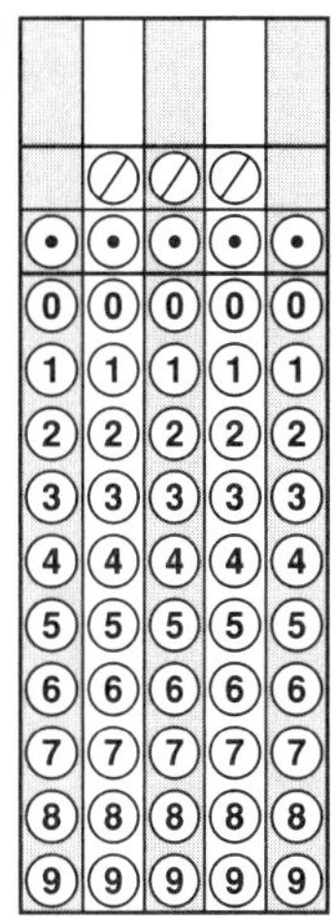

Think, Solve, Explain: Short-Response Problems

For each problem, state the answer and show your work.

9. Miguel is asked to compare the volume of a cube with an edge of 10 cm to a sphere with a diameter of 12 cm.

   Part A Which is greater and by how much? Use $\pi \approx 3.14$.

   Part B Miguel wants to alter the shape of the cube so that it has the same volume as the sphere. If he keeps the base of the box as a square with a side of 10 cm, what must the height be in order to have the same volume as the sphere?

10. The Richter family has a cylindrical storage tank with a diameter of 6 feet and a height of 4 feet. They can buy a new tank with the same height and a diameter 1 foot larger or one that is 1 foot higher and has the same diameter. Show all calculations of the two new capacities and explain which one has a greater volume and by how much.

Section 2 Representing and interpreting situations by using expressions, equations, inequalities, graphs, and formulas

A SOLVING PROBLEMS BY USING ALGEBRAIC EXPRESSIONS, EQUATIONS, AND INEQUALITIES

Algebraic expressions

In some situations, an algebraic expression can be used to solve a problem. For example, suppose that you have a coupon for \$3 off the cost of any item in a store. Then the cost of an item to you can be written as the algebraic expression $x - 3$ where x represents the original cost of the item.

MODEL PROBLEMS

1. A car travels k kilometers per hour. Write an algebraic expression that can be used to find out how far the car has traveled in 7 hours.

   Answer: 7 times k or $7k$

2. After 15 centimeters were cut off of a pipe, g centimeters were left. How long was the original pipe?

   Answer: $15 + g$

TRY THESE

1. One football costs F dollars. Write an algebraic expression to represent the number of footballs that can be bought for \$250.
2. A bowl contains x peanuts and y potato chips. There are 15 calories in each peanut and 17 calories in each potato chip. Write an algebraic expression to represent the total calories in the bowl.

Items 3–6: Represent each of the following algebraically.

3. Number of days in W weeks.
4. Number of feet in K inches.
5. Loss of D dollars from a total of \$250.
6. Number of pieces of paper that each of S students will get if there are 40 pieces.
7. The cost of n candy bars at \$1.50 each.
8. The distance driven at 65 mph for h hours.

9. The number of girls (g) in a class of 35 students with b boys.
10. The cost of b loaves of bread at \$1.25 a loaf and s two-liter bottles of soda at \$0.98 each.

NOTE: For more problems, turn to page 464.

Equations

SOLVING EQUATIONS

Solving an equation involves a process of "undoing." You keep on "undoing" until the variable by itself is equal to a number.

- Undo addition by subtraction: $x + 5 \rightarrow x + 5 - 5 \rightarrow x$
- Undo subtraction by addition: $x - 3 \rightarrow x - 3 + 3 \rightarrow x$
- Undo multiplication by division: $7x \rightarrow \frac{7x}{7} \rightarrow x$
- Undo division by multiplication: $\frac{x}{2} \rightarrow \frac{x}{2} \bullet 2 \rightarrow x$

The order of the "undoing" is very important. *First*, terms added or subtracted to the variable must be "undone." *Next*, you "undo" multiplication or division.

An equation is like a balance scale. To keep it in balance, what is done on one side of the equals sign must also be done on the other side of the equals sign.

EXAMPLE

$2x - 7 = 13$	Equation to be solved.
$2x - 7 + 7 = 13 + 7$	Add 7 to both sides.
$2x = 20$	Simplify.
$\frac{2x}{2} = \frac{20}{2}$	Divide both sides by 2.
$x = 10$	Simplify for solution.

Check the solution by substituting the value of x in the original equation:

$$2x - 7 = 13$$
$$2(10) - 7 = 13$$
$$20 - 7 = 13$$
$$13 = 13$$

When an equation contains sums or differences multiplied by a number or a variable, another important procedure is necessary. This procedure is the use of the **distributive property**.

EXAMPLES OF THE DISTRIBUTIVE PROPERTY

$a(b + c) = ab + ac$

$a(b - c) = ab - ac$

$2(x + 5) = 2x + 10$

$x(y - 3) = xy - 3x$

MODEL PROBLEMS Solve the following equations.

1. $3x + 10 = 16$

Solution:

$3x + 10 - 10 = 16 - 10$	Subtract 10 from both sides.
$3x = 6$	Simplify.
$\frac{3x}{3} = \frac{6}{3}$	Divide both sides by 3.
$x = 2$	Simplify.

Answer: $x = 2$

Check:

$3x + 10 = 16$

$3(2) + 10 = 16$

$6 + 10 = 16$

$16 = 16$

2. $\frac{x}{-4} - 5 = 19$

Solution:

$\frac{x}{-4} - 5 + 5 = 19 + 5$	Add 5 to both sides.
$\frac{x}{-4} = 24$	Simplify.
$\frac{x}{-4} \bullet (-4) = 24 \bullet (-4)$	Muliply both sides by −4.
$x = -96$	Simplify.

Answer: $x = -96$

Check:

$\frac{x}{-4} - 5 = 19$

$\frac{-96}{-4} - 5 = 19$

$24 - 5 = 19$

$19 = 19$

3. $3(x + 2) = 18$

Solution:

$3x + 3(2) = 18$	Use the distributive property.
$3x + 6 = 18$	Simplify.
$3x + 6 - 6 = 18 - 6$	Subtract 6 from both sides.
$3x = 12$	Simplify.
$\frac{3x}{3} = \frac{12}{3}$	Divide both sides by 3.
$x = 4$	Simplify.

Answer: $x = 4$

Check:

$$3(x + 2) = 18$$
$$3(4 + 2) = 18$$
$$3(6) = 18$$
$$18 = 18$$

4. $5x + 3 = 3x - 8$

Solution:

$5x - 3x + 3 = 3x - 3x - 8$	Get the variable on one side of the equation.
$2x + 3 = -8$	Combine.
$2x + 3 - 3 = -8 - 3$	Subtract 3 from both sides.
$2x = -11$	Simplify.
$\frac{2x}{2} = \frac{-11}{2}$	Divide both sides by 2.
$x = \frac{-11}{2}$	Simplify.

Answer: $x = -\frac{11}{2}$ or $-5\frac{1}{2}$

Check:

$$5x + 3 = 3x - 8$$
$$5\left(\frac{-11}{2}\right) + 3 = \left[(3)\left(\frac{-11}{2}\right)\right] - 8$$
$$-27\frac{1}{2} + 3 = -16\frac{1}{2} - 8$$
$$-24\frac{1}{2} = -24\frac{1}{2}$$

5. $3x + 5 + 4x - 3 = 5x + x - 1$

Solution:

$7x + 2 = 6x - 1$ Combine.

$7x - 6x + 2 = 6x - 6x - 1$ Get the variable on one side of the equation.

$x + 2 = -1$ Simplify.

$x + 2 - 2 = -1 - 2$ Subtract 2 from both sides.

$x = -3$ Simplify.

Answer: $x = -3$

Check:

$$3x + 5 + 4x - 3 = 5x + x - 1$$
$$3(-3) + 5 + 4(-3) - 3 = 5(-3) + (-3) - 1$$
$$-9 + 5 - 12 - 3 = -15 - 3 - 1$$
$$-19 = -19$$

TRY THESE Solve each equation for x.

1. $2x + 4 = 12$

2. $3x - 7 = 5x + 3$

3. $4(x + 5) = -8$

4. $\frac{x}{5} - 9 = 12$

5. $3x - 6 = 24$

6. $8(x - 2) = 6x + 10$

7. $\frac{2}{3}x + 3 = 7$

8. $\frac{7x + 2}{4} = 10$

9. $3(x - 5) + 10 = -5$

10. $11x + 8 = x - 14$

NOTE: For more problems, turn to page 465.

APPLICATIONS OF EQUATIONS

Many real-world problems can be solved by setting up and solving an equation. Follow this procedure:

Step 1. Read the problem carefully to determine what is known and what needs to be found.

Step 2. Assign a variable to the unknown quantity.

Step 3. Write an equation. The equation expresses the given information in terms of the variable.

Step 4. Solve the equation for the value of the variable.

Step 5. Check your answer for reasonableness and accuracy.

MODEL PROBLEMS

1. A part-time salesperson receives \$200 per week. She also receives \$5 for every program she sells. Last week she earned \$300. How many programs did she sell?

Solution: Let x = number of programs she sold. Then:

$$5x + 200 = 300$$
$$5x + 200 - 200 = 300 - 200$$
$$\frac{5x}{5} = \frac{100}{5}$$
$$x = 20$$

Answer: She sold 20 programs.

Check:

$$5x + 200 = 300$$
$$5(20) + 200 = 300$$
$$300 = 300$$

2. Danielle bought a skirt and jacket for \$45. The price of the jacket was twice as much as the price of the skirt. How much did the skirt cost?

Solution: Let x = cost of the skirt. Then $2x$ = cost of the jacket. Therefore:

$$x + 2x = 45$$
$$3x = 45$$
$$\frac{3x}{3} = \frac{45}{3}$$
$$x = 15$$

Answer: The skirt cost \$15.

Check:

$$x + 2x = 45$$
$$15 + (2 \bullet 15) = 45$$
$$15 + 30 = 45$$
$$45 = 45$$

Also go back to the original problem to check the solution. The shirt cost \$15 and the jacket cost 2 × \$15 or \$30. Therefore, \$15 + \$30 = \$45.

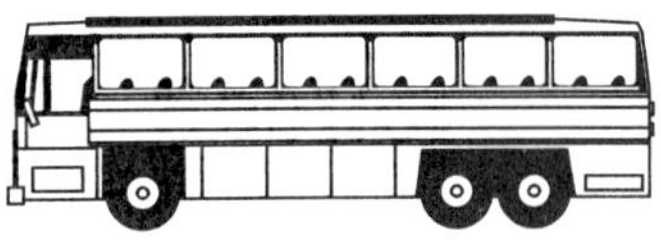

3. A bus company has two types of buses. The smaller bus holds 12 fewer passengers than the larger bus. A club reserves 4 large buses and 5 small buses to seat a total of 336 people. How many passengers will one large bus seat?

Solution: Let x = number of passengers in a large bus.

Then $x - 12$ = number of passengers in a small bus. Therefore:

$$\begin{aligned} 4x + 5(x - 12) &= 336 \\ 4x + 5x - 60 &= 336 \\ 9x - 60 &= 336 \\ 9x - 60 + 60 &= 336 + 60 \\ 9x &= 396 \\ \frac{9x}{9} &= \frac{396}{9} \\ x &= 44 \end{aligned}$$

Answer: A large bus will seat 44 passengers.

Check:

$$\begin{aligned} 4x + 5(x - 12) &= 336 \\ 4(44) + [(5)(44 - 12)] &= 336 \\ 176 + 160 &= 336 \\ 336 &= 336 \end{aligned}$$

Check the answer in the original problem. Since $x = 44$, four large buses hold $4 \bullet 44$ or 176 people. Since $x - 12 = 32$, five small buses hold $5 \bullet 32$ or 160 people. $176 + 160 = 336$.

TRY THESE

1. The sum of the length and width of a rectangle is 44 inches. The length is 3 times the width. Find the width of the rectangle.

2. The special model costs \$2 more than the standard model. Joseph buys 3 standard models and 7 special models. His total bill is \$59. How much does each standard model cost?

3. In a project for a business class project, Tyrone invested three times as much as Andy. At the end of the year, the total net profits were \$328. If profits are shared in the same proportion as the investment, how much money will Andy get?

4. Monique signed up with a cell phone company. Her plan gives her 500 minutes of calls for \$84. Text messages are billed at \$0.05 per message and pictures are billed at \$0.23 per picture. Last month Monique received a bill of \$86.23. If she had no extra minutes and 17 text messages, how many pictures did she send?

5. In the 2006 Winter Olympics in Turin, Italy, team USA won 4 fewer medals than team Germany. The total number of medals won by the two teams was 54. How many medals did each team win?

6. The area of the Chihuahuan Desert in Mexico is approximately 3.24 times the area of the Mojave Desert in California. The sum of their areas is about 593,000 square miles. What is the approximate size of each desert? Round your answers to the nearest integer.

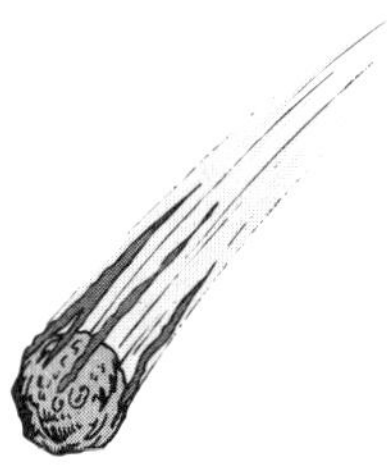

7. The weight of the Hoba West meteorite, located in Namibia, is 4 times the weight of the Campo del Cielo meteorite located in Argentina. Their combined weight is 75 metric tons. What is the weight of each meteorite?

8. An electrician gave the Heart family an estimate of the cost for some electrical services in their house. His rate is \$45 per hour and the cost of the parts is \$140. His total estimate is \$680. How many hours does he estimate for the job?

9. Grandpa Jones is taking his four young grandchildren to see a movie about Sandpaper Sam Suede Pants. A senior citizen ticket is \$7. The total cost for the tickets is \$25. How much is a child's ticket?

10. A piece of rope 18 feet long is to be cut into two pieces so that one piece is 3 feet less than twice the other piece. What is the length of each piece?

Inequalities

The following table lists **signs of inequality**, sometimes called **orders of inequality**.

Sign	Meaning
$<$	less than
$>$	greater than
$\leq$	less than or equal to
$\geq$	greater than or equal to

SOLVING INEQUALITIES

The procedure for solving an inequality is very similar to the procedure for solving an equation. The main difference occurs when both sides of the inequality are multiplied or divided by a negative number. When this step is necessary, the sign of inequality changes direction.

For example, consider this problem: Given that $3 < 4$, multiply both sides of the inequality by -2. If the sign stayed the same, the resulting inequality would be $-6 < -8$, which is false. The sign must change direction in order to have a true statement: $-6 > -8$.

The same is true for the division by a negative. If you divided both sides of $8 > 6$ by -2 and kept the sign the same, the result would be $-4 > -3$, a false statement. To have a true statement, the sign must change direction: $-4 < -3$.

MODEL PROBLEMS

1. Solve the inequality $-2x + 4 < 10$.

Solution:

$-2x + 4 - 4 < 10 - 4$	Subtract 4 from both sides.
$-2x < 6$	Simplify.
$\frac{-2x}{-2} > \frac{6}{-2}$	Divide both sides by -2 and change the direction of the sign of inequality.
$x > -3$	Simplify.

Answer: $x > -3$

Check: You can check by substituting any number greater than -3. Try 0.

$-2x + 4 < 10$

$-2(0) + 4 < 10$

$0 + 4 < 10$

$4 < 10$ (true)

2. Solve the inequality $5x - 3 > 12$.

Solution:

$5x - 3 + 3 > 12 + 3$	Add 3 to both sides.
$5x > 15$	Simplify.
$\frac{5x}{5} > \frac{15}{5}$	Divide both sides by 5. (Sign of inequality does NOT change.)
$x > 3$	Simplify.

Answer: $x > 3$

Check: You can check by substituting any number greater than 3. Try 4.

$5x - 3 > 12$

$5(4) - 3 > 12$

$20 - 3 > 12$

$17 > 12$ (true)

TRY THESE

Solve each inequality.

1. $3y - 6 \le 18$
2. $6x + 1 > -11$
3. $6y + 2 - 8y \ge 14$
4. $8m - 2(2m + 3) < 0$
5. $2x - 1 > 9$
6. $3 - 4c < 27$
7. $3y - 7 \le 5$
8. $5 - 9x > 2$
9. $2x + 3 - 4x < 8$
10. $3(2x - 3) \ge 15$

APPLICATIONS OF INEQUALITIES

To solve a real-world problem that contains an inequality, follow this procedure:

Step 1. Read the problem carefully to determine what is known and what needs to be found.

Step 2. Assign a variable to the unknown quantity.

Step 3. Write an inequality that expresses the given information in terms of the variable.

Step 4. Solve the inequality for the value range of the variable.

Step 5. Check your answer for reasonableness and accuracy.

MODEL PROBLEMS

1. Mr. Manos has paid off \$2,000 of a loan. He still owes \$3,500 or more. What is the least amount that his original loan could have been, before he made the payment?

Solution: Let x = amount of his loan before the payment. Then:

$$x - 2{,}000 \ge 3{,}500$$

$$x - 2{,}000 + 2{,}000 \ge 3{,}500 + 2{,}000$$

$$x \ge 5{,}500$$

Answer: The least possible original loan is \$5,500.

Check:

$$x - 2{,}000 \ge 3{,}500$$

$$5{,}500 - 2{,}000 \ge 3{,}500$$

$$3{,}500 \ge 3{,}500 \text{ (true)}$$

Check your answer in the original problem for reasonableness: \$5,500 − \$2,000 = \$3,500.

2. Ms. Close sells home security systems. She is paid a salary of \$30 per day, plus \$20 for each system she sells. She wants her income to exceed \$150 per day. To do this, how many systems must she sell each day?

Solution: Let x = number of systems she must sell. Then:

$$20x + 30 > 150$$
$$20x + 30 - 30 > 150 - 30$$
$$20x > 120$$
$$\frac{20x}{20} > \frac{120}{20}$$
$$x > 6$$

Answer: She must sell more than 6 systems a day.

Check: You can substitute any value greater than 6. Try 7.

$$20x + 30 > 150$$
$$20(7) + 30 > 150$$
$$140 + 30 > 150$$
$$170 > 150 \text{ (true)}$$

Again, check your answer in the original problem: $30 + 7(20) = 30 + 140 = \$170$. $\$170 > \$150$.

TRY THESE

1. You are comparing two car rental agencies. You find that agency A charges \$260 per week with no charge for mileage. Agency B charges \$200 per week plus 25 cents per mile. Above what mileage does it cost less to rent from agency A?

2. Annual tuition at Wilson State University is \$8,440. Jerry earns no more than \$2,610 per year at his summer job. To earn enough for one year's tuition, what is the least number of summers that he must work?

3. Lydia earns \$8 per hour watching her neighbor's children after school. She is saving her money to buy a \$135 dress for the prom. What is the least number of hours that she will have to work in order to pay for the dress?

4. Bennett, a salesman, earns \$1,200 per month plus a 12% commission on all his sales. What is the minimum amount of sales necessary for him to have a monthly income of \$2,400?

5. A small plane has a maximum take-off weight of 2,200 pounds. The nine people on board weigh an average of 167 pounds each. What is the maximum weight of the luggage the plane can carry?

6. Henry uses the elevator to move cartons of paper to different areas of the company. The elevator has a weight limit of 1,200 pounds. If each box of paper weighs 68 pounds and Henry weighs 196 pounds, what is the maximum number of boxes of paper that he can put on the elevator at one time?
7. Tina's Tiny Cabs take people around the streets of Miami. They charge a base price of \$5.20 plus \$0.25 for each $\frac{1}{8}$ of a mile they drive. What is the maximum number of miles a person can be driven and have the cost be under \$25?
8. Lincoln is paid 1.5 times his hourly rate of \$12 per hour for the time he works over his regular 8-hour day. In order for Lincoln to earn over \$150 per day, how many hours overtime must he work?
9. Tom and Bob work together on a job. Tom earns 1.5 times as much as Bob. If they receive at least \$760 for doing a job, what is the least amount each will earn?
10. Nabih works in a real estate office. He receives a salary of \$600 a month and 8% of all his sales. What is the least amount of sales that he can make if he wants to make more than \$2,500 in a month?

Assessment

Multiple-Choice Problems

1. Which algebraic expression means twice the sum of length l and width w?

   A. $2l + w$ **C.** $l + 2w$

   B. $2 + l + w$ **D.** $2(l + w)$

2. Which algebraic expression means the sum of the squares of leg a and leg b of a right triangle?

   A. $a^2 + b^2$ **C.** $(a + b)^2$

   B. $2a + 2b$ **D.** $2(a + b)$

3. Which algebraic expression means the number of inches in x feet and 5 inches?

   A. $\frac{x}{12} + 5$ **C.** $\frac{x + 5}{12}$

   B. $12x + 5$ **D.** $12(x + 5)$

4. Which algebraic expression means the number of days in 3 less days than y weeks?

   A. $7(y - 3)$ **C.** $7y - 3$

   B. $3 - 7y$ **D.** $7(3 - y)$

5. $7x - 2 < 33$. What is the solution?

   A. $x < 5$ **C.** $x > 5$

   B. $x < \frac{31}{7}$ **D.** $x > \frac{31}{7}$

6. $32 < 3x - 4$. What is the solution?

   A. $x < \frac{28}{3}$ **C.** $x > \frac{28}{3}$

   B. $x < 12$ **D.** $x > 12$

7. $9 - 2x > 15$. What is the solution?

   A. $x > -3$ **C.** $x < -3$

   B. $x > 3$ **D.** $x < 3$

Gridded-Response Problems

8. $4x - 7 = 5$. Solve for x.

9. $5(x + 2) = 2(3x + 1)$. Solve for x.

10. Martin is three times as old as Bill. The sum of their ages is 44 years. How old is Martin?

11. In the last basketball game, sisters Sari and Sandi scored a total of 42 points. Sari scored 3 less than 2 times the number of points Sandi scored. How many points did Sandi score?

12. Mr. Hage is fencing in a rectangular lot. The lot is 5 feet longer than twice its width. Mr. Hage uses 100 feet of fence to enclose the lot. What is the length of the lot?

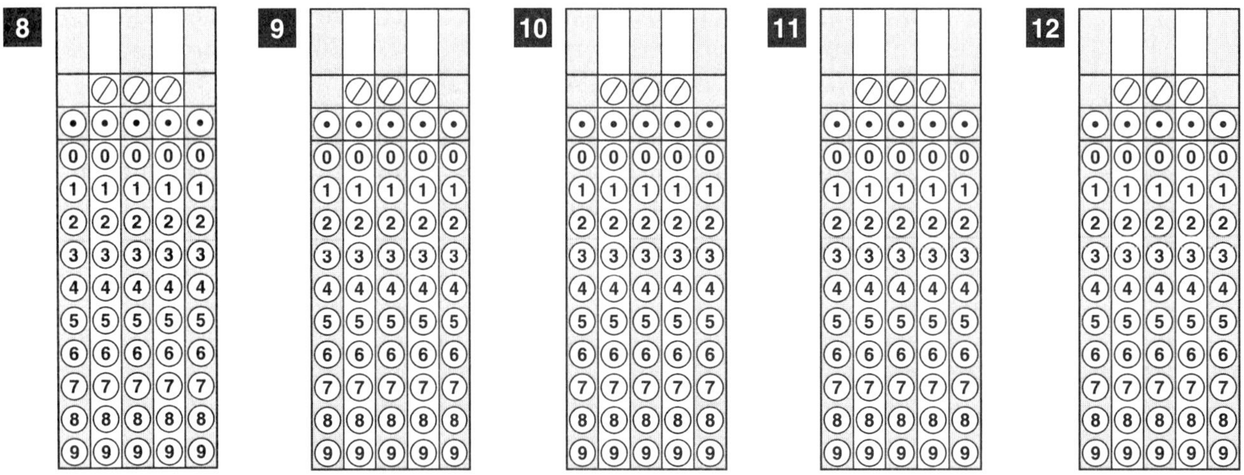

Think, Solve, Explain: Short-Response Problems

13. Solve the following equation, showing your work step by step.

$5(3x - 6) - 2(2x + 3) = 19$

14. Solve the following inequality, showing your work step by step.

$3(2x - 6) - 3(3x + 1) < 21$

15. Maryanna invested a total of $3,000 in two savings accounts. Account A earned 4% simple interest per year and Account B earned 3% simple interest per year. If she earned $102 interest for the year, how much did she invest in each account? Show all computations that lead to your answer.

B SOLVING REAL-WORLD PROBLEMS GRAPHICALLY AND ALGEBRAICALLY WITH SYSTEMS OF EQUATIONS AND INEQUALITIES

Many problems in areas such as business and engineering can be solved by using a system of equations or inequalities. A solution to a system can be found graphically or algebraically.

Systems of linear equations

Solving a system of linear equations means finding the value of each variable that is true in both equations. A solution can be found by graphing or by using algebra.

GRAPHIC SOLUTIONS

A **linear equation in two variables**, when graphed, is a straight line. When two sets of conditions are imposed on the same variables, there are two straight lines. In that case, we have a **system of equations**.

When we graph a system of equations on the same axes, the following situations can occur.

- Lines with **no** point in common: The lines are **parallel**, and the system is called **inconsistent**.
- Lines with **all** points in common: The lines **coincide**, and the system is **consistent but dependent**.
- Lines with **exactly one** point in common: The lines **intersect**, and the system is **consistent and independent**.

When a system is consistent and independent, the point of intersection is the one **solution** to the system. The x-coordinate is the value of x in both equations, and the y-coordinate is the value of y in both equations.

When a system is inconsistent, there is *no* solution. *No* values of x and y are true in both equations.

When a system is consistent but dependent, the number of solutions is infinite. Any values of x and y are true in both equations.

The three kinds of systems are shown below.

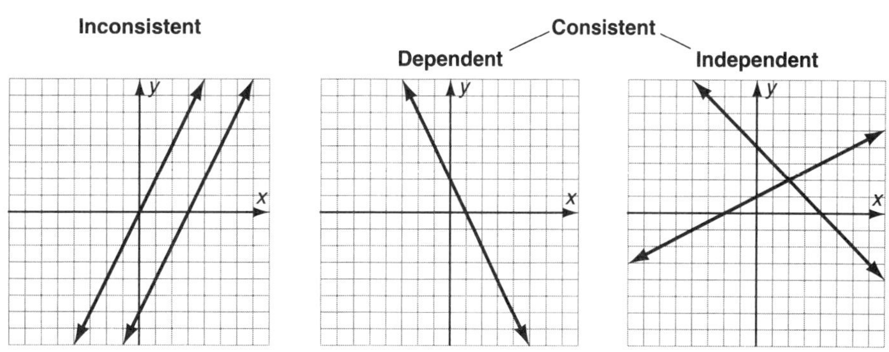

To solve a system of equations graphically, you need to graph each equation. Here is a quick review of methods of graphing a linear equation.

Three-Point Method Choose three different values for the x-coordinate. Using the equation, find the corresponding values of y. Plot the three points on the coordinate axes and connect them with a straight line.

Slope-Intercept Method Write the equation in **slope-intercept form**: $y = mx + b$. In this form, the slope m of the line is the coefficient of x; and b is the **y-intercept**, the point $(0, b)$ where the line crosses the y-axis. Remember that slope $m = \frac{\text{rise}}{\text{run}} = \frac{\text{change in } y}{\text{change in } x}$.

First, plot the y-intercept. Next, find another point by using the ratio of slope starting at the y-intercept. Then connect the two points with a straight line.

Intercept Method Plot the **x-intercept**, the point $(x, 0)$ where the line crosses the x-axis. Plot the y-intercept $(0, y)$. Connect these two points with a straight line.

MODEL PROBLEMS

For each pair of equations, draw the graphs on the same set of axes. State whether the system is inconsistent, consistent but dependent, or consistent and independent. If the system is consistent and independent, give the solution: the coordinates of the intersection.

1. $x + y = 5$

$x + y = 3$

Answer:

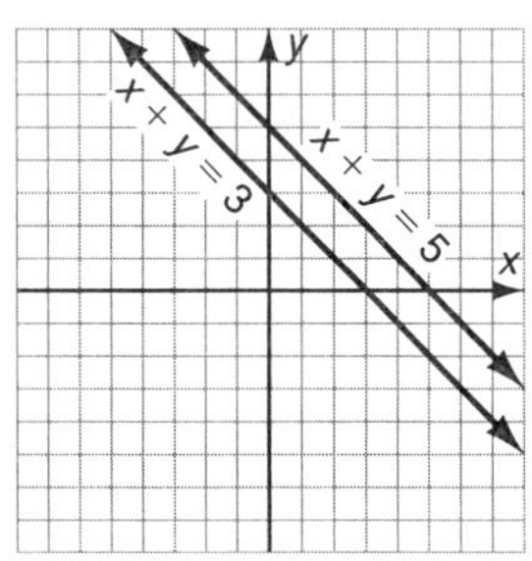

Inconsistent (no solution).

2. $x - y = 2$

$x + 2y = 8$

Answer:

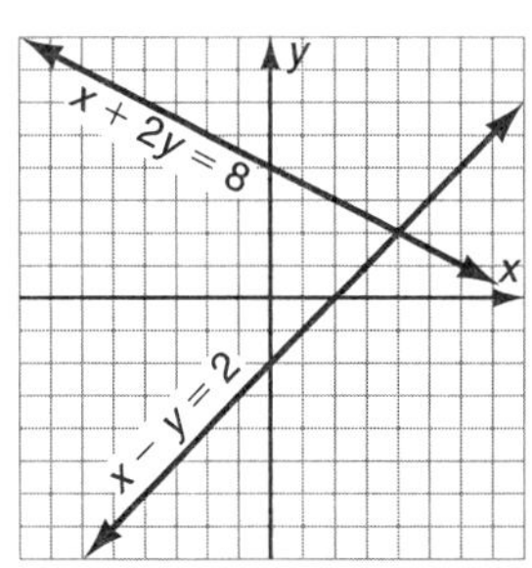

Consistent and independent.

Coordinates of the intersection: (4, 2).

Solution: $x = 4$ and $y = 2$.

3. $x - y = 3$

$2x - 2y = 6$

Answer:

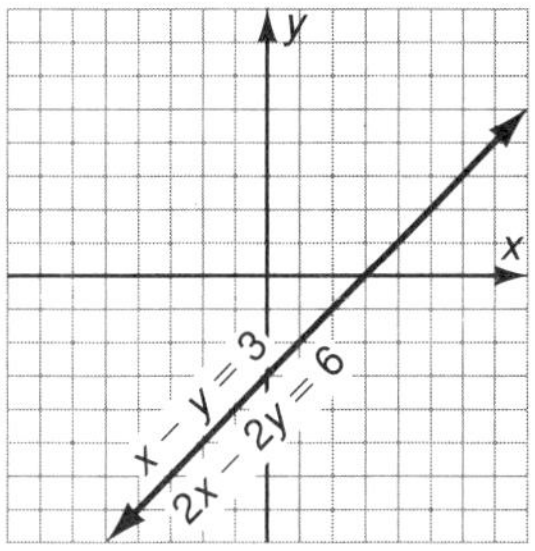

Dependent (infinite number of solutions).

TRY THESE

For each pair of equations, draw the graphs on the same set of axes. State whether the system is inconsistent, consistent but dependent, or consistent and independent. If the system is consistent and independent, give the solution (coordinates of the intersection).

1. $6x - 2y = 9$

$3x - y = 12$

2. $2x + 3y = 5$

$6x + 9y = 15$

3. $5x + y = -4$

$10x + 2y = 4$

4. $3x - y = 9$

$6x - 2y = 0$

5. $y = 2$

$x + 3y = -1$

ALGEBRAIC SOLUTIONS

Solving a system graphically is limited by the accuracy of the graph. For example, if the intersection point is $\left(6\frac{2}{3}, 4\frac{1}{5}\right)$, that cannot be plotted easily on a graph. Solving algebraically avoids this problem. Two ways to solve a system of equations algebraically are (1) by substitution and (2) by elimination.

Substitution Method In this method, we solve one equation in the system for one of the variables, to get its value in terms of the other variable. Then we substitute this value in the other equation and solve that equation for the remaining variable.

EXAMPLE

$x + 4y = 3$

$2x - 3y = 17$

Step 1. Choose one of these equations and solve for one variable in terms of the other. We will solve the first equation for x:

$$x + 4y = 3$$
$$x + 4y - 4y = 3 - 4y$$
$$x = 3 - 4y$$

Step 2. Substitute this value of x in the second equation. This leaves us with only one variable, y:

$$2x - 3y = 17$$
$$2(3 - 4y) - 3y = 17$$

Step 3. Solve this equation for y:

$$2(3 - 4y) - 3y = 17$$
$$6 - 8y - 3y = 17$$
$$6 - 11y = 17$$
$$6 - 6 - 11y = 17 - 6$$
$$-11y = 11$$
$$\frac{-11y}{-11} = \frac{11}{-11}$$
$$y = -1$$

Step 4. Substitute the value of y in the equation that expresses x in terms of y:

$$x = 3 - 4y$$
$$x = 3 - [(4)(-1)]$$
$$x = 3 - (-4) = 3 + 4$$
$$x = 7$$

Answer: Therefore, the solution is $x = 7, y = -1$.
This can be stated as coordinates: $(7, -1)$.

Step 5. Check by substituting the values in both equations of the given system:

$x + 4y = 3$	$2x - 3y = 17$
$7 + [(4)(-1)] = 3$	$2(7) - [(3)(-1)] = 17$
$7 + (-4) = 3$	$14 - (-3) = 17$
$3 = 3$	$17 = 17$

MODEL PROBLEMS

1. Solve the following system algebraically by the substitution method.

$x = 4y$

$2x + y = 18$

Solution:

Step 1. The first equation is already solved for x in terms of y.

$x = 4y$

Step 2. Now we substitute this value of x in the second equation, to get only one variable:

$2(4y) + y = 18$

Step 3. Solve this equation for y.

$$8y + y = 18$$
$$9y = 18$$
$$y = 2$$

Step 4. To find the value of x, substitute this value of y in the equation given for x.

$$x = 4y$$
$$x = 4 \bullet 2$$
$$x = 8$$

Answer: $x = 8, y = 2$ or $(8, 2)$

Step 5. Check.

$x = 4y$	$2x + y = 18$
$8 = 4 \bullet 2$	$2 \bullet 8 + 2 = 18$
$8 = 8$	$16 + 2 = 18$
	$18 = 18$

2. Solve the following system algebraically by the substitution method.

$3x + 2y = 33$

$y - 2x = -8$

Solution:

Step 1. We will solve for y in the second equation.

$$y - 2x = -8$$
$$y = -8 + 2x$$

Step 2. Now we substitute this value of y in the first equation to get only one variable:

$$3x + 2y = 33$$
$$3x + [(2)(-8 + 2x)] = 33$$

Step 3. Solve this equation for x.

$$3x - 16 + 4x = 33$$

$$7x - 16 = 33$$

$$7x - 16 + 16 = 33 + 16$$

$$7x = 49$$

$$\frac{7x}{7} = \frac{49}{7}$$

$$x = 7$$

Step 4. To find the value of y, substitute this value of x in the equation we obtained for y.

$$y = -8 + 2x$$

$$y = -8 + [(2)(7)]$$

$$y = -8 + 14$$

$$y = 6$$

Answer: $x = 7, y = 6$ or $(7, 6)$

Step 5. Check.

$3x + 2y = 33$	$y - 2x = -8$
$3(7) + 2(6) = 33$	$6 - 2(7) = -8$
$21 + 12 = 33$	$6 - 14 = -8$
$33 = 33$	$-8 = -8$

TRY THESE

Solve each system algebraically by the substitution method.

1. $x = 2y$
$3x + 4y = 10$

2. $y = 5x$
$2x - 3y = 39$

3. $x = -3y$
$4x - 2y = 28$

4. $y = x + 2$
$5x - 2y = 5$

5. $3x - y = 15$
$x = 2y + 5$

6. $3x - 4y = 0$
$2x + y = 33$

7. $3x - 2y = 11$
$2x - y = 8$

8. $2x = y - 3$
$3x + 4y = -10$

9. $8x = 2y$

$2x + 3y = 42$

10. $x + 2y = 10$

$2x - y = 5$

11. $x = 2y + 14$

$y - 4x = 0$

12. $6(x + y) = 12$

$x = 6y - 6$

Elimination Method This method eliminates one of the variables by adding two equations. Thus it is sometimes called the **addition method**. Following are three examples.

EXAMPLE 1

$x + y = 4$

$2x - y = 2$

Step 1. Add the two given equations to eliminate one variable, y:

$$\begin{array}{r} x + y = 4 \\ +2x - y = 2 \\ \hline 3x \qquad = 6 \end{array}$$

Step 2. Solve the new equation for the remaining variable, x:

$3x = 6$

$$\frac{3x}{3} = \frac{6}{3}$$

$x = 2$

Step 3. Substitute this value of x in one of the original given equations:

$x + y = 4$

$2 + y = 4$

Step 4. Solve this equation for y:

$2 + y = 4$

$2 - 2 + y = 4 - 2$

$y = 2$

Answer: (2, 2) or $x = 2, y = 2$

Step 5. Check by substituting these values in the other given equation:

$2x - y = 2$

$2(2) - 2 = 2$

$4 - 2 = 2$

$2 = 2$

EXAMPLE 2

When using the elimination method, we sometimes need to multiply one entire equation by a constant in order to eliminate one of the variables.

$3x + 5y = 1$

$x + 4y = 5$

If we just add these, neither variable will be eliminated:

$$\begin{array}{r} 3x + 5y = 1 \\ + \; x + 4y = 5 \\ \hline 4x + 9y = 6 \end{array} \text{ (not helpful)}$$

So we need to do something more. Multiply the second equation by -3:

$$-3(x + 4y = 5) = -3x - 12y = -15$$

Now add this new equation and the first given equation, to eliminate the x variable:

$$\begin{array}{r} -3x - 12y = -15 \\ + \; 3x + 5y = 1 \\ \hline -7y = -14 \end{array}$$

Solve the result for y:

$$-7y = -14 \text{ so } \frac{-7y}{-7} = \frac{-14}{-7} \text{ and } y = 2$$

Substitute the value of y in one of the original given equations, to solve for x:

$$x + 4y = 5$$

$$x + 4(2) = 5$$

$$x = -3$$

Answer: $(-3, 2)$ or $x = -3, y = 2$

EXAMPLE 3

When we want to eliminate a variable by adding the given equations, it is sometimes necessary to multiply *both* given equations by different constants.

$2x + 5y = -5$

$3x - 2y = -17$

Just adding these will not eliminate a variable:

$$\begin{array}{r} 2x + 5y = -5 \\ + \; 3x - 2y = -17 \\ \hline 5x + 3y = -22 \end{array} \text{ (not helpful)}$$

If we multiply the first equation by 2 and the second equation by 5, the y coefficients will be the same number with opposite signs:

$$2(2x + 5y = -5) \rightarrow 4x \underline{\underline{+ 10y}} = -10$$

$$5(3x - 2y = -17) \rightarrow 15x \underline{\underline{- 10y}} = -85$$

Now, adding the two new equations will eliminate the y variable:

$$\begin{array}{r} 4x + 10y = -10 \\ \underline{15x - 10y = -85} \\ 19x \qquad\quad = -95 \end{array}$$

Solve this for x:

$$\frac{19x}{19} = \frac{-95}{19} \text{ so } x = -5$$

Substitute this value of x in one of the original given equations and solve for y:

$$\begin{aligned} 2x + 5y &= -5 \\ 2(-5) + 5y &= -5 \\ -10 + 10 + 5y &= -5 + 10 \\ 5y &= 5 \\ y &= 1 \end{aligned}$$

Answer: $(-5, 1)$ or $x = -5, y = 1$

MODEL PROBLEM

Solve by the elimination method:

$$2x + 3y = 13$$

$$3x - 4y = 1$$

Solution: Multiply the first equation by 3 and the second equation by -2.

$$3(2x + 3y = 13) \rightarrow 6x + 9y = 39$$

$$-2(3x - 4y = 1) \rightarrow -6x + 8y = -2$$

Add:

$$\begin{array}{r} 6x + 9y = 39 \\ \underline{-6x + 8y = -2} \\ 17y = 37 \end{array}$$

Solve for y:

$$y = \frac{37}{17}$$

Substitute this value, $y = \frac{37}{13}$, in one of the original equations and solve for x:

$$3x - 4y = 1$$

$$3x - (4)\left(\frac{37}{17}\right) = 1$$

$$3x - \frac{148}{17} + \frac{148}{17} = 1 + \frac{148}{17} = \frac{165}{17}$$

$$\frac{1}{3}(3x) = \frac{1}{3}\left(\frac{165}{17}\right)$$

$$x = \frac{165}{51} \text{ or } \frac{55}{17} \text{ or } 3\frac{4}{17}$$

Answer: $\left(\frac{55}{17}, \frac{37}{17}\right)$ or $x = \frac{55}{17}, y = \frac{37}{17}$

NOTE: An alternative method, starting at the beginning, is to eliminate y and find the value of x.

TRY THESE

Items 1–8: Solve each system of equations by the elimination method.

1. $6x - y = 11$
$2x + y = 5$

2. $4x + 2y = 26$
$5x - 2y = 1$

3. $2x + 5y = 15$
$3x - 4y = 11$

4. $7x + 2y = 18$
$6x + 14y = 40$

5. $3x + 7y = 6$
$3x - 9y = 6$

6. $7x + 2y = 18$
$2x + 3y = 10$

7. $5x + 3y = 8$
$x - 2y = 4$

8. $4x - 3y = 1$
$3x - 4y = 6$

Items 9 and 10: Solve algebraically by either substitution or the elimination method.

9. $x - y = 5$
$2x + y = 15$

10. $3x + y = 4$
$6x + y = 1$

APPLICATIONS: SYSTEMS OF LINEAR EQUATIONS

Many real-world problems can be solved more easily with a system of equations than with just one equation.

EXAMPLE

Three pills of type A plus 4 pills of type B cost \$39. Five pills of type A plus 7 pills of type B cost \$67. For each type, what is the cost per pill?

Step 1. Assign variables.

Let x = cost per pill for type A

y = cost per pill for type B

Step 2. Write a system of two equations using the given information.

$$3x + 4y = 39$$

$$5x + 7y = 67$$

Step 3. Solve the system graphically or algebraically. Here, we will solve algebraically using the addition method.

$$\begin{aligned} 5(3x + 4y = 39) \rightarrow \quad 15x + 20y &= \quad 195 \\ -3(5x + 7y = 67) \rightarrow -15x - 21y &= -201 \\ \hline -y &= -\ 6 \end{aligned}$$

Therefore: $y = 6$

Step 4. Substitute this value of y in either given equation, and solve for x.

$$3x + 4y = 39$$

$$3x + 4(6) = 39$$

$$3x + 24 - 24 = 39 - 24$$

$$3x = 15$$

$$x = 5$$

Answer: For type A, the cost per pill is \$5.
For type B, the cost per pill is \$6.

Step 5: Check by substituting these values in the given problem:

(3 pills of type A • \$5) plus (4 pills of type B • \$6) cost \$39

\$15 + \$24 = \$39 (OK)

(5 pills of type A • \$5) plus (7 pills of type B • \$6) cost \$67

\$25 + \$42 = \$67 (OK)

MODEL PROBLEMS

1. Two pounds of stew meat and 6 pounds of ground pork cost \$12.30. Three pounds of stew meat and 2 pounds of ground pork cost \$9.70. Find the cost per pound of each kind of meat.

Solution:

Let x = cost per pound of stew meat

y = cost per pound of ground pork

To make the calculation easier, express both costs in cents.

Then $2x + 6y = 1{,}230$

$3x + 2y = 970$

Solve for one variable. Here, we solve for y algebraically by the addition method.

$$\begin{array}{rrcr} 3(2x + 6y = 1{,}230) \rightarrow & 6x + 18y & = & 3{,}690 \\ -2(3x + 2y = 970) \rightarrow & \underline{-6x - 4y} & \underline{=} & \underline{-1{,}940} \\ & 14y & = & 1{,}750 \\ & y & = & 125 \end{array}$$

Substitute to solve for the other variable, x:

$$\begin{aligned} 2x + 6y &= 1{,}230 \\ 2x + 6(125) &= 1{,}230 \\ 2x + 750 - 750 &= 1{,}230 - 750 \\ 2x &= 480 \\ x &= 240 \end{aligned}$$

To write the answer, convert cents back to dollars.

Answer: Ground pork costs \$1.25 per pound. Stew meat costs \$2.40 per pound.

2. Going downstream with the current, a boat can travel 32 miles in 4 hours. The return trip, going upstream against the current, takes 16 hours. What is the rate of the boat in still water with no current? What is the rate of the current? Express the answers in miles per hour (mph).

Solution:

Let r = rate of the boat in still water (mph)

c = rate of the current (mph)

Then:

$r + c$ = rate downstream with the current (mph)

$r - c$ = rate upstream against the current (mph)

Use rate • time = distance or $rt = d$. This gives the following system of equations.

$$4(r + c) = 32 \quad \text{or:} \quad 4r + 4c = 32$$

$$16(r - c) = 32 \quad \text{or:} \quad 16r - 16c = 32$$

We can solve for r algebraically by the addition method.

$$\begin{aligned} 4(4r + 4c = 32) \rightarrow 16r + 16c &= 128 \\ 16r - 16c &= 32 \\ \hline 32r &= 160 \\ r &= 5 \text{ mph} \end{aligned}$$

Substitute and solve for c:

$$\begin{aligned} 4r + 4c &= 32 \\ 4(5) + 4c &= 32 \\ 20 - 20 + 4c &= 32 - 20 \\ 4c &= 12 \\ c &= 3 \text{ mph} \end{aligned}$$

Answer: The rate of the boat in still water is 5 mph. The rate of the current is 3 mph.

TRY THESE

1. The Wood family is considering two plans for home landscaping. Plan 1 uses 8 oak trees and 12 hemlock trees for a total cost of \$520. Plan 2 uses 6 oak trees and 15 hemlock trees for a total cost of \$510. What is the cost of each oak tree and each hemlock tree?

2. Acme Garbage Company has two sizes of containers available for homeowners. The maximum weight for one container is 4 kilograms. The maximum weight for the other container is 3 kilograms. On one garbage collection trip, Acme picked up 30 containers that weighed a total of 100 kilograms. How many containers of each kind were picked up?

3. An airplane flew a distance of 3000 kilometers and then returned. The trip flying against the wind took 6 hours. The trip with the wind took 5 hours. What was the speed of the plane in still air? What was the speed of the wind?

Items 4–8: Solve each problem graphically or algebraically by using a system of equations. If you solve algebraically, you can use the substitution method or the addition method.

4. Two isosceles triangles have perimeters of 8 and 15 centimeters, respectively. They have the same base length, but the **legs** (the nonbase sides) of the larger triangle are twice as long as the legs of the smaller triangle. For each triangle, what is the length of a leg?

5. Jeremy saves quarters and dimes. One night he counts his coins and finds that there are 44 coins with a total value of $6.50. How many of each kind of coin does he have?
6. In an algebra class consisting of freshmen and sophomores, there are 28 students. The number of sophomores is 24 less than 3 times the number of freshmen. How many freshman and how many sophomores are in the class?
7. Tickets for a play cost $8 for adults and $6 for students. A total of 1,125 tickets were sold. The total ticket revenue was $7,700. How many of each kind of ticket were sold?
8. Two friends, Wayne and David, worked at the same store last summer. David earned $5.50 per hour. Wayne earned $6.50 per hour. During one week, together they worked a total of 45 hours and earned a total of $266.50. How many hours did each of them work?

9. Together Liz and Jan made 90 items to be sold at craft fairs. Jan sold her items for $5 each and Liz sold her items for $6.50 each. Together their sales totaled $540. How many items did each sell?
10. For a party, Roberta bought 3 bags of chips and 2 jars of salsa. She spent $14.70 for these items. Jason bought 4 bags of chips and one jar of salsa. He spent $15.85 for these items. Assuming that they both bought the same brands at the same store, how much does 1 bag of chips and 1 jar of salsa cost?

Sysytems of linear inequalities

A **system of inequalities** is simply more than one inequality graphed on the same set of axes. The **solution** is the area that is common to all the inequalities. This area has the shading of each inequality, one shading on top of another. The solution area is sometimes labeled: for instance, it may be labeled S.

GRAPHING SYSTEMS OF LINEAR INEQUALITIES

Systems of linear inequalities are usually solved graphically. Here is a brief a review of graphing a linear inequality.

- First: Graph the inequality as if it were an equation.
- Next: Connect the data points with a dashed line or a solid line, depending on the sign of inequality.

  For $>$ or $<$, use a **dashed line** to show that points on the line are *not* part of the solution set.

  For $\geq$ or $\leq$, use a **solid line** to show that points on the line *are* part of the solution set.

- Finally: Shade the area in which the solutions lie: the area above or below or to the right or left of the line. To find the area where the solutions lie, choose a test point not too near the line. Substitute the coordinates of that chosen point in the inequality. If the result is true, shade the area on that side of the line. If the result is false, shade the area on the other side of the line.

EXAMPLE 1: $y \leq 2x + 3$

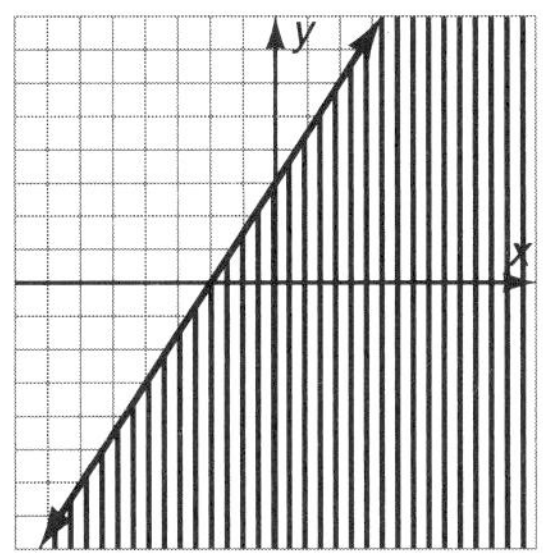

You could choose (5, 0) as a test point:

$y \leq 2x + 3$

$0 \leq 2(5) + 3$

$0 \leq 13$

This is true. Shade on the side that includes point (5, 0).

If the result had been false, you would shade on the other side of the line, the side that does *not* include (5, 0).

EXAMPLE 2: $x > 2$

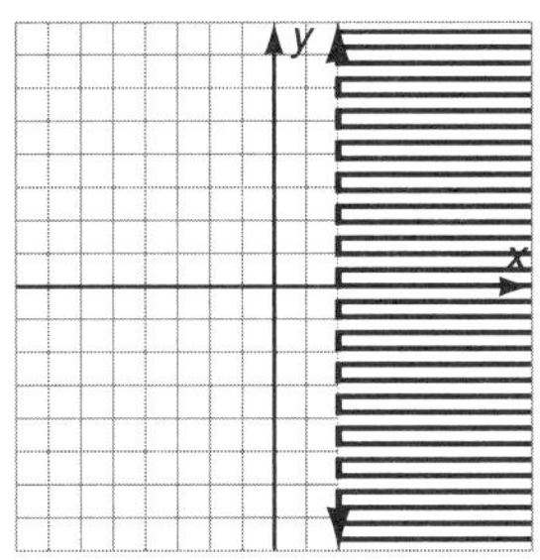

A test point could be (0, 0). Substitute:

$x > 2$

$0 > 2$

This is a false statement. Therefore, shade on the side of the line that does *not* include (0, 0).

If the result had been true, you would shade on the same side of the line, the side that includes the test point.

MODEL PROBLEM Graph this system. Label the solution A.

$y \leq 2x + 3$

$y \geq -2x + 2$

Solution:

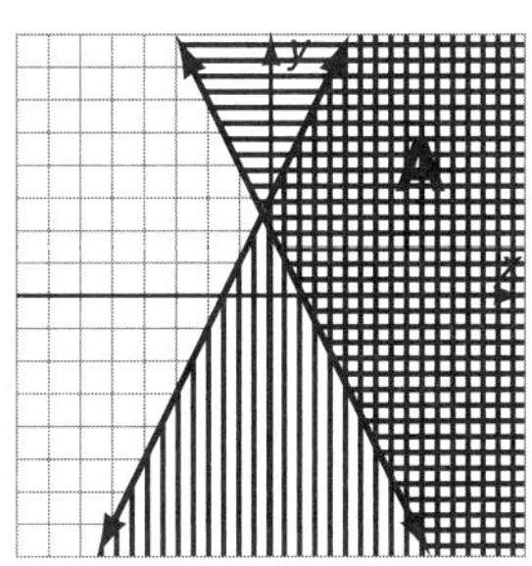

TRY THESE

For each of the following systems, graph the system and label the solution A.

1. $y \geq 2$

$y < x$

2. $y < x + 3$

$y \leq -x + 1$

3. $3x - 2y > 1$

$2x + 3y \leq 18$

4. $2x - 3y \leq 2$

$x + y \leq 6$

5. $3x - y \geq -7$

$3x + y \leq -2$

6. $2x + 3y > 6$

$x - y < 2$

APPLICATIONS: SYSTEMS OF LINEAR INEQUALITIES

A system of linear inequalities is applied to real-world problems in much the same way as a system of equations.

Step 1. Write a system of inequalities using the given information.

Step 2. Graph the inequalities on the same set of axes.

Step 3. Determine the solution. This is the shaded area that the inequalities have in common.

MODEL PROBLEMS

1. The total weight of item A and item B is greater than or equal to 3 pounds but less than or equal to 15 pounds. Using a system of inequalities, graph the solution.

Solution:

Let x = weight of item A

y = weight of item B

Then:

$x + y \geq 3$

$x + y \leq 15$

Graph this system:

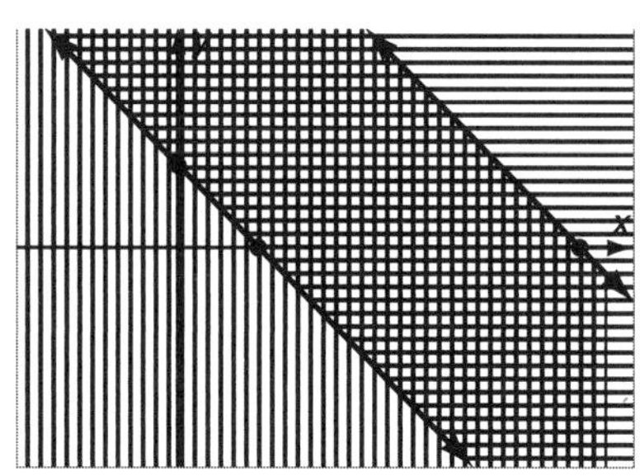

Answer: The solution is the double-shaded area.

2. In basketball, a team earns 2 points when a player makes a basket from inside the 3-point arc and 1 point when a player makes a free throw. If a player scores no more than 20 points in a game, how many 2-point baskets and free throws could the player have made? Show your answer graphically. Would it be possible for the player to have made 9 baskets and 1 free throw? Why or why not?

Solution:

Let x = number of baskets made

y = number of free throws made

Then:

$x \geq 0$

$y \geq 0$

$2x + y \leq 20$

Graph:

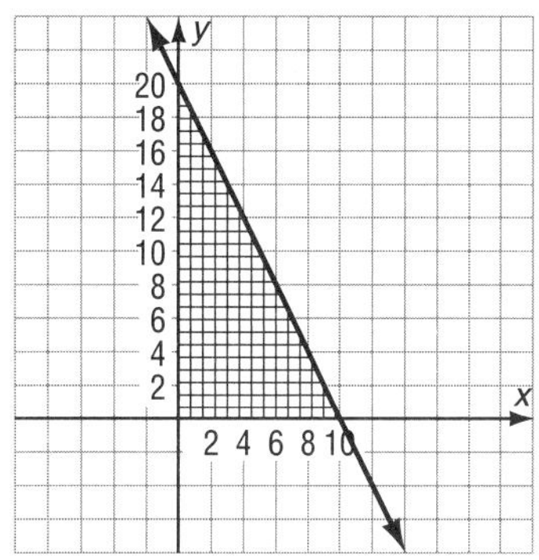

Answer: The double-shaded area contains the possible combinations of baskets and free throws made, such as (5, 2) or (3, 8). Note that the possible answers are only the integer pairs in that area. This is because number of baskets and number of free throws can be expressed only as integers.

(You can't make, say, half a basket or $\frac{3}{5}$ of a free throw.)

TRY THESE

1. The graph below represents the following system of inequalities:

   $y \geq x^2 + 1$

   $y < 4$

   Which of the regions, A, B, C, D, or E, represents the solution of this system? Name the coordinates of one point in the solution set.

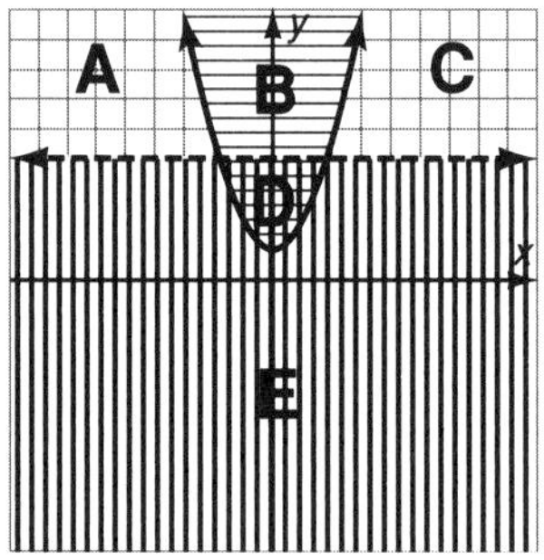

2. The graph below represents the following system of the inequalities:

   $y > 3x - 2$

   $y \leq -2x + 5$

   a. Which region or regions satisfy just the first inequality?

   b. Which region or regions satisfy just the second inequality?

   c. Which region or regions satisfy neither inequality?

   d. Which region or regions satisfy both inequalities?

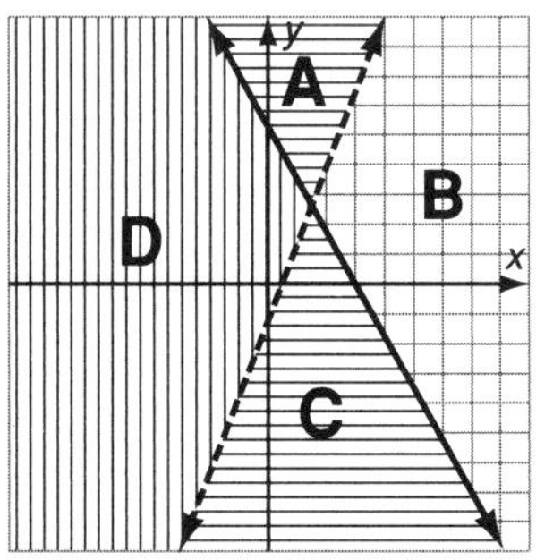

3. The PTA wishes to raise at least \$12,000 at a charity benefit. Ticket prices are either \$20 per person or \$15 per person with at least 500 tickets to be sold at \$20 each. The location of the event can hold at most 800 people. The graph below represents the following inequalities:

$x + y \leq 800$

$20x + 15y \geq 12{,}000$

$x \geq 500$

$y \geq 0$

What region satisfies these conditions?

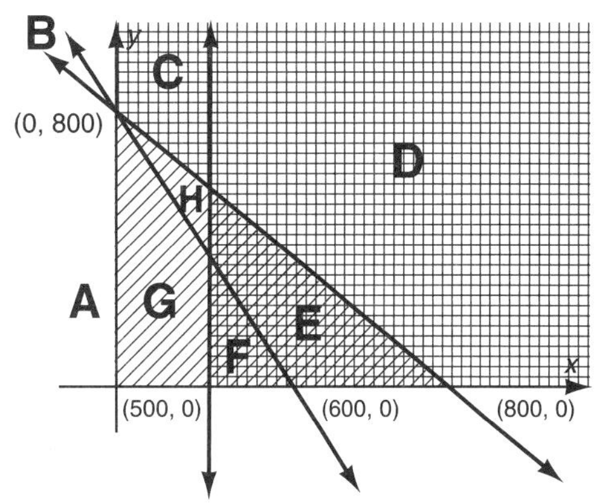

4. A farmer wants to raise chickens and turkeys. He can have NO more than a total of 16 animals and NO more than 12 turkeys. Graph the appropriate inequalities. Show the shaded area where the pairs of integer coordinates representing the chickens and turkeys would fit the conditions of this situation.

5. A small bakery makes pies and cakes. Its oven can hold only 3 cakes at a time, and NO more than a total of 10 items at a time. Graph the appropriate inequalities. Show the shaded area where the pairs of integer coordinates representing number of pies and number of cakes would fit the conditions of this situation.

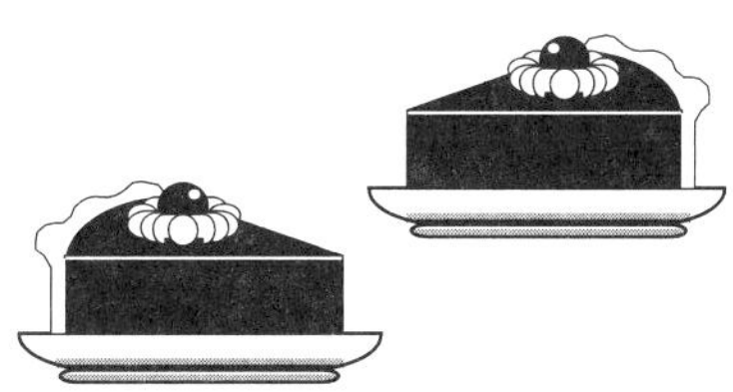

Assessment

Multiple-Choice Problems

1. Gregor has 4.5 times as much money in his savings account as he has in his checking account. The total amount in both accounts is $1,925. How much does he have in his savings account?

A. $1,365

B. $1,555

C. $1,575

D. $1,600

2. In her purse, Tasha has 13 coins that total $1.60. The coins are quarters and dimes. How many dimes does she have?

A. 1 dime

B. 2 dimes

C. 6 dimes

D. 11 dimes

Gridded-Response Problems

3. Keisha bought 3 CDs and 2 DVDs for $85. Last month she bought 5 CDs and 1 DVD for $95. How much does one CD cost if the prices have not changed?

4. A family with 2 adults and 4 children pays $38 to enter Patriot Park for the day. Another family with 4 adults and 3 children pays $51 to enter the park. What is the price for an adult to enter the park?

5. Solve for x:

$9x - 2y = 3$

$2x + 6y = 20$

6. A sign placed along the side of the road must have a perimeter of 144 inches. The length of the sign must be 14 inches longer than its width. What is the length of the sign?

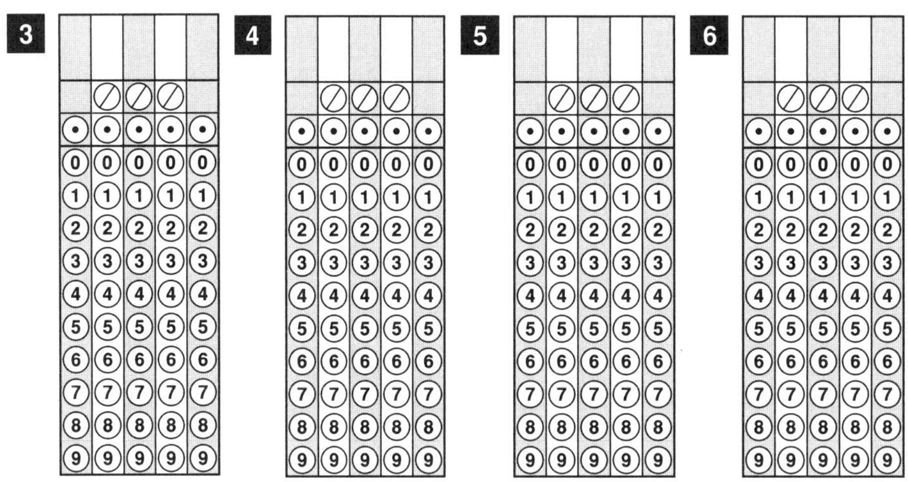

Think, Solve, Explain: Short-Response Problems

7. Shown below are the graphs of $x - y = -3$ and $x + y = 7$.

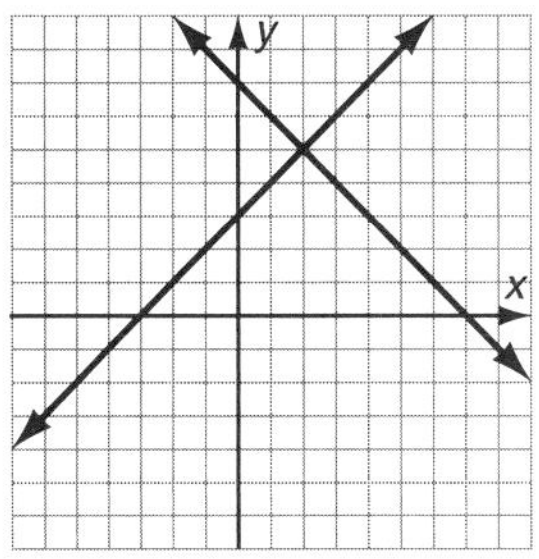

Part A Name the coordinates of a point that satisfies $x - y = -3$ but NOT $x + y = 7$.

Part B Name the coordinates of a point that satisfies $x + y = 7$ but NOT $x - y = -3$.

Part C Name the coordinates of a point that satisfies both equations.

8. Shown below are the graphs of $x - y < -1$ and $x + y < 5$.

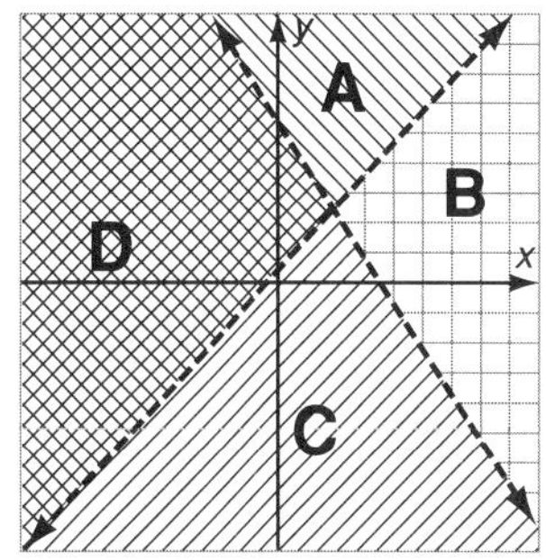

Part A Which of the four regions represents $x - y < -1$ but NOT $x + y < 5$? Name the coordinates of one point in this region.

Part B Which of the four regions represents $x + y < 5$ but NOT $x - y < -1$? Name the coordinates of one point in this region.

Part C Which of the four regions satisfies both inequalities? Name the coordinates of one point in this region.

Part D Which of the four regions satisfies NEITHER inequality? Name the coordinates of one point in this region.

CHAPTER REVIEW

Multiple-Choice Problems

1. Which of the following relations is a function?

   A. (3, 2), (5, 2), (7, 2), (8, 2)
   B. (2, 3), (2, 5), (2, 7), (2, 9)
   C. (1, 2), (2, 5), (2, 1), (5, 1)
   D. (5, 1), (5, 2), (2, 5), (1, 5)

2. Which of the following represents five times the quantity of x and seven is twelve?

   A. $5(x) + 7 = 12$
   B. $5(x - 7) = 12$
   C. $5(x + 7) = 12$
   D. $5x + 7 = 12$

3. Which of the following is the solution to the inequality $8x + 4 \leq 12$?

   A. $x \geq 1$
   B. $x \geq 2$
   C. $x \leq 1$
   D. $x \leq 2$

4. While shopping, DJ buys 3 shirts and a pair of pants for \$84. At the same time, his friend Bryan buys 2 shirts and 2 pairs of pants for \$96. What is the price of 1 shirt?

   A. \$18
   B. \$30
   C. \$36
   D. \$42

Items 5 and 6: A furniture manufacturer can produce at most 19 tables and 30 chairs per day. Each table requires 3 hours of work. Each chair requires 2 hours of work. The maximum number of hours of work that the company can assign is 96. This information can be represented by a set of inequalities (where t = table and c = chair) and a graph:

$$3t + 2c \leq 96$$
$$0 \leq t \leq 19$$
$$0 \leq c \leq 30$$

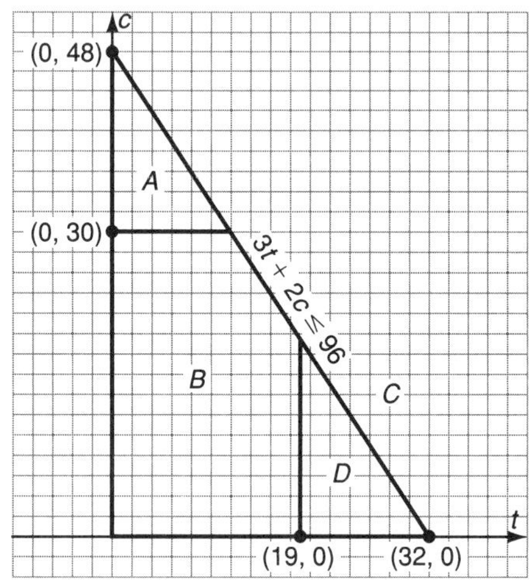

5. Which region of the graph represents the number of tables and chairs the company can produce daily?

   A. region A
   B. region B
   C. region C
   D. region D

6. Which coordinate represents the most possible furniture manufactured?

   A. (0, 30)
   B. (12, 30)
   C. (19, 18.5)
   D. (19, 0)

Gridded-Response Problems

7. The table below represents the function $y = 7 - 2x$. What does y equal when $x = -3$?

x	10	5	1	−3	−8
y	−13	−3	5	?	23

8. $5(x + 2) = 6(2x - 3)$. Solve for x.

9. Alice has 12 coins consisting of dimes and nickels. The total value of the coins is \$0.95. How many dimes does she have?

10. The perimeter of an isosceles triangle is 34 inches. The length of the base is 8 inches. What is the length of one of the legs?

11. What is the next number in the geometric sequence 12, 6, 3, $\frac{3}{2}$, ____?

12. On a display at the store, the cans are stacked in the shape of a pyramid. The top row has 1 can. Each row down has 2 more cans than the previous row. What is the total number of cans in the top 6 rows?

7
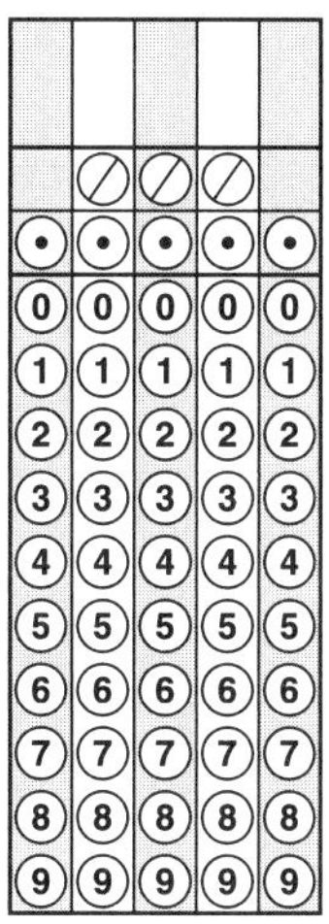

8
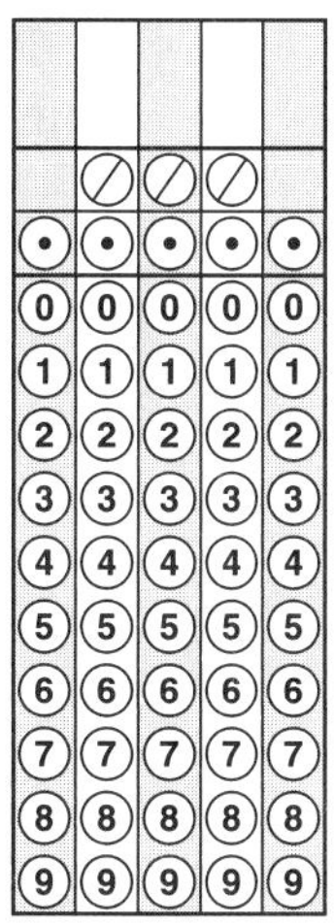

9
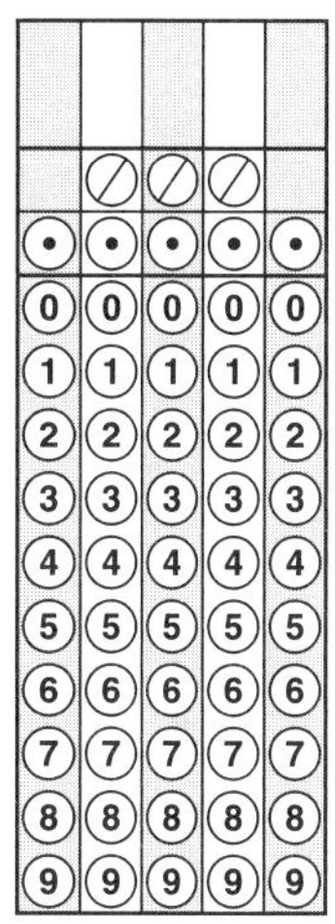

10
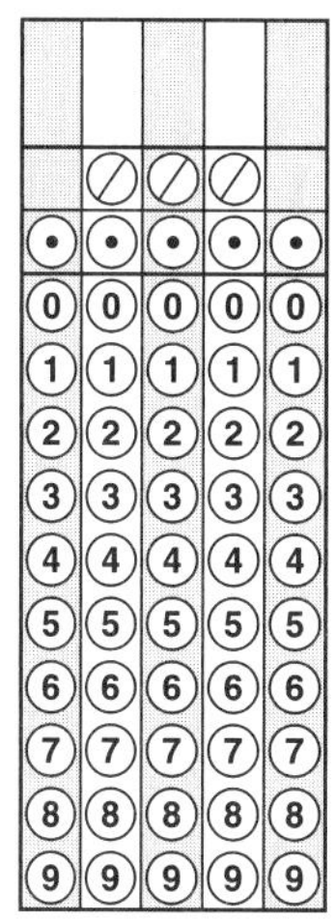

11
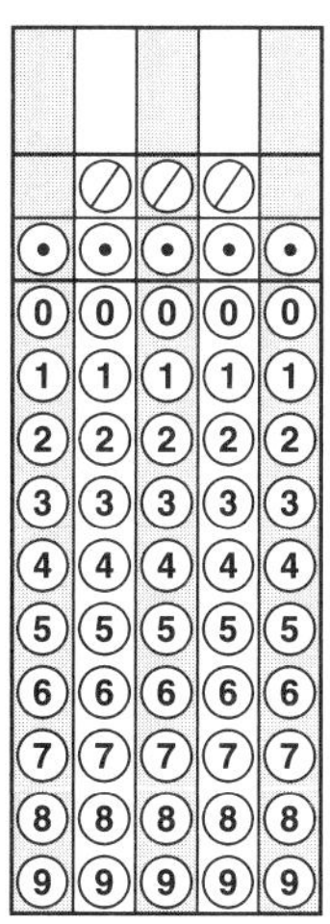

12
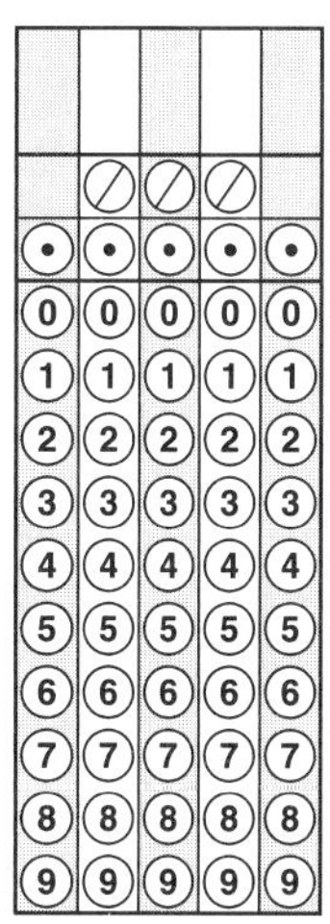

Think, Solve, Explain: Short-Answer Problems

13. Solve the following equation, showing the step-by-step process.

$4(2x - 3) - 3(8x - 1) = 23$

14. Hope had to stop and buy gas two times on her trip to visit her Papa D. She paid \$2.95 a gallon at the first stop and \$2.98 at the second stop. She bought a total of 27 gallons and spent \$80.10. This information can be represented by the following equations where x = the number of gallons purchased at \$2.95 and y = the number of gallons purchased at \$2.98.

$x + y = 27$

$2.95x + 2.98y = 80.10$

Solve the system of equations to find out how many gallons of gas Hope bought at the second stop. Show all your work.

15. The Key Club is going to sell soda at the winter choral concert. They plan to purchase no more than 210 cans of soda. Since more people prefer cola to other brands, they will buy at least twice as much cola. The inequalities below represent this information where x represents the number of cans of cola and y represents the number of cans of non-cola soda.

$x + y \leq 210$

$x \geq 2y$

$x > 0$

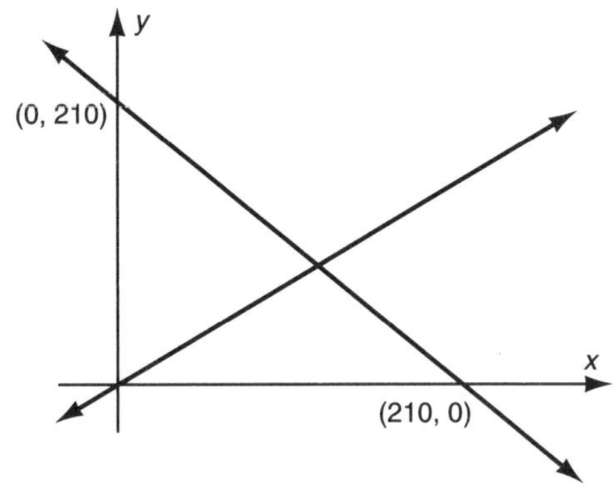

Copy the graph and shade in the area to satisfy the given inequalities and identify the solution set.

CUMULATIVE REVIEW

Multiple-Choice Problems

1. A certain professional baseball player earns \$23,500,000, which includes salary and endorsements. How can this amount be expressed in scientific notation?

   A. 23.5×10^6

   B. 2.35×10^6

   C. 2.35×10^7

   D. 2.35×10^8

2. The circumference of a circle is 14π. What is the area of the circle?

   A. 24.5π

   B. 49π

   C. 98π

   D. 196π

3. Eight years ago, Mr. Cleveland purchased his house for \$180,000. When he had it appraised this year, he found that it had increased in value by 47%. What is his house valued at today?

   A. \$846,000

   B. \$270,000

   C. \$264,600

   D. \$225,400

4. What is the solution set for $4 - 2x > 10$?

   A. $x > 3$

   B. $x > -3$

   C. $x < 3$

   D. $x < -3$

5. Which relation is NOT a function?

   A. (4, 1), (4, 2), (4, 3), (4, 4)

   B. (2, 3), (3, 4), (4, 5), (5, 6)

   C. (1, 4), (2, 4), (3, 4), (4, 4)

   D. (1, 7), (7, 1), (6, −3), (−3, 6)

6. What geometric principle is used to prove the following pair of triangles congruent?

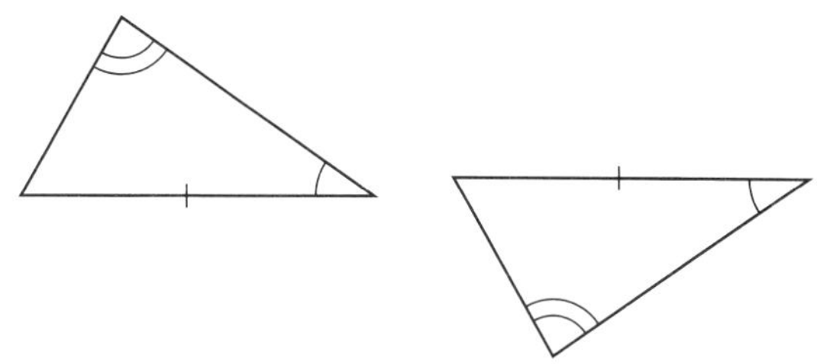

   A. ASA **C.** HL

   B. SAS **D.** AAS

7. Which of the following represents the number of eggs in y dozen and three eggs?

   A. $12 + 3y$

   B. $12y + 3$

   C. $12y + 3y$

   D. $12(y + 3)$

8. Which point is NOT on the parabola $y = x^2 + 2$?

   A. (0, 2) **C.** (2, 6)

   B. (−1, 3) **D.** (1, 4)

9. The figure shown below has how many lines of symmetry?

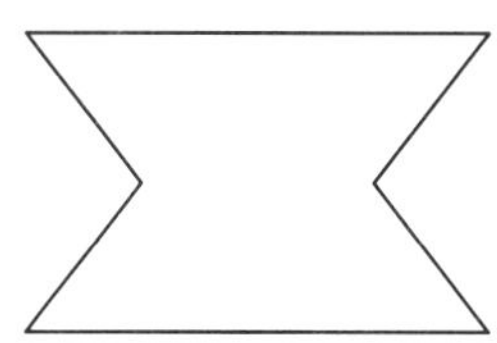

   A. 0 lines **C.** 2 lines

   B. 1 line **D.** 3 lines

10. What are the coordinates of the midpoint of the line segment with endpoints of (1, −8) and (3, 4)?

    A. (4, −4) **C.** (2, 6)

    B. (4, 12) **D.** (2, −2)

Gridded-Response Problems

11. The band formation for the half-time show consists of 5 members in the first row. The second row has 4 more than the first row, the third row has 4 more than the second row, and so forth. How many members will be in the seventh row?

12. What is the surface area of a rectangular box that has dimensions of 5 inches by 14 inches by 3 inches?

13. The vertex angle of an isosceles triangle is 78°. What is the measure of a base angle?

14. What is the slope of a line perpendicular to the line containing the points $(-3, 2)$ and $(0, -2)$?

15. A 6-foot man casts a 10-foot shadow at the same time that his daughter casts a 4-foot shadow. How tall is his daughter?

16. In parallelogram $ABCD$, $\angle D = 55°$. What is the measure of $\angle A$?

17. The radii of two similar cylinders are 1 inch and 2 inches. If the volume of the larger cylinder is 6 cubic inches, what is the volume of the smaller cylinder?

18. On a coordinate graph, one point is at $(-5, 2)$ and another point is at $(0, -10)$. What is the distance between these two points?

19. Joel has a piece of plywood 2 feet by $3\frac{1}{2}$ feet. How many square inches is the top surface of this piece of plywood?

20. Zora has nickels and quarters in her purse. There are 15 coins that total \$2.55. How many quarters does she have in her purse?

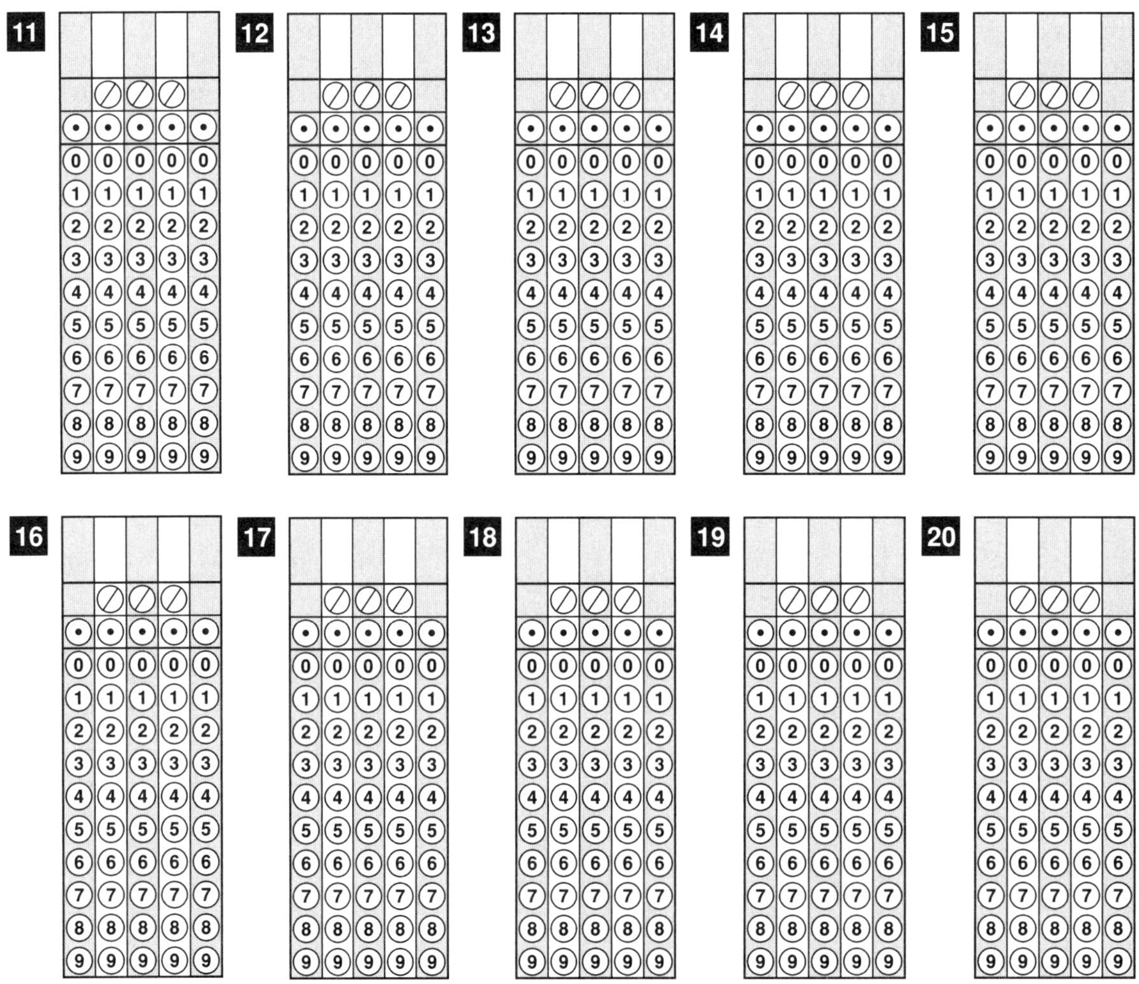

Think, Solve, Explain: Short-Response Problems

21. *ABCD* is a rectangle with $AB = 60$, $BC = 20$, $CD = 2x + y$, and $DA = x - 2y$. Find the values of x and y. Show all necessary work.

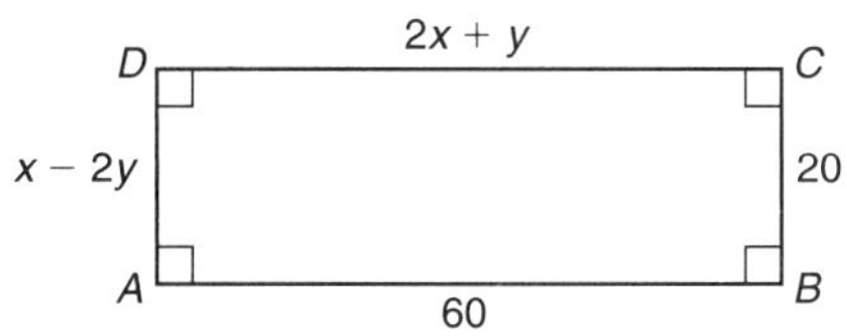

22. Use the graph shown below.

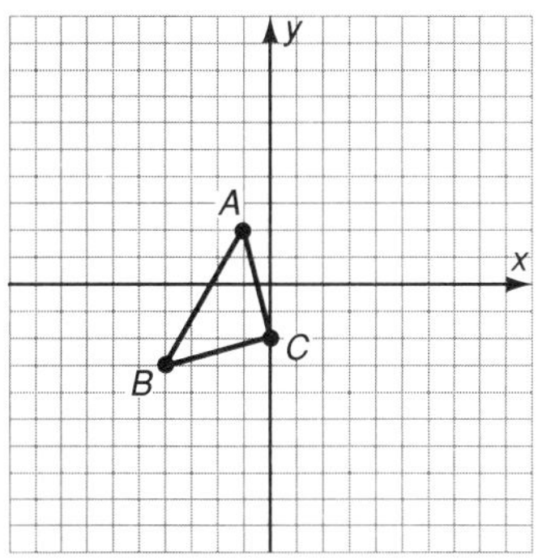

Part A Triangle *ABC* is transformed so that point B' is located at the origin. What will be the coordinates of A' and C'?

Part B Point B' remains the same and the triangle $A'B'C'$ is dilated so that each side is twice as long. What will be the coordinates of A'' and C''?

23. The graph below shows the inequalities $y \leq 3x - 1$ and $y \geq -x + 2$.

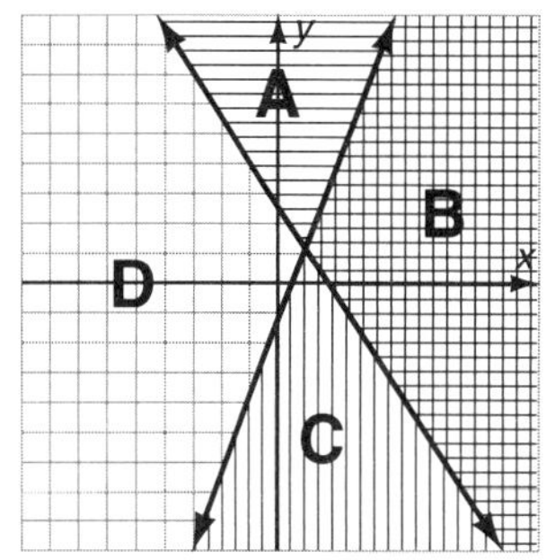

Part A Which region satisfies both inequalities?

Part B Which region satisfies neither inequality?

Part C Which region satisfies $y \geq -x + 2$ but NOT $y \leq 3x - 1$?

24. On his birthday, Pierre's grandmother gave him $2,000. Pierre put part of his money in an account that earns 2% interest and the other part in an account that earns 3% interest. At the end of the year he had earned a total of $45. How much did Pierre put in each account? Show all work necessary to justify your answer.

Think, Solve, Explain: Extended-Response Problem

25. To find the width of a river, a surveying crew used the setup shown below.

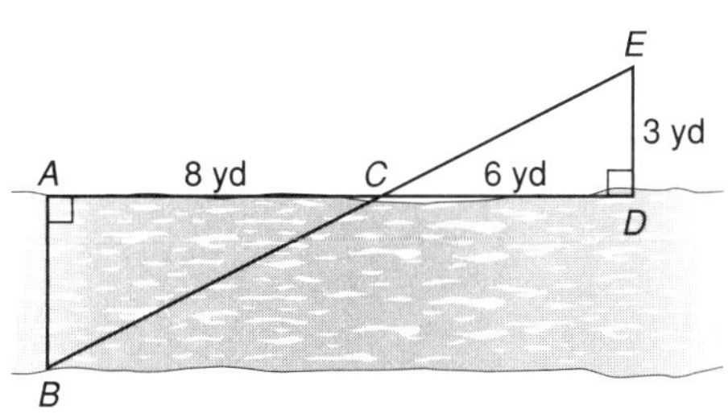

Part A Triangles *ABC* and *DEC* are similar. Explain why this is true, or use a formal geometric proof to show that it is true.

Part B Set up a proportion to find the width *AB* of the river.

Part C Solve the proportion and give the width of the river.

CHAPTER FIVE

Data Analysis and Probability

CHAPTER CONTENTS

Section 1 Understanding and using the tools of data analysis to manage information

Data are related pieces of numerical information. In real-world situations, many problems arise that involve data. Some of the most common examples of numerical data are found in sports, politics, economics, and weather. In certain situations, data may be called "statistics," and **statistics** is also the study of numerical data.

When we describe, summarize, and analyze data, we usually follow four steps:

Step 1. Collect the data.

Step 2. Organize the data into some form of table, chart, or plot.

Step 3. Analyze the data.

Step 4. On the basis of the analysis, draw conclusions or make predictions.

A INTERPRETING DATA THAT ARE ORGANIZED AND DISPLAYED IN TABLES, CHARTS, AND PLOTS

When data are collected, they are usually random and unorganized. To make any statement or draw any conclusion, we usually need to **organize**, or rearrange, the data in an orderly setup. **Tables**, **charts**, and **plots** are common tools for organizing and displaying data.

Tables

Tables are very helpful for organizing data. In a table, data are presented in **columns** (reading down) and **rows** (reading across). One kind of table is a table of frequencies.

EXAMPLE

A teacher records the following grades for a class:

90, 93, 86, 76, 65, 91, 100, 74, 75, 82, 81, 77, 93, 91, 99, 64, 82, 86, 88, 78, 82, 95, 67, 88, 80, 70, 62, 99, 71, 85

These grades are much easier to analyze when we put them into a table of frequencies:

Grade interval	Tallies	Frequency
90–100	~~||||~~ ||||	9
80–89	~~||||~~ ~~||||~~	10
70–79	~~||||~~ ||	7
60–69	||||	4
50–59	none	0

To construct a table of frequencies from ungrouped data, we choose **intervals** of equal length. The data are then grouped within those intervals. Almost always, we want to keep the number of intervals somewhere between 5 and 15, because the grouping will not be effective if there are too many or too few intervals. Then, for each interval, we **tally** the data that fall into it, as shown in the table above. Each vertical line | is a tally indicating one piece of data. For every fifth tally we use a slanted line to indicate a group of five. **Frequency** is the number of tallies in each interval.

NOTE: In working with data, a helpful first step is to write them in order from least to greatest or from greatest to least.

MODEL PROBLEMS

1. These data are weights of packages, in kilograms. Organize them in a table of frequencies.

   70, 43, 48, 72, 53, 81, 76, 54, 58, 64, 51, 53, 75, 62, 84, 67, 72, 80, 88, 65, 60, 43, 53, 42, 57, 61, 55, 75, 82, 71

   Answer:

Interval	Tallies	Frequency
80–89 kg	𝍸	5
70–79 kg	𝍸 ||	7
60–69 kg	𝍸 |	6
50–59 kg	𝍸 |||	8
40–49 kg	||||	4

Use your table to answer the following questions.

2. Which weight interval contains the most packages?

   Answer: 50–59 kilograms

3. How many packages weigh less than 70 kilograms?

   Answer: 18 packages. These are the packages in the bottom three intervals: 6 + 8 + 4 = 18.

TRY THESE

1. Here are heights, in centimeters, of 18 students:

   149, 155, 157, 158, 158, 160, 162, 162, 162, 163, 166, 170, 171, 171, 173, 175, 175, 180

   a. Construct a table, using intervals with a length of 10.

   b. How many students are less than 160 centimeters tall?

   c. In which interval or intervals do we find the greatest number of students?

2. These data are the number of hours a group of teenage students spent listening to music:

   3, 4, 4, 5, 6, 8, 9, 10, 11, 12, 13, 15, 16, 17, 17, 18, 18, 19, 19, 20, 20, 20, 22, 25, 25, 27, 28, 31, 32, 36

   a. Construct a table, using intervals with a length of 5.

   b. How many students listened for 10 hours or less?

   c. Which time interval contains the LEAST number of students?

3. The following data represent the number of points received by each of the students in Ms. Martin's class:

   83, 92, 71, 64, 85, 94, 63, 75, 91, 68, 85, 81, 85, 98, 58

   a. Construct a table using intervals beginning with 50 and intervals of length 10.

   b. Which interval had the greatest number of students?

   c. How many students have 80 points or greater?

   d. How many students have less than 70 points?

4. Liam surveyed the people leaving a fast-food restaurant. They were asked the amount of money spent per person. The data below represent these amounts:

   \$1.89, \$3.25, \$2.28, \$4.50, \$3.89, \$2.95, \$5.10, \$4.28, \$6.35, \$4.10, \$2.10, \$3.80, \$3.14, \$3.62, \$3.01

   a. Construct a table beginning with 0 and intervals of length 100 cents.

   b. Which interval contains the most values?

   c. How many people spent \$4.00 or more?

   d. How many people spent between \$2.00 and \$4.00?

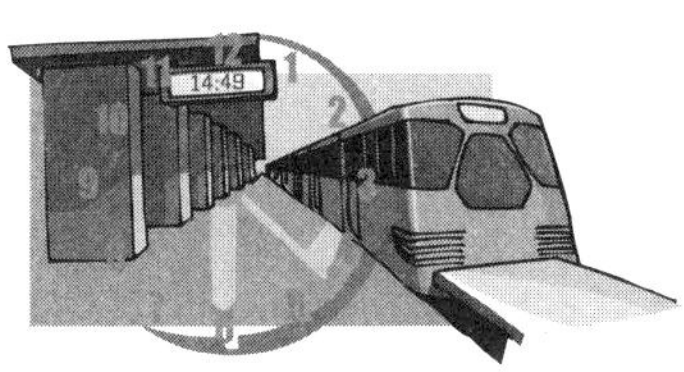

5. The following data represent the number of minutes that trains were late at one stop recorded over a period of three days:

   0, 5, 2, 10, 8, 6, 2, 0, 8, 6, 9, 12, 14, 1, 3

   a. Construct a table beginning with 0 and intervals of length 3 minutes.

   b. Which interval has the greatest frequency?

   c. Which interval has the least frequency?

   d. Based on this survey, what percent of the trains arrived 6 or fewer minutes late?

Bar graphs

Another way to organize data is to use a **bar graph**. A bar graph has two **scales**, or **axes**: one horizontal and one vertical. Rectangular bars are used to represent the information. The bars can be vertical (going upward) or horizontal (going from left to right).

Here are guidelines for constructing a bar graph.

- Begin all the bars on the same line.
- Make all the bars the same width.
- Put equal space between the bars.
- If both scales are numerical, try to begin each scale at zero.
- On each axis, the scale intervals must be **uniform**, that is, evenly spaced.
- Define each axis by giving it a **label**.
- Give your bar graph a **title**.

Below is an example of a **vertical bar graph**. It shows the number of students who earned each of five grades (A, B, C, D, and F) in geometry for one semester.

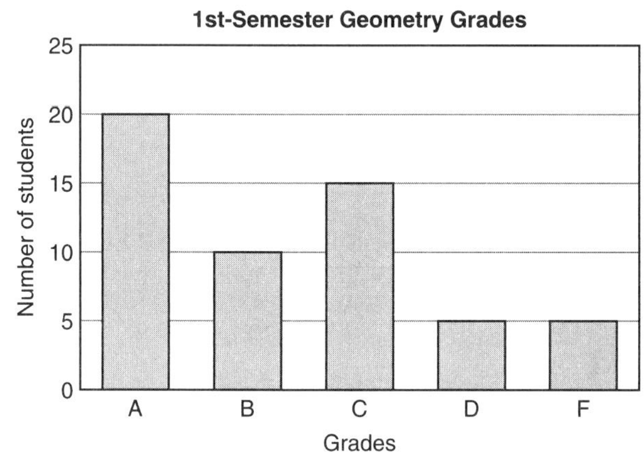

In a vertical bar graph like this one, the horizontal axis identifies the category, which can be a numerical quantity or a named quantity. The vertical axis is a numerical amount or a frequency.

If the axes are reversed, the graph becomes a **horizontal bar graph**:

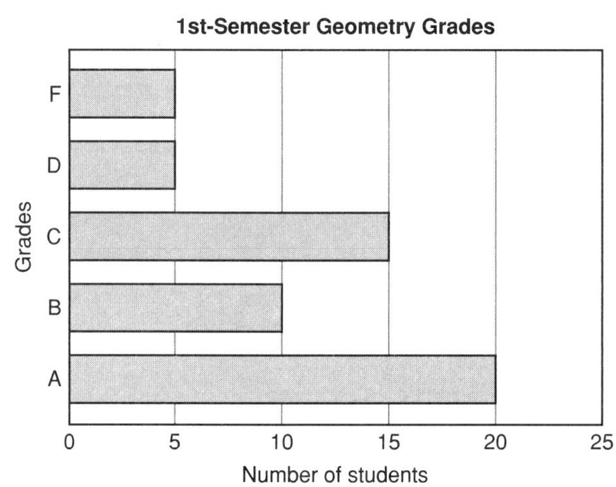

MODEL PROBLEMS

One hundred adults responded to a survey about the importance of extracurricular activities at their child's school. Following are the results:

39 said that extracurricular activities were "extremely important" (category 1)

41 said "very important" (category 2)

14 said "fairly important" (category 3)

3 said "not important" (category 4)

3 had "no opinion" (category 5)

1. Use these data to construct a vertical bar graph.

Answer: First, organize your data. Here, a table is used to organize the given information.

Category	Response	Number responding
1	Extremely important	39
2	Very important	41
3	Fairly important	14
4	Not important	3
5	No opinion	3

Then construct the graph.

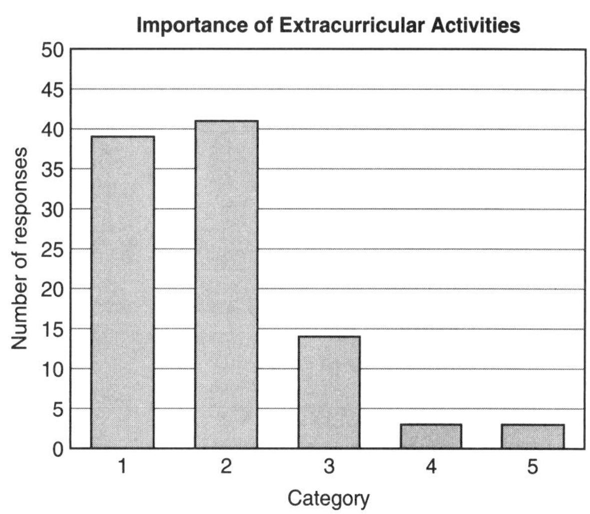

2. In the graph above, which of these categories has the greatest frequency?

   A. category 1

   B. category 2

   C. category 3

   D. category 4

   Answer: B (category 2, "very important")

3. What percent of the parents felt that extracurricular activities had at least some importance?

   A. exactly 39 percent

   B. exactly 41 percent

   C. approximately 50 percent

   D. more than 75 percent

   Answer: D (more than 75 percent)

TRY THESE

1. N-R-G Cereal Company makes 5 kinds of breakfast cereal. The percentage of protein in each kind is: oatmeal, 6%; rice, 2%; raisin, 2%; wheat, 4%; and high-protein, 10%.

   a. Construct a vertical bar graph displaying these data.

   b. Using your graph, determine how much greater the percentage of protein is in high-protein cereal than in rice cereal.

   c. Construct a horizontal bar graph using the same data.

2. The following information represents the number of cars that were sold at Bud's Used Cars over an eight-week period.

Week	1	2	3	4	5	6	7	8
Number of cars sold	2	5	1	6	3	4	2	5

   a. Construct a vertical bar graph displaying these data.

   b. What was the total number of cars sold during the 8-week period?

   c. What percent of the total cars sold were sold during week 4?

   d. Compare the total sales of the first 4 weeks with the total sales of the last 4 weeks.

3. The following data concern the operations performed by the doctors at an eye, ear, nose, and throat hospital during the first quarter of the year.

Type	Eye	Ear	Nose	Throat
Number of operations	56	88	36	48

a. Construct a vertical bar graph displaying these data.

b. What percent of the operations deal with the ear?

c. What is the percent difference in the number of operations for the ear than for the nose?

To display data that are directly related, we can use a **double bar graph** instead of using two separate graphs.

4. The graph below represents the average number of hours spent in after-school activities by students in a tenth-grade math class.

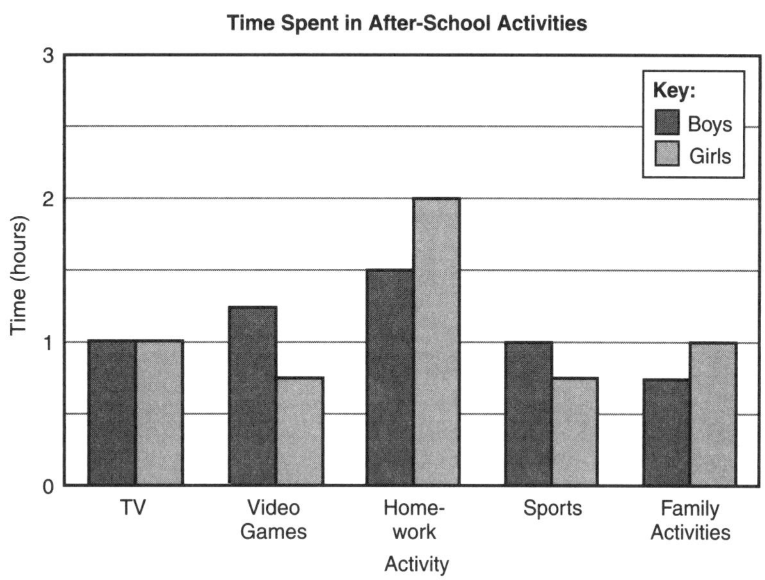

a. On what activity did the girls and boys spend the same amount of time?

b. How much more time than the girls did the boys spend playing video games.

c. Which group spent more time on the combined activities of homework and sports? How much more time did they spend?

5. Here are data on rainfall in Martin and Palm Beach counties.

Rainfall (Inches)

Month	Palm Beach	Martin
1	2	3
2	3	3
3	1	4
4	2	3.5
5	6	4
6	10	4

a. Construct a multiple bar graph for these data.

b. What was the difference in the amounts of rainfall in the two counties during the third month?

c. During which month did the two counties have the same amount of rainfall?

d. During which month was there the greatest difference between the counties in amounts of rainfall?

Circle graphs

A **circle graph** is a good way to display data when you need a visual comparison. The complete circle represents a whole amount or an entire group. We can think of this whole, or write it, as 100 percent. Then each category is some portion smaller than 100 percent. The categories are shown as parts, or **sectors**, of the circle.

To construct a circle graph, follow these steps:

Step 1. Look at the value of each category. Compare this value with the whole, to get a fraction or a percent. Each fraction or percentage will be one sector of the circle.

Step 2. To determine the size of each sector, multiply 360 degrees (the entire circle) by the fraction or percent you calculated. The product tells you the measure of the angle that you will draw in the center of the circle. Round each product to the nearest degree, and make sure that all the angles add up to 360 degrees.

Step 3. Draw a circle. Then use a protractor to measure the central angles and mark off the sectors.

Step 4. Label each sector.

Step 5. Give your circle graph a title.

MODEL PROBLEMS Marcia's weekly budget is $50. She spends it as follows: $10 for gasoline; $20 for clothes; $5 for entertainment; $15 for dance lessons.

1. Construct a circle graph for these data.

Answer:

First calculate the percentages of the whole circle ($50, 360°).

Gasoline $= \frac{10}{50} = \frac{1}{5} = 20\%$ and $\frac{1}{5}$ or $20\% \bullet 360° = 72°$

Clothes $= \frac{20}{50} = \frac{2}{5} = 40\%$ and $\frac{2}{5}$ or $40\% \bullet 360° = 144°$

Entertainment $= \frac{5}{50} = \frac{1}{10} = 10\%$ and $\frac{1}{10}$ or $10\% \bullet 360° = 36°$

Dance lessons $= \frac{15}{50} = \frac{3}{10} = 30\%$ and $\frac{3}{10}$ or $30\% \bullet 360° = 108°$

Then draw the circle graph:

Marcia's Budget ($50 Total)

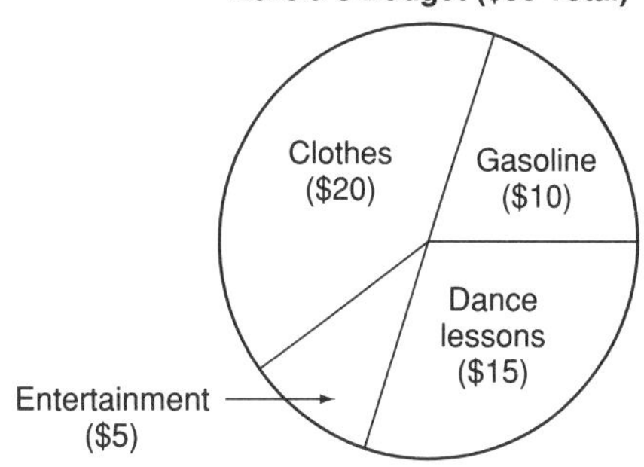

2. What percentage does Marcia spend on gas and clothes combined?

A. 10%

B. 30%

C. 50%

D. 60%

Answer: D (20% + 40% = 60%)

3. What percentage is NOT spent on entertainment?

A. 90%

B. 72%

C. 45%

D. 10%

Answer: A (100% − 10% = 90%)

TRY THESE

Items 1–3: Jo Smith owns a small store. He says that of every dollar he receives in sales, 60 cents goes for the cost of the product, 15 cents goes for his employees' salaries, 10 cents goes for operating costs, and 15 cents is profit.

1. Construct a circle graph to display this information.
2. Last month Jo received $150,000 in revenue. How much profit did he make?
3. How much of last month's revenue was used for employee salaries and operating costs?

Items 4–6: Use the circle graph below.

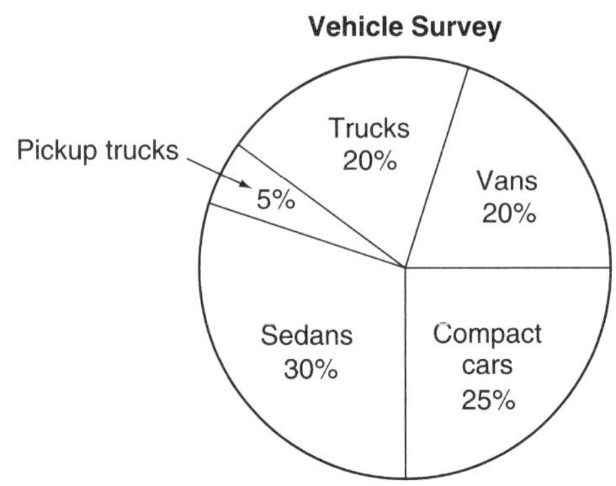

If 2,500 vehicles pass the survey point:

4. Predict how many of these vehicles will be sedans.
5. Predict how many of these vehicles will NOT be vans.
6. How many of the vehicles that passed were cars?

Items 7–10: The table below shows the scores on the AP exams of students from Excalibur High School.

Score	5	4	3	2	1
Number of students	36	62	70	54	18

7. Construct a circle graph to display this information.
8. What percent of the students received a 5?
9. A score of 3 or higher is considered passing. What percent of the students earned a passing score?
10. Display the data in a vertical bar graph.

Line graphs

Sometimes we have one type of data and the value can increase, decrease, or stay the same. In that case, a good way to display the data is to use a **line graph** (sometimes called a **frequency polygon**).

In a line graph, numerical data are represented as points on a graph grid. These points are connected with line segments (sometimes called **broken lines**). A falling line shows a decrease in value. A rising line shows an increase. If there is no change in the data, that is shown by a horizontal line.

To construct a line graph:

- Plot your data as points on the grid.
- Connect the points with line segments.
- Label the horizontal axis and the vertical axis.
- Give your line graph a title.

Below is an example of a line graph. It displays the average high temperatures in one city, Miami, over a 6-month period.

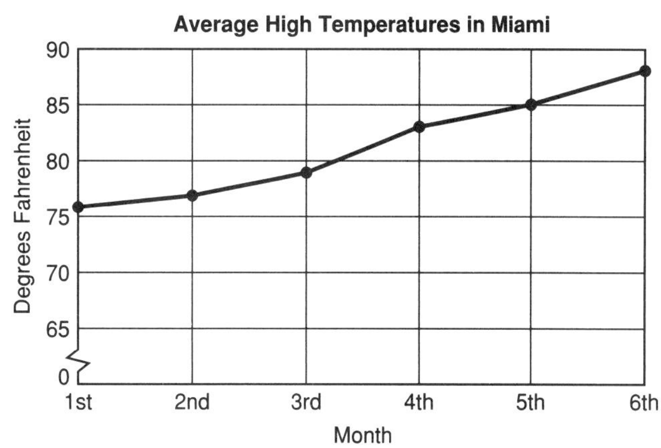

MODEL PROBLEMS

1. Refer to the line graph above, showing temperatures in Miami.

a. What is the approximate difference in temperatures between the first and sixth month?

Answer: About 88° − 76° ≈ 12 degrees

b. Does the temperature change more than 5° between any two consecutive months?

Answer: No. The greatest change in temperature between any two consecutive months is approximately 4° (between months 3 and 4).

2. The table gives cash dividends per share of XYZ stock from 1996 to 2006.

a. Construct a line graph for these data.

Year	Dividend (cents per share)	Year	Dividend (cents per share)
1996	14	2002	30
1997	12	2003	38
1998	15	2004	51
1999	20	2005	30
2000	15	2006	40
2001	15		

Answer:

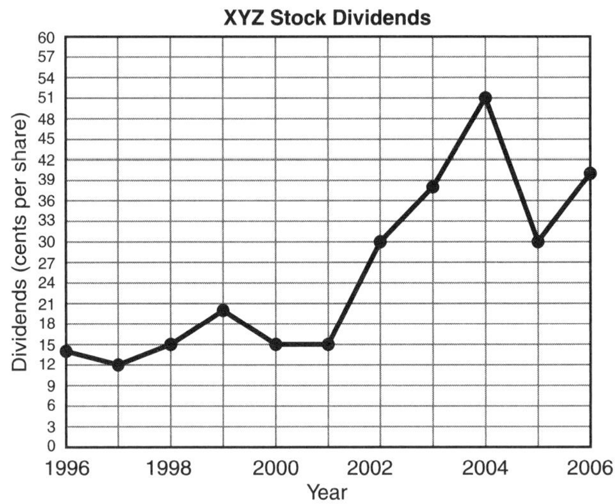

b. Using the line graph above, answer the following questions.

(1) In what year was the lowest dividend paid?

Answer: 1997

(2) Between what two consecutive years did the amount of the dividend stay the same?

Answer: 2000–2001

(3) Between what two consecutive years did the amount of the dividend decline most sharply?

Answer: 2004–2005

(4) What is the difference between the amounts of the highest and lowest dividends paid?

Answer: $51 - 12 = 39$ cents

To compare two or more sets of data, we can use a **multiple line graph**.

3. The following table shows data for the average high temperatures for the first six months of the year in two cities: Miami and Pensacola.

a. Construct a double line graph for these data.

Average High Temperatures (degrees Fahrenheit)

Month	Miami	Pensacola
1	76	62
2	77	64
3	78	68
4	83	77
5	85	84
6	88	92

REMEMBER: In a multiple line graph, the axes must represent the same facts for all the data sets.

Answer:

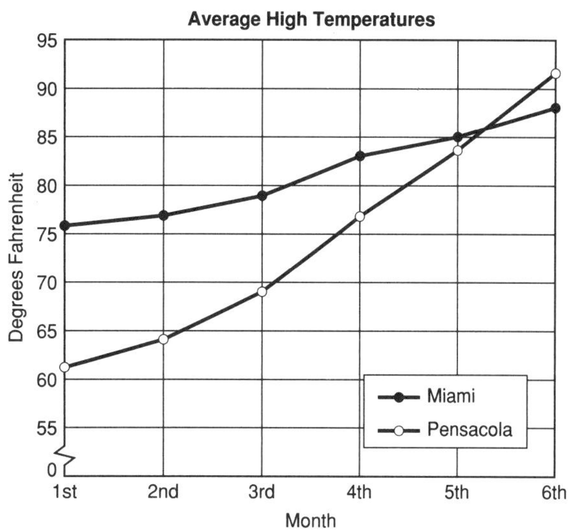

b. Between what two months are the temperatures in the two cities most similar?

Answer: Between the fifth and sixth months

c. Between what two months is there the greatest difference in temperatures?

Answer: Between the first and second months

d. During which month does Pensacola have a higher average temperature than Miami?

Answer: During the sixth month

TRY THESE

1. Here are data on absences at a company.

Absences, Central Manufacturing Company

January	February	March	April	May	June
125	200	250	200	75	100

a. Construct a single line graph for these data.

b. How many more absences were there in March than in February?

c. Between which two consecutive months was the increase in number of absences greatest?

d. During which months were there the same number of absences?

2. Here are data on average temperatures in three cities: Atlanta, Chicago, and Los Angeles (L.A.).

Average Temperature (Degrees Fahrenheit)

City	Jan.	Feb.	Mar.	Apr.	May	June	July	Aug.	Sept.	Oct.	Nov.	Dec.
Atlanta	43	45	52	61	70	76	78	77	72	63	52	45
Chicago	24	26	35	47	58	67	72	72	65	54	40	29
L.A.	55	56	58	59	62	66	70	71	69	65	61	57

a. Construct a multiple line graph for these data.

b. During which months is the temperature higher in Los Angeles than in the other cities?

c. During which months is the temperature higher in Chicago than in the other cities?

d. During which month is there the greatest difference between the highest and lowest temperature?

3. The following table shows the number of tomatoes on 50 tomato plants in Becky's garden.

Tomatoes Grown in Becky's Garden

Number of tomatoes per plant	6	7	8	9	10	11	12	13
Frequency (number of plants)	6	2	5	8	14	9	4	2

a. Construct a single line graph for these data.

b. What percent of the plants produced more than 9 tomatoes per plant?

c. What is the total number of tomatoes produced?

4. The table below displays the heights, in inches, of two students over a period of 6 years.

Students' Heights (Inches)

Grade	7	8	9	10	11	12
John	64	66	68	72	74	75
Harold	64	67	70	71	72	73

   a. Construct a double line graph to display the data.

   b. Which student had the largest one-year difference in height and when did this occur?

   c. What was the greatest height difference between Harold and John when Harold was taller? During what grade did this occur?

5. The table below displays the number of e-mails received by three students during one school week.

E-mails Received in One Week

Student	Monday	Tuesday	Wednesday	Thursday	Friday
Danielle	6	4	7	2	5
Adam	4	6	3	5	2
Kahlil	7	1	4	3	6

   a. Construct a multiple line graph for these data.

   b. On which days did Kahlil receive more e-mails than the other students?

   c. On which days did Adam receive more e-mails than Danielle and Kahlil?

   d. Who received the most e-mails over the five-day period?

Scatter plots or scattergrams

Suppose that we want to graph the heights and weights of a group of people. Since both height and weight are variables, we use the phrase **bivariate data**, meaning that there are two variables. Bivariate data are best displayed on a **scatter plot** or **scattergram**.

To construct a scatter plot, follow these steps:

Step 1. Draw two perpendicular axes. In the example of height and weight, the horizontal or x-axis could represent weight and the vertical or y-axis could represent height.

Step 2. Scale both axes so that all the data will fit.

Step 3. Plot the data on the graph. Each data point represents both an x-value and a y-value, so its coordinates are (x, y). In our example, the coordinates of a point are (weight, height).

Step 4. Do NOT connect the points. This is because each point represents a particular fact. In our example, the "fact" is one person.

Step 5. After you plot all the points, look at them to see if there is a trend, a pattern. If the points form a pattern that tends to rise, we say that there is a **positive correlation**. If the points form a pattern that tends to fall, we say that there is a **negative correlation**. If the points do not show any organized pattern, there is **no correlation**.

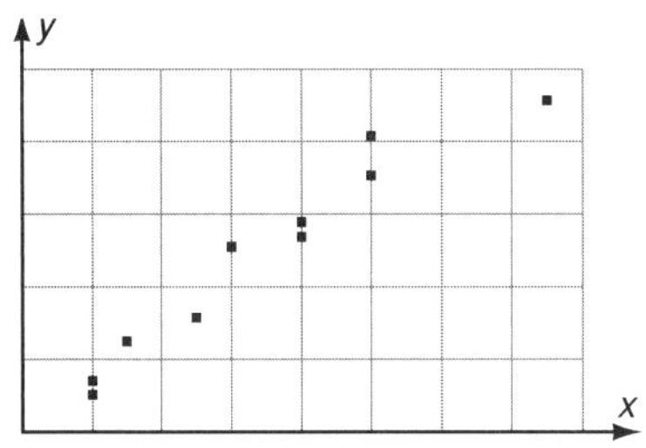

Positive correlation

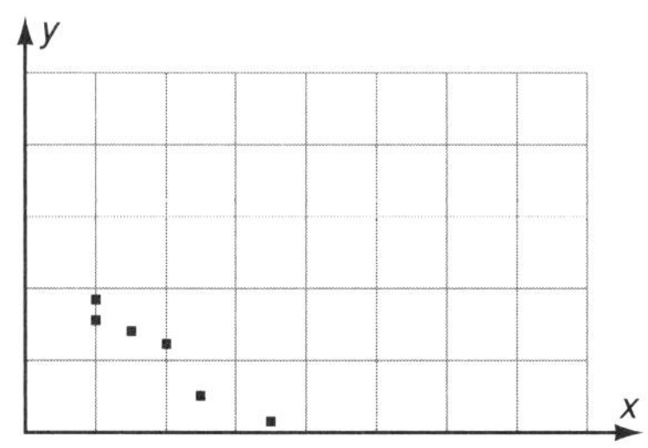

Negative correlation

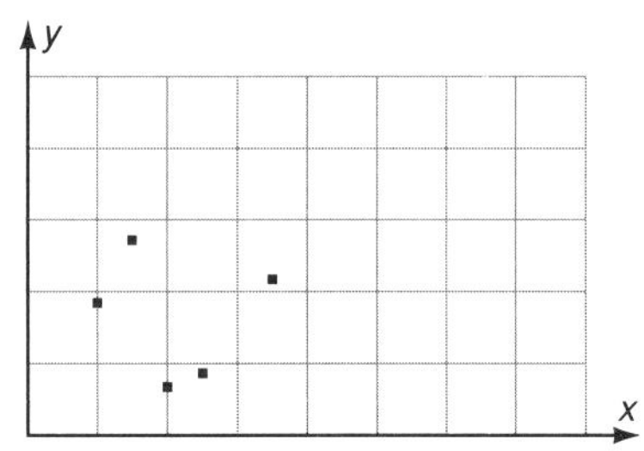

No correlation

EXAMPLE

Here are weights and heights displayed as a table and as a scatter plot.

Weight (pounds)	Height (inches)	Weight (pounds)	Height (inches)
130	63	155	64
120	62	165	70
195	73	125	63
145	63	160	69
165	68	190	71

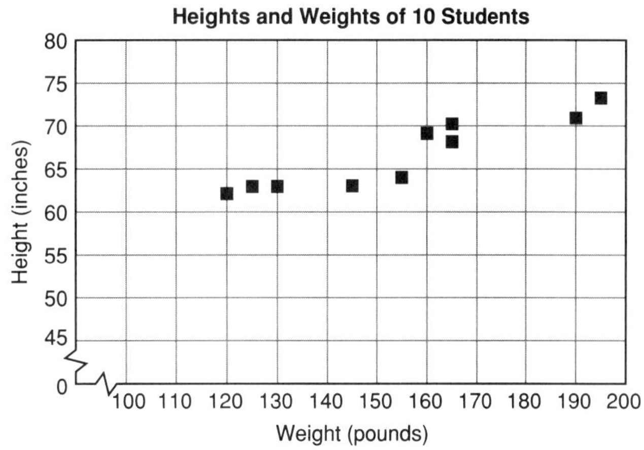

Notice that the scales do not have to be the same, but they must be evenly (uniformly) spaced on each axis. In this example, there seems to be a positive correlation between the heights and weights of these students. As weight increases, height also increases.

MODEL PROBLEM

Ten portable stereos were rated on a scale of 0 to 100, with 100 being "perfect." Prices and ratings of these 10 stereos are as follows:

Price (dollars)	Rating	Price (dollars)	Rating
$120	73	$140	59
180	65	170	64
70	61	100	52
90	55	75	56
200	41	130	65

Part A Construct a scatter plot for these data.

Answer:

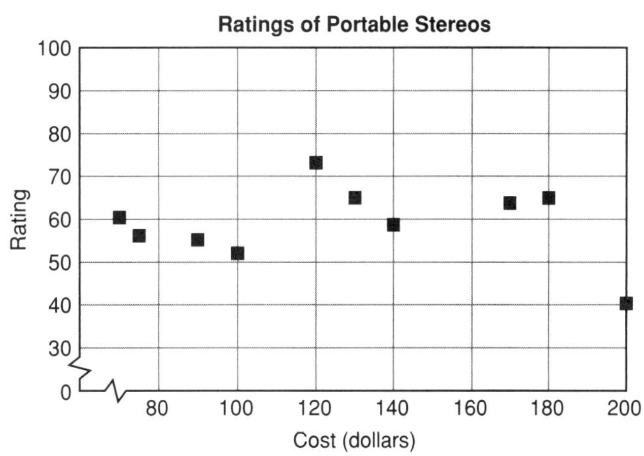

Part B Is there a correlation? If so, is it positive or negative?

Answer: There seems to be no correlation between cost and rating.

TRY THESE

1. Construct a scatter plot using the following information. Is there a correlation? If so, is it positive or negative?

Student	Homework (minutes)	Television (minutes)
1	30	60
2	90	90
3	90	0
4	75	120
5	60	150
6	75	90
7	45	60
8	60	0
9	0	210
10	45	30

2. The data displayed below show the football scores of the Pittsburgh Steelers (2006 Super Bowl champions) and their opponents for the 2005 season.

Game	Steelers	Opponents	Game	Steelers	Opponents
1	34	7	11	7	26
2	27	7	12	31	38
3	20	23	13	21	9
4	24	22	14	18	3
5	17	23	15	41	0
6	27	13	16	35	21
7	20	19	17	31	17
8	20	10	18	21	18
9	34	21	19	34	17
10	13	16	20	21	10

a. Construct a scatter plot using the scores. Label the vertical axis Steelers and the horizontal axis Opponents.

b. Draw a line on the graph connecting the points where the teams would have the same scores. For example, connect the points (7, 7) and (14, 14) and so forth.

c. Is there correlation in the data?

d. Based on the graph, what was the Steelers' win-loss record for the season?

e. What was the largest margin of points in a Steelers victory?

f. What was the largest margin of points in a Steelers loss?

3. The scatter plot below shows the length of completed passes in football compared to the yards gained after the catch. The players are identified by letters.

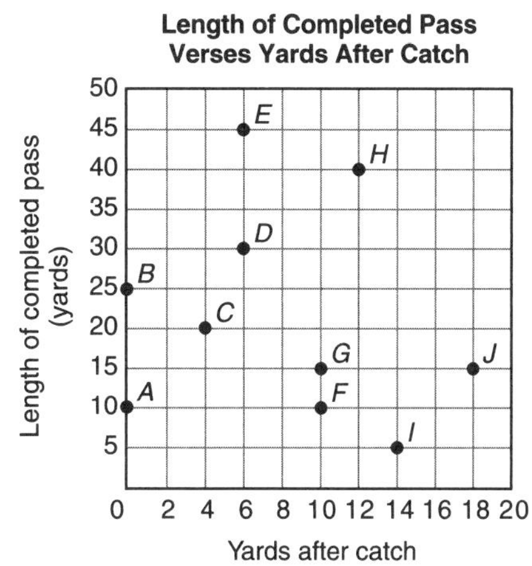

a. Which player caught a pass for the same distance as the yards gained after the catch? What was the distance of each?

b. What was the greatest amount of yards gained after a catch and by which player?

c. Which player made the biggest gain (catch and yards after the catch)? What was the total yards gained?

d. Is there a correlation in the data?

4. The following table lists items from a McDonald's menu with grams of protein and grams of total fat per 100 grams.

Menu Item	Protein	Total Fat
Chicken Nuggets®	10	10
Double Cheeseburger	25	23
Filet-O-Fish®	14	18
Big Mac®	25	30
McChicken® Sandwich	15	16
Big N' Tasty®	27	26
Double Quarter Pounder® with Cheese	47	40

a. Construct a scatter plot using the vertical axis for protein and the horizontal axis for total fat.

b. Is there a correlation?

c. What is the smallest amount of total fat?

d. What is the largest amount of protein?

5. The scatter plot shown below displays the number of 2-point and 3-point baskets scored by a girls' basketball team. The players are identified by letters.

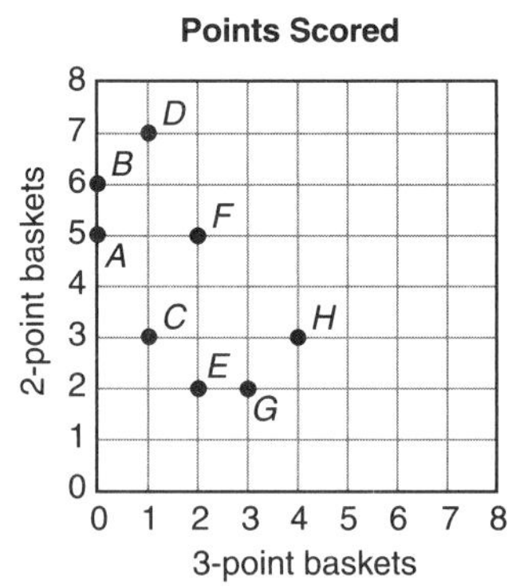

a. Which players scored the least points?

b. Which player scored the most points?

c. Which player scored the most 2-point baskets?

Venn diagrams

Venn diagrams are another way to display data. A **Venn diagram** is a picture that uses circles and rectangles to represent sets.

A **set** can be defined as a collection of **elements** for which there is a method of determining whether an element is or is not a member of the set. The collection "all positive odd integers" is an example of a set. The number 5 would be a member of this set, but 74 would not be a member. Some sets, such as "the set of all integers between 2 and 3," have no members. A set with no members is called an **empty set**.

In a Venn diagram, separate circles are used for sets that do not have any members in common. For example, if you use a Venn diagram to show the set *A* of juniors in your school and the set *B* of seniors in your school, it would look like this:

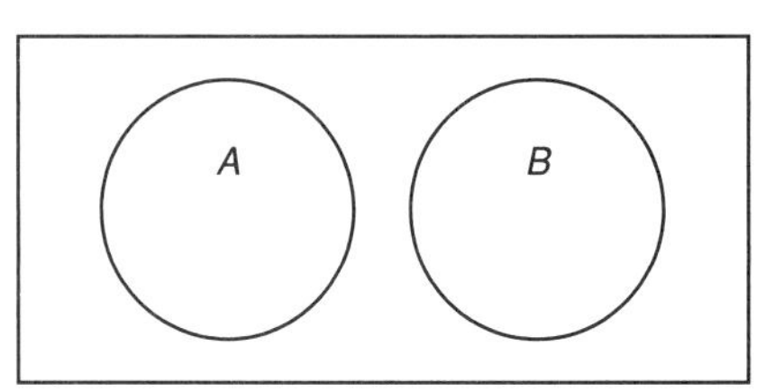

There are no juniors who are also seniors and no seniors who are also juniors, so the circles do not overlap.

Overlapping circles are used to indicate two or more sets that have some but not all elements in common. For example, if set *A* is {*a, b, c, d, e, f*} and set *B* is {*d, e, f, g, h, i*}, the circles will overlap. The intersection of the circles will represent members *d, e,* and *f.*

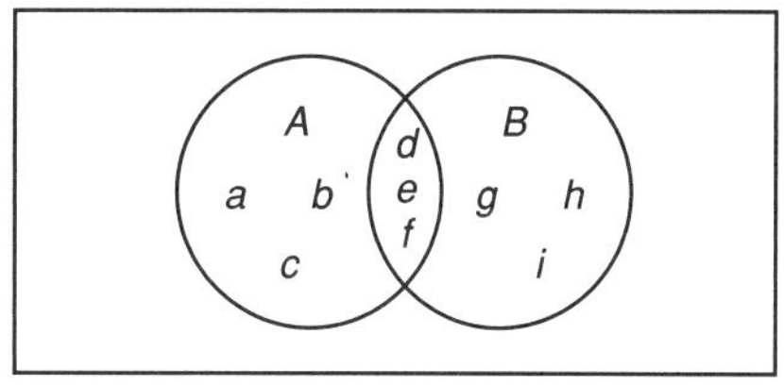

If one circle is totally within another circle, we say that the smaller circle is a **subset** of the larger circle. All the elements represented by the smaller circle are contained in the larger circle. For example, if A = blue, red, green, yellow and B = red, blue, then B is a subset of A.

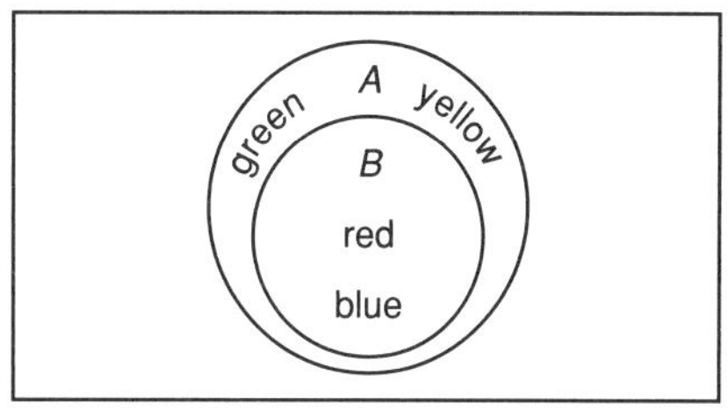

Venn diagrams are shaded to show different operations on sets. Remember that an **intersection of sets** represents the elements the sets have in common. For example, in the Venn diagram below, 20 juniors play football, 15 juniors play basketball, and 6 juniors play both. The shaded area is the intersection.

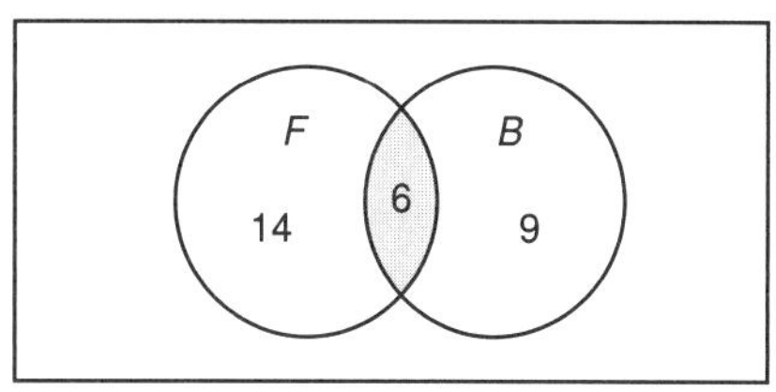

A **union of sets** represents all the elements in the sets. For example, in a sophomore class, 30 students take Spanish, 55 students take French, and 15 students take German. Six students take Spanish and French, 2 students take German and French, 1 student takes German and Spanish, and no student takes all three. This is illustrated in the following Venn diagram. The shaded area is the union of the sets.

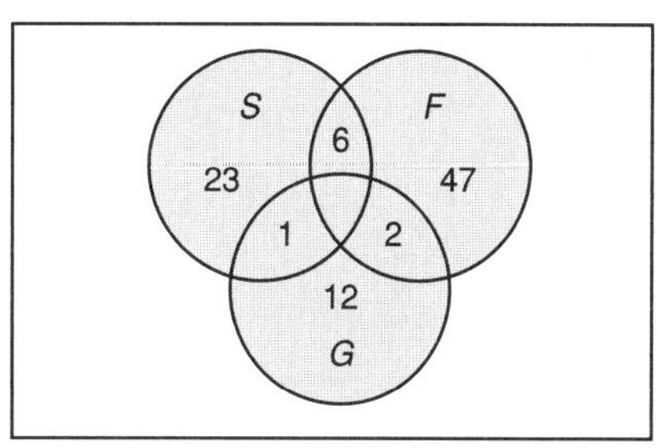

The **complement of a set** is all elements that are *not* in the set. A complement can be indicated by a prime sign (′). For example, if set A is all the even integers, its complement A' is the set of all numbers that are not even integers. In the Venn diagram below, the shaded area A' is the complement of A.

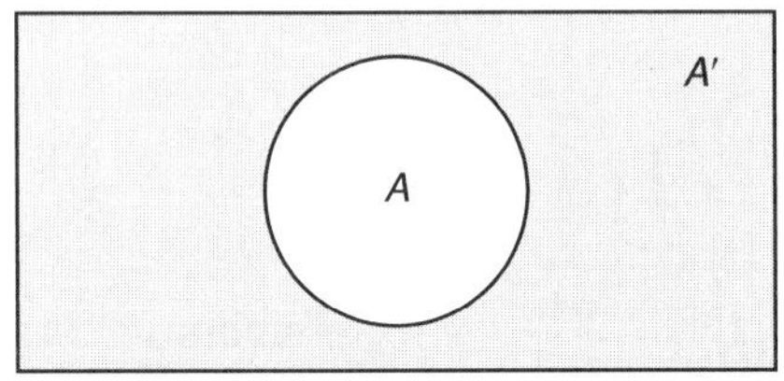

MODEL PROBLEMS

This Venn diagram shows the number of students, out of a class of 30, who earned an A in mathematics (M) and an A in English (E).

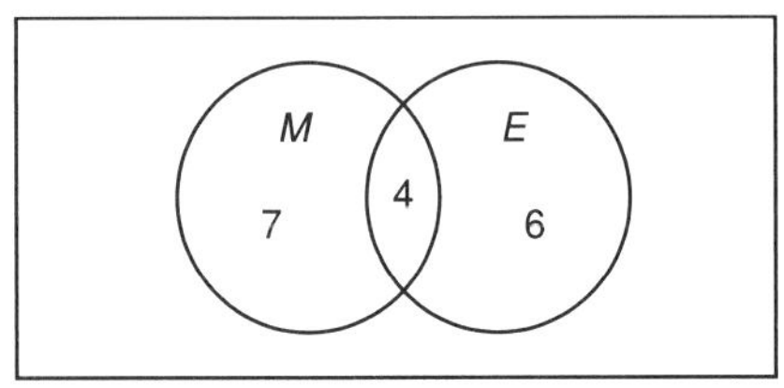

1. How many students earned an A in mathematics?

   Answer: 11 (since $7 + 4 = 11$)

2. How many students earned an A in English?

   Answer: 10 (since $6 + 4 = 10$)

3. How many students earned an A in both mathematics and English?

   Answer: 4 (the number shown in the intersection of the sets)

4. How many students did *not* earn an A in mathematics and did *not* earn an A in English?

   Answer: 13, since $30 - (7 + 4 + 6) = 13$

5. How many students received an A in mathematics but *not* in English?

   Answer: 7

6. How many students received an A in English but *not* in mathematics?

   Answer: 6

TRY THESE

1. This Venn diagram shows salad toppings used for 100 salads in a cafeteria. B = bacon, CH = cheese, and CR = croutons.

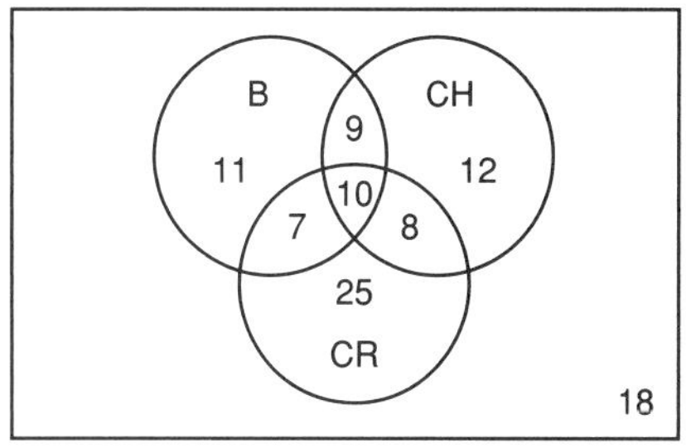

   a. How many salads had just bacon as a topping?

   b. How many salads had just bacon and cheese?

   c. How many salads had just croutons and bacon?

   d. How many salads had all three toppings?

   e. How many salads had *no* topping?

2. There are three models of computers in a school library: model A, model B, and model C. This Venn diagram shows the number of students who are able to use each type of computer.

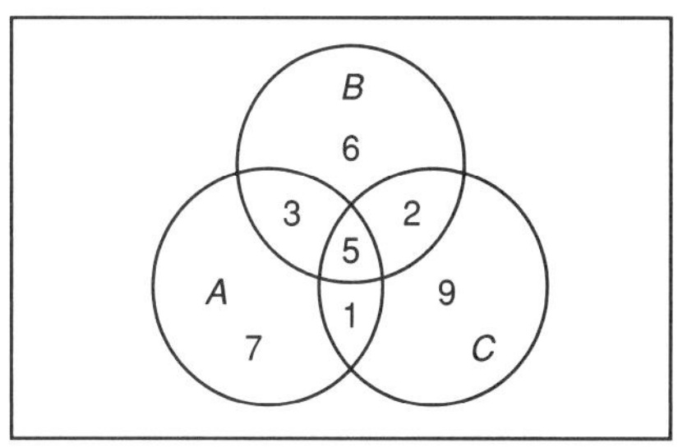

   a. How many students can use model A?

   b. How many students can use model C?

   c. How many students can use both A and B?

   d. How many students can use all three models?

   e. All together, how many students can use the computers?

3. The Venn diagram below shows the results of a survey taken at Multi-Language High School. Students were asked whether they were studying Spanish, French, or Italian.

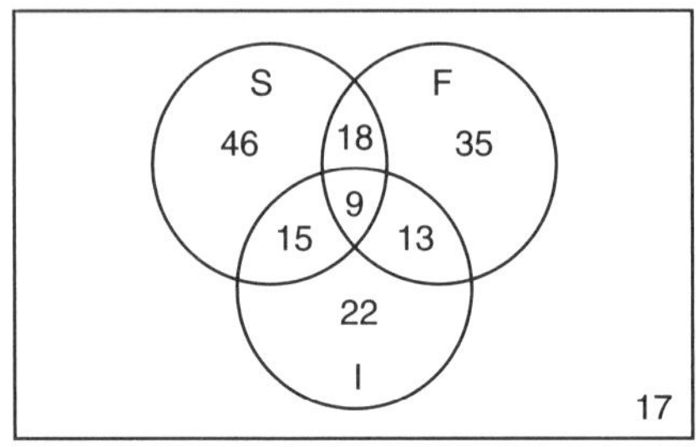

a. How many students were surveyed?

b. How many students study all three languages?

c. How many students study French and Italian?

d. How many students study only French?

e. What percent of the students do *not* study Spanish or Italian?

4. Ms. Pollins conducted a survey about the technology in her students' homes. The items in question are BlackBerrys, iPods, and cell phones. The results of the survey are shown in the Venn diagram below.

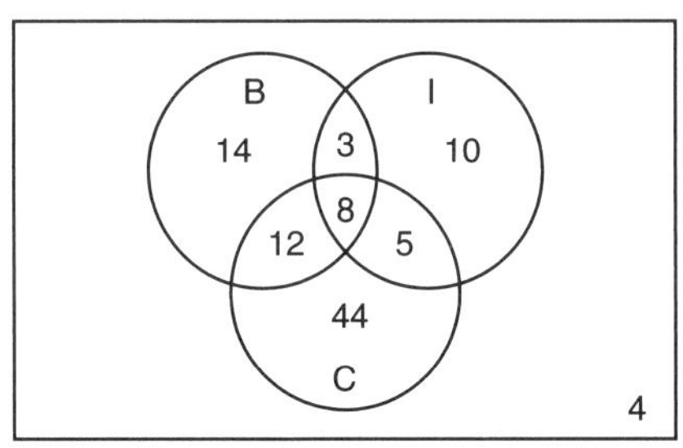

a. How many families have none of the three items?

b. How many families have BlackBerrys and cell phones but no iPods?

c. How many families have either iPods or cell phones but no BlackBerrys?

d. What percent of those surveyed have all three items?

5. The members of the FFA are getting ready for the county fair. The Venn diagram shown below displays the kinds of animals that the members are going to show at the fair.

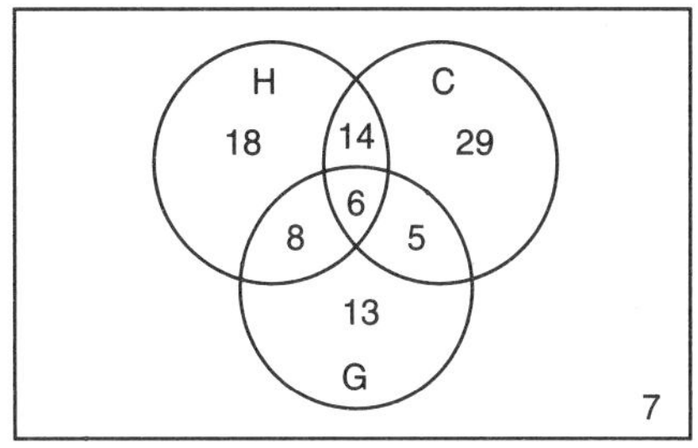

a. How many members are going to show two different kinds of animals?

b. How many members are going to show goats (G)?

c. How many members are going to show just one kind of animal?

d. How many members are not going to show any animals?

Stem-and-leaf plots

Some types of data can be displayed in a diagram called a **stem-and-leaf plot**. Stem-and-leaf plots are used for one-variable numerical data, such as grades on an algebra test. The **stem** is the higher place value (such as tens or hundreds). The **leaf** is the lower place value (such as units).

EXAMPLE

For the data below, a stem-and-leaf plot would be set up as follows.

Test grades: 85, 88, 75, 76, 96, 65, 95, 62, 95, 77, 88, 74, 91, 80, 66, 70, 72, 89, 81, 82, 89

Stem	Leaf
6	2 5 6
7	0 2 4 5 6 7
8	0 1 2 5 8 8 9 9
9	1 5 5 6

Key: $8|2 = 82$

The stem represents the tens digit and the leaf represents the units digit. For instance, the top row represents 62, 65, and 66 from the data. Notice that the leaves are put in order from smallest to largest.

We can draw conclusions from this diagram. For instance, there were more grades in the 80s than in any other interval, since the leaf whose stem is 8 is longest. The highest grade was 96. The lowest grade was 62. The middle grade can be found by looking at the 21 pieces of data in order and choosing the eleventh piece of data. It is 81.

MODEL PROBLEM

Average Daily Temperatures (°F) During March

Stem	Leaf
4	5 6 8
5	2 2 4 7 9 9
6	0 0 2 3 3 4 4 7 7 7 9
7	1 2 3 3 5 6 9
8	0 1 1 2

Key: 5|4 = 54°F

Part A What was the highest daily temperature in March?

Answer: 82°F

Part B What was the lowest daily temperature in March?

Answer: 45°F

Part C On how many days was the temperature in the 60s?

Answer: 11 days

Part D What temperature occurred most often?

Answer: 67°F

TRY THESE

1. This stem-and-leaf plot shows the heights of the 12 members of the Suncoast basketball team.

Stem	Leaf
17	5 8
18	0 0 1 4 4 4 5 5 8
19	2

Key: 18|1 = 181 centimeters

a. How tall is the tallest member of the team?

b. What is the most common height?

c. How many players are shorter than 181 centimeters?

2. The stem-and-leaf plot below shows the annual salaries of the 12 employees of a small computer company.

Stem	**Leaf**
3	1 3
4	4 5 7
5	8 9
6	2 5
7	1
8	
9	2
10	
11	5

Key: 6|2 = $62,000

a. What is the highest salary?

b. Which range of salaries includes the most employees?

c. What is the difference between the highest salary and the lowest salary?

3. The stem-and-leaf plot below displays the ages of people staying at a small motel on one particular night.

Stem	**Leaf**
0	2 5 9
1	5 6
2	7
3	4 5
4	
5	9 9
6	1
7	2

Key: 3|4 = 34 years old

a. What is the age of the youngest person staying at the motel?

b. How many people are under the age of 35?

c. There is a senior discount for anyone over the age of 55. How many people qualify for the discount?

4. The stem-and-leaf plot below shows the number of pull-ups students in coach Sutton's class were able to do.

Stem	Leaf
0	0 0 3 5
1	2 2 2 4 8
2	3 5
3	7
4	1

Key: 2|3 = 23 pull-ups

a. How many students' results were recorded?

b. What was the most common number of pull-ups done?

c. How many students did *not* do at least 25 pull-ups?

5. The stem-and-leaf plot below displays the number of points the boys' basketball team scored in each game this season.

Stem	Leaf
3	6
4	6
5	1 2 4 4 5 6
6	0 1 1 1 4 6 8
7	1

Key: 5|1 = 51 points

a. How many games did they play this season?

b. What is the range between the highest and lowest number of points scored?

c. They won all the games in which they scored over 51 points. What was their win-loss record?

Histographs or histograms

A histogram is like a vertical bar graph, but there are some differences between them. In a **histogram**, or **histograph**, both axes are numerical. The vertical axis measures the frequency of a particular variable. The horizontal scale is usually given as intervals. On a histogram (unlike a bar graph), there is no space between the bars.

EXAMPLE

A histogram can be useful for displaying data such as the following grades.

Test grades: 85, 88, 75, 76, 96, 65, 95, 62, 95, 77, 88, 74, 91, 80, 66, 70, 72, 89, 81, 82, 89

The histogram shows the number of test grades in each 10-point interval:

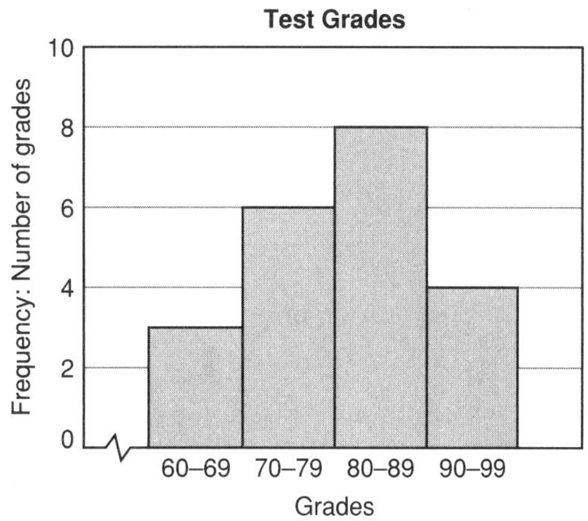

In this histogram, the first bar represents the three grades that are in the interval 60–69, the second bar represents the six grades in the interval 70–79, the third bar represents the eight grades in the interval 80–89, and the fourth bar represents the four grades in the interval 90–99.

MODEL PROBLEM

This histogram shows the speeds of vehicles passing a checkpoint.

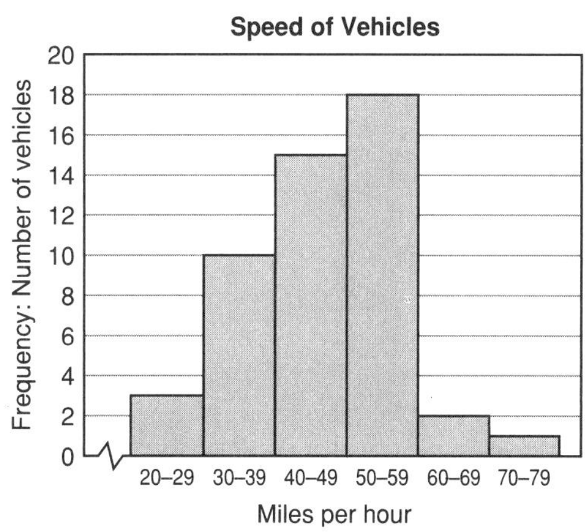

a. How many vehicles were traveling between 30 and 39 miles per hour?

Answer: 10 vehicles

b. The speed limit on this road is 50 miles per hour. How many vehicles were traveling at less than the speed limit?

Answer: 3 + 10 + 15 = 28 vehicles

c. What is the lowest possible speed at which one of these cars could have been traveling?

Answer: 20 miles per hour

d. How many vehicles passed the checkpoint?

Answer: $3 + 10 + 15 + 18 + 2 + 1 = 49$ vehicles

TRY THESE

1. Jessica conducted a survey of shoppers entering a mall on a Saturday morning in February. She asked each shopper his or her age. This histogram shows the results.

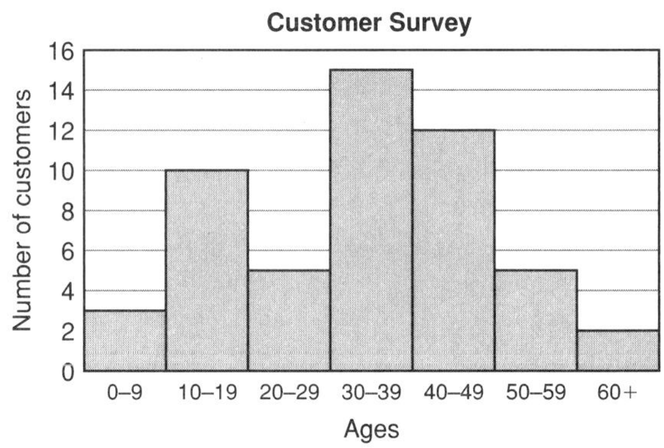

a. How many people over age 60 came into the mall?

b. What age group had the largest number of people?

c. Do you think the results would have been different if Jessica had done her survey on a Wednesday? Explain your answer.

2. The histogram below shows the number of employees living at certain distances (in miles) from their place of employment.

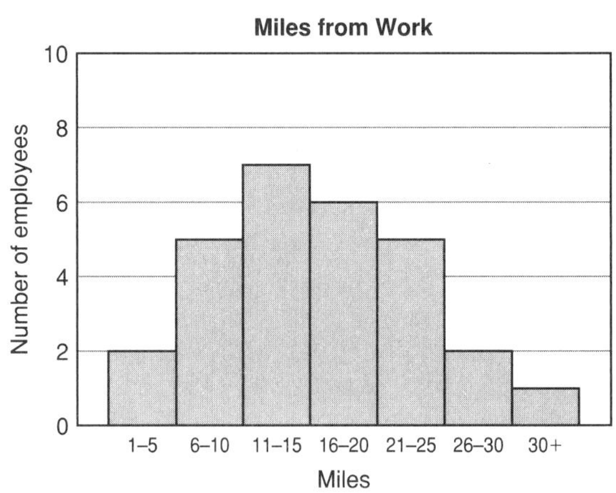

a. How many employees are included in these data?

b. How many employees live 11 or more miles from their workplace?

c. From this graph, can you determine how long it takes the employees in each category (interval) to get to work? Explain your answer.

3. The students in Ms. Nefflen's class recorded the lengths of earthworms prior to their dissection. The results, in millimeters, are shown in the histogram below.

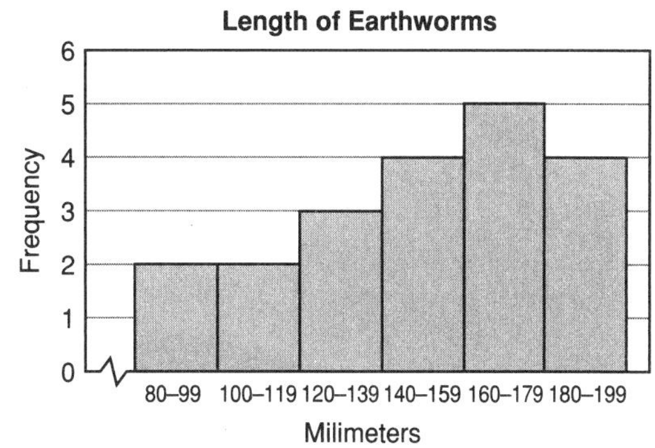

a. How many earthworms were measured?

b. Which interval contained the most earthworms?

c. What percent of the worms were in the interval found in part b?

4. The histogram below shows the results of the times in seconds of the students' 400-meter run.

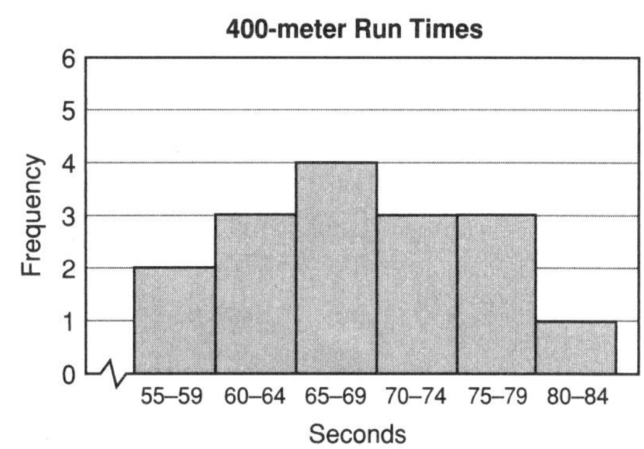

a. How many students were timed in the race?

b. To receive a physical fitness award, students had to run the 400-meter race in less than 75 seconds. How many students qualified for the award?

c. Using the histogram, are you able to tell the fastest time? Explain why or why not.

5. The histogram below shows the heights, measured in centimeters, of the members of the football team.

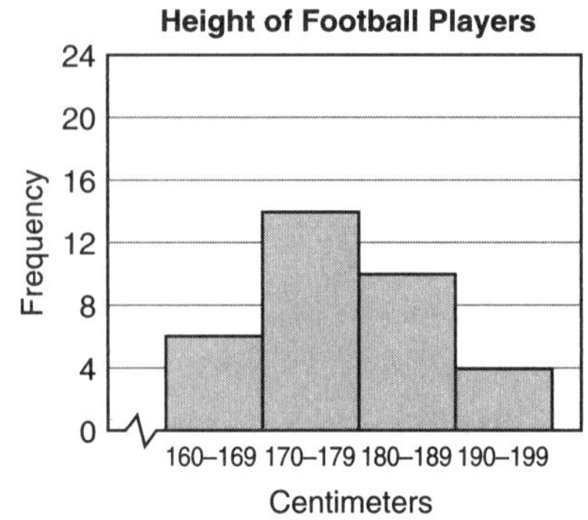

a. How many people are on the football team?

b. How many team members are taller than 169 centimeters?

c. What percent of the players are between 170 and 179 centimeters tall?

Step graphs

In some situations, we use a special type of line graph called a step graph. A **step graph** is the graph of a **step function**, and it looks like a series of steps. To indicate each set of values, we draw a line segment with a solid circle to represent an endpoint and an open circle to represent a missing endpoint.

EXAMPLE

The price of a train ticket increases by $2 every 3 years. From the first time the tickets were sold until 1995, the price of a ticket was $3. At the beginning of 1996, the ticket price jumped to $5. At the beginning of 1999, the price became $7, and so forth.

To put this another way, the price of a ticket remains the same throughout one 3-year period. Then it increases by $2 and stays at the new value for another 3 years, and so on. The step graph below represents this situation.

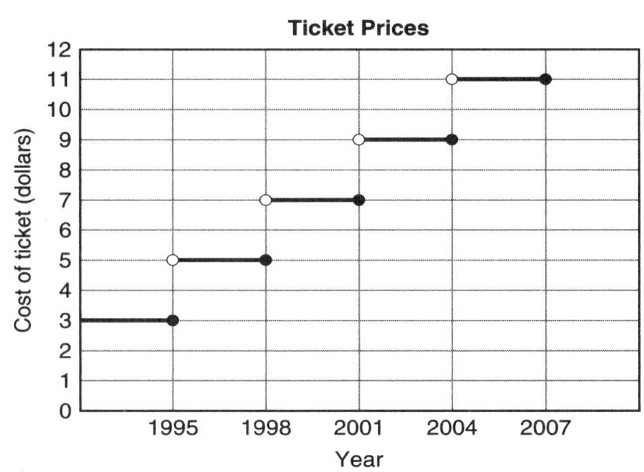

Question: What was the price of the ticket in 2002?

Answer: $9

MODEL PROBLEM The cost of sending a letter to South America depends on the weight. The step graph shows the costs. What was the cost for a letter weighing $5\frac{1}{4}$ ounces?

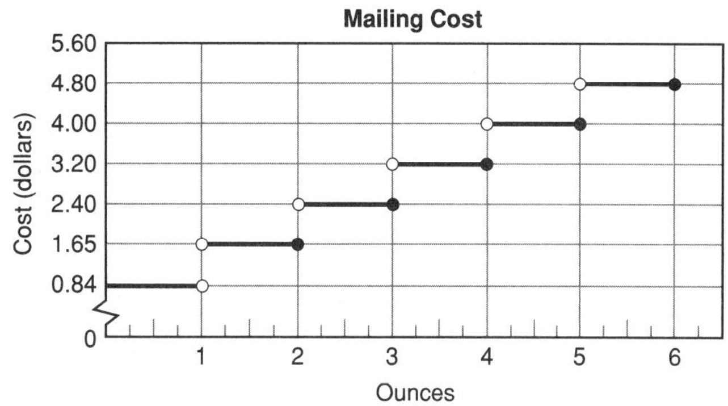

Answer: $4.80

TRY THESE

1. The step graph below is based on the income tax tables for 2005. If Reuben earned $33,425 in 2005, how much was his income tax?

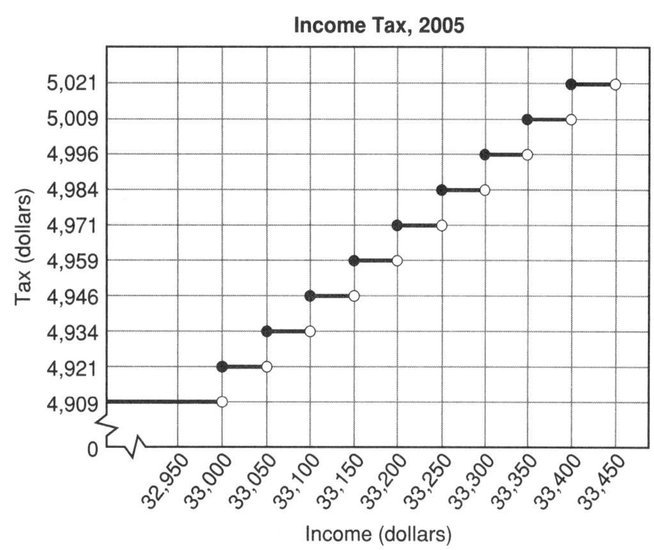

2. The cost of renting a private swimming pool for a two-hour party is based on the number of people who attend. You must have one lifeguard for every 15 people at the party. The charge for the pool with one lifeguard is $80. For each 15 additional people you need another lifeguard at $20 per lifeguard. The step graph representing the charges is shown below.

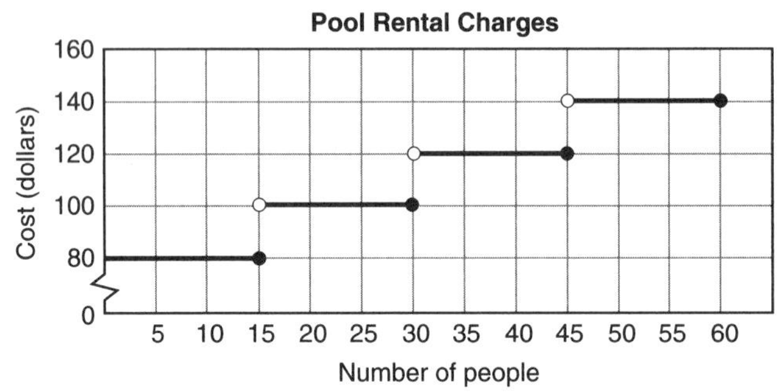

a. How much does it cost for a party with 40 people?

b. How much would it cost for a party with 85 people?

3. Barrie is ordering invitations for her wedding. The cost for the first 50 invitations is $3.50 per invitation. The price then decreases to $3.00 per invitation for the next 50 invitations. It then decreases to $2.50 each for any in excess of 100 invitations. The step graph representing these charges is shown below.

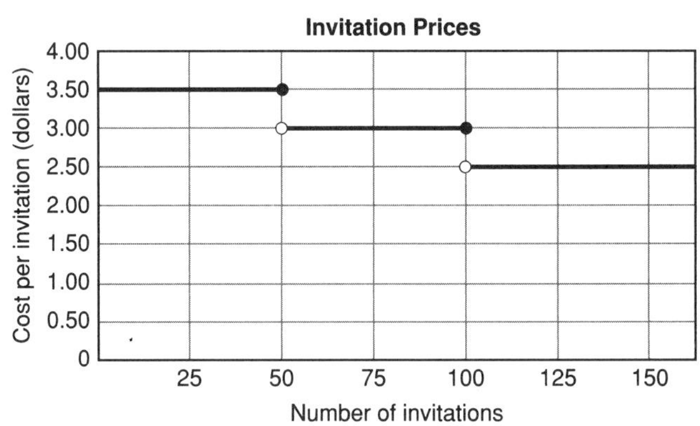

a. How much will it cost for 50 invitations?

b. How much will it cost for 125 invitations?

Assessment

Multiple-Choice Problems

Items 1 and 2: Use the graph below.

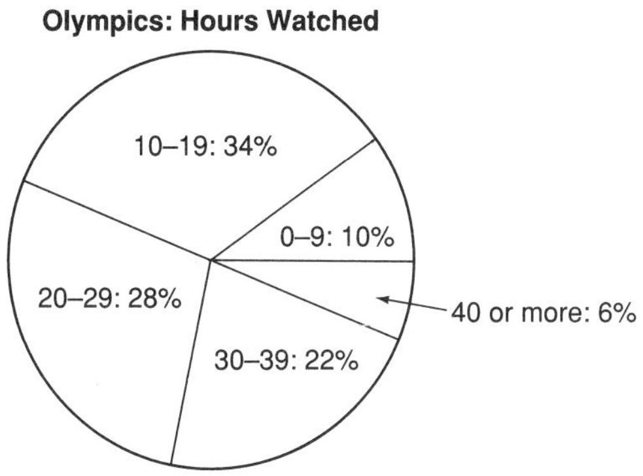

1. In a survey, 50 people were asked how many hours they watched the Olympics on television. Of the people surveyed, how many watched LESS than 30 hours?

   A. 5 **C.** 28

   B. 14 **D.** 36

2. If 120 people were surveyed, approximately how many could be expected to say that they had watched more than 40 hours?

   A. 3 **C.** 10

   B. 7 **D.** 13

Items 3 and 4: Use the graph below.

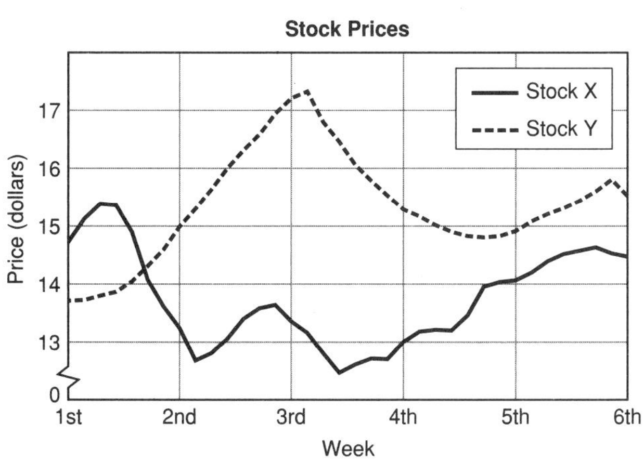

3. During which week did the price of stock X hit its low?

   A. 1 **C.** 3

   B. 2 **D.** 4

4. During which week was the price of stock X the same as the price of stock Y?

   A. 1 **C.** 3

   B. 2 **D.** 4

Items 5 and 6: Use the following graph.

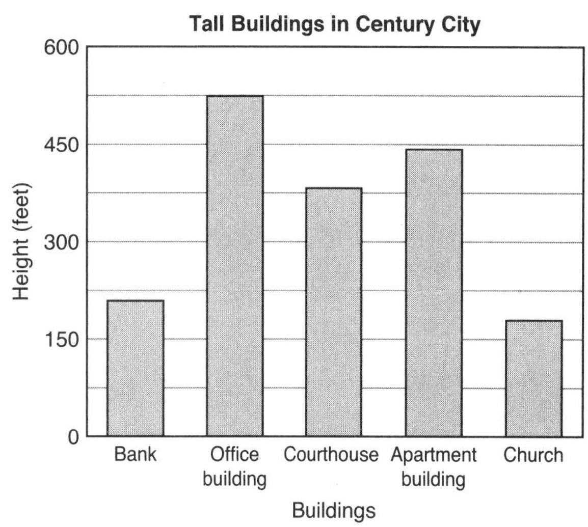

5. To the nearest 25 feet, what is the height of the courthouse?

   A. 350 feet **C.** 400 feet

   B. 375 feet **D.** 440 feet

6. What is the approximate difference in height between the courthouse and the office building?

   A. 100 feet **C.** 150 feet

   B. 125 feet **D.** 175 feet

Items 7–10: Use the following graph.

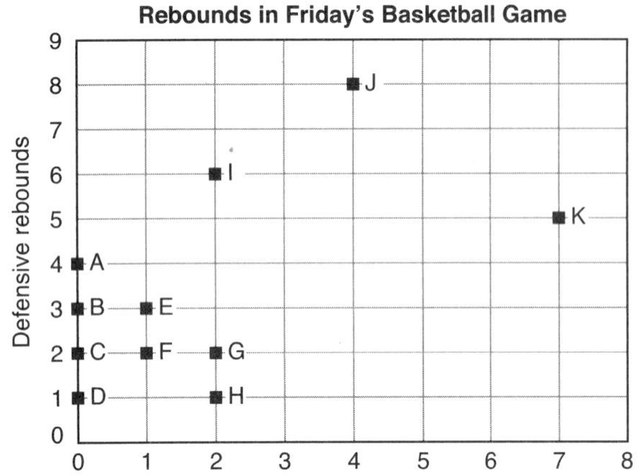

7. Which player had more offensive rebounds than defensive rebounds?

   A. player D
   B. player I
   C. player J
   D. player K

8. Which player had more defensive rebounds than offensive rebounds?

   A. player H
   B. player I
   C. player G
   D. player K

9. How many players had no offensive rebounds?

   A. 2 **C.** 6
   B. 4 **D.** 7

10. Which player had the same number of each kind of rebound?

   A. player G
   B. player H
   C. player E
   D. player F

Items 11–13: Use the following stem-and-leaf plot.

Grades on an Algebra Test

Stem	Leaf
10	0 0
9	0 2 4 7 7 7
8	1 3 3 6 7 8 8 9
7	0 4 8 8 9
6	2 3 5
5	4

Key: 8|3 = 83 percent

11. How many students took the test?

   A. 10 **C.** 25
   B. 15 **D.** 30

12. What score occurred most often?

   A. 100 **C.** 88
   B. 97 **D.** 78

13. If a failing grade is any grade less than 70 percent, how many students failed the test?

   A. 1 **C.** 4
   B. 3 **D.** 9

Items 14 and 15: Use the table below.

Resident Population of Florida in 2000

Age (years)	Population (thousands)
Less than 5	946
5–14	2,089
15–24	1,942
25–59	7,460
60–84	3,214
Greater than 84	331

14. In 2000, how many residents were between the ages of 15 and 24?

   A. 1,942 **C.** 194,200
   B. 19,420 **D.** 1,942,000

15. The total population in 2000 was 15,982,000. Approximately what percent of the population was over 59 years of age?

A. 20% **C.** 25%

B. 22% **D.** 68%

Items 16 and 17: Use the graph below.

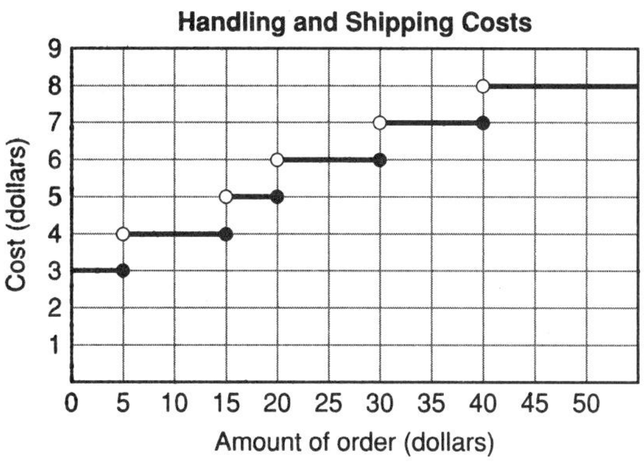

16. If 3 items are ordered that cost $7.50, $18.95, and $1.18, what is the shipping and handling cost?

A. $3 **C.** $5

B. $4 **D.** $6

17. If the shipping and handling cost is $5, which could NOT be the amount of the order?

A. $15.75

B. $16.95

C. $19.98

D. $22.05

Gridded-Response Problems

Items 18–20: Use the Venn diagram. The letters identify radio stations.

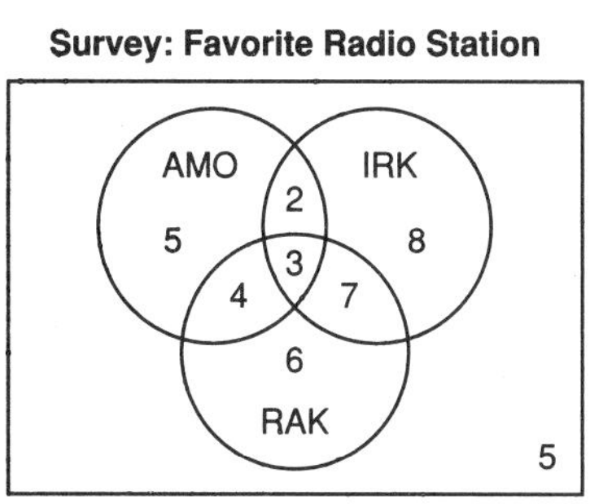

18. How many of those surveyed liked just AMO and RAK?

19. How many of those surveyed liked all three radio stations?

20. How many people were surveyed?

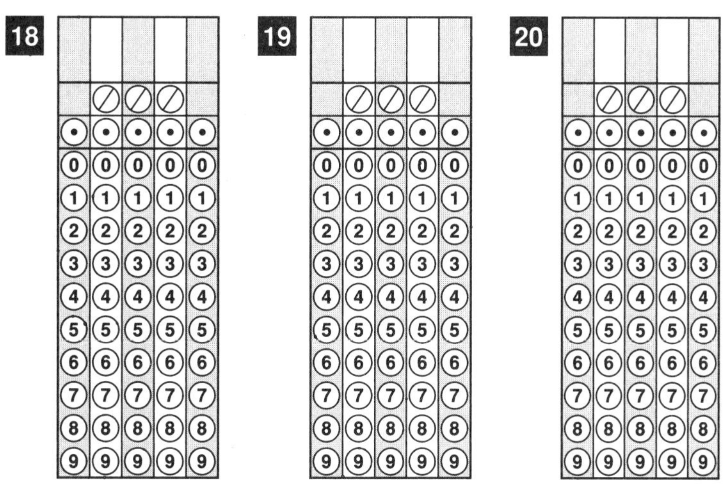

Think, Solve, Explain: Extended-Response Problems

21. You measure the heights of 15 high school students, with the following results (in inches): 63, 68, 69, 71, 60, 63, 66, 57, 70, 67, 66, 72, 61, 61, 70

Part A Construct a table to show the frequencies of these heights. Use intervals of 4 inches, starting with 57 inches. (Thus the first interval is 57–60.)

Part B How many students are less than 65 inches tall?

Part C Suppose that you measured the heights of 15 middle school students. Describe how your frequency table would differ from the table for the high school students.

22. From the income that Ellen earns from her after-school job she budgets 30% for lunches, 30% for clothes, and 15% for entertainment. She saves the rest for college.

Part A Construct a circle graph representing her budget.

Part B If she earns $560 a month, how much does she spend on clothes and entertainment combined?

Part C The first 6 months she earned $560 per month and the second half of the year she received a raise and earned $600 per month. How much will she save for college this year? Show all work.

23. Amanda, who works for a lawn service, is conducting a survey about the size of homeowners' lawns. At a mall, she asks 30 people at random to classify their lawns as follows.

"No lawn"

"Very small" = less than 30 square feet

"Small" = less than 300 square feet (but at least 30 square feet)

"Medium" = less than 1,200 square feet (but at least 300 square feet)

"Large" = greater than 1,200 square feet

Five people say they have no lawn at all; 3 people have very small lawns; 8 people have small lawns; 10 people have medium lawns; and 4 people have large lawns.

Part A Construct a bar graph to display this information. For the horizontal axis, use the verbal categories (words), NOT square footage.

Part B What percent of the people surveyed have less than 300 square feet of lawn?

Part C On another day, Amanda surveys 40 people at the same mall. Given the information from the original survey, predict how many people will say they have lawns greater than 300 square feet. Show the mathematics involved in making an accurate prediction.

Part D The lawn service targets people who have lawns greater than 300 square feet. Assume that the people surveyed at the mall are a representative sample. Estimate how many "targets" (potential customers) there would be among 1,000 people.

24. A TV station ran a fund-raising telethon to benefit victims of a hurricane. Phone calls were accepted between 5 P.M. and midnight. The calls were tallied at the end of each hour. The table shows the number of calls received.

Telethon Calls

Time (P.M.)	Number of calls
5–5:59	200
6–6:59	340
7–7:59	540
8–8:59	820
9–9:59	860
10–10:59	600
11–11:59	180

Part A Draw a line graph representing these data.

Part B The fund-raisers' goal was 3,500 calls by the end of the evening. How close were they to that goal?

Part C During which time period did they receive the largest percentage of calls? What was that percentage?

25. The box score from an Orlando Magic and Miami Heat game listed the players' points and the amount of time played. (The time has been rounded to the nearest minute.) This information is shown below.

Magic Players' Points vs. Time Played

Orlando Magic	Minutes played	Points scored
Francis	37	15
Howard	37	13
Dooling	31	15
Turkoglu	30	15
Garrity	24	0
Hill	20	12
Battie	16	6
Kasun	15	9

Heat Players' Points vs. Time Played

Miami Heat	Minutes played	Points scored
Haslem	42	10
Wade	40	38
Walker	39	12
Williams	31	10
O'Neal	31	25
Payton	24	2
Mourning	16	6
Anderson	13	2

Part A Display these data on two scatter plots, one per team. Use minutes as the horizontal axis and points as the vertical axis.

Part B On the basis of your scatter plots, does one set of data show a better positive correlation than the other? If so, which one? Explain your reasoning.

B CALCULATING AND DETERMINING MEANINGFUL MEASURES OF CENTRAL TENDENCY (MEAN, MEDIAN, AND MODE) AND DISPERSION IN THE FORM OF RANGE

When data are displayed in an organized way, we can **analyze** them. This section reviews ways of analyzing data.

Measures of central tendency

We can often analyze a set of data by calculating a single value: a **mean**, a **median**, or a **mode**. Mean, median, and mode are called **measures of central tendency**. Which one of these three is used depends on the situation, the data, and the desired results.

MEAN

The **mean**, or **arithmetic mean**, is found by adding all the elements in a set of data and then dividing the sum by the number of elements. This is commonly known as finding an "average."

EXAMPLE

If the data are 24, 36, 21, 27, 21 and 33, they can be rearranged as 21, 21, 24, 27, 33, 36. The mean is calculated as follows:

$$\frac{21 + 21 + 24 + 27 + 33 + 36}{6} = \frac{162}{6} = 27$$

The mean of these data is 27.

NOTE: When you calculate a mean, the data can be in random order. However, it is helpful to get into the habit of always arranging data in an orderly way, from least to greatest or vice versa.

REMEMBER: If you are finding a mean with a scientific or graphing calculator, be sure to use parenthesis in the numerator.

MEDIAN

The **median** is the middle of a set of data. To find a median, first put the data in numerical order from greatest to least or least to greatest. Then, how we calculate the median depends on whether the data set has an odd or even number of elements (members).

EXAMPLE 1

If the number of data is odd, the median is the middle number. For instance, if the data are 20, 24, 32, 35, and 42, the number of elements, 5, is odd. Therefore the middle number is the third number, 32. The **median** of this set is 32.

EXAMPLE 2

If the number of data is even, the median is halfway between the two middle elements. For instance, if the data are 23, 26, 37, 43, 46, and 57, the number of elements, 6, is even. Therefore the median is halfway between the third element and the fourth element. Halfway between 37 and 43 is:

$$\frac{37 + 43}{2} = \frac{80}{2} = 40$$

The **median** of this data set is 40.

MODE

The **mode** is the number that occurs most often in a set of data. A data set can have no mode, one mode, or more than one mode.

- A set has **no mode** if no value occurs more than the others.
- A set has **one mode** if exactly one number appears more than any other number.
- If a set has two numbers that occur most frequently, it is called **bimodal**.
- A set can also have three or more modes.

EXAMPLES

The set of numbers 32, 45, 24, 35, 17, 25, 56, 35 (which can be rearranged as 17, 24, 25, 32, 35, 35, 45, 56) has a mode of 35, since 35 is repeated more times than any other number.

The set of numbers 2, 5, 6, 7, 3, 3, 6 (which can be rearranged as (2, 3, 3, 5, 6, 6, 7) is bimodal because 3 and 6 both occur twice, and no other value occurs twice or more than twice.

The set of numbers 12, 31, 35, 46, 16, 29, 55 (which can be rearranged as 12, 16, 29, 31, 35, 46, 55) has no mode, because no number is repeated.

MODEL PROBLEMS

1. Find all three measures of central tendency—mean, median, and mode—for this data set:

5, 0, 11, 1, 14, 3, 4, 7, 11

Answer: First arrange the data in order.

0, 1, 3, 4, 5, 7, 11, 11, 14

$$\text{Mean} = \frac{0 + 1 + 3 + 4 + 5 + 7 + 11 + 11 + 14}{9} = \frac{56}{9} = 6.\overline{2}$$

Median = 5. There is an odd number of elements, 9. Therefore, the median, the middle, is the fifth element:

0, 1, 3, 4, 5, 7, 11, 11, 14

Mode = 11. The only repeated element is 11:

0, 1, 3, 4, 5, 7, 11, 11, 14

2. Rodger earned the following scores on 20-point quizzes during one semester, 16, 18, 12, 14, 10, 15, 19, 16, 15, 19, 20, 16, 12, 17, 15, 11, 14, 17. Find the mean, median, and mode of these scores.

Answer: First put the data in order.

10, 11, 12, 12, 14, 14, 15, 15, 15, 16, 16, 16, 17, 17, 18, 19, 19, 20

Mean:

$$\frac{10 + 11 + 12 + 12 + 14 + 14 + 15 + 15 + 15 + 16 + 16 + 16 + 17 + 17 + 18 + 19 + 19 + 20}{18}$$

$$= \frac{276}{18} = 15.\overline{3}$$

Median = 15.5. There is an even number of elements, 18. So the median is the value halfway between the two middle numbers, 15 and 16:

10, 11, 12, 12, 14, 14, 15, 15, $\underline{15}$, $\underline{16}$, 16, 16, 17, 17, 18, 19, 19, 20

Calculate: $\dfrac{15 + 16}{2} = \dfrac{31}{2} = 15.5$

Modes = 15 and 16. There are two modes because two numbers appear more than any other number. The numbers 15 and 16 appear three times each:

10, 11, 12, 12, 14, 14, $\underline{15}$, $\underline{15}$, $\underline{15}$, $\underline{\underline{16}}$, $\underline{\underline{16}}$, $\underline{\underline{16}}$, 17, 17, 18, 19, 19, 20

3. Find the median and the mode of the data in this stem-and-leaf plot:

Stem	**Leaf**
5	5 7 8
6	1 3 7 8
7	0 4 6 6

Key: 7|4 represents 74

Answer:

Median = 67. These data are already in numerical order. There are 11 elements. Therefore, the middle number is the sixth number, which is 67.

Mode = 76, because 76 is the only number that is repeated.

TRY THESE

1. Find the mean, median, and mode of the following set of data.

100, 181, 135, 125, 150, 172, 114, 135

2. Find the median and the mode of the data in this stem-and-leaf plot.

Stem	Leaf
5	2 4 6
6	4 5
7	0 2 6 6 9
8	1 5 7 7 7

Key: 7|2 represents 7.2

3. Six different stores sell the same jogging jacket for different prices: \$31.95, \$26.30, \$29, \$32.50, \$25.25, and \$38. Find the:

   a. mean price

   b. median price

4. Following are the scores earned by juniors on the last test in their math course. Find the mean, median, and mode. Round your answers to the nearest tenth.

   85, 98, 68, 88, 100, 91, 77, 75, 85, 66, 50, 86, 83, 95

5. The morning news reported the following predicted high temperatures for the locations in its viewing area: 92°, 86°, 89°, 89°, 82°, 91°, 88°, 83°, 94°, 86°, 88°, 85°, 87°, 89°, 87°. Find the mean, median, and mode of the temperatures.

REPRESENTATIVE AND MISLEADING MEASURES

Sometimes, it can be misleading to use a particular measure of central tendency to describe a specific set of data. For example, suppose that the salaries of the employees of a small company are \$65,000, \$29,000, \$29,000, \$28,500 and \$28,000. The *mean* of these salaries is \$35,900. But the mean is certainly not indicative of the employees' "average" salary. Here, the *median* of the salaries, \$29,000, is a better indication than the mean. In this case, we would say that the median is more **representative** of the data, or that it is a better **representation** of the data.

In this example, the mean is not a very good representation because the set contains an **outlier**: a number that lies a large distance from the rest of the data. An outlier affects the mean, giving a misleading average. A high outlier (such as the \$65,000 salary here) pulls the mean up. A low outlier pulls the mean down. Following is another illustration of an outlier.

EXAMPLE

The ages of the mothers of all the people in a kindergarten classroom were listed. Then the mean and the median were calculated. The mean was much higher than the median, because the data included the age of the teacher's mother, who was much older than the children's mothers. The teacher's mother's age was an outlier, pulling the mean up.

The *mode* is probably the least reliable measure of central tendency. In many sets of data, the element that occurs most frequently is close to the center, but this is not always the case. Sometimes, the mode can occur toward the lowest or highest values of the set.

EXAMPLE

Set I: 3, 6, 7, 8, 8, 10 **Set II**: 2, 2, 9, 10, 22

In set I, the mode is 8, the median is 7.5, and the mean is 7. In this set, the three measures of central tendency are very close, and each measure can represent the data.

But in set II, the mode is 2 whereas the median is 9 and the mean is 9. Thus for set II, the mode would not be a good measure to describe the data.

MODEL PROBLEMS

1. In a chemistry experiment, a student gets the following readings: 2, 4, 3, 5, 3, 30. If the student wants to determine the "average" reading, should he or she find the mean or the median?

Answer: The student should find the median, because 30 is an outlier that will pull the mean up. Therefore, the mean would not be representative of the data.

2. Suppose that you own a hat shop. For one particular hat, you sold several sizes last year but this year you want to order only one size. To decide which size to order, you examine last year's sales figures for this hat, which show how many of each size were sold. Should you find the mean, the median, or the mode of last year's data?

Answer: You should find the mode, since that was the size sold most often.

TRY THESE

1. Mr. Smith has five math classes, with the following numbers of students:

Class	Number of students
Algebra II	25
Algebra II	30
Trigonometry	27
Trigonometry	28
Integrated Math	12

a. Find the mean, median, and mode for the data.

b. Which of these measures is most indicative of the average class size? Explain your reasoning.

2. Ten families in a certain neighborhood have the following annual incomes: $28,600, $38,400, $35,300, $40,700, $36,500, $39,900, $38,400, $95,000, $36,000, $38,300.

    a. Find the mean, median, and mode.

    b. Which of these measures is NOT indicative of the average salary? Why?

3. Pablo is reviewing his electric bills for the year. They are as follows:

January	$180	July	$210
February	$200	August	$215
March	$175	September	$40
April	$170	October	$190
May	$185	November	$182
June	$182	December	$194

    a. Find the mean, median, and mode.

    b. Which of these measures is NOT indicative of the average electric bill? Why?

4. The members of the Interact Club and its sponsors contributed the following amounts of money to buy presents for children in need: $3, $2.10, $5, $1, $4, $1.90, $20, $2.50, $20, $1.50, $3.25, $0.75, $1.10, $2.25, $1.65.

    a. Find the mean, median, and mode.

    b. Which measure is the most indicative of the average contribution? Why?

5. Candy's Chocolate Confectionary recorded the sales for the four weeks in February.

    Week 1: $1,800

    Week 2: $5,200

    Week 3: $1,700

    Week 4: $1,650

    a. Find the mean, median, and mode for the four weeks.

    b. Which of these measures is NOT indicative of the average sales per week? Why?

Measures of variation

The mean, median, and mode measure the central tendency of a set of data. Another measure of data that is helpful in statistics is **variation**.

RANGE AND QUARTILES

The measure that describes the “spread” of a set of data is called the **range** of the set. The range is the difference between the greatest value and the least value in the data. The least and greatest numbers are sometimes called the **extremes**. Thus the range can also be described as the difference between the extremes.

EXAMPLE

If the largest number in a data set is 100 and the least number is 64, the range is 100 – 64, or 36.

If a data set is divided into four equal parts, we say that the set is expressed in **quartiles**. There are three quartiles:

- The **median** is the quartile that divides the data in half.
- The middle number of the lower half is called the **lower quartile**.
- The middle number of the upper half is called the **upper quartile**.

The difference between the upper quartile and the lower quartile is called the **interquartile range**. This is another measure of the variation in the data.

EXAMPLE

Lower quartile = 33 ↓ (between 33 and 33); Median = 68 ↓ (between 65 and 71); Upper quartile = 78 ↓ (between 74 and 82)

25 25 33 ↓ 33 65 65 ↓ 71 73 74 ↓ 82 84 162

Here, the median, 68, is halfway between the two middle numbers, which are 65 and 71.

The lower quartile, the middle point of the bottom half of the data, is 33.

The upper quartile, the middle point of the upper half, is 78.

The interquartile range is the difference between 78 and 33: $78 - 33 = 45$.

The range of the data is the largest number, 162, minus the smallest number, 25: $162 - 25 = 137$.

BOX-AND-WHISKER PLOTS

This information about variation in data can be displayed in a type of graph called a **box-and-whisker plot**. To construct a box-and-whisker plot, we use five numbers, called the **five-number summary**:

(1) Least value

(2) Greatest value

(3) Lower quartile

(4) Upper quartile

(5) Median

These five values are placed on an appropriately scaled number line. The upper quartile, the lower quartile, and the median are each marked with a vertical line. The least value and the greatest value (the extremes) are each marked with a dot. The vertical lines are connected to form the **box**. Then, a line segment, called a **whisker**, is drawn from each end of the box to the dot on that side.

Here are two examples of box-and-whisker plots.

EXAMPLE 1

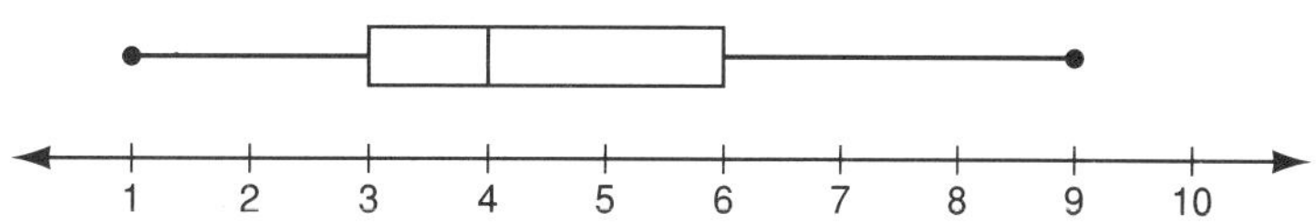

In this box-and-whisker plot, the five-value summary is:

Least value = 1

Greatest value = 9

Lower quartile = 3

Upper quartile = 6

Median = 4

EXAMPLE 2

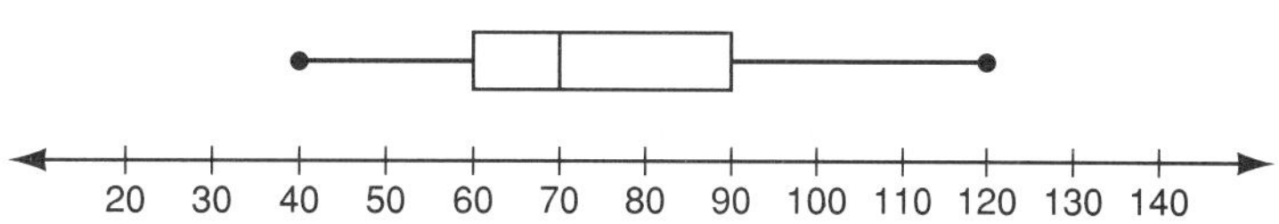

Here the five-number summary is:

Least value = 40

Greatest number = 120

Lower quartile = 60

Upper quartile = 90

Median = 70

The range, the difference between the extremes, is 120 − 40 = 80.

The interquartile range, the upper quartile minus the lower quartile, is 90 − 60 = 30.

Dividing data into quartiles can also be thought of as representing the data in **percents**. For instance, if we use percentages in Example 2, the percent of data located in each part of the box-and-whisker plot would be as follows:

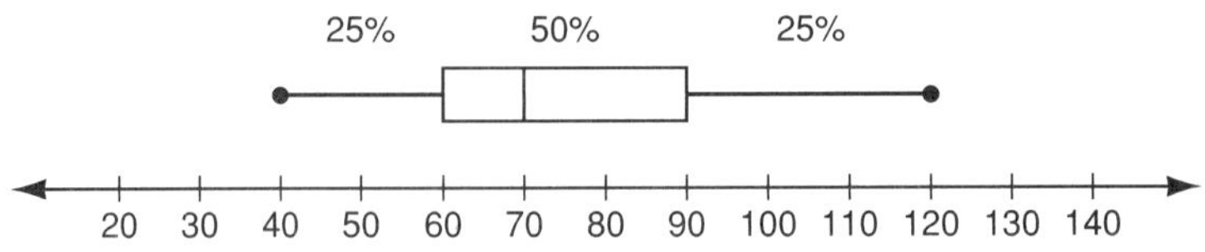

In a box-and-whisker plot:

- 25 percent of the data lie between the lower extreme and the lower quartile.
- 50 percent of the data lie between the lower quartile and the upper quartile.
- 25 percent of the data lie between the upper quartile and the upper extreme.

MODEL PROBLEMS

1. A city in New England recorded the following data. Find the range, median, upper and lower quartiles, and interquartile range.

Month	Days below 32°F
November	6
December	21
January	20
February	15
March	8

Answer: First arrange the data in order as 6, 8, 15, 20, 21.

Range = 21 − 6 = 15

Median = 15

Lower quartile $= \frac{6 + 8}{2} = \frac{14}{2} = 7$

Upper quartile $= \frac{20 + 21}{2} = \frac{41}{2} = 20.5$

Interquartile range = 20.5 − 7 = 13.5

2.

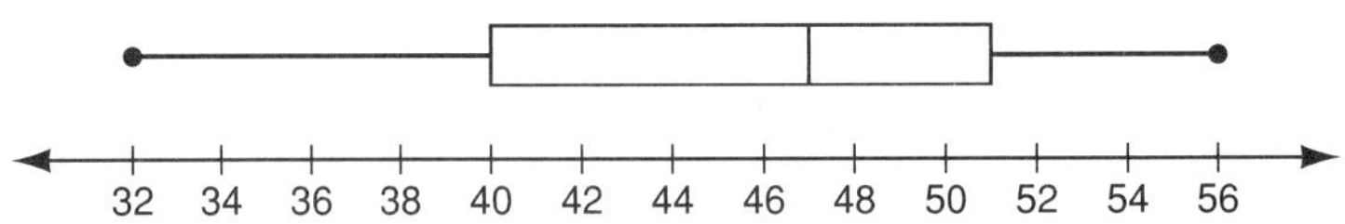

a. What is the median of the data in this box-and-whisker plot?

Answer: 47

b. What is the lower extreme?

Answer: 32

c. What is the range?

Answer: 24 (the difference of 56 and 32)

d. What percent of the data lie between 32 and 51?

Answer: 75 percent. Explanation: 25 percent lie between the lower extreme and the lower quartile; and 50 percent lie between the lower quartile and the upper quartile. Therefore, 75 percent lie between the lower extreme (32) and the upper quartile (51).

TRY THESE

1. A race car had the following speeds in five different races: 158, 159, 150, 180, and 172 miles per hour. Find the median and the range of these speeds.

2. Find the range of the data in this stem-and-leaf plot:

Stem	**Leaf**
18	2 3 7
19	1 6
20	0 5 8
21	3 7 9

Key: 19|6 represents 196

3.

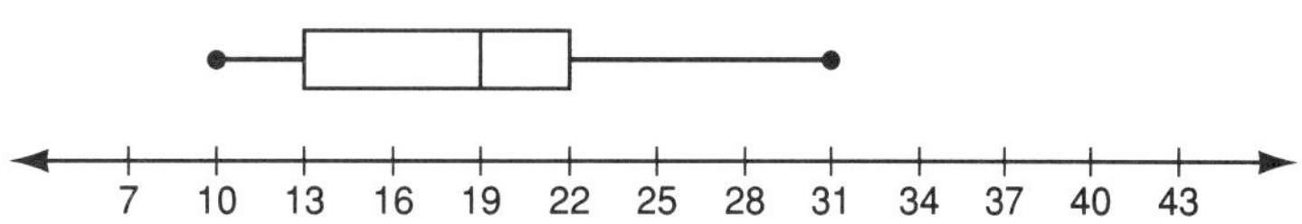

a. What is the range of the data in this box-and-whisker plot?

b. What percent of the data lie between 13 and 22?

4. The box-and-whisker plot below shows the scores on the final exam in history.

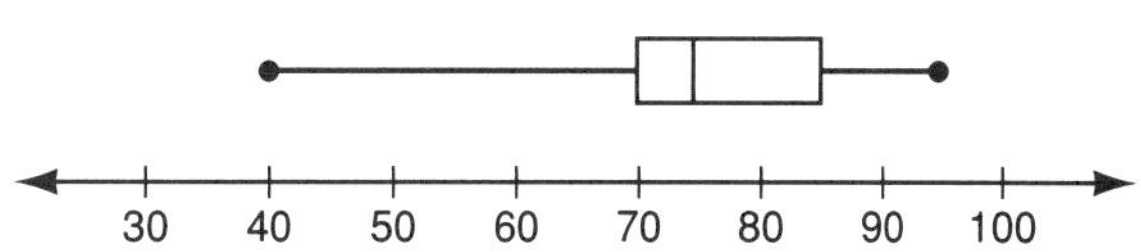

a. What is the range of the scores on the test?

b. What is the median score?

c. If 70 is the lowest passing score, what percent of the students passed?

Items 5–10: The box-and-whisker plots below show the ratings for three networks during a week in February. A rating of 10.3 means that out of every 100 houses with televisions, 10.3 households were watching a particular network at that time.

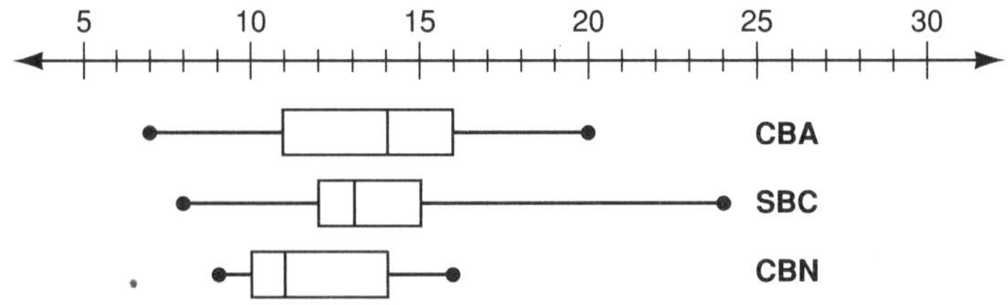

5. What is the median rating of each network?

6. What is the interquartile range for each network?

7. What is the rating range of each network?

8. If winning is determined by the highest rated show, which network would win?

9. If winning is determined by the highest median rating, which network would win?

10. Which appears to be the biggest loser of the three networks?

Assessment

Gridded-Response Problems

Items 1 and 2: After 10 games, the scores of a bowling team were 138, 144, 164, 150, 216, 154, 156, 182, 151, and 164.

1. Find the mean of the bowling team's scores.

2. Find the median of the bowling team's scores.

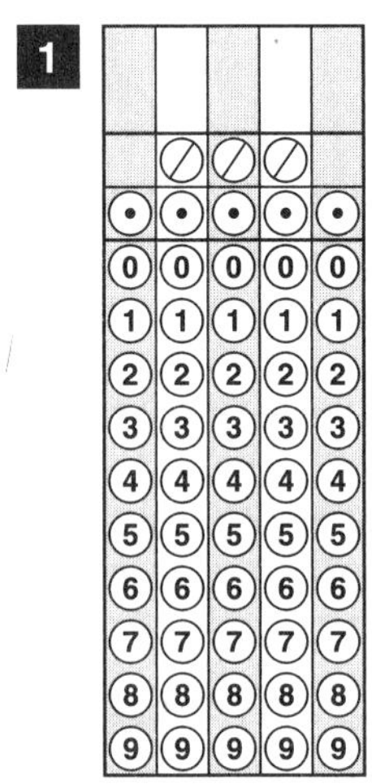

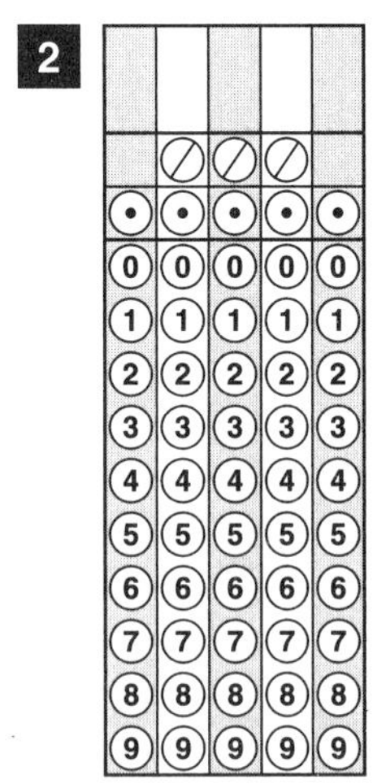

Items 3–5: Brian and Stacy are testing the strength and durability of disks. To do this, they drop each disk from a height of 40 inches to see if it breaks. The disks are grouped in sets of 20. After dropping 10 sets, Brian and Stacy record the following data

Set	Number broken
1	12
2	8
3	13
4	12
5	14
6	11
7	9
8	10
9	10
10	12

3. What is the mean of the data in the table above?

4. What is the median of these data?

5. What is the mode of these data?

Items 6 and 7: The fire department in Center City kept a log of the number of fire-related calls for one year, by day of the week. At the end of the year the results were shown in the table below.

Day	Number of calls
Sunday	89
Monday	62
Tuesday	50
Wednesday	57
Thursday	67
Friday	112
Saturday	94

6. To the nearest integer, what is the mean number of calls per day?

7. To the nearest integer, what is the median of these data?

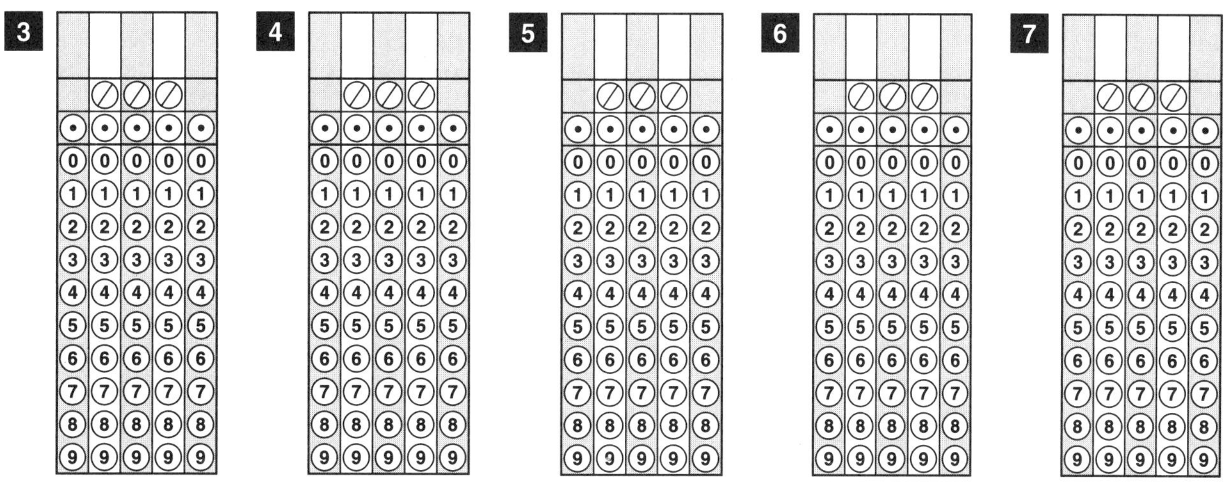

Multiple-Choice Problems

Items 8–10: At a gymnastics meet, nine judges rated the athletes on a 10-point scale, with 1 being the lowest rating and 10 the highest. For the floor exercise, Li received the following marks: 4, 5, 5, 5, 6, 7, 6, 8, 5.

8. What is the approximate mean of Li's marks?

A. 5

B. 5.67

C. 25.5

D. 51

9. What is the median of Li's marks?

A. 5

B. 5.67

C. 6

D. 7

10. What is the mode of Li's marks?

A. 4

B. 5

C. 6

D. 7

Items 11 and 12: The following table shows the number of desks in 10 rooms of Palm High School.

Room	Desks
202	32
203	30
210	37
212	28
214	29
304	33
306	31
342	35
350	32
Gym	95

11. What is the median number of desks in these rooms?

A. 28.2

B. 32

C. 42.4

D. 142

12. What is the mean number of desks?

A. 32

B. 38.2

C. 64

D. 142

Items 13–15: The workers at the Steel City Grill receive the following weekly salaries: \$385, \$400, \$560, \$400, \$450, \$600, \$385, \$400, \$800, \$1,000.

13. What is the mean salary?

A. \$400

B. \$425

C. \$538

D. \$550

14. What is the median salary?

A. \$400

B. \$425

C. \$538

D. \$550

15. What is the mode of the salaries?

A. \$385

B. \$400

C. \$425

D. \$538

Items 16–18: This box-and-whisker plot represents prices of toasters.

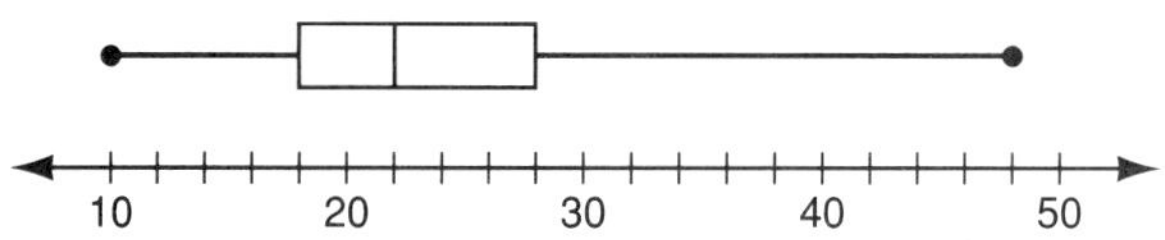

16. What is the highest price?

A. \$22

B. \$28

C. \$48

D. \$50

17. What is the median price?

A. \$18

B. \$22

C. \$23

D. \$28

18. What percent of the toasters cost between \$18 and \$28?

A. 25%

B. 50%

C. 60%

D. 75%

Items 19 and 20: The scout troop collected money for hurricane relief in front of the local grocery store. During the first hour they received the following donations: \$0.50, \$2.50, \$1, \$4, \$2, \$5, \$3, \$2.50, \$1, \$2, \$20, \$20, \$3, \$20, \$0.50.

19. Which measure of central tendency best describes the average donation?

A. mean

B. median

C. mode

D. range

20. If two more people donated \$3 each, which one of the measures of central tendency would have the greatest change?

A. mean

B. median

C. mode

D. range

Section 2 Using concepts of probability and statistics to identify patterns in data and make predictions from data

A PROBABILITY

People often think that probability theory is applied mostly in games of chance, but in fact probability is also involved in genetics, astronomy, manufacturing, insurance, and many other fields. Public opinion pollsters, for instance, apply probability when they report the results of a survey.

To review the basic vocabulary of probability theory, we will use a 6-sided solid cube. One face of the cube has 1 dot, the second face has 2 dots, the third face has 3 dots, and so on up to 6 dots. This is commonly known as a **number cube**, a **die**, or a **fair die**. The plural of "die" is **dice**, and we often refer to a **pair of dice**.

- **Experiment**: Toss the die and see which face lands up, that is, how many dots land faceup.
- **Outcome**: Number of dots on the face that lands up.
- **Equally likely**: Each number, 1, 2, 3, 4, 5, and 6, has an equal chance of landing faceup. (This is what we mean by a "fair" die.)
- **Sample space**: All possible outcomes of an experiment, the set {1, 2, 3, 4, 5, 6}.
- **Event**: Any subset of the sample space, for example, {2}.

The **probability** of an event A is defined as the number of successful outcomes divided by the total number of all possible outcomes. It is written as a fraction:

$$P(A) = \frac{\text{number of successful outcomes}}{\text{total number of possible outcomes}}$$

Suppose we want to calculate the probability of getting a 4 in our example of an experiment with a die. There are 6 possible outcomes: {1, 2, 3, 4, 5, 6}. There is only 1 successful outcome, because there is only 1 way to get a 4: the face with 4 dots must land up. Because we have defined the die as "fair," getting a 4 is just as likely as getting any other number, such as a 1 or a 5. Therefore, the probability of getting a 4 is:

$$P(4) = \frac{\text{number of successful outcomes}}{\text{total number of possible outcomes}} = \frac{1}{6}$$

If the number of successful outcomes in an experiment is zero, then the probability is 0. For example, the probability of getting a 7 when tossing a die is zero, since there is no way to get a successful outcome:

$$P(7) = \frac{\text{number of successful outcomes}}{\text{total number of possible outcomes}} = \frac{0}{6} = 0$$

If the number of successful outcomes is the same as the total number of all possible outcomes, the probability is 1. For example, the probability of getting a number less than 7 when rolling a die is 1, because there are 6 ways to get this outcome:

$$P(N < 7) = \frac{\text{number of successful outcomes}}{\text{total number of possible outcomes}} = \frac{6}{6} = 1$$

MODEL PROBLEMS

1. One side of a coin shows a person's face; this side is called **heads**. The other side is called **tails**. When you flip a dime, what is the probability of getting tails?

   Answer: There are two possible outcomes, heads or tails: {H, T}. There is only one way to get tails. As with dice, we assume that a coin is "fair," so heads and tails are equally likely. Therefore, the probability of tails is:

$$P(\text{T}) = \frac{\text{number of successful outcomes}}{\text{total number of possible outcomes}} = \frac{1}{2}$$

2. When you roll a die, what is the probability of getting a number (N) less than 5?

   Answer: There are six possible outcomes: {1, 2, 3, 4, 5, 6}. There are 4 ways to get a number less than 5: {1, 2, 3, 4}. Therefore, the probability of getting a number less than 5 is:

$$P(N < 5) = \frac{\text{number of successful outcomes}}{\text{total number of possible outcomes}} = \frac{4}{6} = \frac{2}{3}$$

TRY THESE

1. When you toss a die, what is the probability of getting an even number?
2. When you toss a die, what is the probability of getting a number greater than 1?
3. When you toss a die, what is the probability of getting a number greater than 4?
4. When you toss a die, what is the probability of getting a number greater than 6?

Items 5–10: You roll a 10-sided die with one of the numbers 1 through 10 on each side.

5. What is the probability of getting a number less than 10?
6. What is the probability of getting a number less than 4?

7. What is the probability of getting a number greater than 2?
8. What is the probability of getting a 3?
9. What is the probability of getting a 0?
10. What is the probability of getting a number less than 11?

B SIMPLE EVENTS

When an experiment consists of one event, such as rolling one die or tossing one coin, the outcome is called a **simple event**. Here are three examples of calculating probability for simple events.

EXAMPLE 1

When we roll a die, what is the probability of getting an odd number? This event will be called event A. What is the probability of getting a number less than 4? This event will be called event B. What is the probability of event A *and* B?

To determine this, we must look at the sample space: {1, 2, 3, 4, 5, 6}. Event A involves the subset of odd numbers: {1, 3, 5}. Event B involves the subset of numbers less than 4: {1, 2, 3}. The probability of event A *and* B is the *intersection* of these subsets, or {1, 3}. Thus:

$$P(A \text{ and } B) = \frac{2}{6} = \frac{1}{3}$$

EXAMPLE 2

Suppose we want to know the probability of A *or* B: getting an odd number *or* a number less than 4. This probability is the *union* of the two subsets, or {1, 2, 3, 5}. Thus:

$$P(A \text{ or } B) = \frac{4}{6} = \frac{2}{3}$$

REMEMBER: When we express a union of sets, we must be sure *not* to repeat an element that appears in both sets. In this case, the 1 and the 3 appear in both sets, but each appears just once in the union.

EXAMPLE 3

When we roll a die, the probability of getting a 2 is $\frac{1}{6}$. What is the probability of *not* getting a 2?

In rolling a die, the outcome must be either a 2 or not a 2. This is certain. Therefore, the probability of getting a 2 *or* not getting a 2 is 1. Therefore, to find the probability of not getting a 2, we subtract the probability of getting a 2 from 1:

$$P(\text{not } 2) = 1 - \frac{1}{6} = \frac{5}{6}$$

Another way to find the probability of not getting a 2 is to calculate the probability of all the other possible events. When we roll a die, there are five ways *not* to get a 2: {1, 3, 4, 5, 6}. Therefore, the probability of not getting a 2 is $\frac{5}{6}$.

In an experiment, the sum of the probabilities of all the possible events must equal 1. Therefore, a **formula** for the probability of "not *A*" is:

$$P(\text{not } A) = 1 - P(A)$$

MODEL PROBLEMS

Items 1–3: A child's bank contains 7 quarters (Q), 5 dimes (D), 12 nickels (N), and 16 pennies (P).

1. Without looking at the contents, the child reaches in and takes out one coin. Find the probability that the coin is a:

a. dime

b. quarter

c. penny or a nickel

d. quarter or a dime

Answers: The sample space consists of 7 + 5 + 12 + 16 = 40 coins. Therefore, the probabilities are:

a. $P(\text{D}) = \frac{5}{40} = \frac{1}{8}$

b. $P(\text{Q}) = \frac{7}{40}$

c. $P(\text{P or N}) = \frac{16}{40} + \frac{12}{40} = \frac{28}{40} = \frac{7}{10}$

d. $P(\text{Q or D}) = \frac{7}{40} + \frac{5}{40} = \frac{12}{40} = \frac{3}{10}$

2. What is the probability that the child will NOT select a quarter?

Answer: $P(\text{Q}) = \frac{7}{40}$. Therefore, $P(\text{not Q}) = 1 - \frac{7}{40} = \frac{33}{40}$.

Another way to calculate this is to sum the probabilities of all the other outcomes:

$$P(\text{not Q}) = \frac{5}{40} + \frac{12}{40} + \frac{16}{40} = \frac{33}{40}$$

3. What is the probability of selecting a quarter *or* a dime *or* a nickel *or* a penny?

Answer: This probability is the sum of the probabilities of all the possible outcomes. Therefore:

$$P(\text{Q or D or N or P}) = \frac{7}{40} + \frac{5}{40} + \frac{12}{40} + \frac{16}{40} = \frac{40}{40} = 1$$

You could simply answer "1" without calculating, since in any experiment the sum of the probabilities of all possible outcomes must be 1.

4. Given the set of numbers {1, 2, 3, 6, 7, 11}, what is the probability of selecting a number that is odd *and* greater than 6?

Answer: This probability is an intersection of subsets. The sample space consists of six numbers. The subset of odd numbers is {1, 3, 7, 11}. The subset of numbers greater than 6 is {7, 11}. The intersection of these two subsets is {7, 11}. Thus:

$$P(\text{odd and} > 6) = \frac{2}{6} = \frac{1}{3}$$

5. Given the spinner shown, find the probability that in one spin the pointer will land on a 1.

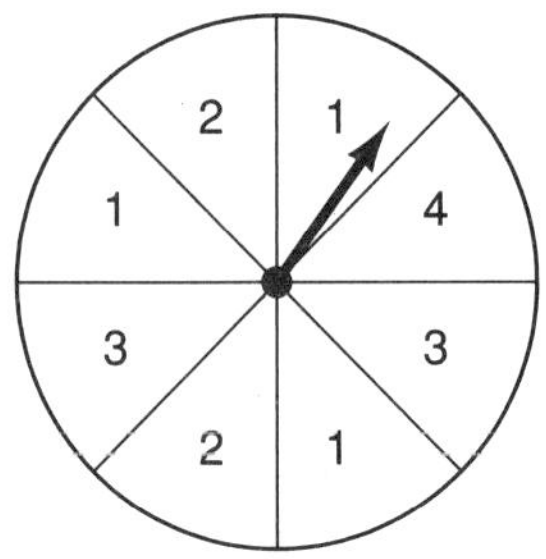

We assume that a spinner (like a die or a coin) is "fair." Thus there is an equal chance of landing on each sector. We also assume that the sectors are of equal size (unless a problem states otherwise).

Answer: There are 8 sectors on which the pointer can land, so the sample space is 8. Since three sectors are marked "1," there are 3 successful outcomes. Therefore, the probability is:

$$P(1) = \frac{\text{number of successful outcomes}}{\text{total number of possible outcomes}} = \frac{3}{8}$$

TRY THESE

Items 1 and 2: Use the spinner from the model problem above.

1. What is the probability that the pointer will land on a 2?

2. What is the probability that the pointer will NOT land on a 4?

Items 3 and 4: A jar contains 25 marbles: 5 red, 7 blue, 2 black, 2 white, and 9 yellow. Without looking, you take 1 marble.

3. Find the probability of selecting:
   a. a red marble
   b. a black marble
   c. a white or a yellow marble
   d. a red or a blue marble
4. What is the probability of NOT choosing a blue marble?
5. Given the set of numbers {1, 2, 4, 7, 9, 12}, what is the probability of selecting a number that is even *and* less than 7?
6. Given the spinner shown, predict how many times the pointer will land on red if the spinner is spun 84 times.

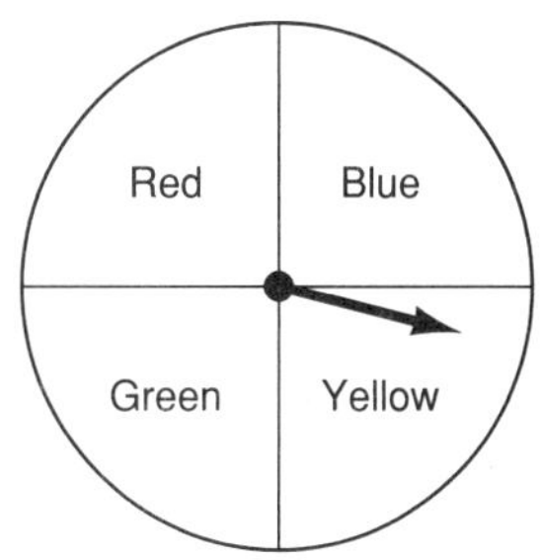

Items 7–12: In the game of Scattergories, a person rolls a 20-sided die with 20 different letters on it. All letters in the alphabet are used except Q, U, V, X, Y, and Z.

7. What is the probability that a person will roll an A, E, I, or O?
8. What is the probability that a person will roll a letter with NO curves in its capital letter form?
9. What is the probability that a person will roll a letter that is composed of straight lines and curves?
10. What is the probability that a person will roll a U?
11. What is the probability of getting a vowel and a letter in the alphabet before G?
12. What is the probability of getting a vowel or a letter in the alphabet before G?

C ODDS

People often talk about odds. We say, “What are the odds that it will rain today?” or “The odds of my passing this test are good.”

In the study of probability, odds are expressed in one of two ways. First, we can find the **odds in favor of an event**: the odds that an event will occur. Second, we can find the **odds against an event**: the odds that the event will *not* occur.

Let A stand for an event. The probability of A is:

$$P(A) = \frac{\text{number of successful outcomes}}{\text{total number of possible outcomes}}$$

The odds in favor of A are calculated as follows:

$$\text{Odds of } A = \frac{\text{probability that } A \text{ will occur}}{\text{probability that } A \text{ will not occur}} = \frac{P(A)}{P(\text{not } A)}$$

The odds against A are calculated as:

$$\text{Odds against } A = \frac{\text{probability that } A \text{ will not occur}}{\text{probability that } A \text{ will occur}} = \frac{P(\text{not } A)}{P(A)}$$

REMEMBER: To find $P(\text{not } A)$, you can use two methods. The first method is to subtract $P(A)$ from one: $1 - P(A)$. The second method is to find the sum of the probabilities of all the other events, all the events that are not A.

EXAMPLE 1

A jar contains 5 candies: 3 red and 2 green. You choose 1 candy without looking. What are the odds in favor of getting a red candy?

There are 3 chances that the event "red" will occur and 2 chances that the event "green" will occur. Therefore, the odds in favor of red are:

$$\text{Odds of red} = \frac{\text{probability that red will occur}}{\text{probability that red will not occur}} = \frac{P(\text{red})}{P(\text{not red})}$$

$$= P(\text{red}) \div P(\text{not red}) = \frac{3}{5} \div \frac{2}{5} = \frac{3}{5} \bullet \frac{5}{2} = \frac{15}{10} = \frac{3}{2}$$

We express this answer by saying that the odds in favor of choosing red are 3 to 2.

EXAMPLE 2

One of three students—Ann, Bill, and Cory—will be chosen at random to lead a group discussion. What are the odds that Ann will be selected? What are the odds against Ann being selected?

The probability that Ann will be selected is $\frac{1}{3}$, so the probability that she will *not* be selected is $1 - \frac{1}{3} = \frac{2}{3}$. The odds *in favor* of selecting Ann are calculated as:

$$\frac{P(A)}{P(\text{not } A)} = \frac{\frac{1}{3}}{\frac{2}{3}} = \frac{1}{3} \div \frac{2}{3} = \frac{1}{3} \bullet \frac{3}{2} = \frac{3}{6} = \frac{1}{2}$$

Therefore, the odds in favor of selecting Ann are 1 to 2. We can also say "1 in 2" or "1 out of 2."

An alternative method: There is 1 way to select Ann and 2 ways to *not* select Ann. Therefore, the odds in favor of selecting Ann are 1 to 2.

The odds *against* selecting Ann are:

$$\frac{P(\text{not } A)}{P(A)} = \frac{2}{3} \div \frac{1}{3} = \frac{2}{3} \bullet \frac{3}{1} = \frac{6}{3} = \frac{2}{1}$$

We say that the odds against selecting Ann are 2 to 1.

An alternative method: There are 2 ways not to select Ann and 1 way to select Ann. Therefore, the odds against selecting Ann are 2 to 1.

MODEL PROBLEMS

1. There are 20 people in a room: 13 males (M) and 7 females (F). What are the odds that one person chosen at random will be a male?

Answer: $P(M) = \frac{13}{20}$ and $P(\text{not } M) = 1 - \frac{13}{20} = \frac{7}{20}$.

Therefore, the odds in favor of choosing a male are:

$$\frac{P(M)}{P(\text{not } M)} = \frac{13}{20} \div \frac{7}{20} = \frac{13}{20} \bullet \frac{20}{7} = \frac{13}{7}$$

2. You enter a contest in which the probability of winning the prize is $\frac{2}{15}$. What are the odds in favor of winning?

Answer: The probability $P(\text{not } W)$ of not winning is $\frac{13}{15}$.

Therefore, the odds in favor of winning (W) are:

$$\frac{P(W)}{P(\text{not } W)} = \frac{2}{15} \div \frac{13}{15} = \frac{2}{15} \bullet \frac{15}{13} = \frac{2}{13}$$

3. Shirts and pants are hanging on a rack in a dark closet. You reach in and take one item at random. In this situation, it has been determined that the odds of choosing a shirt are $\frac{2}{9}$. If you choose 44 items in this way, predict the number of shirts you will get.

REMEMBER: Odds of shirt $= \frac{P(S)}{P(\text{not } S)} = \frac{2}{11} \div \frac{9}{11} = \frac{2}{11} \bullet \frac{11}{9} = \frac{2}{9}$

Answer: Since $2 + 9 = 11$, the given odds imply that out of every 11 items chosen, 2 will be shirts. Calculate:

$$\frac{2}{11} \bullet 44 = \frac{88}{11} = 8$$

Therefore, you can predict that you will get 8 shirts.

TRY THESE

1. You choose one number from the numbers 1, 2, 3, 4, and 5. What are the odds in favor of choosing an even number?
2. Students in Mr. McDonnell's history class are being chosen at random to give oral reports. There are 28 students, of which 15 are girls. What are the odds against a boy being the first person chosen to give a report?
3. In one class of 30 students, the odds that a student will go to college outside of the state of Florida are 1 out of 5. Predict how many students will go to college outside of Florida.

Items 4 and 5: Choose one letter from the word FLORIDA.

4. What are the odds in favor of choosing a vowel?
5. What are the odds against choosing the letter A?

Items 6 and 7: There are 6 sealed envelopes containing different amounts of money. Only 2 of the envelopes contain more than $50. Triston must select one envelope.

6. What are the odds that he will select an envelope that contains less than $50?
7. What is the probability that he will select an envelope that contains more than $50?
8. The weather forecaster reports that the probability of rain today is $\frac{4}{10}$. What are the odds that it will NOT rain today?

Items 9 and 10: Use the spinner shown below.

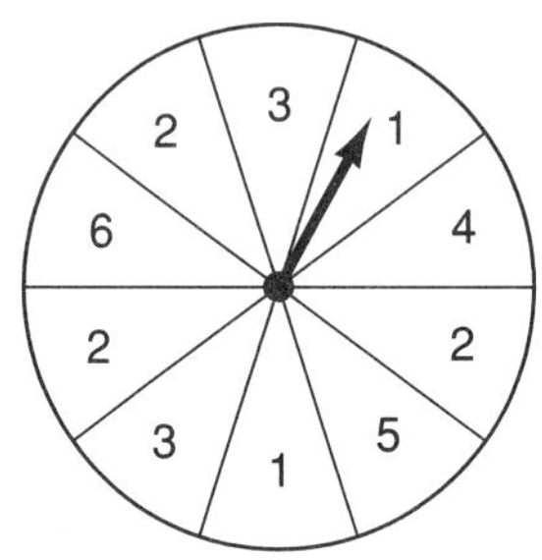

9. What are the odds of the pointer landing on either a 1 or a 2?
10. If the pointer is spun 20 times, using probability, how many times can you expect it to land on a 2?

D THE COUNTING PRINCIPLE

So far, we have reviewed probabilities involving a single activity. Most probability problems, however, deal with situations in which there are two or more activities. A familiar example is rolling a pair of dice. In determining the probability of more than one event, the **counting principle** can be very helpful.

EXAMPLE 1

At an ice cream party, three flavors of ice cream are available: vanilla (V), chocolate (C), and strawberry (S). Along with the ice cream, there are two choices of toppings: fudge (F) and marshmallow (M). This is a situation with two events or two activities: flavor and topping.

To determine the sample space of all possible outcomes in such a situation, we can use a **tree diagram**:

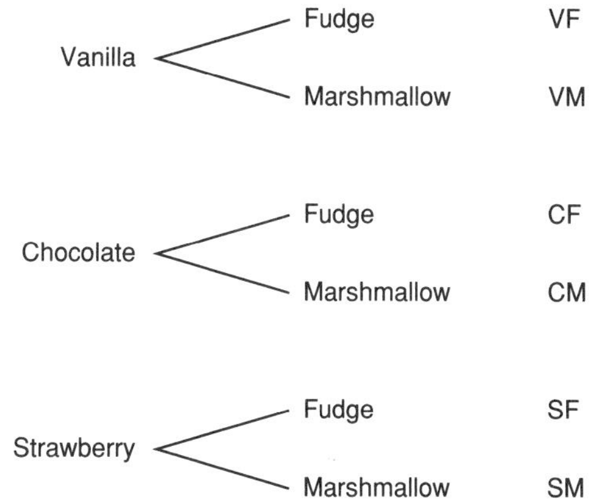

This diagram shows 6 possibilities. According to the counting principle, the sample space contains:

(3 flavors) $\times$ (2 toppings) $= 3 \times 2 = 6$ possible outcomes

EXAMPLE 2

Sometimes a situation involves multiple events because an activity is repeated. For instance, we can toss a coin three times. Here we can find all possible outcomes by using the following tree diagram. H = heads and T = tails:

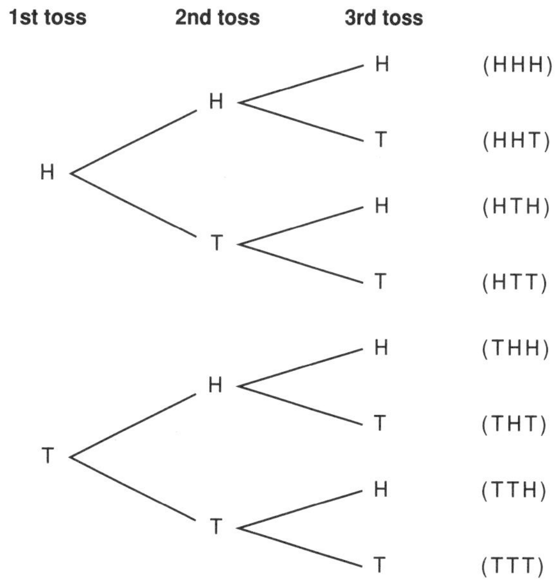

According to the counting principle, there are 8 possible outcomes. Each time the coin is tossed there are 2 possibilities, and we toss the coin 3 times. Therefore, we calculate $2 \times 2 \times 2 = 8$.

MODEL PROBLEMS

1. A school cafeteria sells two kinds of pizza: pan (P) and thin (T). For each kind, three toppings are available: pepperoni (P), sausage (S), or extra cheese (C). You are about to buy pizza at this cafeteria. Draw a tree diagram to determine the number of possible choices you have.

   Answer: The sample space consists of $2 \times 3 = 6$ possible choices.

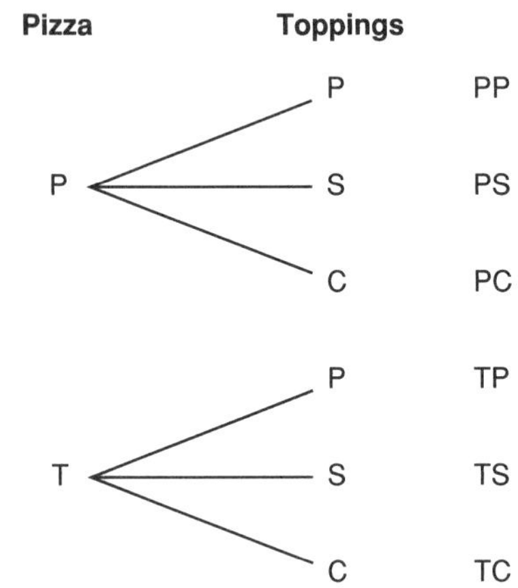

2. To demonstrate dances from different countries, the members of a culture club want to form couples consisting of a boy and a girl. The club has 17 members: 9 girls and 8 boys. How many different couples can be formed? Use the counting principle.

   Answer: $9 \times 8 = 72$ different couples

TRY THESE

1. There are 3 true-or-false questions on a quiz. What are the possible sets of answers for these 3 questions? Use a tree diagram. (Let T represent true and F represent false.)

2. You have won a car in a raffle. The car dealer will let you choose from 3 colors, 2 types of seats, and 5 option packages. How many variations are available to you? Use the counting principle.
3. A box has three marbles: red (R), blue (B), and green (G). One marble is selected, its color is noted, then it is put back into the box. A second marble is selected and its color noted. Draw a tree diagram to show all possible outcomes.
4. A box has three marbles: red (R), blue (B), and green (G). One marble is selected, its color is noted, then it is NOT put back into the box. A second marble is selected and its color noted. Draw a tree diagram to show all possible outcomes.
5. At a county fair, one of the games requires that you pick a door and then pick a curtain behind the door. There are 3 doors and 4 curtains behind each door. Use a tree diagram to show all possible choices.

E INDEPENDENT EVENTS

Suppose that you roll a pair of dice and a 3 comes up on the red die while a 1 comes up on the green die. In this case, the 1 on the green die has absolutely nothing to do with the 3 on the red die. In other words, the probability of getting either of these two outcomes has nothing to do with the probability of getting the other outcome. Outcomes like these are called **independent events**.

We can use the counting principle to determine the probability that two or more independent events will all occur at the same time. In this case, for instance, the probability of a 3 on the red die is $\frac{1}{6}$, and the probability of a 1 on the green die is $\frac{1}{6}$. The probability that we will get a 3 on the red die and a 1 on the green die is:

$$\frac{1}{6} \times \frac{1}{6} = \frac{1}{36}$$

REMEMBER: In order to use the counting principle to find a probability, the events must be independent of one another.

MODEL PROBLEM

Assume that being born on any day of the week is equally likely. You choose 2 students in your class at random. What is the probability that both of them were born on a Monday?

Answer: These two events are independent, since the day one student was born has nothing to do with the day the other was born.

There are 7 days in the week, so the probability that the first student was born on Monday is $\frac{1}{7}$, and the probability that the second student was born on Monday is also $\frac{1}{7}$. The probability that both were born on Monday is:

$$\frac{1}{7} \bullet \frac{1}{7} = \frac{1}{49}$$

TRY THESE

Items 1–3: Joe rolls 1 die. Jane tosses 1 coin. Jeff chooses 1 sock at random from a drawer containing only 6 brown socks and 6 green socks. Find each of the following probabilities.

1. Joe will get a 5 *and* Jane will get heads *and* Jeff will choose a green sock.
2. Joe will get an odd number *and* Jane will get heads *and* Jeff will get a red sock.
3. Joe will get a 2 or a 3 *and* Jane will get heads or tails *and* Jeff will get a brown sock.

4. A fair coin is tossed twice. Find the probability that the results are a head and a tail in either order.

Items 5 and 6: A single die is rolled twice.

5. What is the probability of getting two odd numbers?
6. What is the probability of getting two numbers that both exceed 4?

Items 7 and 8: It has been said that the probability that South Florida will be hit by a hurricane in any single year is $\frac{5}{19}$.

7. What is the probability that South Florida will be hit by a hurricane in 3 consecutive years?
8. What is the probability that South Florida will NOT be hit by a hurricane in 2 consecutive years?
9. What is the probability of a family having 3 boys in a row?
10. What is the probability of a family having 3 boys in a row or 2 girls in a row?

F COMPOUND EVENTS

A **compound event** consists of two or more independent events. Examples of compound events are tossing four coins, choosing a dessert and a beverage, or rolling a die and spinning an arrow.

Using the counting principle for compound events

In some situations, the counting principle can be used to find the probability of a compound event, provided that the events making up the compound event are independent.

EXAMPLE

A spinner has 8 equal sectors numbered 1 through 8. The arrow is spun 2 times. What is the probability that the arrow will land on the 3 both times?

The probability of landing on the 3 once is $\frac{1}{8}$. Landing on the 3 a second time is independent of what happens the first time. Therefore, we can use the counting principle:

$P(3, \text{then } 3) = \frac{1}{8} \times \frac{1}{8} = \frac{1}{64}$

Finding the sample space for compound events

We cannot always use the counting principle to find the probability of a compound event. In some problems we need to determine the sample space, all the possible outcomes, before we can calculate a probability.

To find the probability of a compound event in such a situation, we follow three steps:

Step 1. Find the number of elements in the sample space.

Step 2. Count the number of elements in the sample space that will fit the given situation.

Step 3. Set up a ratio to express the probability:

$$\frac{\text{number of elements that fit the situation}}{\text{number of elements in the sample space}}$$

One way to determine the sample space in these problems is by using a tree diagram, as the following examples illustrate.

EXAMPLE 1

A family has three children. What is the probability that there are 2 girls and 1 boy? In this case, the counting principle will not give us the probability because we need to examine the sample space. We can display the sample space as in the following tree diagram:

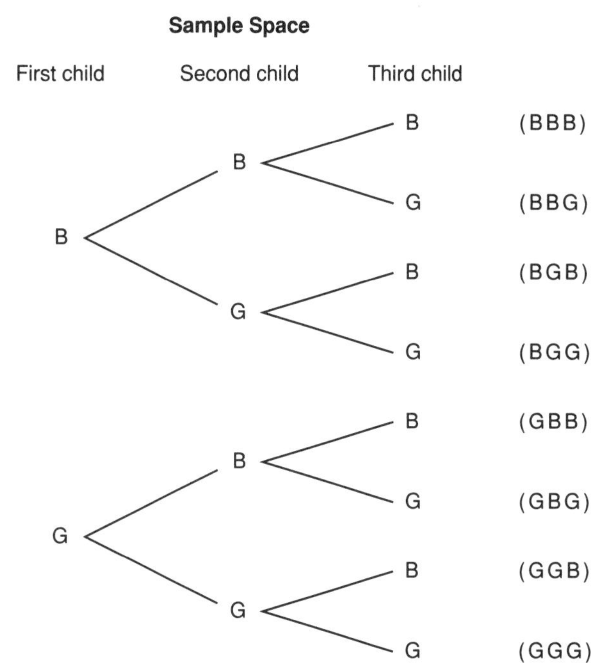

The tree diagram shows a sample space of 8 possibilities. Of these 8 possibilities, 3 contain 2 girls and 1 boy. Therefore, the probability of 2 girls and 1 boy is:

$$\frac{\text{number of elements that fit the situation}}{\text{number of elements in the sample space}} = \frac{3}{8}$$

EXAMPLE 2

Using the tree diagram above, find the probability that there will be at least 1 girl in the family. When we look at the outcomes in the tree diagram, we see that 7 elements of the sample space have at least 1 girl. Therefore, the probability is:

$$\frac{\text{number of elements that fit the situation}}{\text{number of elements in the sample space}} = \frac{7}{8}$$

MODEL PROBLEMS

Items 1–4: A pair of dice is tossed.

1. Set up a diagram of all possible outcomes. How many possible outcomes are there?

Answer: There are 36 possible outcomes. Diagram:

Die II \ Die I	1	2	3	4	5	6
1	1, 1	1, 2	1, 3	1, 4	1, 5	1, 6
2	2, 1	2, 2	2, 3	2, 4	2, 5	2, 6
3	3, 1	3, 2	3, 3	3, 4	3, 5	3, 6
4	4, 1	4, 2	4, 3	4, 4	4, 5	4, 6
5	5, 1	5, 2	5, 3	5, 4	5, 5	5, 6
6	6, 1	6, 2	6, 3	6, 4	6, 5	6, 6

2. What is the probability that the same number will come up on both dice?

Answer: The same number on both dice occurs 6 times in the diagram: 1, 1; 2, 2; 3, 3; 4, 4; 5, 5; and 6, 6. Therefore, the probability is $\frac{6}{36} = \frac{1}{6}$.

3. What is the probability that the sum of the numbers will be 5?

Answer: The sum of 5 occurs 4 times in the diagram: 1, 4; 2, 3; 3, 2; and 4, 1. Therefore, the probability is $\frac{4}{36} = \frac{1}{9}$.

4. What is the probability that both numbers will be even?

Answer: In the diagram, 2 even numbers occur 9 times: 2, 2; 2, 4; 2, 6; 4, 2; 4, 4; 4, 6; 6, 2; 6, 4; and 6, 6. Therefore, the probability is $\frac{9}{36} = \frac{1}{4}$.

5. In a contest a person chooses one door, A, B, or C, and then behind that door the contestant chooses a prize or money. The given probabilities of the events are shown on the branches of the tree diagram below.

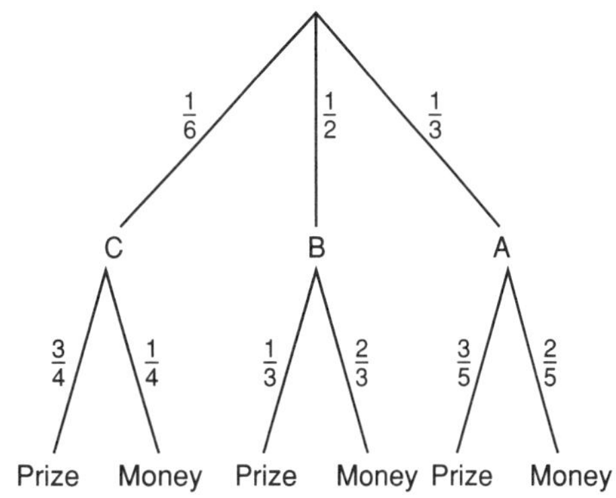

What is the probability of a contestant choosing door B and then the prize?

Answer: The probability of choosing door B is $\frac{1}{2}$. The probability of choosing the prize behind door B is $\frac{1}{3}$. The probability of choosing door B and then the prize is $\frac{1}{3} \bullet \frac{1}{2} = \frac{1}{6}$.

TRY THESE

Items 1–3: Toss a pair of dice.

1. What is the probability that you will get a sum of 11?

2. What is the probability of getting a sum of 1?

3. What is the probability of getting a difference of 1?

Items 4 and 5: Three coins are tossed simultaneously.

4. What is the probability that all three will land tails up?

5. What is the probability that any two coins land heads up and the other coin lands tails up?

Items 6–8: A coin and a die are tossed.

6. What is the probability that the coin will land heads up and the die will be a 2?

7. What is the probability that the coin lands tails up and the die is an even number?

8. What is the probability that the coin will NOT land heads up and the die will be a number less than 5?

Items 9 and 10: The spinner shown below is spun two times.

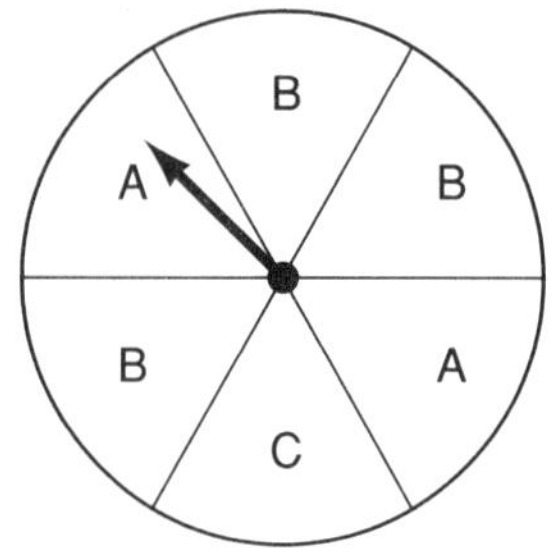

9. What is the probability that both spins will cause the pointer to land on A?

10. What is the probability that one spin will cause the pointer to land on C and the other spin will cause the pointer to land on B?

11. Use the tree diagram in Model Problem 5 to find which door has the highest percentage of people going for the prize.

G DEPENDENT EVENTS

Tom has 15 coins in his pocket: 2 quarters, 4 dimes, 4 nickels, and 5 pennies. He takes out 1 coin at a time and does not put any coins back into his pocket. What is the probability that the first 2 coins he takes will be nickels? In this situation, the outcomes are not independent, because once Tom has taken out the first coin, the number of coins remaining in his pocket changes.

Outcomes like these are called **dependent events**, and problems like this one are called selection **without replacement**.

EXAMPLE

As an example, we will solve the problem of Tom's coins. What is the probability that Tom will get 2 nickels on the first two tries?

If he is going to get 2 nickels on the first two tries, the first coin he gets must be a nickel. The probability that this first coin will be a nickel is $\frac{4}{15}$.

He does not put the first nickel back, so this leaves only 14 coins in his pocket, and only 3 of those are nickels. Thus the probability that the second coin will be a nickel is $\frac{3}{14}$.

The probability that both coins will be nickels is:

$$P(\text{nickel, then nickel}) = \frac{4}{15} \times \frac{3}{14} = \frac{12}{210} = \frac{2}{35}$$

Suppose that before taking out the second coin, Tom replaces the first coin. In that case, the two events (taking one coin and then taking a second coin) would be independent. Such a problem is called selection **with replacement**. Then the probability that both coins will be nickels is:

$$\frac{4}{15} \times \frac{4}{15} = \frac{16}{225}$$

NOTE: If a problem does not specify "with replacement" or "without replacement," you must look at the nature of the situation and decide which condition would apply.

MODEL PROBLEMS

Items 1 and 2: A jar contains 20 jelly beans: 3 red (R), 7 green (G), 4 yellow (Y), and 6 purple (P). Marni takes out 1 jelly bean at random, eats it, and then takes a second jelly bean.

1. What is the probability that the first jelly bean was yellow and the second jelly bean was purple?

Answer: Since Marni eats the first jelly bean, this is a problem without replacement. The events are dependent. We assume that the first jelly bean (the one she ate) was yellow. That leaves 19 jelly beans in the jar, and 6 of these are purple. The probability of drawing the first (yellow) jelly bean is:

$$P(\text{Y}) = \frac{4}{20}$$

The probability of drawing the second (purple) jelly bean is:

$$P(\text{P}) = \frac{6}{19}$$

So the probability of drawing first the yellow and then the purple is:

$$P(\text{Y, then P}) = \frac{4}{20} \times \frac{6}{19} = \frac{24}{380} = \frac{6}{95}$$

2. What is the probability that the two jelly beans were both green?

Answer: Since Marni eats the first jelly bean, this is still a problem without replacement. The probability of choosing green first is:

$$P(\text{G first}) = \frac{7}{20}$$

This leaves 19 jelly beans in the jar, of which 6 are green. So the probability of choosing a green second is:

$$P(\text{G second}) = \frac{6}{19}$$

The probability of two greens is:

$$P(\text{G, then G}) = \frac{7}{20} \times \frac{6}{19} = \frac{42}{380} = \frac{21}{190}$$

3. In a container there are 5 pencils, each a different color. What is the probability that the red one will be selected on the second try if the first one selected is not returned to the container?

Answer: Since there is only one pencil of each color, we must assume that the first one was not red. There are four possibilities of "not red," so the probability for the first one is:

$$P(\text{not red}) = \frac{4}{5}$$

Now, only 4 pencils are left in the container, of which 1 is red. The second time, red is selected. The probability of this event is:

$$P(\text{red}) = \frac{1}{4}$$

Therefore, the probability of selecting red on the second try is:

$$P(\text{not red, then red}) = \frac{4}{5} \times \frac{1}{4} = \frac{4}{20} = \frac{1}{5}$$

TRY THESE

Items 1 and 2: In a drawer are 1 red belt, 2 brown belts, and 4 black belts.

1. Without looking, you take 2 belts out of the drawer, 1 at a time. The first belt you get is black. You do not put the first belt back into the drawer. What is the probability that the second belt will be brown?

2. Without looking, you take 1 belt out of the drawer. You replace it. Then you take out another belt. What is the probability that the first belt was red and the second belt was black?

Items 3 and 4: Many Colors is a school singing group consisting of 6 members. The group's costume is a solid-color T-shirt. All the singers wear the same size shirt, but each shirt is a different color. The shirts are kept in a box.

3. When it is time for the concert, what is the probability that the second shirt taken out of the box will be blue?

4. What is the probability that the third shirt will be blue and the fourth shirt will be green?

Items 5–7: In a box of 15 chocolates, all shaped the same, there are 4 chocolate-covered cherries, 7 cocoanut filled, and 4 solid chocolates. One piece is selected, eaten, and then a second piece is selected.

5. What is the probability of selecting 2 chocolate-covered cherries in a row?

6. What is the probability of first selecting a cocoanut-filled chocolate and then a solid chocolate?

7. What is the probability of first selecting a solid chocolate and then a chocolate-covered cherry?

Items 8–10: A picnic basket contains 5 cans of apple juice, 7 cans of grape juice, and 3 cans of orange juice. You reach into the basket and randomly select 3 cans of juice.

8. What is the probability that all 3 cans are grape juice?
9. What is the probability that 2 cans are orange juice and 1 can is apple juice?
10. What is the probability that NONE of the cans are orange juice?

Assessment

Multiple-Choice Problems

Items 1 and 2: Refer to the illustration.

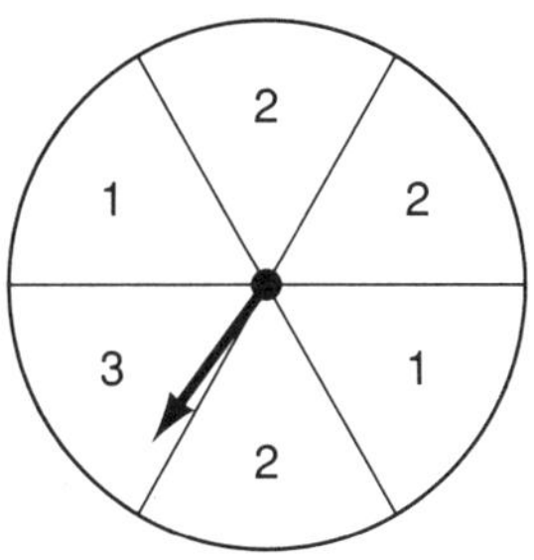

1. What is the probability that this spinner will land on a 1?

   A. $\frac{1}{3}$ **C.** $\frac{2}{3}$

   B. $\frac{1}{2}$ **D.** $\frac{3}{4}$

2. If you spin the spinner 90 times, how many times would you expect to get a 2?

   A. 1

   B. 3

   C. 15

   D. 45

Items 3 and 4: A basket contains the following sandwiches: 2 ham, 6 beef, 4 turkey, 5 chicken, and 3 tuna. Each sandwich is wrapped in aluminum foil with NO label saying what kind it is. You are choosing a sandwich.

3. What is the probability that you will NOT choose a ham sandwich?

   A. $\frac{1}{10}$ **C.** $\frac{3}{4}$

   B. $\frac{2}{5}$ **D.** $\frac{9}{10}$

4. What is the probability that you will choose a poultry sandwich?

   A. $\frac{9}{20}$ **C.** $\frac{1}{5}$

   B. $\frac{1}{4}$ **D.** $\frac{3}{20}$

5 The odds that a machine part will be defective are 1 to 15. Approximately how many defective parts would be expected in a box of 480 parts?

   A. 25 **C.** 35

   B. 30 **D.** 40

Items 6–8: Jar X contains 6 balls: 2 blue and 4 red. Jar Y contains 8 balls: 4 red and 4 blue. You draw 1 ball at random from each jar.

6. What is the probability that you will get a blue ball from each jar?

A. $\frac{1}{2}$ **C.** $\frac{1}{6}$

B. $\frac{1}{3}$ **D.** $\frac{1}{10}$

7. What is the probability that you will draw a red ball from jar X and a blue ball from jar Y?

A. $\frac{1}{10}$ **C.** $\frac{1}{3}$

B. $\frac{1}{6}$ **D.** $\frac{1}{2}$

8. Jefferson reaches into jar X and pulls out a ball. He then places that ball in jar Y. Next, Suzuki reaches into jar Y and pulls out a ball. What is the probability that the ball taken from jar X is blue and the ball taken from jar Y is blue?

A. $\frac{5}{9}$ **C.** $\frac{1}{3}$

B. $\frac{7}{18}$ **D.** $\frac{5}{27}$

9. When you roll a single die, what is the probability of getting a 2 or 5?

A. $\frac{1}{6}$ **C.** $\frac{1}{3}$

B. $\frac{2}{9}$ **D.** $\frac{5}{6}$

10. If you roll a pair of dice, what is the probability of getting a sum of 5?

A. $\frac{1}{9}$ **C.** $\frac{1}{18}$

B. $\frac{1}{12}$ **D.** $\frac{1}{36}$

Items 11–13: Using the two spinners below, there is an equally likely chance of the pointer landing on any sector.

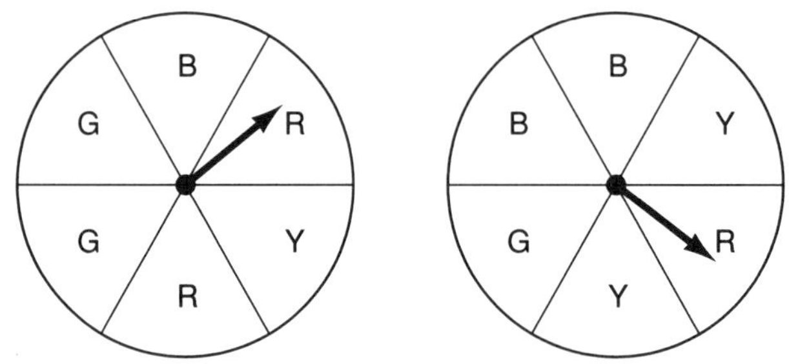

11. What is the probability of both pointers landing on red?

A. $\frac{1}{18}$ **C.** $\frac{1}{3}$

B. $\frac{1}{4}$ **D.** $\frac{1}{2}$

12. What is the probability of one pointer landing on green and the other pointer landing on yellow?

A. $\frac{1}{18}$ **C.** $\frac{5}{36}$

B. $\frac{1}{9}$ **D.** $\frac{1}{2}$

13. What is the probability of neither pointer landing on blue?

A. $\frac{1}{9}$ **C.** $\frac{5}{9}$

B. $\frac{4}{9}$ **D.** $\frac{7}{9}$

Gridded-Response Problems

14. When you roll a pair of dice, what is the probability of getting a sum of more than 5?

15. A math class consists of 35 students: 18 boys and 17 girls. The teacher calls on students at random to answer questions. What are the odds that a girl will be asked the first question?

16. Ms. Weber puts the names of her 10 debating students into a box. Speeches will be given in the order in which the names are chosen. What is the probability that John's name will be chosen third?

17. If we use the letters in the word PROBABILITY, what is the probability that a letter chosen at random will be an I or a B?

18. You choose 1 letter at random from the letters in the word GRADUATION. What are the odds of selecting a consonant?

19. At the bakery there are 6 different kinds of cookies: chocolate chip, sugar, oatmeal, peanut butter, raisin, and butterscotch. They are sold in packages, each package containing two different kinds. What is the total number of different combinations in which these cookies could be packaged?

20. A group of students are participating in a probability experiment. They were given the choice of oranges or grapes. Half of the students chose oranges and half chose grapes. They were then instructed to make a second choice of apples or grapefruits. Of those who chose grapes, three-fourths chose apples. Of those who chose oranges, three-fourths chose grapefruits. This information is displayed on the tree diagram shown below. What is the probability that a person chosen at random will choose grapes first and then a grapefruit?

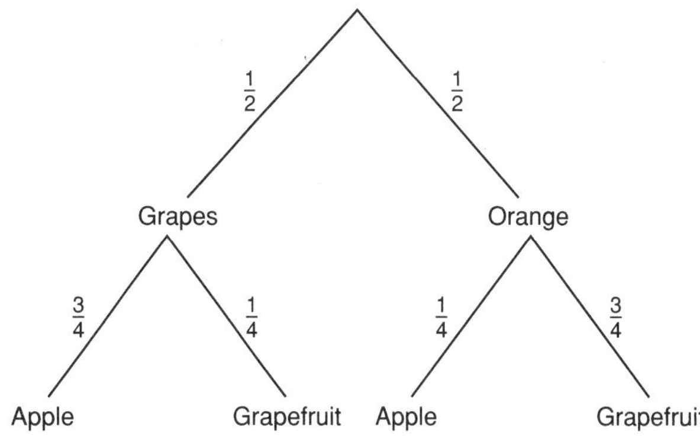

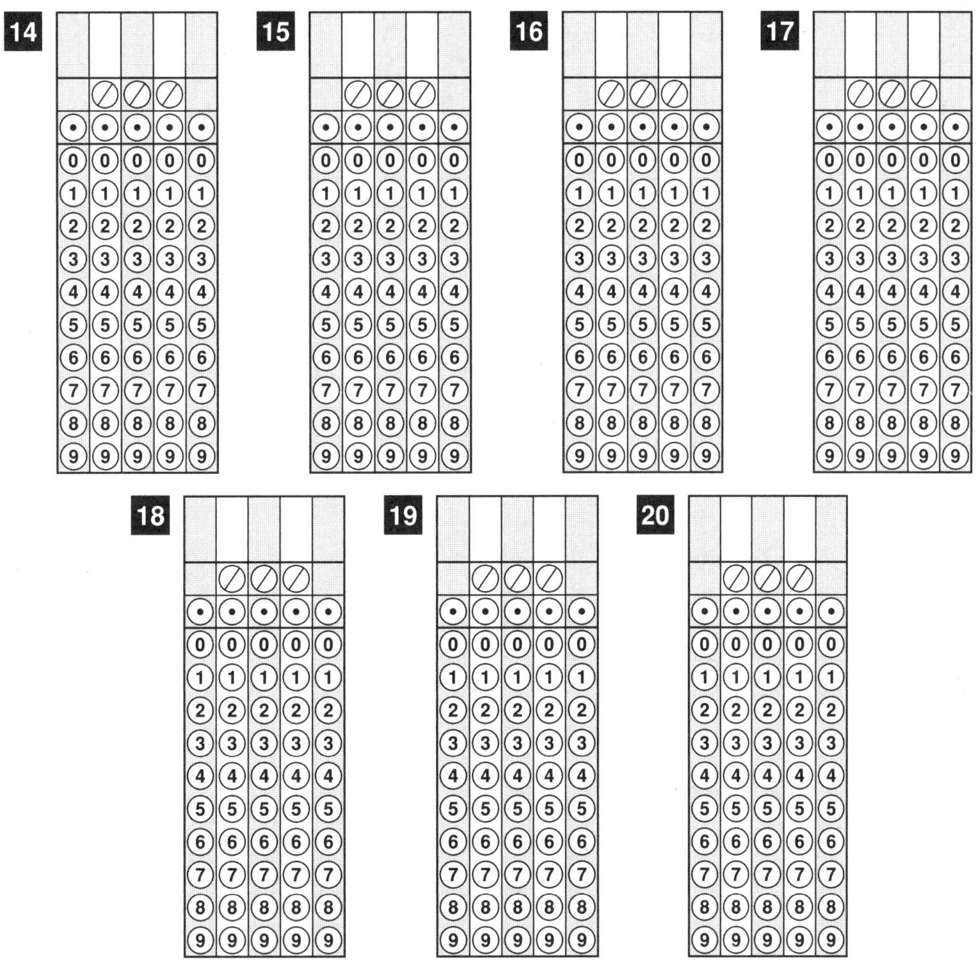

Section 3 Using statistical methods to make inferences and valid arguments about real-world situations

A READING, ANALYZING, AND INTERPRETING DISPLAYED DATA

In our reviews so far, data were displayed and measured. How does this information relate to real-world needs and situations? The problems below give examples of real-world applications.

MODEL PROBLEMS

A new high school was being built in Mason County. Some of the students who had been going to another high school would be transferred to this new high school. In order to choose the school colors for the new high school, the students being transferred were asked to vote on four possibilities that had been selected by a committee.

Seniors did not vote, since they would not be attending the school the next year. Eighth-grade students did not vote, because they were coming from many middle schools.

The table lists the results of the voting.

Number of Votes

Class	Green and gold	Red and blue	Blue and silver	Black and gold
Freshmen	44	43	36	27
Sophomores	24	20	31	45
Juniors	40	12	25	33
Total votes	108	75	92	105

1. To the nearest unit, what percent of the juniors voted for the winning colors?

Answer: 36 percent

The highest number of votes (108) was for green and gold, so these were the winning colors. The total number of juniors who voted was 110. Of these, 40 voted for green and gold. Calculate the percent: $\frac{40}{110} = 0.\overline{36} \approx 36\%$.

2. Which class did NOT prefer the winning choice?

Answer: Sophomores. Their highest vote (45) was for black and gold.

3. To the nearest unit, what percent of the total votes were cast by the freshmen?

Answer: 39 percent

The total number of votes was 380. The total number of votes cast by freshmen was 150. Calculate the percent: $\frac{150}{380} \approx 39\%$.

TRY THESE

Use the data in the Model Problems on page 390.

1. To the nearest unit, what percent of the total voters selected red and blue?
2. What was the difference in the number of votes between the freshmen who voted for black and gold and the juniors who voted for black and gold?
3. What percent of the voters were sophomores?
4. How many votes were cast by the class that cast the most votes?
5. What type of graph could be used to best display these data? Explain your reasoning.

MODEL PROBLEMS

When data are collected, the results can be displayed in a bar graph. Central High School had a fund-raiser that involved selling candy bars over a 4-week period. The results are shown in this bar graph.

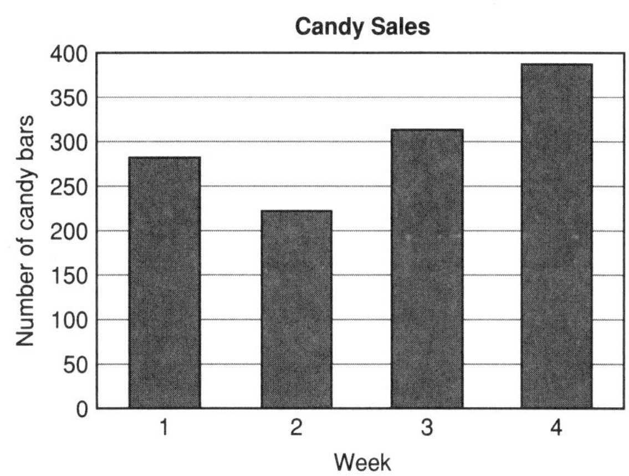

1. Approximately how many more candy bars were sold in week 3 than in week 2?

   Answer: Approximately 90 more, since $310 - 220 = 90$.

2. Approximately what was the number of candy bars sold during the 4-week period?

   Answer: Approximately 1,200, since $280 + 220 + 310 + 390 = 1{,}200$.

TRY THESE

Use the graph above about candy sales.

1. Determine the difference in sales between the first and fourth weeks.
2. What percent of the total sales occurred during the fourth week?

3. What do you think would explain the dip in sales during week 2 and the increases in weeks 3 and 4?

MODEL PROBLEMS

The brains of animals represent different percentages of total body weight, depending on the animal. This table shows the average total body weight and average brain weight of several animals.

Brain and Body Weights

Animal	Average total body weight (pounds)	Average brain weight (pounds)	Brain weight as percent of body weight
Alligator	550	0.0375	0.01%
Lion	500	0.65	0.1%
Porpoise	425	5	1.2%
Ostrich	325	0.11	0.03%
Wolf	75	0.425	0.6%
Baboon	50	0.475	1%
Human	150	3.625	2.4%
Chimpanzee	150	1	0.67%

1. On the basis of the data in the table, which statement is true?

   A. Animals that have approximately the same total body weight also have the same brain weight.

   B. The heavier the animal, the heavier the brain.

   C. The size of the brain is determined by the size of the animal.

   D. Of the animals listed, the porpoise has the heaviest brain.

   Answer: D. The porpoise has a brain weighing 5 pounds, the heaviest listed in this table.

2. What is the difference in percent of brain weight to body weight between the heaviest animal and the lightest animal?

   Answer: 0.99 percentage point. The figure for the heaviest animal listed, the alligator, is 0.01%. The figure for the lightest animal listed, the baboon, is 1%. Subtract: 1% − 0.01% = 0.99 percentage point.

TRY THESE

Items 1 and 2: Use the table about brain weights in the Model Problem above.

1. On the basis of the data in the table, which statement is FALSE?

   A. Of the animals listed, the alligator has the smallest percentage of brain weight to body weight.

B. The average total body weight of the human is 3 times that of the baboon.

C. The lion and the baboon have the same percent of brain weight to body weight.

D. The brain weight of the alligator is less than the brain weight of the ostrich.

2. How much heavier is the baboon's brain than the wolf's brain?

Items 3–6: The U.S. Census Bureau published the following information on October 19, 2004.

Population of Florida

Subject	1990	2000
Total population	12,937,926	15,982,378
Male	6,261,719	7,797,715
Female	6,676,207	8,184,663
Age		
Under 5 years	849,596	945,823
5 to 9 years	814,118	1,031,718
10 to 14 years	748,355	1,057,024
15 to 19 years	799,279	1,014,067
20 to 24 years	870,546	928,310
25 to 59 years	5,807,563	7,460,343
60 to 84 years	2,838,359	3,213,806
85 years and over	210,110	331,287
Median age	36.3	38.7
Under 18 years	2,866,237	3,646,340
Male	1,468,433	1,870,986
Female	1,397,804	1,775,354
Urban and Rural Residence		
Total population	12,937,926	15,982,378
Urban population	10,970,445	14,274,392
Rural population	1,967,481	1,707,986
Farm population	47,436	38,567

3. What is the percent of increase in Florida's population from 1990 to 2000?

4. How many more people over the age of 59 lived in the state of Florida in 2000 than in 1990?

5. What is the percent of decrease in farm population from 1990 to 2000?

6. Which gender, under the age of 18, had the greater percent of increase from 1990 to 2000?

B IDENTIFYING LIMITATIONS OF STATISTICAL TECHNIQUES IN MAKING INFERENCES AND VALID ARGUMENTS

Two popular sayings about statistics are, "You can prove anything with numbers" and "Statistics lie." The problem is not in the statistics or numbers themselves, but in how data are gathered, displayed, or interpreted. Common problems with statistics include bias in gathering data, misleading displays of data, such as graphs, and misleading use of "averages." Misleading use of "averages" was discussed in Section 1B.

Bias in data collection

When a statistical experiment is performed and facts involving numbers are gathered, the experimenters or researchers must be careful to avoid what is called **bias**. Bias can arise in several ways.

In a survey, for instance, the group being surveyed should be representative, or typical, of the whole population. If the people being surveyed are *not* representative, the data in the survey will be considered biased. For example, suppose that you conduct a survey to find out how many hours a day students spend on homework. If you survey only fifth graders, your conclusion may be valid for fifth graders, but it will *not* necessarily be valid for all students.

Bias can also exist when researchers or experimenters ask sensitive questions. People may give an answer that they think is socially correct, or they may give the answer that they think the interviewer wants to hear. For example, suppose that a police officer, armed and in full uniform, takes a survey asking people's opinion about the honesty of the police force. The results of that survey might be very different from those of a survey conducted by an unarmed police officer wearing civilian clothes. The results might also be different if someone who is not a police officer conducted the survey. In a case like this, it can be said that bias exists because of the way a survey was taken.

MODEL PROBLEMS

1. Simeen surveys all the students in her trigonometry class and finds that 95 percent of them like block scheduling. She concludes that 95 percent of the students in the school like block scheduling. Is this conclusion valid? Explain why or why not.

   Answer: Her conclusion is not valid, because the results of her survey of one class (trigonometry) are not necessarily valid for the entire school. Since she has surveyed only one class, the whole population (the whole school) is underrepresented. Also, the one class she has surveyed is a higher-level math class, so it probably would not be representative, or typical, of the entire school.

2. A company that makes computer CDs has manufacturing plants in three cities. The plant in Boca Raton makes 60 percent of the CDs. The plants in Deland and Fort Myers each make 20 percent of the CDs.

   An inspector checking for quality finds that over several months, each plant produces about 350 defective CDs per week. The inspector concludes that all three plants are producing work at the same standard of quality.

   Should we agree or disagree with the inspector's conclusion? Explain your answer.

   Answer: We should disagree. In fact, the plant in Boca Raton seems to be producing higher-quality work. Boca Raton makes 3 times as many CDs as either of the other two plants, and yet it produces no more poor-quality CDs than the others.

TRY THESE ✔

1. A company is giving its annual banquet, and a committee of 25 employees has been selected to choose a speaker. Of the company's 21 employees in management, 7 were selected to be on the committee. Of the 12 clerical employees, 4 were selected. Of the 42 assembly-line employees, 14 were selected. Is this committee representative of the employees? Why or why not?

2. The school board plans to relocate the high school to the other side of town. People opposed to the move held a protest rally outside the school district headquarters. A reporter from the local newspaper surveyed the people at the rally. Based on his survey, he reported that the community is 100 percent against building the new high school across town. Is this conclusion valid? Why or why not?

3. Students at a school are voting on a new school motto. Two choices are offered: motto A and motto B. This table shows the results of the vote.

Motto Votes Cast

Class	Motto A	Motto B
Seniors	200	10
Juniors	90	110
Sophomores	80	160
Freshmen	100	120
Total	**470**	**400**

After all the votes are counted, motto B is chosen as the new school motto. What would you conclude is the reasoning behind this decision?

4. The school newspaper wanted to determine who the students consider to be the best teacher at Palm Tree High School. For five days, they set up a table outside the cafeteria and asked students to cast a vote for the best teacher. Records were kept as to who had already voted and by the end of the week 89 percent of the students had voted. In their next edition it was reported that Ms. Pla was the best teacher. Is this conclusion valid? Why or why not?

5. Last Friday Adrianna surveyed the students at her lunch table as to their favorite school lunch. Seventy-five percent of the students surveyed said that the meal of hamburgers and French fries was their favorite. She concluded that the meal of hamburgers and French fries was the most popular meal that the cafeteria served. Could the fact that every Friday the cafeteria serves hamburgers and French fries have any effect on her results? Was her conclusion valid? Why or why not?

Misleading displays of data

We often see data displayed in graphs in newspapers, magazines, and advertisements. These graphs can sometimes be misleading. For instance, a graph that has no scale or an incorrect scale can lead to an incorrect interpretation of the data.

To get accurate information from graphs, we need to look carefully at the scales. The scale that is used can make a major difference in how we interpret a graph.

EXAMPLE 1

The graph below seems to show a great increase in sales.

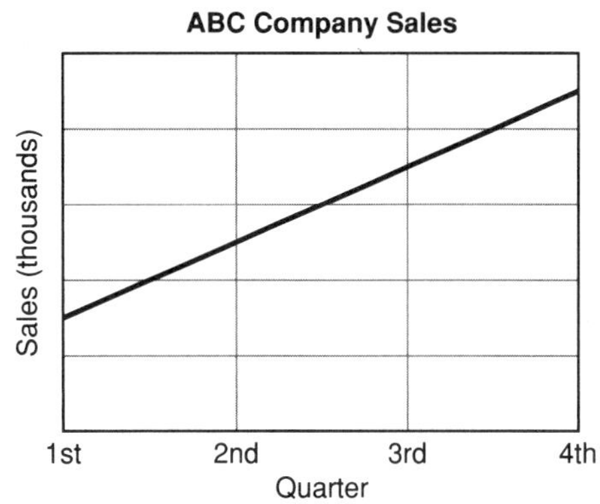

Notice, however, that this graph has no vertical scale, so we really have no idea how much sales have increased.

Now look at the same graph *with* a vertical scale:

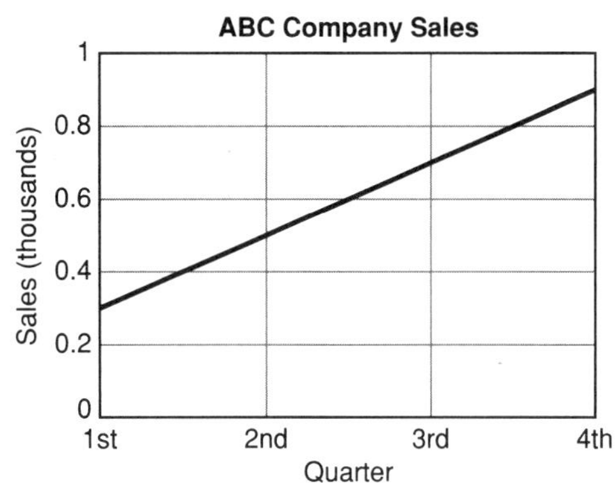

The vertical scale shows that the increase in sales was really very small.

EXAMPLE 2

Each student in a science class kept track of the average daily temperature for 5 days and then graphed the results. Two students, Ahmed and Bette, produced the two graphs below. The graphs show exactly the same data.

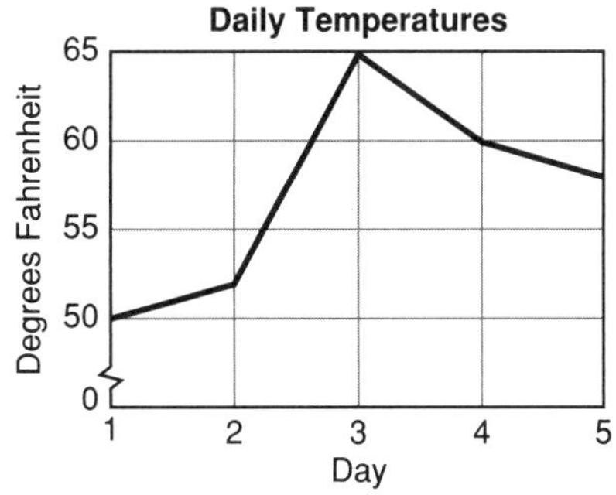

Ahmed's graph

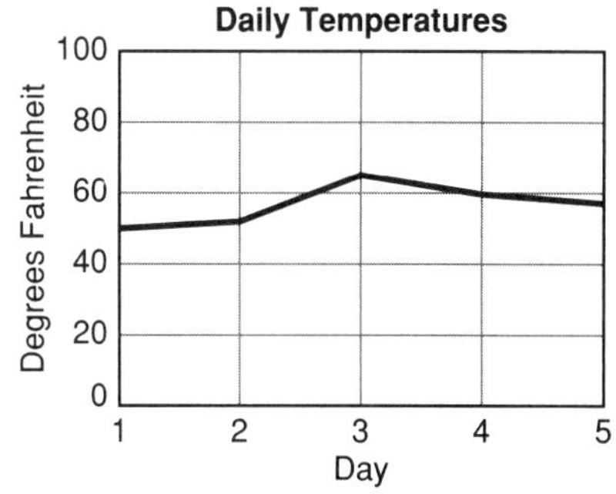

Bette's graph

The students were asked to decide, from their graphs, if the temperature had varied much during the 5 days. On the basis of his graph, Ahmed would probably say that the temperature had varied greatly. But on the basis of her graph, Bette would probably say that the temperature did *not* vary much. The reason for the difference in the graphs, and the different interpretations, is that Ahmed and Bette used different vertical scales.

MODEL PROBLEM This graph is misleading. Identify each error that makes it misleading and state what should be done to correct the error and clarify the graph.

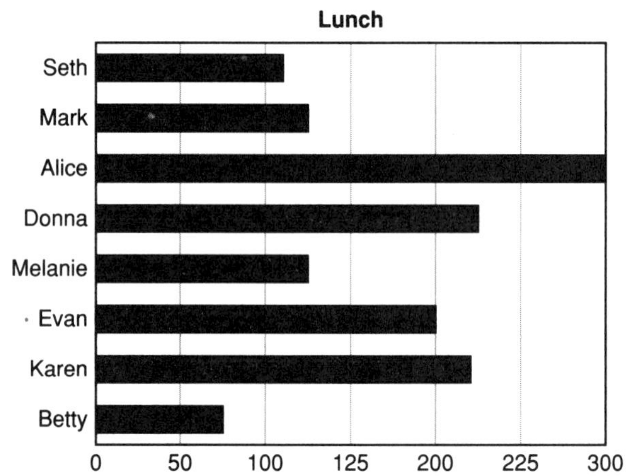

Answer:

- Error: Title does not give enough information. "Lunch" is too vague.

  Correction: Write a more specific title, such as "Calories Eaten at Lunch."

- Error: Vertical axis has no label, so we don't know who these people are.

  Correction: Give the vertical axis a label, such as "Juniors" or "Students" or "Children."

- Error: Horizontal axis has no label, so we don't know what these numbers mean.

  Correction: Give the horizontal axis a label, "Calories" or "Number of Calories."

- Error: Horizontal scale is not uniform—the intervals are unequal.

  Correction: Make the intervals equal. For example, the intervals could be 0, 50, 100, 150, 200, 250, and 300. Or they could be 0, 25, 50, 75, 100, etc.

TRY THESE

Items 1–4: The graphs below are misleading. Identify each error and state what should be done to correct it and clarify the graph.

1.

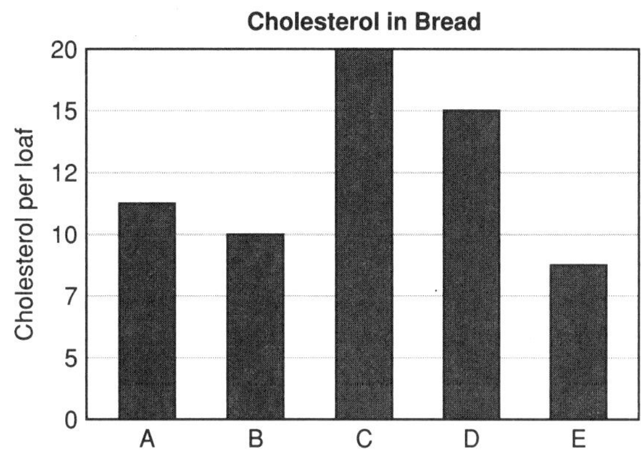

2.

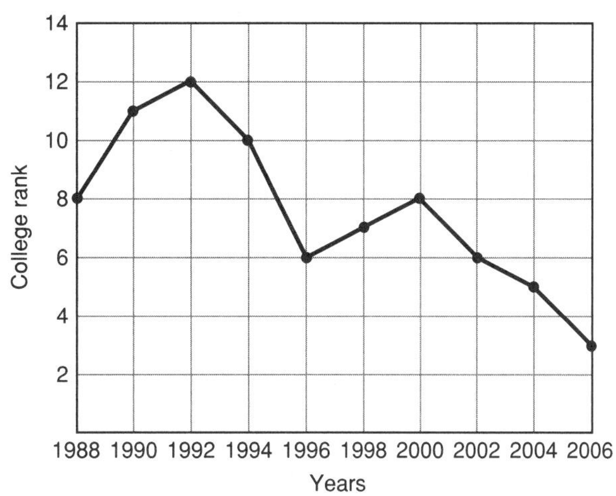

3.

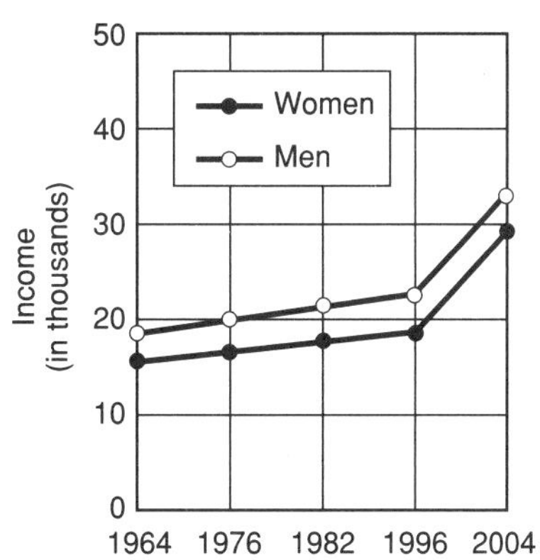

4. **Reasons for Not Turning in Homework**

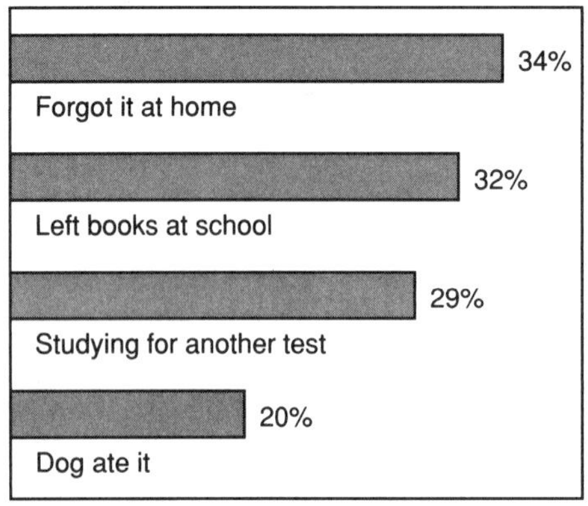

5. Families were surveyed concerning the remote controls for their entertainment center. These are the results of the survey: 45% never misplace them, 37% misplace them 1 to 3 times a week, 16% misplace them 4 to 8 times a week, and 2% misplace them more than 8 times a week. These data are displayed in the circle graph below. Identify any errors in the graph.

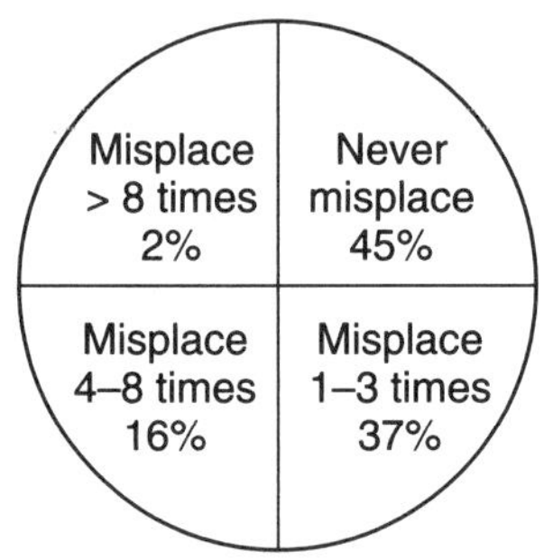

6. Jason was asked to make a stem-and-leaf plot of the high temperatures of 15 cities in South Florida. The data were as follows: 86°, 88°, 82°, 94°, 88°, 94°, 96°, 88°, 92°, 91°, 83°, 88°, 90°, 86°, 95°. Jason's plot is shown below.

Stem	Leaf
8	2 3 6 8
9	0 1 2 4 5

He reported to the class that the median temperature was 90°. Identify any errors.

Assessment

Multiple-Choice Problems

Items 1–4: A vacant lot was given to a city, and the city conducted a survey to find out how the residents wanted the lot to be used. Each person surveyed was asked to vote for one of eight uses. The city chose the use that received the most votes in the survey. The following table shows the results of the city's survey.

Use	Votes	Use	Votes
Open park	8	Basketball court	30
Playground	19	Swimming pool	18
Rose garden	5	Tennis court	12
Baseball field	22	Band shell	6

1. What percent of those surveyed voted for the winning use?

   A. 47% C. 25%

   B. 30% D. 12%

2. Suppose that the lot could have two uses: one sport and one nonsport. According to the survey, what nonsport use would be preferred?

   A. rose garden C. baseball field

   B. band shell D. open park

3. There are approximately 10,000 residents in this city. If the sample surveyed was representative, how many residents would be expected to prefer the basketball court?

   A. 2,500 C. 25,000

   B. 3,000 D. 30,000

4. Which one of the following statements is true?

   A. Twice as many people surveyed prefer a swimming pool to a band shell.

   B. Twenty-four percent of the people surveyed prefer a rose garden.

   C. A tennis court is the fourth most popular choice.

   D. Twenty-five percent of the people surveyed preferred a basketball court.

5. What is wrong with the graph shown below?

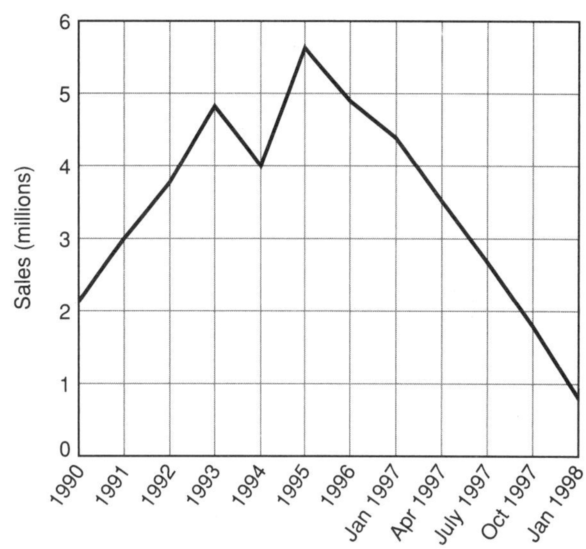

   A. Title is incomplete.

   B. Horizontal scale is not consistent.

   C. Vertical scale is labeled incorrectly.

   D. Nothing is wrong with the graph.

Items 6 and 7: The table below compares the annual sales of 5 florists.

Flourist	Annual Sales
A	$32,000
B	$29,000
C	$26,000
D	$38,000
E	$34,000

6. Florist D displayed the data on the graph below? What is misleading about the graph?

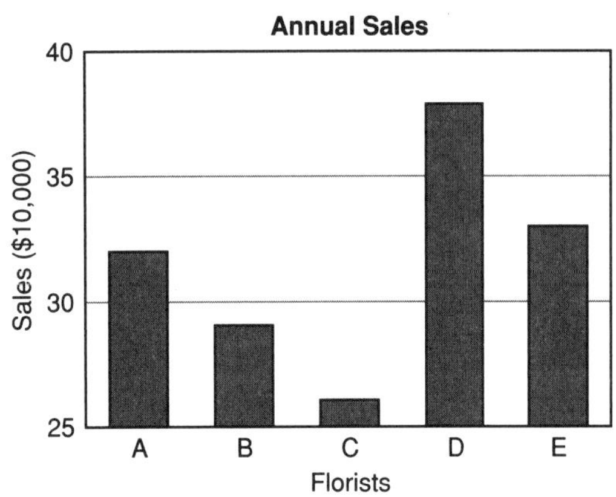

A. The columns are too narrow.

B. The vertical scale is not consistent.

C. The differences in heights have been exaggerated.

D. Nothing is misleading.

7. Florist C displayed the data on the graph below? What is misleading about the graph?

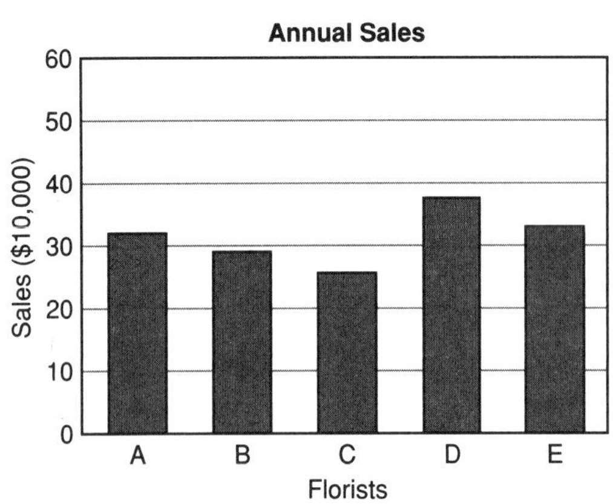

A. Columns are all the same width.

B. Horizontal title is incorrect.

C. Vertical scale is not consistent.

D. Nothing is misleading.

Items 8–10: On the TV show *U.S. Statue*, people were asked to call in and vote for their favorite U.S. statue. The results of the voting are shown in the table below. The categories are age ranges and the votes are in thousands.

	A	B	C	Total votes
6–12	157	125	216	498
13–19	322	418	512	1,252
20–45	524	871	695	2,090
> 45	352	492	316	1,160
Total by statue	1,355	1,906	1,739	

8. Based on the information in the table, which of the following statements is true?

A. The age group of 13–19 cast more votes for the winning statue.

B. Two statues each received over one-third of the votes.

C. The over 45 age group cast fewer votes than the 20–45 age group because the show was on after their bedtime.

D. The winning statue received more votes than the other two combined.

9. The number of votes cast for the winning statue by the 20–45 age group is what percent of the total of all votes cast?

A. 17% C. 42%

B. 29% D. 46%

10. What percent of the 6–12 age group cast their vote for the winning U.S. statue?

A. 3%

B. 7%

C. 10%

D. 25%

CHAPTER REVIEW

Multiple-Choice Problems

Items 1–3: In a vending machine there are 10 gum balls: 3 red, 2 blue, 4 yellow, and 1 white.

1. A customer who puts in a coin gets 1 gum ball, at random. What is the probability that the customer will get a yellow gum ball?

   A. $\frac{4}{5}$

   B. $\frac{3}{5}$

   C. $\frac{4}{10}$

   D. $\frac{1}{10}$

2. A customer puts in a coin and gets a gum ball, then puts in a second coin and gets a second gum ball. What is the probability that the first gum ball will be blue and the second gumball will be red?

   A. $\frac{1}{15}$

   B. $\frac{4}{15}$

   C. $\frac{8}{15}$

   D. $\frac{13}{15}$

3. A customer puts in a coin. What is the probability of getting a blue gum ball or a white gum ball?

   A. $\frac{1}{5}$

   B. $\frac{1}{10}$

   C. $\frac{3}{5}$

   D. $\frac{3}{10}$

Items 4–6: Marsha, Tanya, and Julia each bowled four games in a charity event. The scores are displayed in the Venn diagram below.

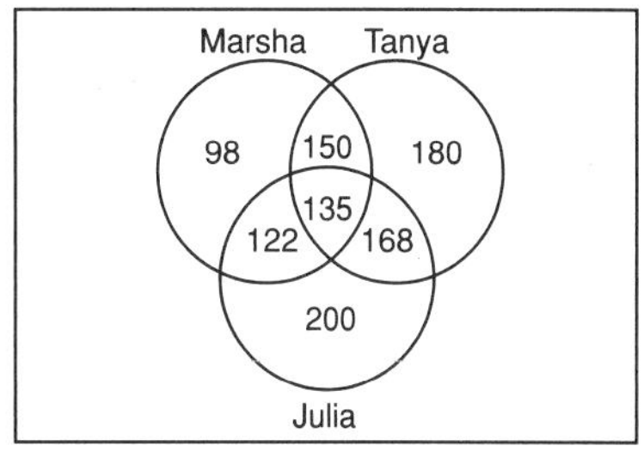

4. What score do they all have in common?

   A. 122

   B. 135

   C. 150

   D. 168

5. What is Marsha's highest score?

   A. 200

   B. 180

   C. 150

   D. 98

6. What is the highest score that Tanya and Julia have in common?

   A. 135

   B. 168

   C. 180

   D. 200

7. The office manager kept track of the number of absences of the employees for the year. The results are as follows: 7, 6, 8, 4, 7, 3, 2, 6, 5, 3, 7, 8, 0, 2, and 1. Which measure of the data is the lowest?

   A. mean

   B. median

   C. mode

   D. range

Gridded-Response Problems

Items 8–10: Derrick volunteers as an aid at the elementary school. He prepared a back-to-back stem-and-leaf plot of the heights of boys and girls in Mr. Hale's class.

Heights of Boys and Girls

Girls		Boys
9 8 3 2	3	1 2 3 5 6 8 9
6 6 4 3	4	3 6 9
7 6 5 2 1	5	1 2 3 3 4

Key: 7|5 = 57 inches (left leaves)
3|2 = 32 inches (right leaves)

8. What is the difference between the median height of the boys and the median height of the girls?

9. What is the difference between the mode height of the boys and the mode height of the girls?

10. What is the difference between the range of the heights of the boys and the range of the heights of the girls?

Items 11 and 12: Mr. Onder put his students' names on cards. Each period he shuffles the cards and then calls on students in the order of the cards. There are 8 girls in a class of 20 students.

11. If a boy is called on first, what are the odds that it will be Darryl Yetter? Give your answer in fraction form.

12. On the next day, Mr. Onder shuffles the cards again. What is the probability that Lucy Brown will be chosen second? Give your answer in fraction form.

Items 13 and 14: The hospital keeps records of the ages of all people who come to the emergency room. Last Tuesday evening, the data recorded were displayed on the box-and-whisker plot shown below.

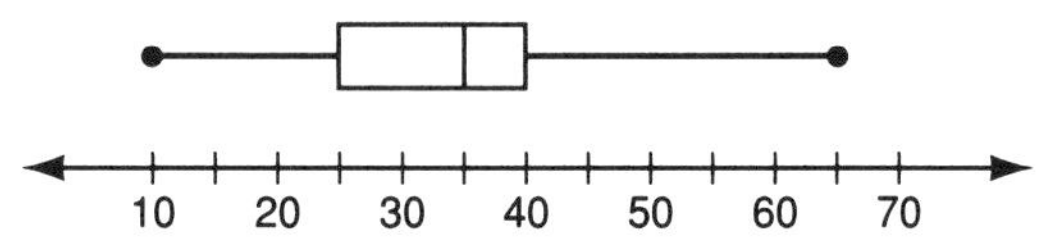

13. Fifty percent of those using the emergency room were under what age?

14. What percent of those using the emergency room were between the ages of 40 and 65?

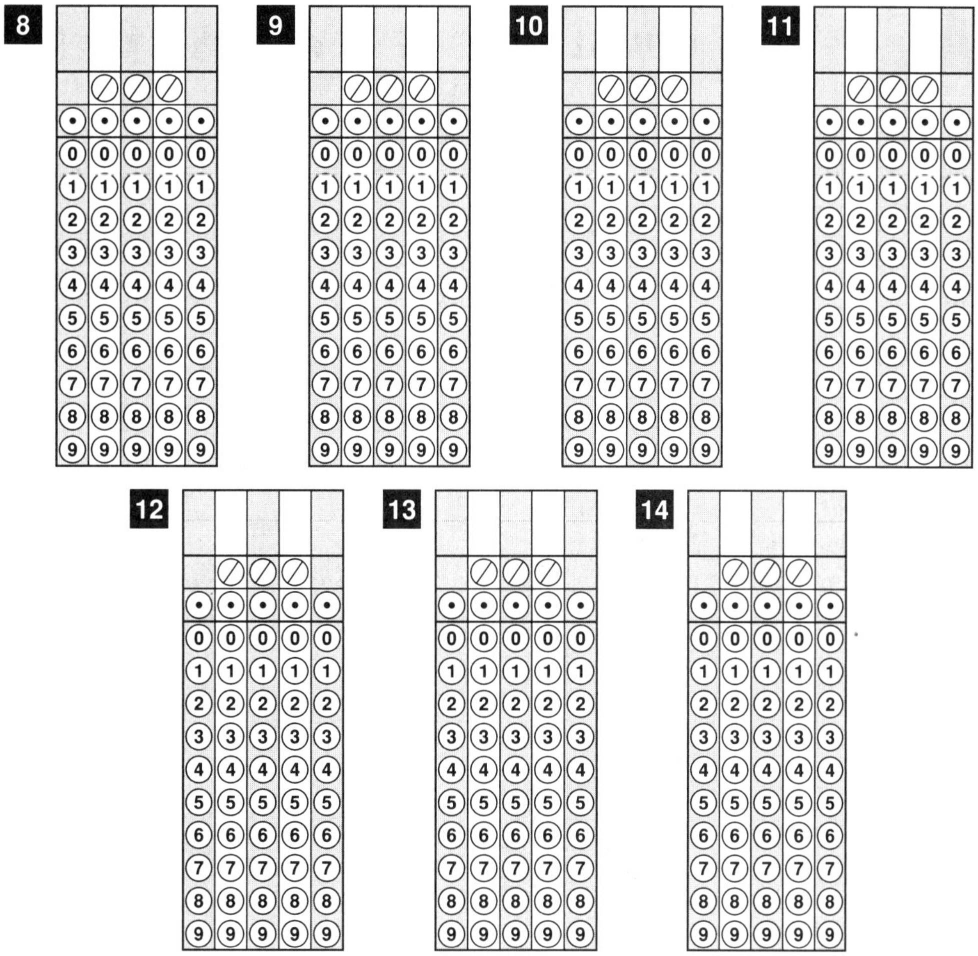

Think, Solve, Explain: Extended-Response Problem

15. Students in the auto shop class checked on the gas mileage (miles per gallon) of their cars in January. They checked the gas mileage again in May. The points on the graph represent the miles per gallon for each student's car in January and May.

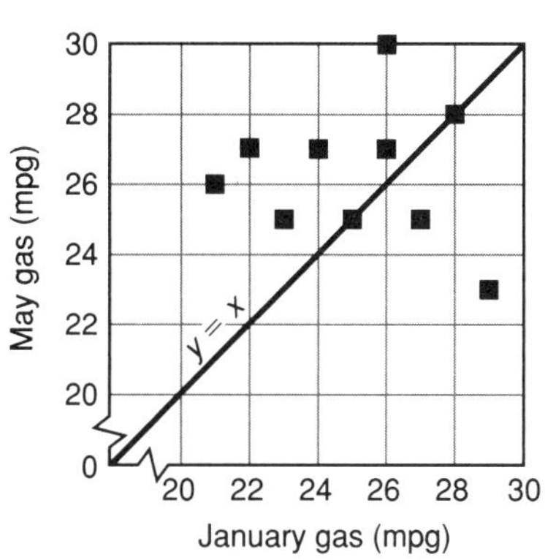

Part A What is the mean of the gas mileage in January?

Part B What is the mean of the gas mileage in May?

Part C What is the significance of the points that lie on the line $y = x$?

Part D Overall, what conclusions can you draw concerning the gas mileage of the students' cars in January and May?

CUMULATIVE REVIEW

Multiple-Choice Problems

1. Bananas sell for \$0.78 per pound. What is the cost of a bunch of bananas that weigh 2 pounds 8 ounces?

A. \$1.56

B. \$1.64

C. \$1.95

D. \$2.18

2. The company's budget allows for 0.3% to be used for TV advertising. What is the fractional equivalent of this percent?

A. $\frac{3}{10}$

B. $\frac{3}{100}$

C. $\frac{3}{1,000}$

D. $\frac{3}{10,000}$

3. In 2006, the population of the world is approximately 6.5 billion. How can this number be expressed in scientific notation?

A. 6.5×10^{10} **C.** 6.5×10^{8}

B. 6.5×10^{9} **D.** 6.5×10^{7}

4. Paul and Bruce recorded the time they spent riding their bicycles. Paul left at 8:00 A.M. and rode at 20 miles per hour. Bruce left at 9:00 A.M. and rode at 22 miles per hour. Both stopped riding at 3:00 P.M. Which statement best describes their ride?

A. Bruce rode 14 miles farther than Paul.

B. Paul rode 14 miles farther than Bruce.

C. Bruce rode 8 miles farther than Paul

D. Paul rode 8 miles farther than Bruce.

5. The indoor track is in the shape of a rectangle with a semicircle on each end. The rectangle is 5 yards by 15 yards. To the nearest yard, what is the perimeter of the track?

A. 40 yards

B. 46 yards

C. 53 yards

D. 61 yards

6. Triangle *ABC* is similar to triangle *DEF*. Which statement is false?

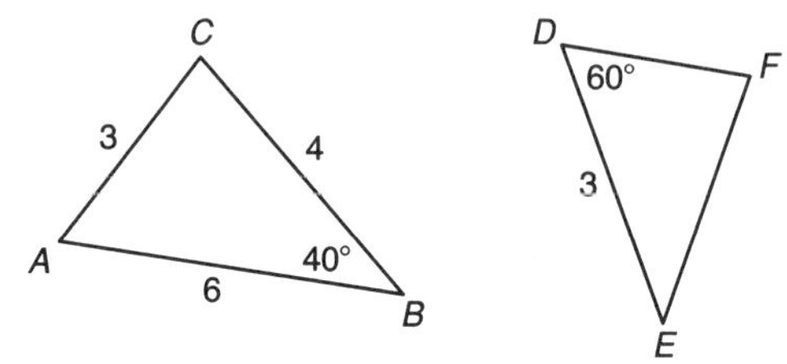

A. $\angle C = 60°$

B. $FE = 2$

C. $\angle A = 60°$

D. $\angle F = 80°$

7. The following box-and-whisker plot represents the prices of dresses shown in a catalogue.

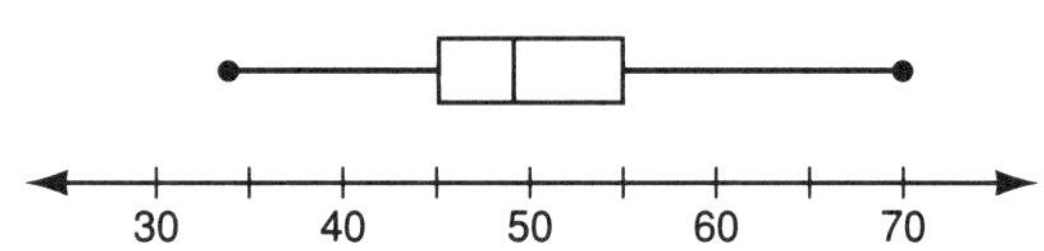

Twenty-five percent of the dresses are below what cost?

A. \$34

B. \$45

C. \$49

D. \$55

8. Which graph shows the relationship of time to height above the ground when a person rides on a Ferris wheel?

A.

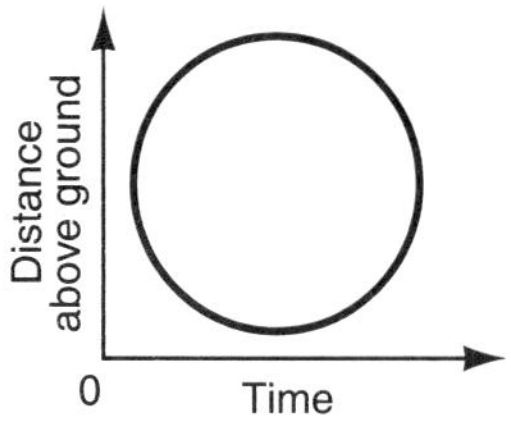

C.

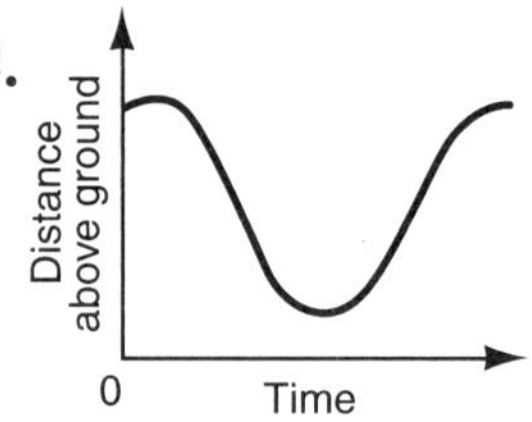

B.

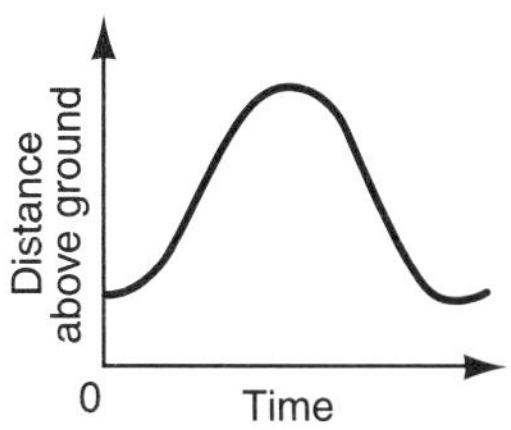

D.

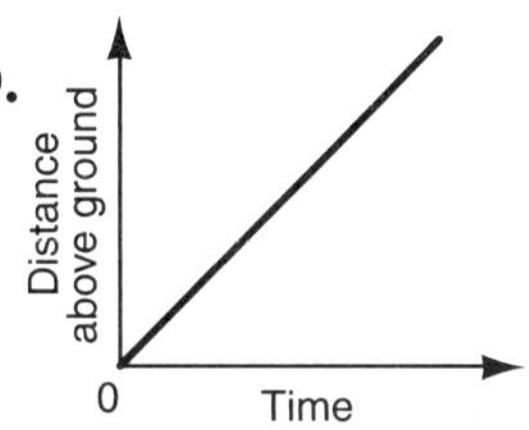

9. *ABCD* is a trapezoid. What is the measure of angle *D*?

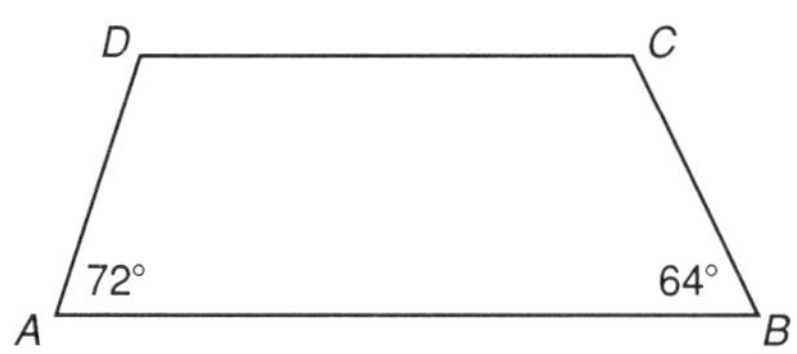

A. 18° **C.** 108°

B. 26° **D.** 116°

10. How would the following statement be expressed algebraically?

Seven less than twice a number x is forty-two.

A. $7 - 2x = 42$ **C.** $2 - 7x = 42$

B. $2x - 7 = 42$ **D.** $7x - 2 = 42$

11. What is the solution to the inequality $6 - 5x \leq -19$?

A. $x \leq 2.6$ **C.** $x \leq 5$

B. $x \geq 2.6$ **D.** $x \geq 5$

12. In November the student council collects canned goods for a food bank. The class that brings in the most canned items receives a pizza party. Ms. Dorff-Pennea's class won when 8 students each brought in 1 can, 8 students brought in 2 cans each, 8 students brought in 4 cans each, and Ms. Dorff-Pennea brought in 40 cans. Which of the following best describes the average number of cans brought in per person?

A. mean **C.** mode

B. median **D.** range

Gridded-Response Problems

13. If the spinner below is spun 30 times, how many times would it be expected to land on a 3?

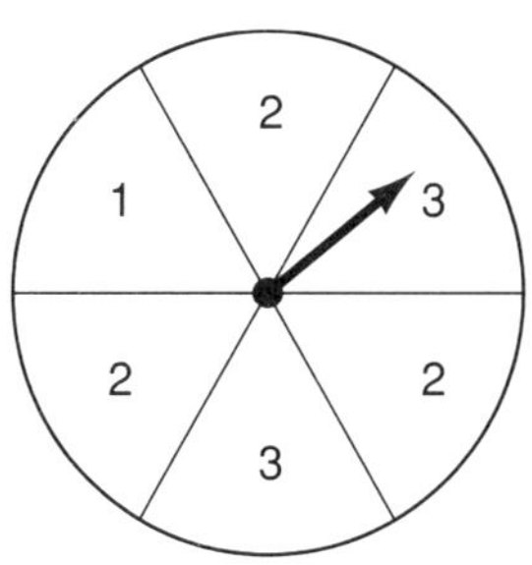

14. Evaluate $\dfrac{2x^2 - 3xy}{5y}$ when $x = 7$ and $y = 2$.

13

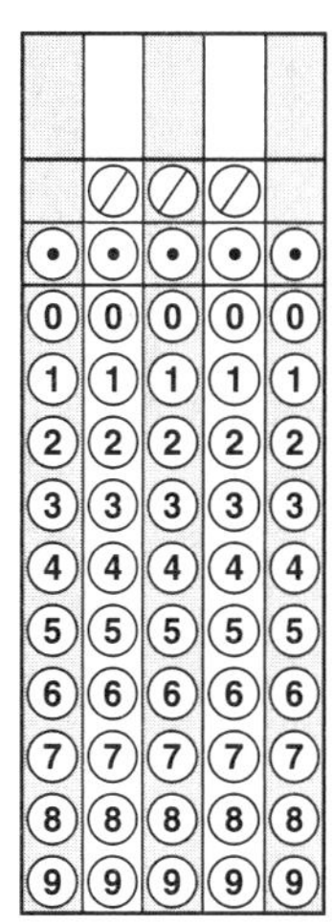

14

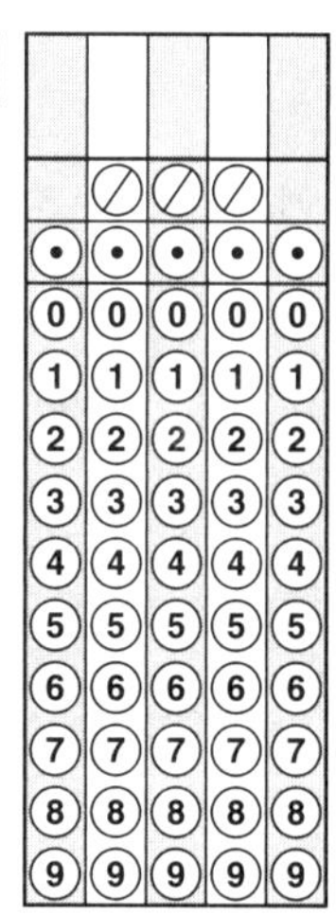

15. The base of a ladder is placed 7 feet from the house. The ladder reaches up to a window 24 feet above the ground. How long is the ladder?

16. Stamps, each 1 inch square, are printed on a sheet that is 2 feet by 3 feet. How many stamps can be printed on the sheet?

17. What is the slope of the line passing through point A (7, 4) and point B (2, −3)?

18. On a map with a scale of 2 inches = 26 miles, two towns are located 9.5 inches apart. What is the distance, in miles, between the two towns?

19. $3(2x - 5) = 12(x - 8)$. What is the value of x?

20. Mikala went to Appliances-a-Plenty to price steam irons. The prices are displayed on the stem-and-leaf plot below. What is the median price of the steam irons?

Stem	Leaf
2	2 4 6 7
3	1 2 2 3 4
4	2 5 7
5	8 9

Key: 2|2 = $22

21. John bought 3 shirts and 2 pairs of pants for $73. Buying the same priced items, Marty bought 2 shirts and 3 pairs of pants for $82. This problem can be represented by the equations below where s equals the price of one shirt and p equals the price of one pair of pants. What is the price of one shirt?

$3s + 2p = 73$

$2s + 3p = 82$

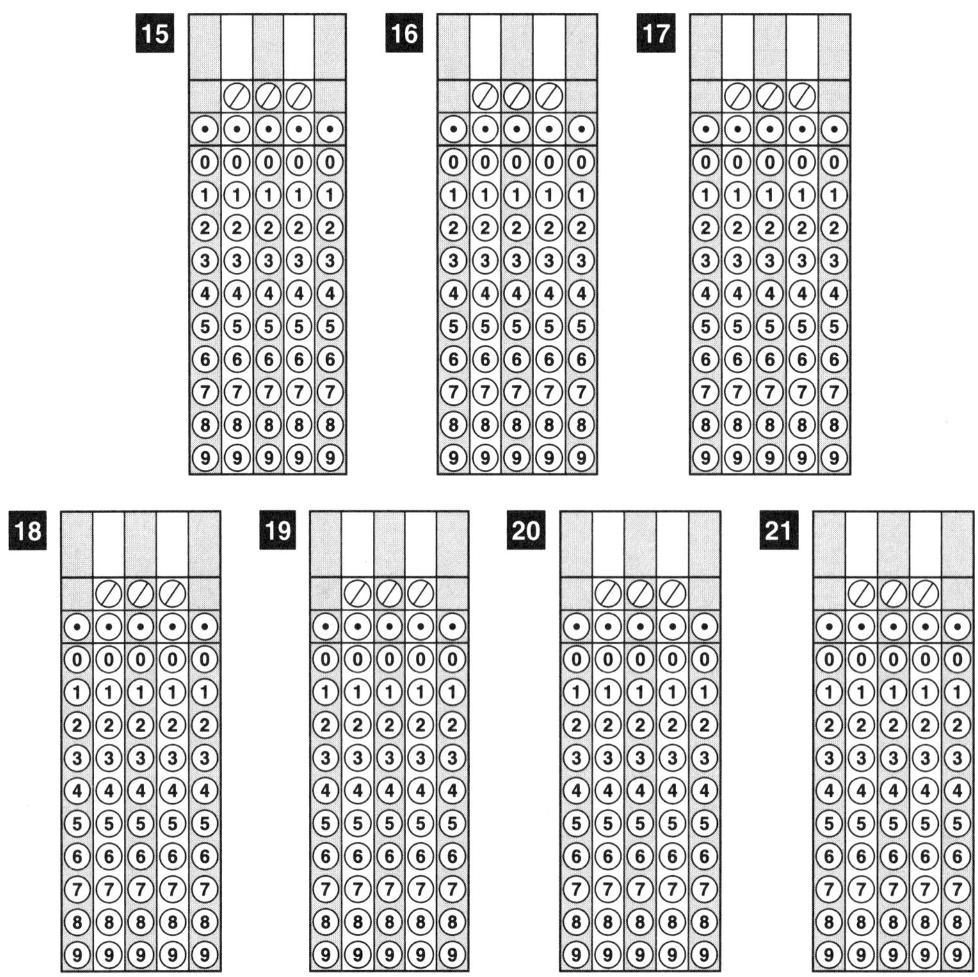

Think, Solve, Explain: Short-Response Problems

22. Two water pipes have the same length. Their diameters are 8 inches and 10 inches. They are to be replaced with one pipe that has the same length and the same capacity as the two pipes together. What size diameter is needed for the new pipe? Express your answer to the nearest hundredth. Show all necessary work.

23. The legs (equal sides) of an isosceles triangle are each 5 inches longer than the base. The perimeter of the triangle is 37 meters.

Part A Write an equation that can be used to find the lengths of the sides of the triangle.

Part B Solve the equation.

Part C Find the lengths of all three sides of the triangle.

Think, Solve, Explain: Extended-Response Problems

24. The surface area of the larger of two cones is 4 times larger than the surface area of the smaller cone.

Part A The volume of the larger cone is 16π cubic inches. What is the volume of the smaller cone?

Part B The volume of a cylinder with a 2.5-inch radius is the same as the volume of the larger cone. What is the height of the cylinder? Show all necessary work.

25. The Smith Company has been in business for 4 years. Every year its profits have increased, as shown in this table.

Year	Profits
1	\$230,000
2	\$275,000
3	\$320,000
4	\$370,000

Part A Construct a line graph to display these data.

Part B Use your graph to predict the profits in years 5 and 6 if the trend shown continues.

ANSWER SHEET: TEST 1

NOTE: When you answer performance tasks on your own paper, use graph paper if necessary.

PART I

1 Ⓐ Ⓑ Ⓒ Ⓓ

2 Ⓕ Ⓖ Ⓗ Ⓘ

3 Ⓐ Ⓑ Ⓒ Ⓓ

4 Ⓕ Ⓖ Ⓗ Ⓘ

5

6

7 Use your own paper.

8 Ⓐ Ⓑ Ⓒ Ⓓ

9 Ⓕ Ⓖ Ⓗ Ⓘ

10 Ⓐ Ⓑ Ⓒ Ⓓ

11 Ⓕ Ⓖ Ⓗ Ⓘ

12 Ⓐ Ⓑ Ⓒ Ⓓ

13 Ⓕ Ⓖ Ⓗ Ⓘ

14

15

16

17

18 Use your own paper.

19 Ⓐ Ⓑ Ⓒ Ⓓ

20 Ⓕ Ⓖ Ⓗ Ⓘ

21 Ⓐ Ⓑ Ⓒ Ⓓ

22

23

24

25 Use your own paper.

26 Ⓕ Ⓖ Ⓗ Ⓘ

27 Ⓐ Ⓑ Ⓒ Ⓓ

28 Ⓕ Ⓖ Ⓗ Ⓘ

29

30

PART II

1 Ⓐ Ⓑ Ⓒ Ⓓ
2 Ⓕ Ⓖ Ⓗ Ⓘ
3 Ⓐ Ⓑ Ⓒ Ⓓ
4 Ⓕ Ⓖ Ⓗ Ⓘ
5 Ⓐ Ⓑ Ⓒ Ⓓ
6 Ⓕ Ⓖ Ⓗ Ⓘ
7 Ⓐ Ⓑ Ⓒ Ⓓ
8 Use your own paper.
9
10
11
12 Ⓕ Ⓖ Ⓗ Ⓘ
13 Ⓐ Ⓑ Ⓒ Ⓓ
14 Ⓕ Ⓖ Ⓗ Ⓘ
15 Use your own paper.
16
17
18
19
20 Ⓐ Ⓑ Ⓒ Ⓓ
21 Ⓕ Ⓖ Ⓗ Ⓘ
22 Ⓐ Ⓑ Ⓒ Ⓓ
23 Use your own paper.
24
25
26
27
28 Ⓕ Ⓖ Ⓗ Ⓘ
29 Ⓐ Ⓑ Ⓒ Ⓓ
30 Ⓕ Ⓖ Ⓗ Ⓘ

TEST 1

PART I

Multiple-Choice Problems

1. The number 4.2×10^2 is located between what two numbers?

   A. 419 and 421 **C.** 4 and 5

   B. 41 and 43 **D.** 0 and 1

2. The electric company attached a steel cable to a pole for support. The engineer used the Pythagorean theorem to determine that the length of the cable is $7\sqrt{11}$ feet. Which of the following numbers is closest to the cable's length?

   F. $23\frac{1}{4}$ feet **H.** $23\frac{2}{7}$ feet

   G. $23\frac{1}{5}$ feet **I.** $23\frac{2}{9}$ feet

3. Which of the following expressions is equivalent to $3x + 7(x - 2) - 6x(x + 4)$?

   A. $-6x^2 - 9$

   B. $-6x^2 - 14x - 14$

   C. $-6x^2 + 34x - 2$

   D. $-6x^2 + 34x - 14$

4. To celebrate the school's 10th anniversary, the principal ordered a big cake. In the drawing below, each block in the grid represents a piece of cake that can feed approximately 18 people. How many people will the cake feed?

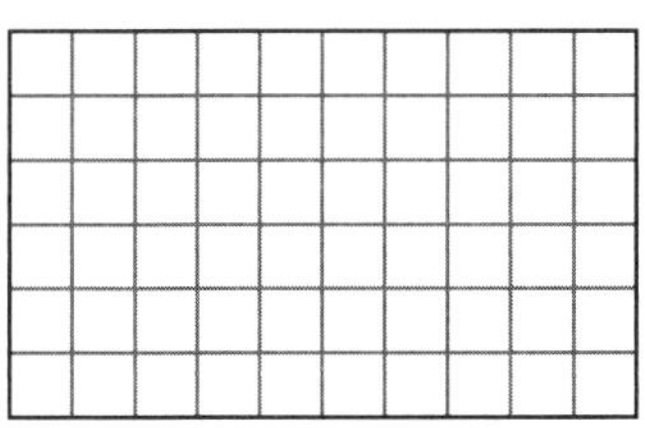

   F. 950

   G. 1,075

   H. 1,250

   I. 1,400

Gridded-Response Problems

5. In order to regain a larger market share of auto sales, a car company decides to reduce the price of one of its cars from \$22,350 to \$19,221. What is the percent of decrease?

6. Jeron's doctor wants him to drink eight 10-ounce glasses of water a day for the next 5 days. At the end of the 5 days, how many gallons of water will he have consumed?

Think, Solve, Explain: Short-Response Problem

7. A water-filled cone with a base radius of 8 inches and a height of 9 inches is poured into a cylinder. The water level rises to a depth of 4 inches. What is the radius of the base of the cylinder expressed to the nearest integer? Show all necessary work.

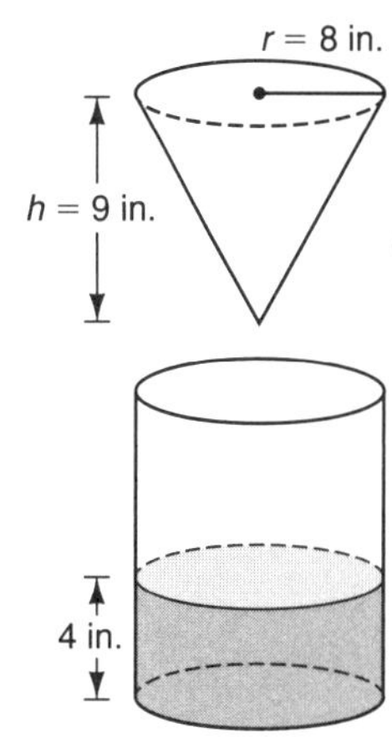

Multiple-Choice Problems

8. In the drawing below, what type of angle is $\angle x$?

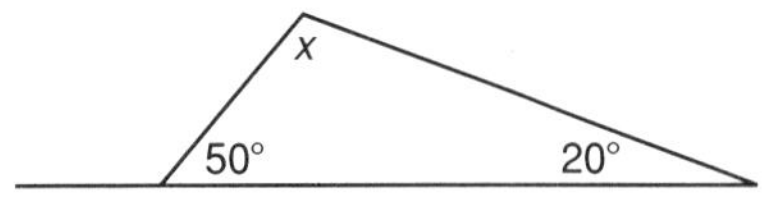

A. acute

B. right

C. obtuse

D. straight

9. The figure below is composed of a rectangle, a square, and an equilateral triangle. What is the perimeter of the compound figure?

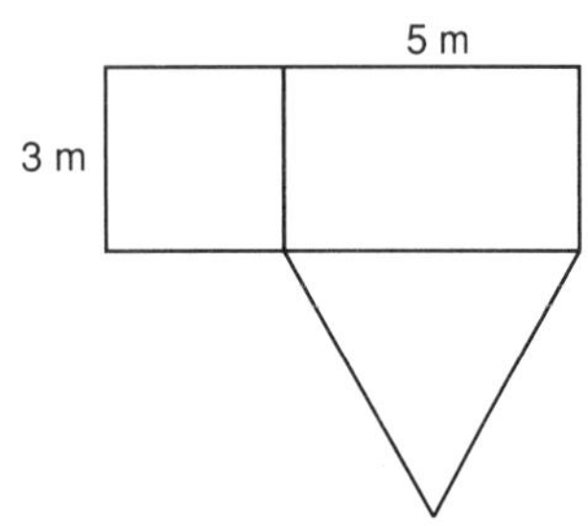

F. 21 meters

G. 27 meters

H. 32 meters

I. 35 meters

10. Shanghai, China, is located at 31.2° N latitude and London, UK, is located at 51.5° N latitude. What is the angle measurement between these two cities?

A. 20.3°

B. 31.2°

C. 51.5°

D. 82.7°

11. What is the midpoint of the line segment with endpoints of $(-2, -7)$ and $(4, 5)$?

F. $(-3, -1)$

G. $(1, -1)$

H. $(1, -6)$

I. $(-3, -6)$

12. In rectangle $ABCD$ shown below, diagonal $DB = 15$ and $BC = 11$. What is the length of AB?

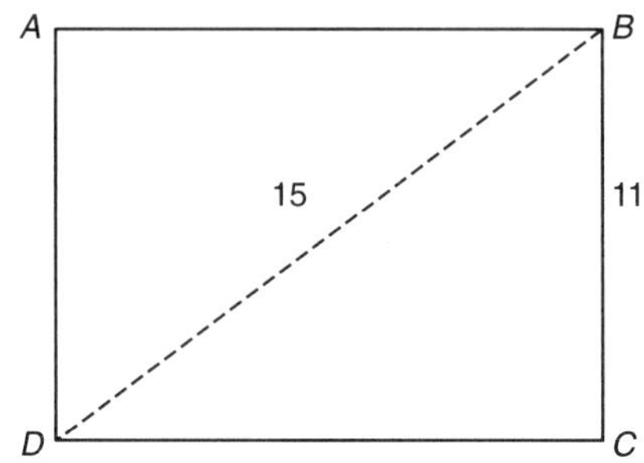

A. 3.7

B. 5.1

C. 10.2

D. 18.6

13. Quadrilateral $MNPQ$ is a rhombus and $MN = 2x + 1$, $QN = 3x - 11$, and $MQ = x + 13$. What is the value of x?

F. 1

G. 2.4

H. 6

I. 12

Gridded-Response Problems

14. In the figure shown below $\angle ACB = 60°$. What is the measure of $\angle DCE$?

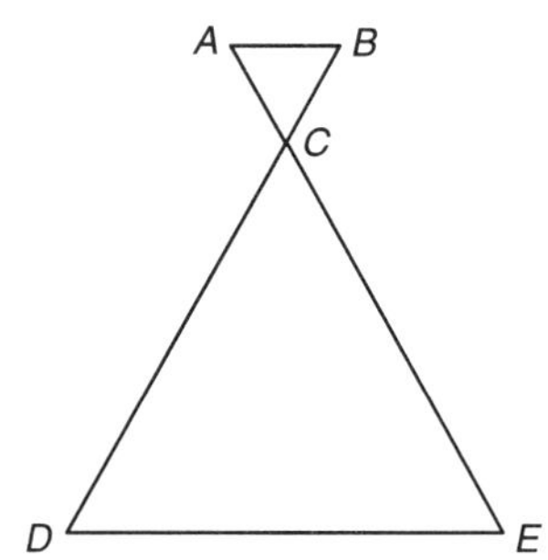

15. Q is a point on $\overline{PR}$ such that $PQ{:}QR = 2{:}3$. If $QR = 9$, what is the length of PR?

16. Using the triangle below, what is the length of side AB? Round the answer to the nearest hundredth.

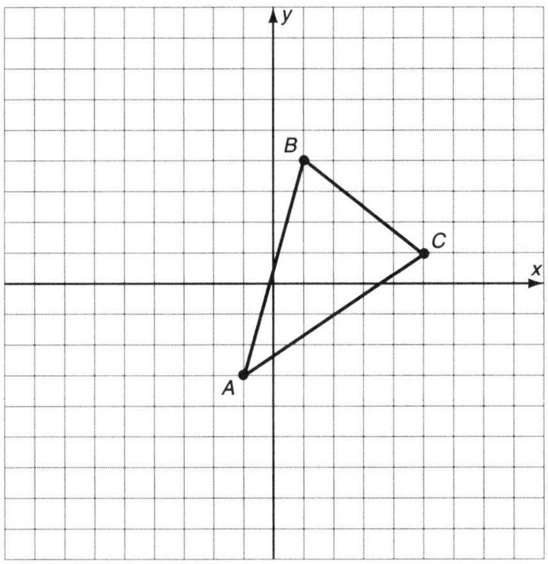

17. The drawing of a living room has a scale of 1 inch = 2 feet. In the drawing, the length of the fireplace is 2.5 inches. What is the actual length of the fireplace?

Think, Solve, Explain: Extended-Response Problem

18.

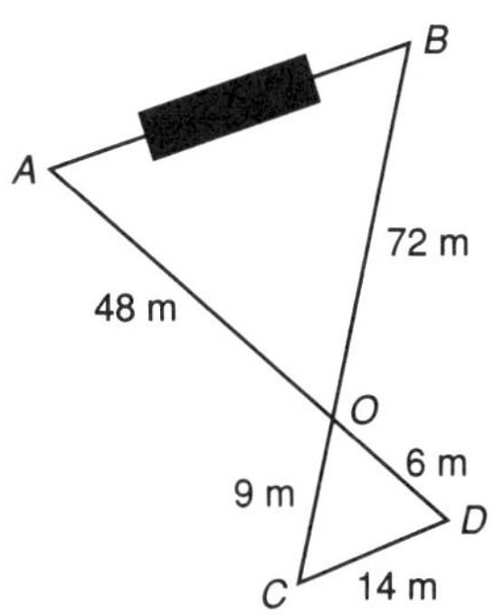

Part A Explain why $\triangle ABO \sim \triangle DCO$.

Part B Set up a proportion to find the distance between A and B.

Part C Solve the proportion to find the length of AB.

Multiple-Choice Problems

19. Jeri has a window that is 3 feet wide by 4 feet high. She wants to replace it with a window that is one foot wider and has twice as much area. What must be the new dimensions of the window?

A. 3 feet by 6 feet

B. 3 feet by 8 feet

C. 4 feet by 6 feet

D. 4 feet by 8 feet

20. In the Fibonacci sequence, the first number is repeated, then each number after that is the sum of the two previous numbers: 1, 1, 2, 3, 5, 8, ... What is the sum of the first ten terms?

F. 88 **H.** 143

G. 89 **I.** 232

21. A dress designer is offering stores a purchase deal on two lines of clothes. One deal is the purchase of 8 dresses and 3 sweaters for \$286. The other deal is the purchase of 5 dresses and 6 sweaters for \$253. This information can be represented by the system of equations, where d represents the cost of a dress and s represents the cost of a sweater.

$8d + 3s = 286$

$5d + 6s = 253$

What is the cost of one dress and one sweater?

A. dress \$26 and sweater \$26

B. dress \$29 and sweater \$18

C. dress \$23 and sweater \$34

D. dress \$36.20 and sweater \$12

Gridded-Response Problems

22. In the function $f(x) = x^2 + 1$, what is the value of $f(2)$?

23. In the geometric sequence $-\frac{1}{512}, \frac{1}{64}, -\frac{1}{8}, 1, -8, \ldots$, what is the eighth term?

24. The company uses a freight elevator to move the auto parts from the production floor to the loading dock. The freight elevator has a 2,000-pound limit. If the operator weighs 210 pounds and each box weighs 74 pounds, what is the maximum number of boxes, along with the operator, that the elevator can hold?

Think, Solve, Explain: Short-Response Problem

25. The teacher asked the class to draw a graph representing a child swinging on a swing. Jan and Don both turned in their graphs. Which one is correct? Explain your choice and make sure to include the reasons why the other one is incorrect.

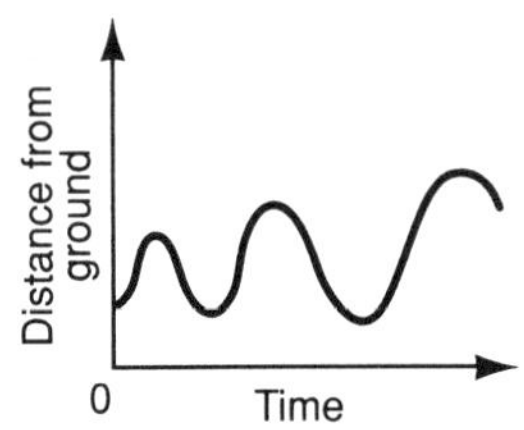

Jan's graph

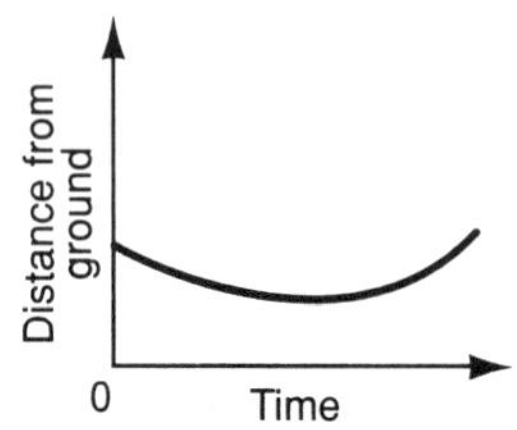

Don's graph

Multiple-Choice Problems

26. The scores of the ten top golfers after the first round of the state tournament are as follows: 73, 80, 78, 76, 73, 78, 75, 79, 75, and 73. What is the mode of the scores?

    F. 73

    G. 75

    H. 78

    I. It has no mode.

27. Jamie surveyed all the people in the lawn and garden center at High's store on a Saturday afternoon. He reported to his boss that out of the 50 people surveyed, 40 have a gas lawn mower. Based on his survey, he concluded that the majority of people in town own a gas lawn mower. Which best describes his conclusion?

    A. Not a valid conclusion because of the location of the survey.

    B. Not a valid conclusion because it should have been done by a professional survey company.

    C. Not a valid conclusion because the sample size is too small.

    D. Valid conclusion.

28. Natasha has a bag with 3 red marbles, 2 blue marbles, and 1 green marble. She reaches in and selects a marble at random, notes the color, puts the marble back in the bag, and selects another marble. What is the probability that she will select 2 red marbles or 2 blue marbles?

    F. $\frac{13}{36}$

    G. $\frac{8}{30}$

    H. $\frac{1}{4}$

    I. $\frac{1}{9}$

Gridded-Response Problems

29. In a class of 15 students, three of the 7 boys have the first name of Robert. What are the odds that a Robert will be chosen first to answer a question?

30. The histogram at the right displays the scores on the last American history exam. What percent of the students received a passing grade of 70% or higher? Express the answer to the nearest whole percent.

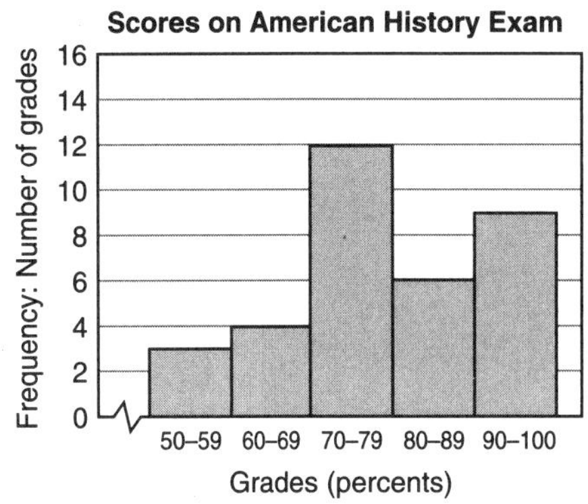

PART II

Multiple-Choice Problems

1. During one week the price of gasoline changed on a daily basis. Sunday, regular unleaded sold for \$3.15 a gallon. On Monday the price increased by 5 cents a gallon. On Tuesday there was an increase of 3 cents a gallon, while on Wednesday the price decreased by 4 cents a gallon. On Thursday the price increased by 7 cents, on Friday it decreased by 2 cents, and on Saturday it increased by 1 cent. At the end of the week, what was the price of a gallon of regular unleaded gasoline?

A. \$3.22 **C.** \$3.31

B. \$3.25 **D.** \$3.37

2. In a basketball game, Marcia made x three-point baskets and y two-point baskets. Which of the following expressions represents her total points scored?

F. $(3 \bullet 2)(x + y)$ **H.** $3y + 2x$

G. $(3 + 2)(x + y)$ **I.** $3x + 2y$

3. The four members of the Keller family are going to Pasta Prime restaurant for dinner. The special on the menu states that a family of four can eat for \$25. Mr. Keller knows that he must pay a 6% sales tax on the cost of the meal. He also pays a 15% tip on the meal before the tax is added. What will be his total cost?

A. \$26.50 **C.** \$30.25

B. \$28.75 **D.** \$30.50

4. What is the value of $\dfrac{2 - 17 \bullet 5 - 3 \bullet 5}{3^2 + 11}$?

F. $-\dfrac{60}{17}$

G. $-\dfrac{9}{2}$

H. $-\dfrac{49}{10}$

I. $-\dfrac{98}{17}$

5. William's grandmother cuts a piece of pie with a central angle of 60°. What is the length of arc x if the radius of the pie is 3 inches?

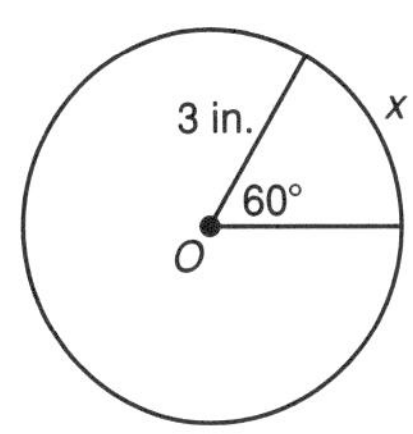

A. 0.789 inch

B. 1.57 inches

C. 3.14 inches

D. 6.28 inches

6. The area in front of Martha's house is in the shape of a trapezoid with an area of 33 square feet. The parallel sides are 5 feet and 7 feet. What is the distance between these sides?

F. 2.75 feet **H.** 4 feet

G. 3 feet **I.** 5.5 feet

7. Lucas and his sister Sylvia each use approximately 8 ounces of milk on their cereal. If this is the only use of the milk, approximately how many days will a gallon of milk last?

A. 16 days **C.** 8 days

B. 12 days **D.** 6 days

Think, Solve, Explain: Short-Response Problem

8. In order to increase the number of parking spaces available, the electric company has agreed to move its pole located at point A so that the poles are in the shape of a square.

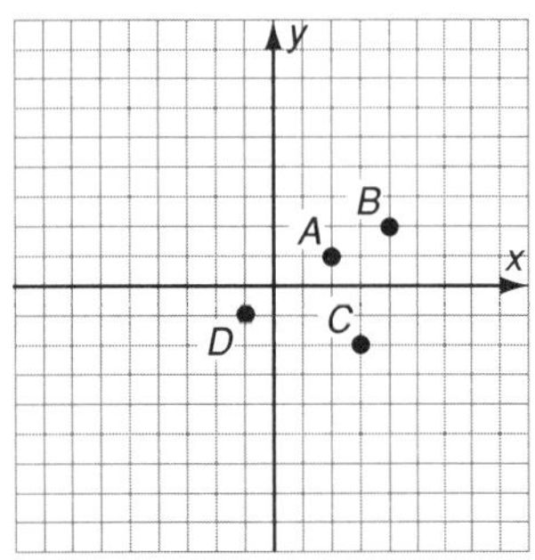

Part A Plot the new location for the pole and write its new coordinates. Label the new location A'.

Part B Use the definition or properties of a parallelogram to verify that the poles form a square. You must use the slopes of the sides, the lengths of the sides, or both to verify your answer.

Gridded-Response Problems

9. Alexandria is completing some calculations for her science project. What is the value of the following expression?

$(-3.1)^2 - (0.4)^3$

10. What is the perimeter of an equilateral triangle that has a height of 8 centimeters and an area of 14 square centimeters?

11. What is the volume of a sphere that has a surface area of 154 square meters? Use $\pi \approx \frac{22}{7}$ and round your answer to the nearest tenth.

Multiple-Choice Problems

12. Two angles are complementary. The measure of the larger angle is five times the measure of the smaller angle. What is the measure of the larger angle?

F. 15° **H.** 75°

G. 30° **I.** 150°

13. Pentagon $ABCDE \cong$ pentagon $FGHIJ$. The measure of $\angle BCD$ is equal to the measure of which of the following angles?

A. $\angle CDE$ **C.** $\angle FGH$

B. $\angle GHI$ **D.** $\angle ABC$

14. A line AB contains the points A $(-3, 2)$ and B $(4, -1)$. What is the slope of the line that is perpendicular to line AB?

F. $-\frac{7}{3}$ **H.** $\frac{3}{7}$

G. $-\frac{3}{7}$ **I.** $\frac{7}{3}$

Think, Solve, Explain: Extended-Response Problem

15. Mr. Check, the gym teacher, is checking on the physical fitness of 10 of his students. He prepared a scatter plot to show the number of push-ups completed in September and then again in December. The points represent the number of push-ups of each student.

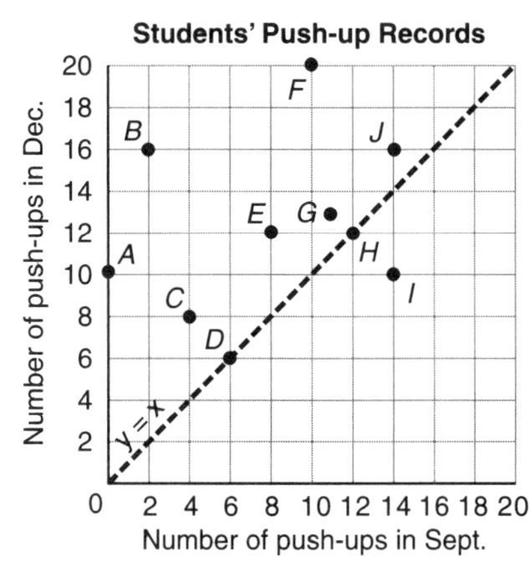

Part A Determine the mean number of push-ups done in September. Show the calculations you used to get your answer.

Part B What can be said about students *D* and *H*?

Part C Which student showed the greatest improvement? Explain.

Part D Mr. Check feels that his students improved. Is he correct? Explain why or why not.

Gridded-Response Problems

16. In the drawing below, what is the length of the median *EF*?

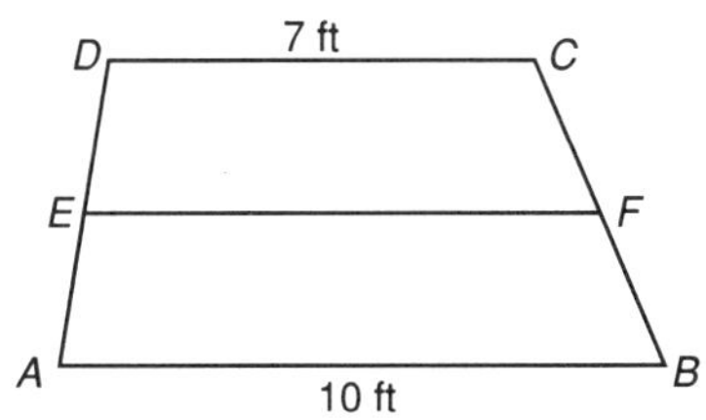

17. In the drawing below, the diameter of the circle is 20 centimeters and angle $CAB = 60°$. What is the length of *AC*?

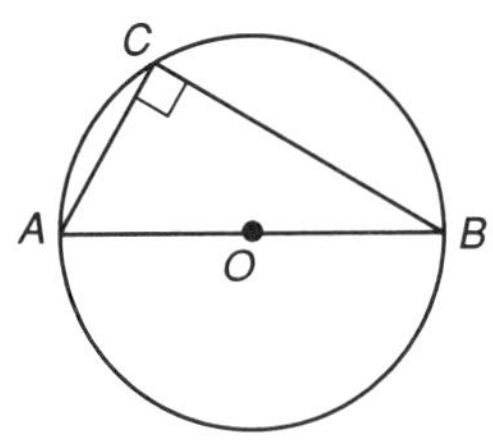

18. Cylinders I and II shown below are similar figures. The surface areas are in the ratio of 9 to 4. If the volume of cylinder II is 64 cubic inches, what is the volume of cylinder I?

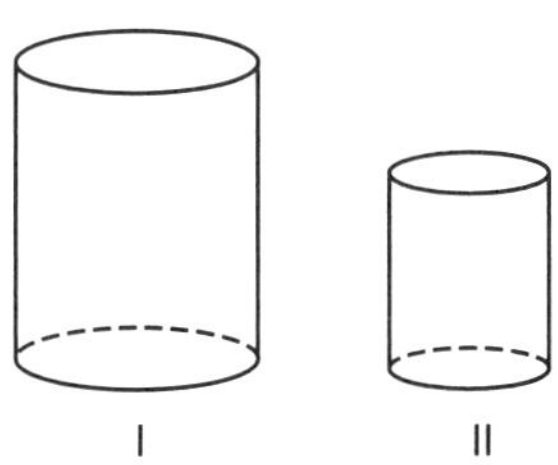

19. The movie theater has been showing the movie *Pi Makes the World Go Round*. The table shows the attendance for the past few weeks. The movie will no longer be shown when the attendance drops below 150 people. If the pattern continues, how many more weeks will the movie be shown?

Week	Attendance
1	500
2	440
3	380
4	320

Multiple-Choice Problems

20. Which of these is not a function?

A. $x + y = 3$

B. $x = y$

C. $x = 3$

D. $y = 3$

21. David and Kat rowed their canoe upstream against the current. They can row x miles per hour in still water, and the current's average rate is y miles per hour. It took 2 hours to row $1\frac{1}{8}$ miles. Which of the following represents this information?

F. $1\frac{1}{8}(x + y) = 2$

G. $1\frac{1}{8}(x - y) = 2$

H. $2(x + y) = 1\frac{1}{8}$

I. $2(x - y) = 1\frac{1}{8}$

22. It took Skyler $2\frac{1}{2}$ hours to drive home at an average speed of 60 miles per hour. Which of these graphs best represents Skyler's drive home?

A.

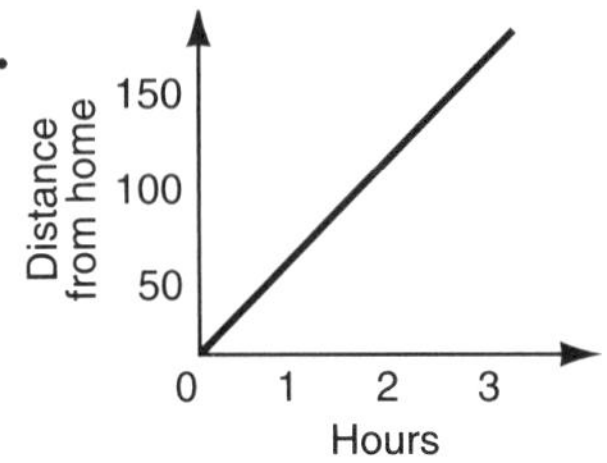

B.

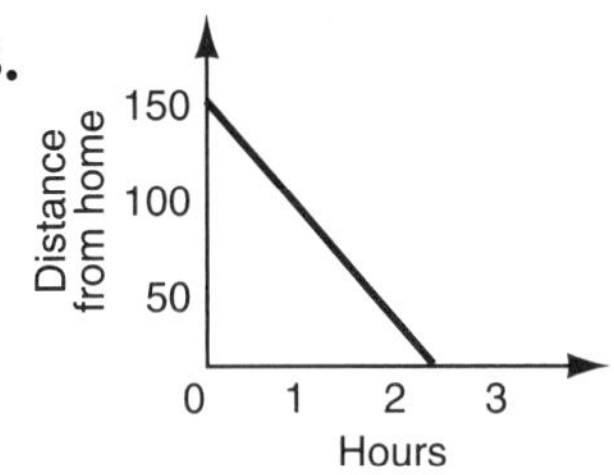

C.

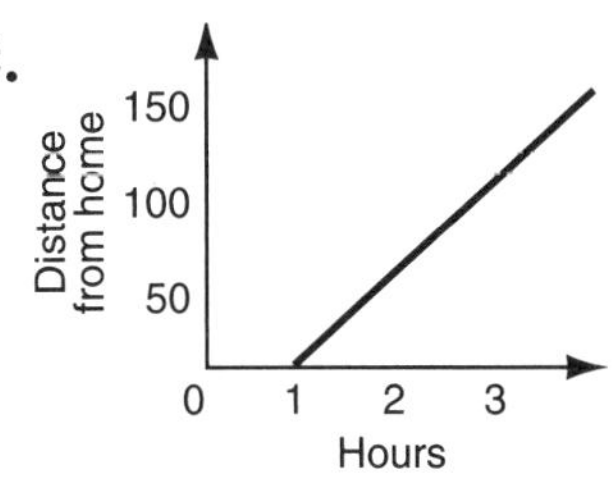

D.

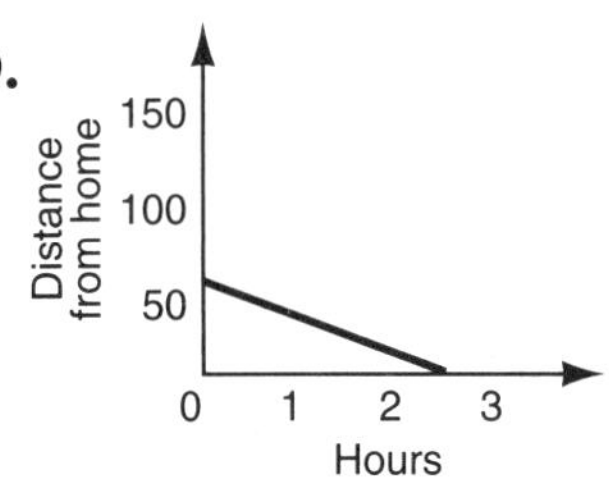

Think, Solve, Explain: Short-Response Problem

23. Ann has a right circular cylinder vase with dried flowers sitting on her shelf. The vase is 8 inches tall and has a diameter of 4 inches. She saw a similar vase in the store that also has a height of 8 inches, but the diameter is half the diameter of her vase. What is the ratio of the volume of her vase to the vase in the store? Show all necessary work.

Gridded-Response Problems

24. The geometric sequence of $-1, 3, -9, 27, \ldots$, has a common ratio of -3. What is the sum of the first six terms?

25. What is the value of x in the equation below?

$$\frac{1}{2}(x + 16) = 3(x + 2)$$

26. What is the smallest value of x in the solution set of the inequality shown below?

$$6 - 3x \leq -15$$

27. What is the median of the following data set: 0, 0, 1, 2, 3, 5, 8, 11, 13, 13, 13, 14, 15, 15?

Multiple-Choice Problems

28. The salaries per week of people who work at the Sandwich Shoppe are \$75, \$220, \$89, \$84, \$89, \$75, \$75, \$79, \$80, and \$84. The owner wants to hire another employee. She advertised that the average salary is \$95. Which measure of central tendency is she using?

F. mean

G. median

H. mode

I. range

29. A table of the products of two numbers is shown below. What is the probability that the product of two numbers, one odd and one even, selected at random will be less than 11?

x	**0**	**2**	**4**
1	0	2	4
3	0	6	12
5	0	10	20

A. $\frac{4}{9}$ **C.** $\frac{7}{9}$

B. $\frac{5}{9}$ **D.** $\frac{8}{9}$

30. The Venn diagram displays the participation in sports of girls at Alligator High School. What percent of those surveyed play all three sports?

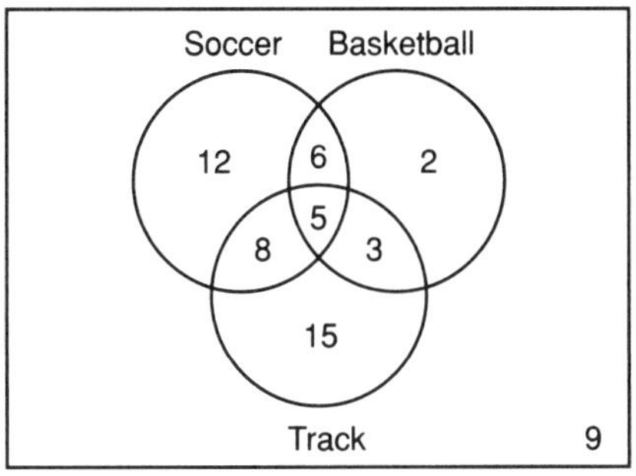

F. 8%

G. 10%

H. 37%

I. 43%

ANSWER SHEET: TEST 2

NOTE: When you answer performance tasks on your own paper, use graph paper if necessary.

PART I

1 Ⓐ Ⓑ Ⓒ Ⓓ

2 Ⓕ Ⓖ Ⓗ Ⓘ

3 Ⓐ Ⓑ Ⓒ Ⓓ

4

5

6

7 Ⓕ Ⓖ Ⓗ Ⓘ

8 Ⓐ Ⓑ Ⓒ Ⓓ

9 Ⓕ Ⓖ Ⓗ Ⓘ

10 Ⓐ Ⓑ Ⓒ Ⓓ

11 Ⓕ Ⓖ Ⓗ Ⓘ

12 Use your own paper.

13

14

15

16 Ⓐ Ⓑ Ⓒ Ⓓ

17 Ⓕ Ⓖ Ⓗ Ⓘ

18 Ⓐ Ⓑ Ⓒ Ⓓ

19 Use your own paper.

20 Ⓕ Ⓖ Ⓗ Ⓘ

21 Ⓐ Ⓑ Ⓒ Ⓓ

22 Ⓕ Ⓖ Ⓗ Ⓘ

23 Use your own paper.

24

25

26

27

28 Ⓐ Ⓑ Ⓒ Ⓓ

29 Ⓕ Ⓖ Ⓗ Ⓘ

30 Ⓐ Ⓑ Ⓒ Ⓓ

PART II

1

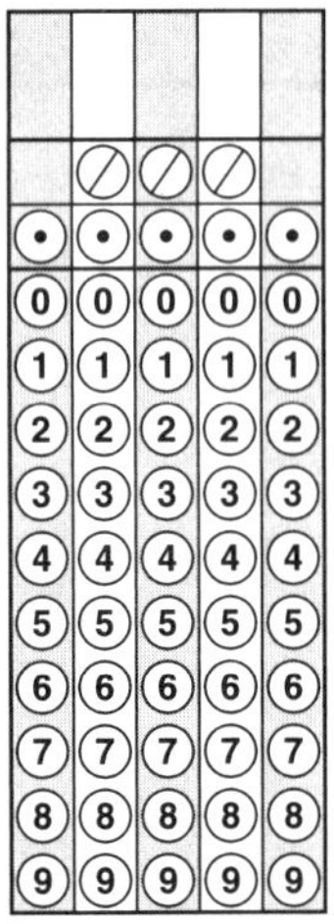

2

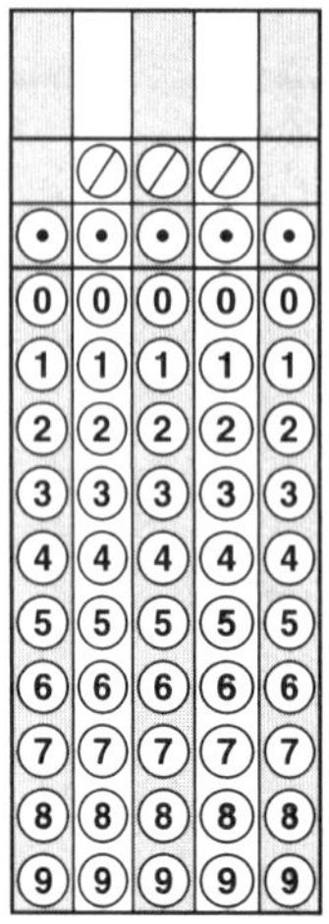

3

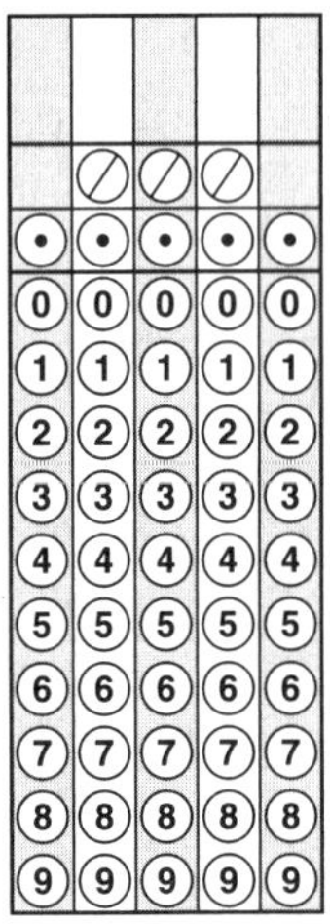

4 Ⓐ Ⓑ Ⓒ Ⓓ

5 Ⓕ Ⓖ Ⓗ Ⓘ

6 Use your own paper.

7 Ⓐ Ⓑ Ⓒ Ⓓ

8 Ⓕ Ⓖ Ⓗ Ⓘ

9 Ⓐ Ⓑ Ⓒ Ⓓ

10 Ⓕ Ⓖ Ⓗ Ⓘ

11 Ⓐ Ⓑ Ⓒ Ⓓ

12

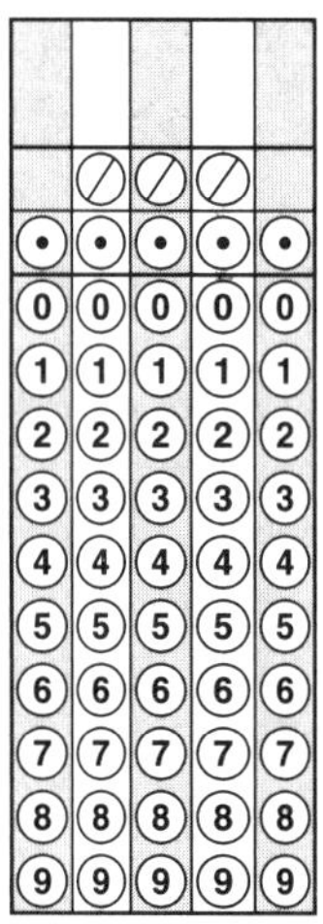

13

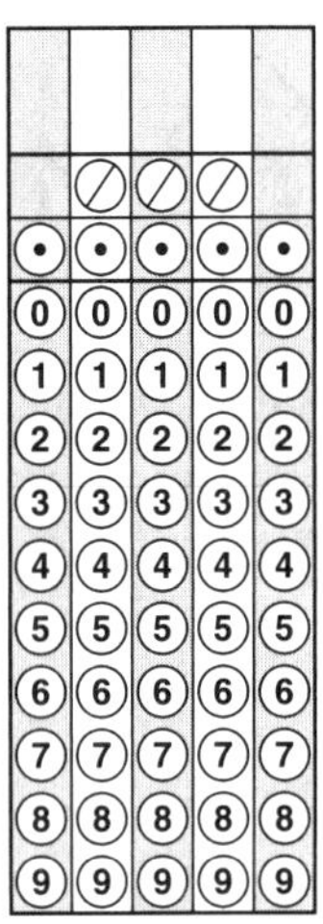

14

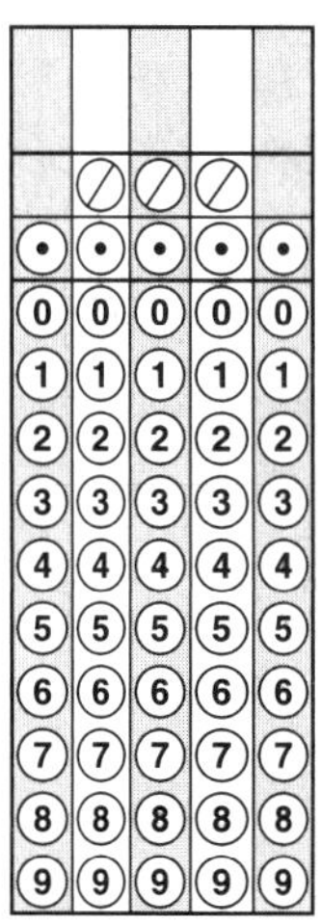

15 Ⓕ Ⓖ Ⓗ Ⓘ

16 Ⓐ Ⓑ Ⓒ Ⓓ

17 Use your own paper.

18

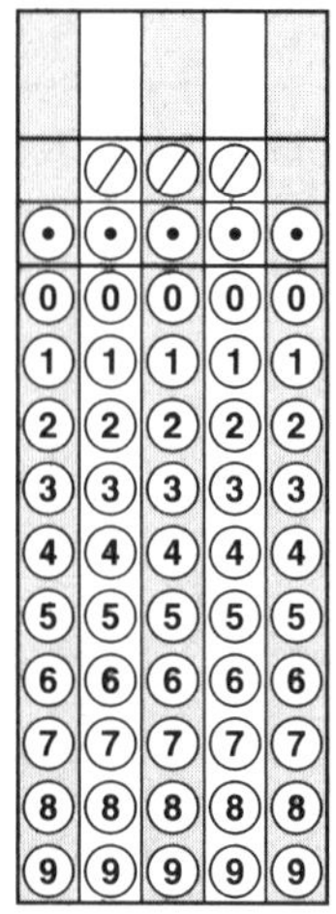

19

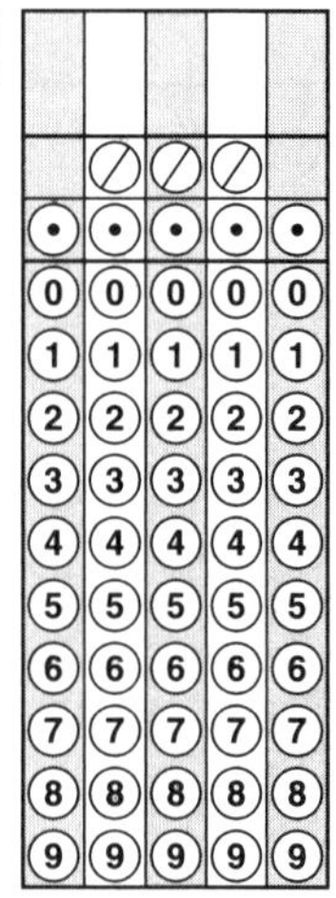

20 Ⓕ Ⓖ Ⓗ Ⓘ

21 Ⓐ Ⓑ Ⓒ Ⓓ

22 Ⓕ Ⓖ Ⓗ Ⓘ

23 Use your own paper.

24

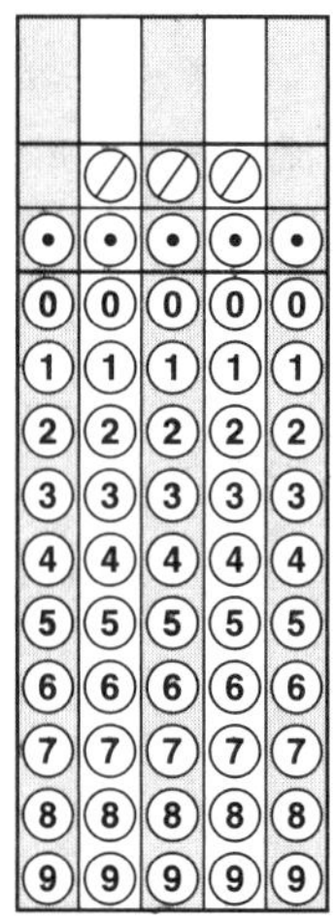

25

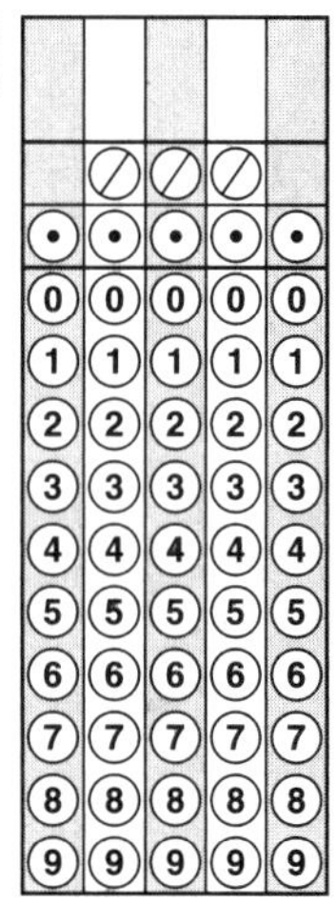

26

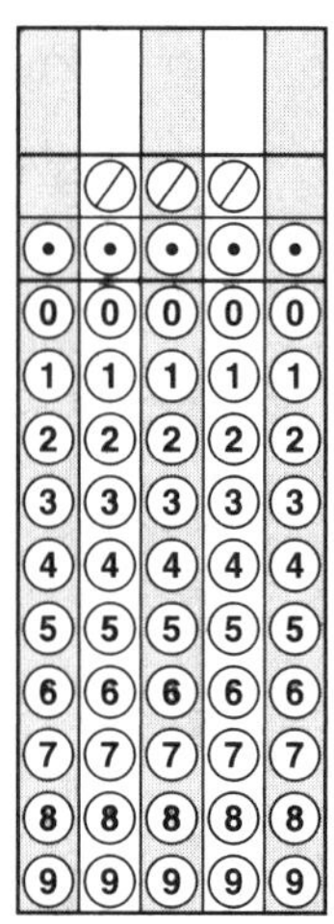

27 Ⓐ Ⓑ Ⓒ Ⓓ

28 Ⓕ Ⓖ Ⓗ Ⓘ

29 Ⓐ Ⓑ Ⓒ Ⓓ

30 Ⓕ Ⓖ Ⓗ Ⓘ

TEST 2

PART I

Multiple-Choice Problems

1. How many degrees are in arc x?

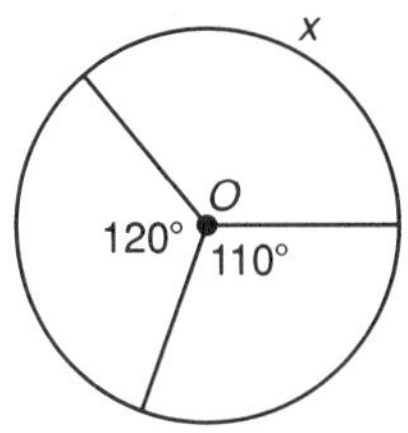

A. 90° **C.** 180°

B. 130° **D.** 230°

2. Byron wants to put tile in $\frac{1}{4}$ of the entryway in his house. How many square feet of tile will he need?

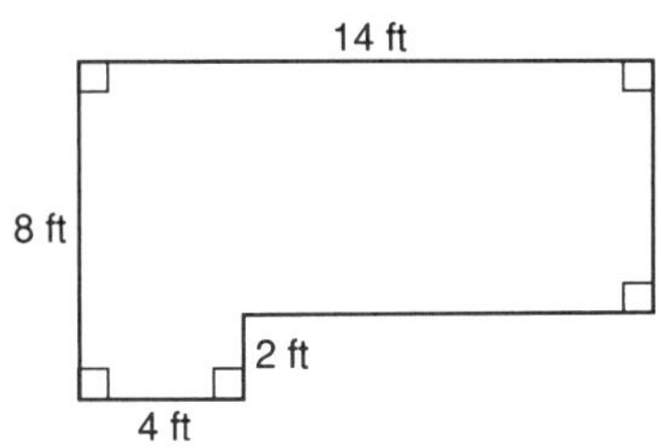

F. 23 square feet

G. 28 square feet

H. 92 square feet

I. 112 square feet

3. Melissa's favorite recipe requires 3 ounces of lemon concentrate. The recipe serves 4 people. How many quarts of lemon concentrate will she need for a party of 64 people?

A. 1 quart **C.** 2.25 quarts

B. 1.5 quarts **D.** 6 quarts

Gridded-Response Problems

4. Gladys types 140 words in 2.5 minutes. What is her rate in words per minute?

5. An orange has a diameter of 10 centimeters. Marshall cuts the orange in half, forming a hemisphere (half a sphere). What is the total surface area of the orange half? Use $\pi \approx 3.14$.

6. A right triangle is shown below. The ratio of the base of the triangle to the hypotenuse is $\frac{3\sqrt{6}}{6\sqrt{2}}$. What is the value of $\frac{3\sqrt{6}}{6\sqrt{2}}$? Round your answer to the nearest thousandth.

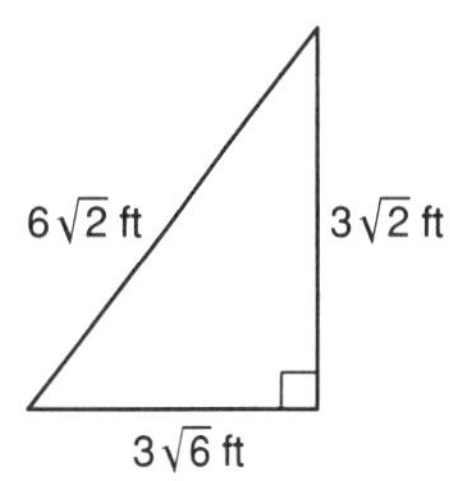

Multiple-Choice Problems

7. Each ship that passes through the Panama Canal requires about 52 million gallons of water to help it cross from the Atlantic to the Pacific Ocean. If 27 ships move through the canal, approximately how many gallons of water are needed?

F. 1.4×10^8

G. 1.4×10^9

H. 7.9×10^8

I. 7.9×10^9

8. After the Jacksonville Jaguars made the 2006 playoffs, shirts were selling for 20% more than the previous price. If the shirts sold for \$36 each after the playoffs, what was the original price of a shirt?

A. \$25 **C.** \$30

B. \$28.80 **D.** \$34

9. If $x = -3$, $y = \frac{1}{2}$, and $z = 2$, what is the value of $3x - 4z + 4y^2$?

F. -16 **H.** 2

G. -1 **I.** 5

10. Sean took his family to the Royal Dragon restaurant. The bill was $75 before the tax was added. Sean had a coupon for $10 off. A 6% tax is added to the bill after the coupon is deducted. Sean leaves a 20% tip based on his final bill including tax. How much did it cost to take his family out for dinner?

A. $95.40 **C.** $82.68

B. $94.50 **D.** $78.32

11. *ABCD* is a rectangle with *AB* = 12 inches, *BC* = 5 inches, and *AC* = 13 inches. How long is *BD*?

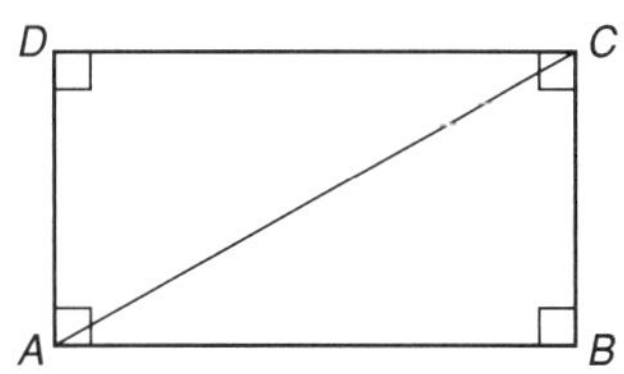

F. 5 inches **H.** 13 inches

G. 12 inches **I.** 17 inches

Think, Solve, Explain: Extended-Response Problem

12. During a sale, Ms. Chin was asked to reduce the price of certain items. On Wednesday morning she marked an $80 dress down 10%. On Thursday morning she marked the price of the same dress down another 20%. On Friday morning she marked the price of the same dress down another 40%.

Part A Determine the price of the dress on Wednesday, Thursday, and Friday.

Part B What is the total percentage change of the dress from Tuesday to Friday?

Part C Make a line graph to show the price of the dress at the start of each day from Tuesday through Friday. Be sure to include a title for the graph, a label for each axis, appropriate and consistent scales, and the line itself.

Gridded-Response Problems

13. A 9-foot board is used to help support a palm tree until its roots get established. If the board forms a 30° angle with the trunk of the tree, how far from the base of the tree is the bottom of the board?

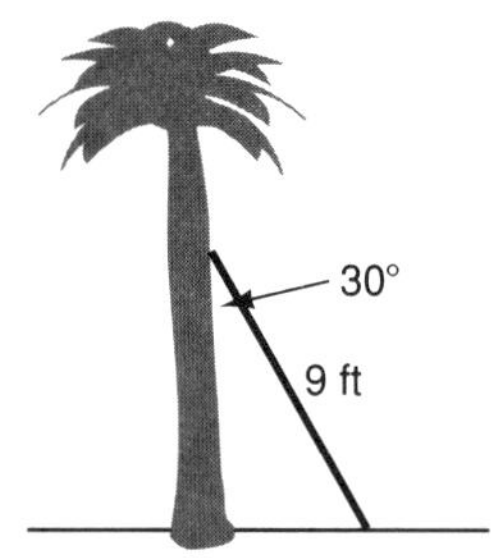

14. In the drawing below, $\triangle ABC \sim \triangle DEF$. How long is side *DF*?

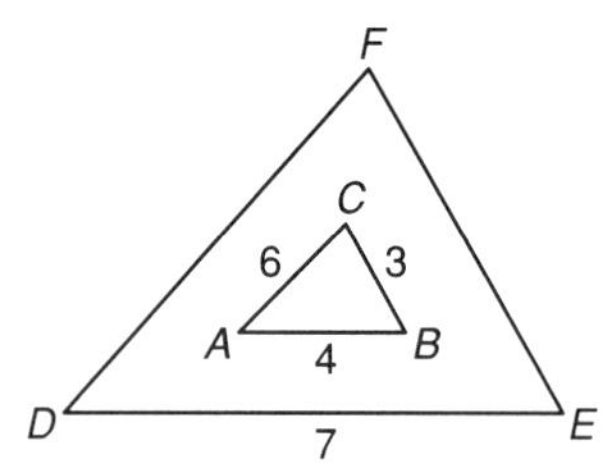

15. What is the slope of a line that contains the points (1, 6) and (−3, −3)?

Multiple-Choice Problems

16. Pentagon *ABCDE* ~ pentagon *RSTVW*. What is the ratio of the area of *ABCDE* to *RSTVW*?

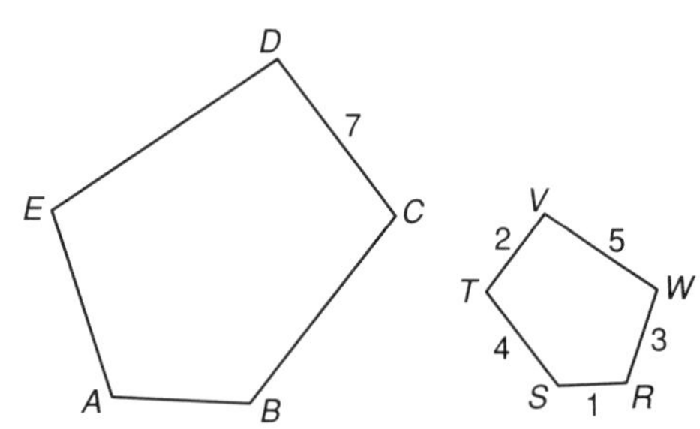

A. $\frac{7}{2}$ **C.** $\frac{49}{4}$

B. $\frac{7}{5}$ **D.** $\frac{49}{25}$

17. In the drawing below, $\overline{DE} \parallel \overline{AB}$. Which of the following proportions can be used to find $\overline{EB}$?

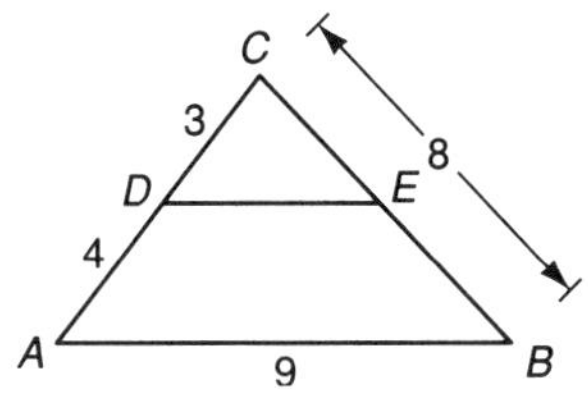

F. $\frac{3}{4} = \frac{EB}{8}$

G. $\frac{4}{9} = \frac{EB}{8}$

H. $\frac{3}{7} = \frac{EB}{8}$

I. $\frac{4}{7} = \frac{EB}{8}$

18. Which of the following shows the result of $\triangle ABC$ reflected first over the x-axis and then over the y-axis?

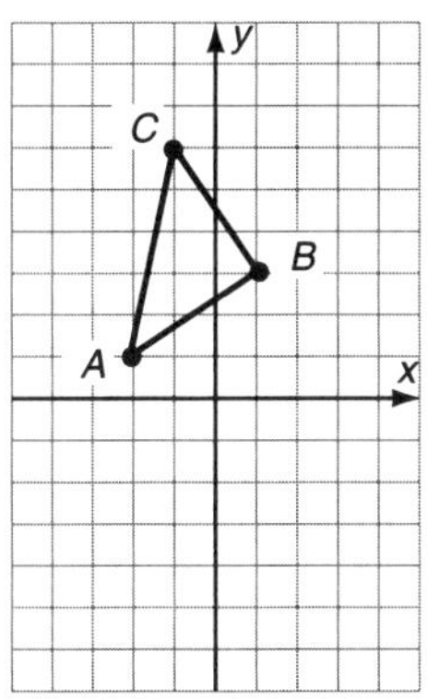

A.

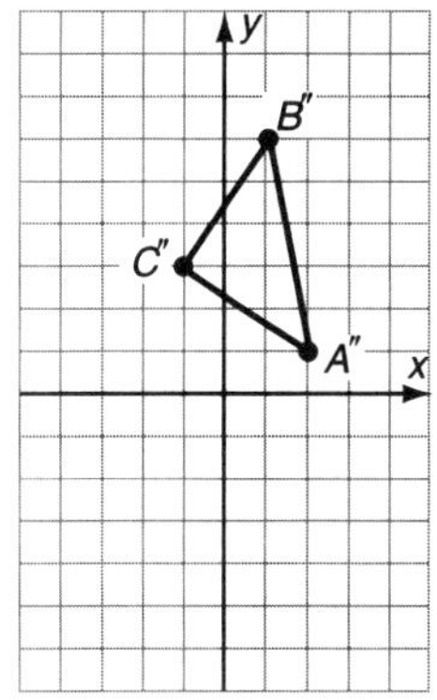

B.

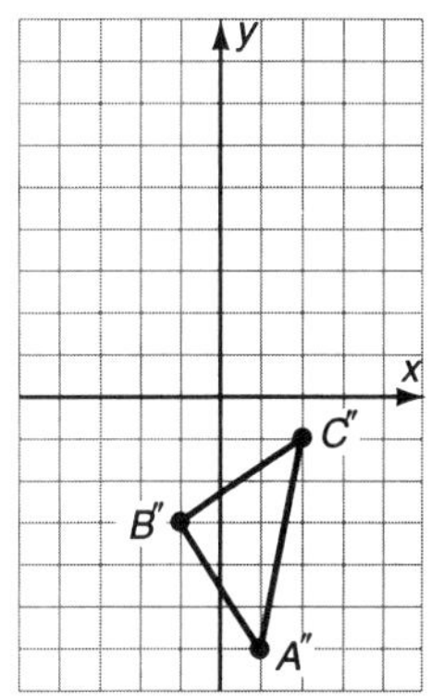

C.

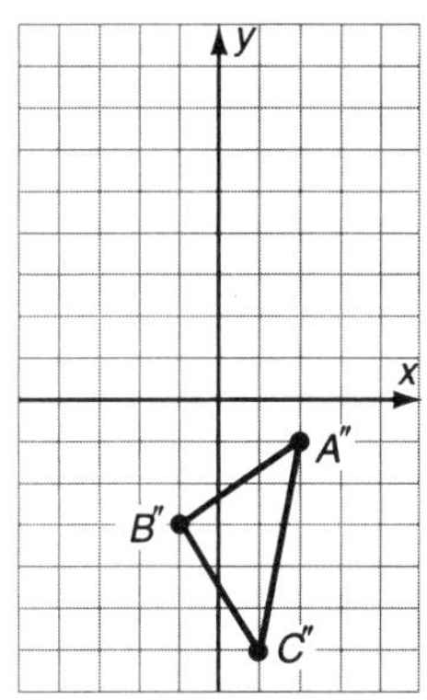

D.

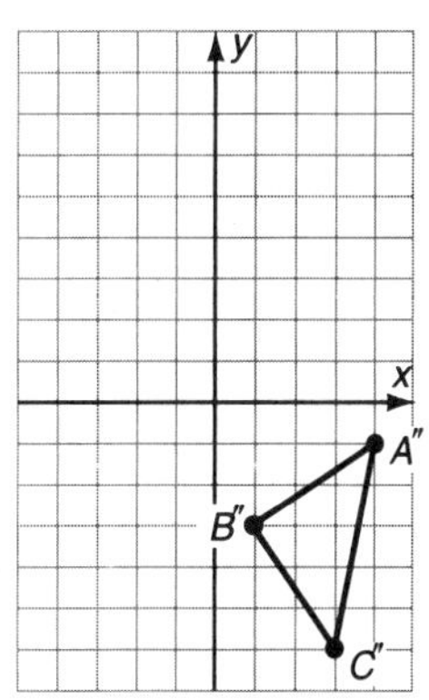

Think, Solve, Explain: Short-Response Problem

19. Use the drawing below.

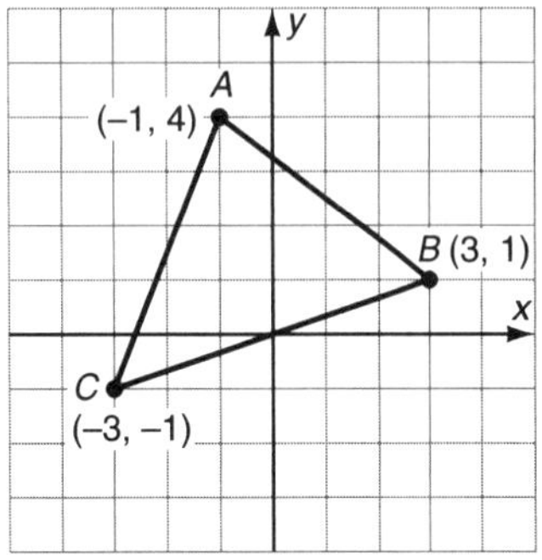

Part A Find the slope m of each side of the triangle in this graph.

Part B Find the length d of each side of the triangle.

Part C What type of triangle is shown? You must use the slopes of the sides, the lengths of the sides, or both to help verify your answer.

Multiple-Choice Problems

20. Which of the following represents the expression seven less than the sum of twice a number and eight?

F. $\left(\frac{1}{2}x + 8\right) - 7$

G. $(2x + 8) - 7$

H. $7 - (2x + 8)$

I. $7 - (2x - 8)$

21. Jacquet is participating in an exercise program that requires increasing by 4 the number of sit-ups done each day. The first day she did 5 sit-ups. On which day of the program will she do 25 sit-ups?

A. day 4

B. day 5

C. day 6

D. day 25

22. On the first day of the craft fair Malika sold 3 dolls and 4 puppets for more than \$90. On the second day she sold 5 dolls and 2 puppets for less than \$115. This information can be represented by the following inequalities where d represents the selling price of one doll and p represents the selling price of one puppet. The following inequalities are shown in the graph below.

$3d + 4p > 90$
$5d + 2p < 115$

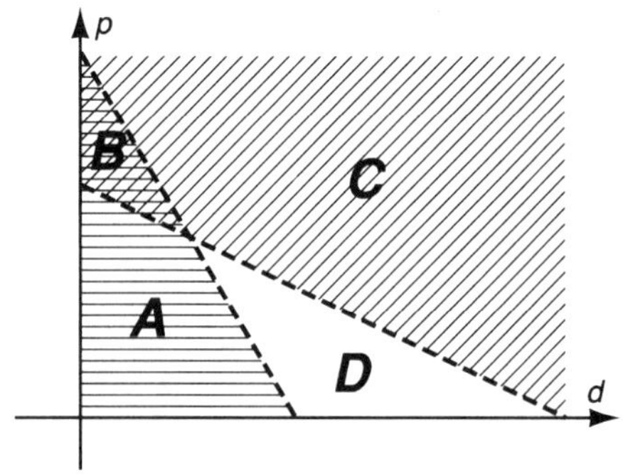

Which region of the graph represents the possible selling prices of both the dolls and the puppets?

F. region *A* **H.** region *C*

G. region *B* **I.** region *D*

Think, Solve, Explain: Short-Response Problem

23. On Wednesday the value of a certain stock was \$20 per share. By the end of the day on Thursday the price per share had gone up 40%. At closing on Friday the value had gone down 10% from the value on Thursday.

Part A Determine the value of the stock at closing on Thursday and Friday. Show all work necessary to justify your answer.

Part B Find the total percent change from Wednesday to Friday.

Gridded-Response Problems

24. The first term of a geometric sequence is 1 and the common ratio is 2. What is the sum of the first six terms of the sequence?

25. In the function $f(x) = \frac{1}{3}x - 2$, what does $f(6)$ equal?

26. What is the value of x in the following equation?

$3(2x + 4) = 2(x + 10)$

27. At the Dog Stand, Rochelle has her choice of a hot dog, cheese dog, or sausage. She has a choice of 2 different types of rolls and a choice of 3 different sauces. How many ways can she choose 1 meat, 1 roll, and 1 sauce?

Multiple-Choice Problems

28. The Dollar Sergeant Store is well known for selling products at an inexpensive price. Their prices are displayed on the stem-and-leaf plot below. What is the range of prices at the store?

Stem	Lea
0	8 9
1	2 5 6
2	4 5 7 9
3	5 8 9 9

Key: 2|4 = $2.40

A. $2.50

B. $3.00

C. $3.10

D. $3.90

29. Nicholas prepared a circle graph displaying the percent of time spent on activities when not in school. What, if anything, is wrong with his circle graph?

Nicholas's Out-of-School Activities

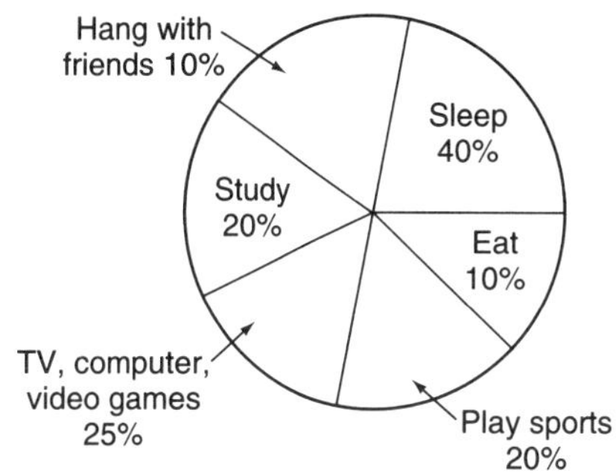

F. He should have presented the data in a bar graph.

G. He should have presented the data in a table.

H. The percentages add up to more than 100%.

I. Nothing is wrong with the graph.

30. Use the graph below.

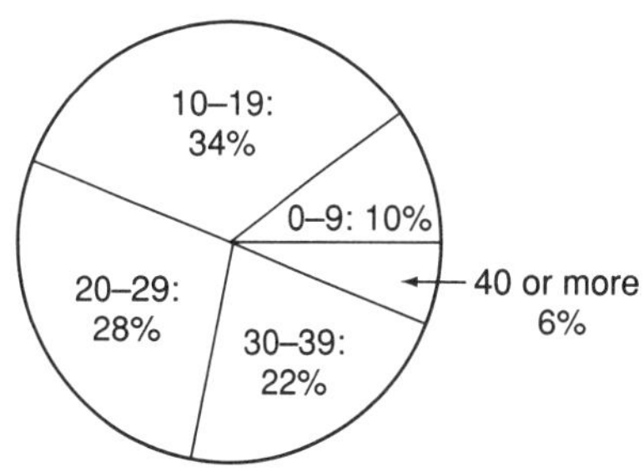

In a survey, 50 people were asked how many hours they watched the Olympics on television. Of the people surveyed, how many watched LESS than 30 hours?

A. 5

B. 14

C. 28

D. 36

PART II

Gridded-Response Problems

1. When a point on a merry-go-round makes half a revolution, it covers a distance of 157 feet. What is the radius of the merry-go-round? Use ≈ 3.14.

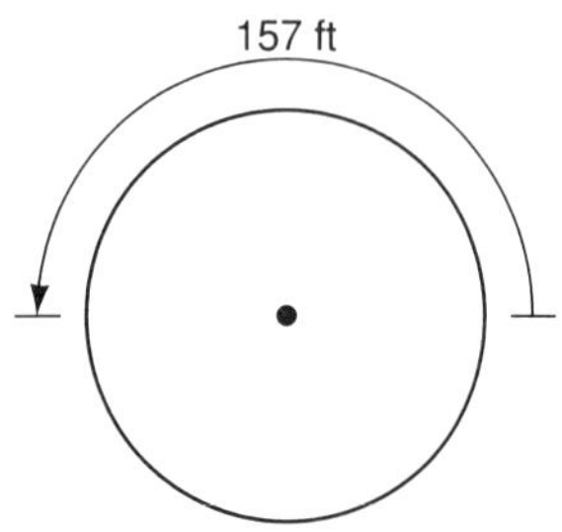

2. Oceanic dolphins can swim, on average, at a rate of 15 mph. What is this rate in feet per second?

3. Last year, 20 people tried out for the cheerleading squad. This year, 32 people tried out. What is the percent of increase in the number of people who tried out this year over last year?

Multiple-Choice Problems

4. What is the area of the shaded region of the square picture frame shown below? Each of the small openings is a square with a side that measures 3 inches.

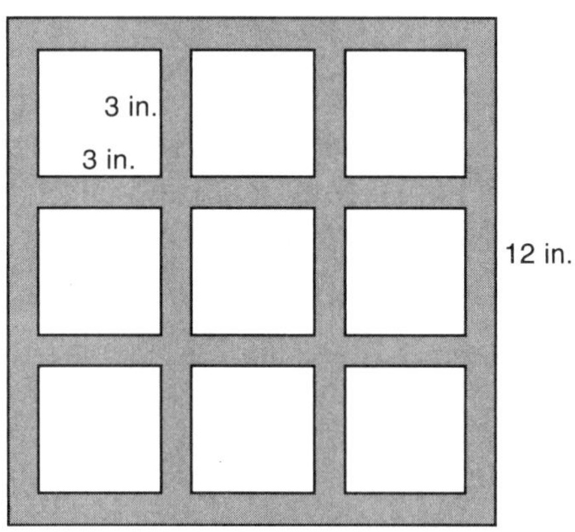

A. 63 square inches

B. 81 square inches

C. 133 square inches

D. 144 square inches

5. What is the length of a side of a square that has the same area as a circle with a diameter of 14 meters?

F. 30.8 meters **H.** 15.4 meters

G. 24.8 meters **I.** 12.4 meters

Think, Solve, Explain: Short-Response Problem

6. A square pyramid with a height of 5 feet and a base edge of 24 feet is shown below.

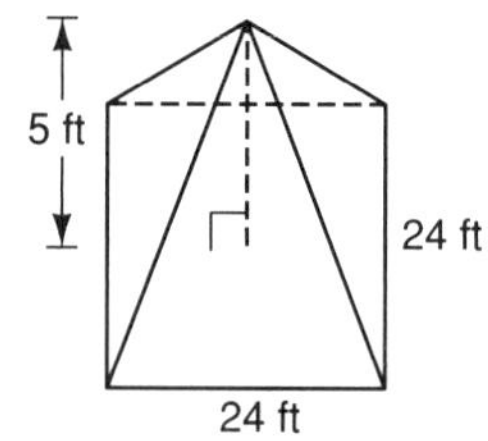

Part A What is the slant height of a face?

Part B What is the total surface area of the pyramid including the base? Show all work necessary.

Multiple-Choice Problems

7. Between what two numbers does 4% of 36 lie?

A. 0 and 1 **C.** 8 and 10

B. 1 and 2 **D.** 14 and 15

8. The following table displays the weekly profit and loss statements of a small pizza shop during the four weeks in February.

Week 1	Profit	$1,253
Week 2	Loss	$352
Week 3	Loss	$210
Week 4	Profit	$1,563

At the end of February, what was the total profit for the month?

F. $252 **H.** $2,674

G. $2,254 **I.** $3,378

9. The game commission wants to estimate the number of deer in a wilderness area. They counted 4 deer in a 0.5-square-mile area as shown by the shaded area on the grid. The total wilderness area is outlined. If the deer are evenly distributed throughout the area, how many deer are predicted to be in the entire area?

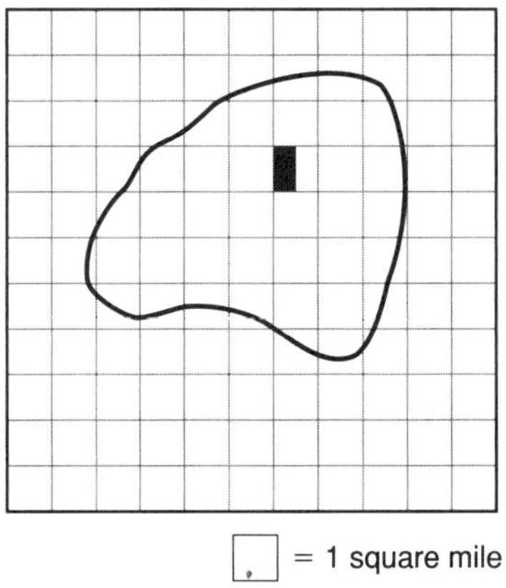

A. 200 deer **C.** 240 deer

B. 220 deer **D.** 280 deer

10. One method of naming the year in the Chinese calendar is the stem-branch method. The following years have been named jia-zi: 1804, 1864, 1924, and 1984. If n represents any jia-zi year, which of the following represents the next jia-zi year?

F. $n - 60$ **H.** $60n$

G. $n + 60$ **I.** $\frac{n}{60}$

11. The dotted line below is the graph of a line that is perpendicular to the vertical line. What is the slope of the dotted line?

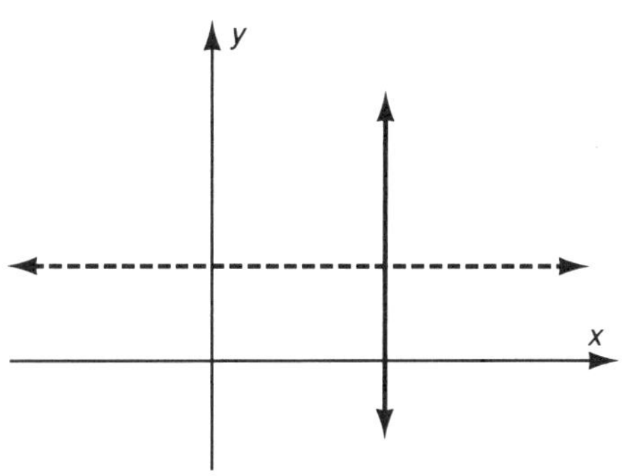

A. −1 **C.** 1

B. 0 **D.** no slope

Gridded-Response Problems

12. Four is the geometric mean between 5 and what number?

13. In the figure below, $\angle BCD$ is an exterior angle of triangle ABC. How many degrees are in $\angle A$?

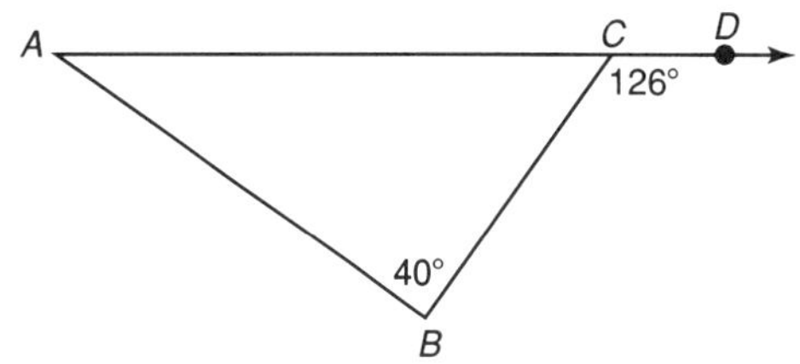

14. On a map, three cities are located in the shape of a right triangle. The distance from point C to point B is 5 centimeters. The distance from point B to point A is 12 centimeters. The map has a scale of 1 cm = 24 km. What is the actual distance from point A to point C in kilometers?

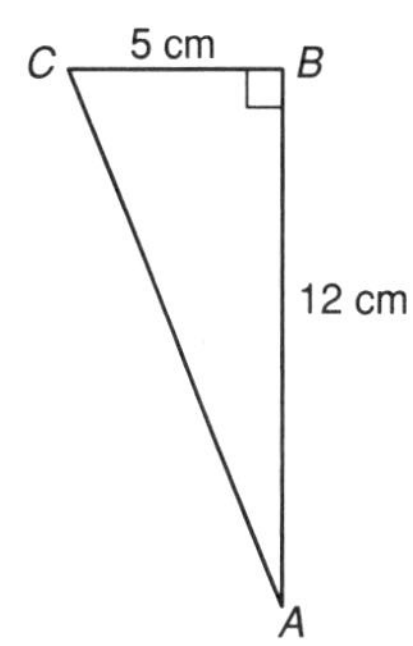

Multiple-Choice Problems

15. Which of the following shows a reflection of the letter **N**?

F.  **H.** 

G.  **I.** 

16. Point B is the midpoint of $\overline{AC}$. The coordinates of point A are (3, 8) and the coordinates of point B are (5, 10). What are the coordinates of point C?

A. (4, 9) **C.** (7, 12)

B. (6, 16) **D.** (8, 18)

Think, Solve, Explain: Extended-Response Problem

17. A 9-meter by 12-meter rectangle is divided into 3 rectangles and 2 trapezoids as shown in the drawing below. *HIFG* and *BCDK* are each one-ninth the size of *ACEG*.

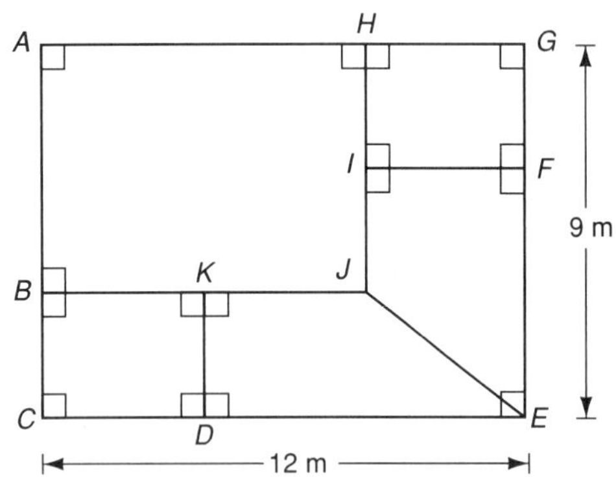

Part A Find the lengths of $\overline{BC}$, $\overline{KJ}$, $\overline{BK}$, and $\overline{IJ}$.

Part B Prove that the area of *ACEG* is 108 square meters by finding the sum of all the parts. Show the calculations that lead to your conclusion.

Gridded-Response Problems

18. The box-and-whisker plot below displays the prices of shirts at the LFN shop outside of Amateur Players Stadium. What percent of the shirts cost less than $20?

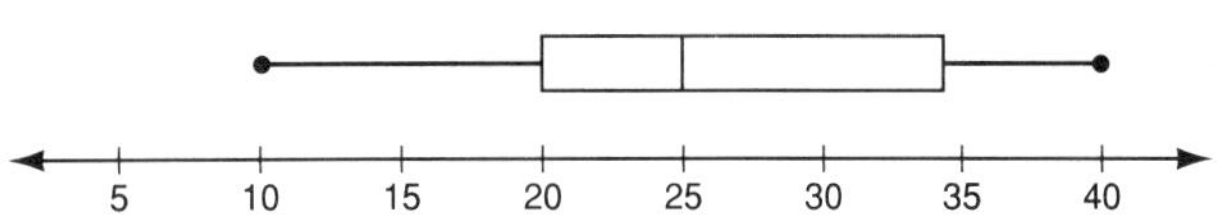

19. The table below shows the points scored this season by the kicker for a high school football team. What was the total number of points scored by the kicker?

Extra points (1 point)	𝍸 𝍸 III
Field goals (3 points)	𝍸 I
Touchdowns (6 points)	II

Multiple-Choice Problems

20. The students in Mr. Lightner's class were having a discussion about younger siblings. They made a bar graph to display the number of younger siblings in the students' families. What is the total number of younger siblings of all the members of the class?

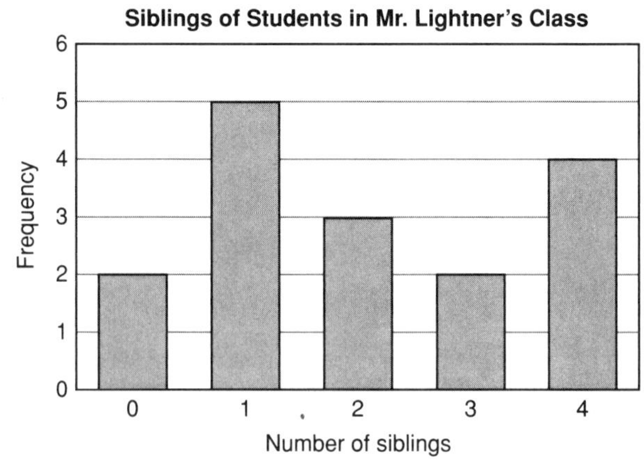

F. 4

G. 16

H. 33

I. 35

21. The numbers 3, 4, 8, and 10 have a mean of 6.25 and a median of 6. Which number, when included with these numbers, will cause the data to have a mean of 7 and a median of 8?

A. 8

B. 9

C. 10

D. 12

22. Suppose the odds of getting a hit in a softball game are 3 to 5. If Jasmine bats 40 times, how many hits would she be expected to get?

F. 10

G. 15

H. 24

I. 25

Think, Solve, Explain: Short-Response Problem

23. The system of the inequalities:

$x - y < 2$

$x + 2y > -1$

$y < 2$

is shown in the graph below.

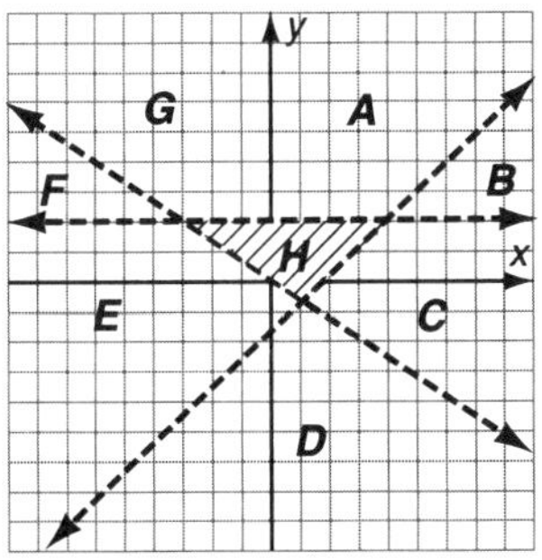

Part A Which region contains points that satisfy all three inequalities?

Part B Which regions satisfy the inequality $y < 2$?

Part C Which regions do NOT satisfy the inequality $x - y < 2$?

Gridded-Response Problems

24. What is the sum of the first 8 terms of the arithmetic sequence with a first term of 3 and a common difference of 2?

25. The Lopez family spends 25% of their budget on food. Last year their income was $52,000. This year their income increased to $60,000. How much more will they spend on food if they intend to spend the same percentage as last year?

26. What is the value of x in the following system of equations?

$3x + 4y = -10$

$x - y = 20$

Multiple-Choice Problems

27. A water lily covers an area of 5 square inches. These lilies double in the area covered every week. How much area will one water lily cover in 6 weeks?

A. 25 sq in. **C.** 160 sq in.

B. 30 sq in. **D.** 15,625 sq in.

28. If $f(x) = x^2 - x + 1$, what is $f(3)$?

F. 4 **H.** 10

G. 7 **I.** 13

29. Which of the following graphs represents the speed in relationship to time when a child is going down a slide?

A.

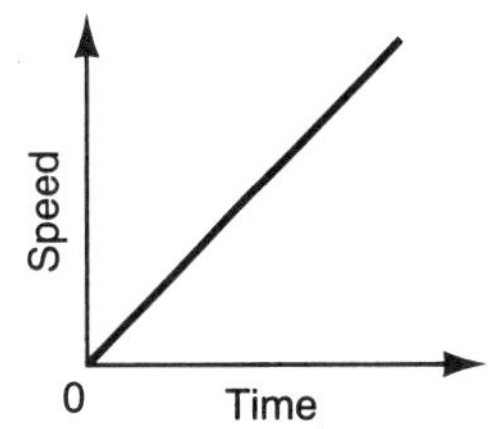

B.

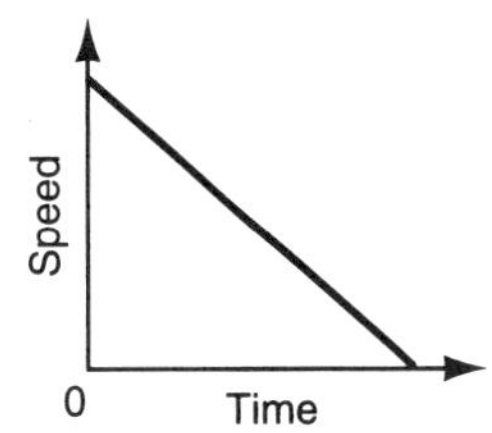

C.

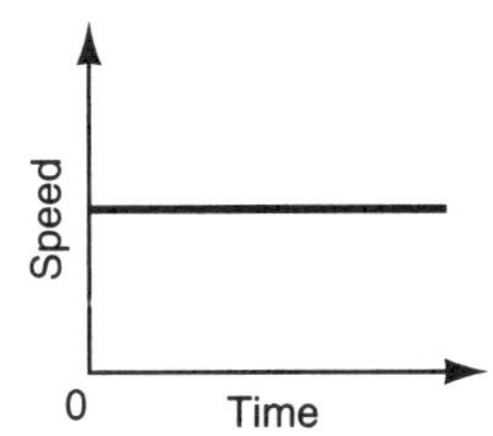

D.

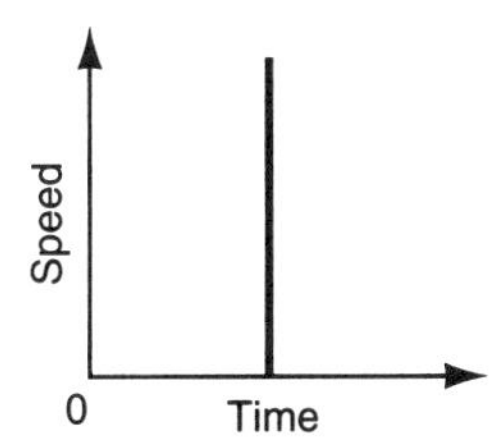

30. In the inequality $-2(x - 3) > 4(x + 2)$ what are the values of x?

F. $x < -\frac{1}{3}$ **H.** $x > -\frac{1}{3}$

G. $x < 3$ **I.** $x > 3$

ANSWER SHEET: TEST 3

NOTE: When you answer performance tasks on your own paper, use graph paper if necessary.

PART I

1

2

3

4

5 Ⓐ Ⓑ Ⓒ Ⓓ

6 Ⓕ Ⓖ Ⓗ Ⓘ

7 Ⓐ Ⓑ Ⓒ Ⓓ

8 Ⓕ Ⓖ Ⓗ Ⓘ

9 Ⓐ Ⓑ Ⓒ Ⓓ

10 Use your own paper.

11 Ⓕ Ⓖ Ⓗ Ⓘ

12 Ⓐ Ⓑ Ⓒ Ⓓ

13 Ⓕ Ⓖ Ⓗ Ⓘ

14 Ⓐ Ⓑ Ⓒ Ⓓ

15 Ⓕ Ⓖ Ⓗ Ⓘ

16 Ⓐ Ⓑ Ⓒ Ⓓ

17 Use your own paper.

18

19

20

21

22

23 Ⓕ Ⓖ Ⓗ Ⓘ

24 Ⓐ Ⓑ Ⓒ Ⓓ

25 Use your own paper.

26 Ⓕ Ⓖ Ⓗ Ⓘ

27 Ⓐ Ⓑ Ⓒ Ⓓ

28 Ⓕ Ⓖ Ⓗ Ⓘ

29

30

PART II

1 Ⓐ Ⓑ Ⓒ Ⓓ

2 Ⓕ Ⓖ Ⓗ Ⓘ

3 Ⓐ Ⓑ Ⓒ Ⓓ

4

5

6

7 Ⓕ Ⓖ Ⓗ Ⓘ

8 Ⓐ Ⓑ Ⓒ Ⓓ

9 Ⓕ Ⓖ Ⓗ Ⓘ

10 Use your own paper.

11 Ⓐ Ⓑ Ⓒ Ⓓ

12 Ⓕ Ⓖ Ⓗ Ⓘ

13 Ⓐ Ⓑ Ⓒ Ⓓ

14 Ⓕ Ⓖ Ⓗ Ⓘ

15 Ⓐ Ⓑ Ⓒ Ⓓ

16 Ⓕ Ⓖ Ⓗ Ⓘ

17 Ⓐ Ⓑ Ⓒ Ⓓ

18 Use your own paper.

19

20

21

22

23 Use your own paper.

24 Ⓕ Ⓖ Ⓗ Ⓘ

25 Ⓐ Ⓑ Ⓒ Ⓓ

26 Ⓕ Ⓖ Ⓗ Ⓘ

27 Ⓐ Ⓑ Ⓒ Ⓓ

28

29

30

TEST 3

PART I

Gridded-Response Problems

1. Ben uses the function $f(t) = -4.9t^2 + 120$ to determine the position of a ball after it has been dropped from the top of a building. He lets t represent the time since the ball left his hand. How far above the ground is the ball 2 seconds after leaving his hand?

2. Ms. Hauber has a series of posters hanging on the walls of her classroom. The first poster shows 1 ball, the second poster pictures 4 balls, the next one has 9 balls, the next 16, and so forth. What is the sum of the number of balls in the first seven posters?

3. On some steep roads in North Carolina, signs display the grade or steepness of the road. A 5 percent grade means that the road changes 5 feet vertically for every 100 feet of horizontal distance traveled. With a 5 percent grade, what is the vertical change, in feet, if the horizontal distance is $\frac{1}{2}$ of a mile?

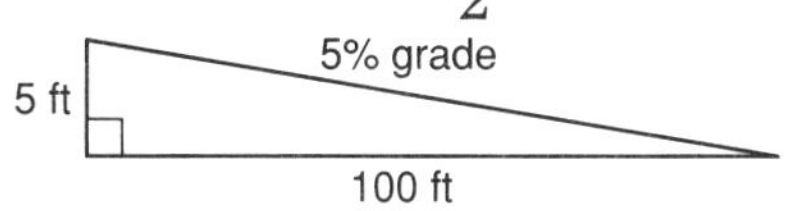

4. Circle I has a radius of 2 centimeters and circle II has a radius of 3 centimeters. What is the ratio of the area of circle II to the area of circle I?

Multiple-Choice Problems

5. In the drawing below $\overline{AC}$ is an altitude to the hypotenuse $\overline{BD}$. How long is $\overline{AD}$?

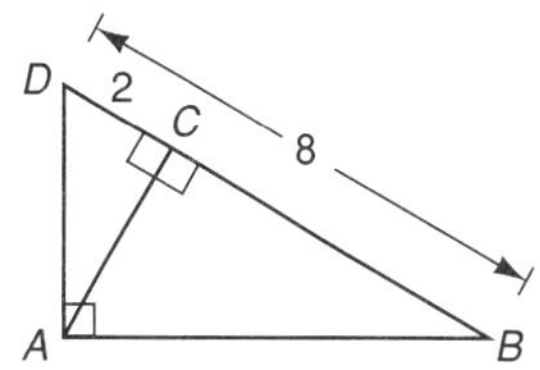

A. 2

B. 4

C. 8

D. 16

6. In the inequality, what is the value of x?

$3(x - 2) - 5(x + 3) \geq -19$

F. $x \leq 14$

G. $x \leq -1$

H. $x \geq -1$

I. $x \geq 14$

7. What are the coordinates of the vertex of the parabola shown below?

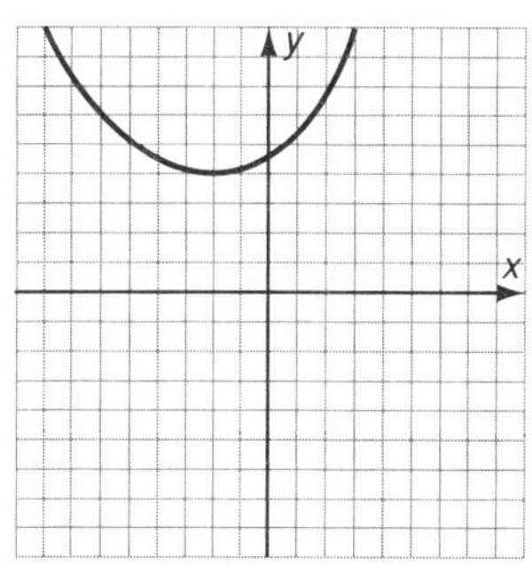

A. (2, 4)

B. (−2, 4)

C. (4, −2)

D. (2, −4)

8. Which of the following choices shows a dilation from triangle I to triangle II?

F.

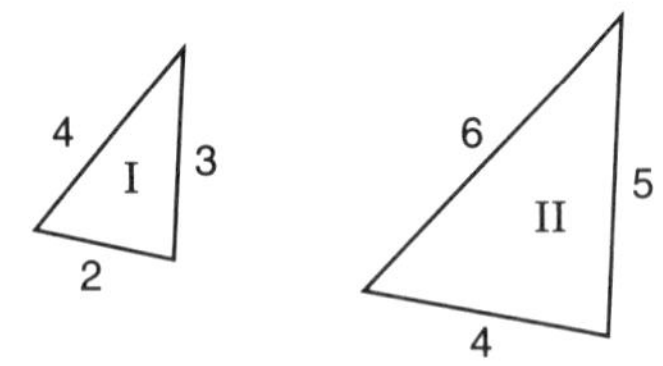

G.

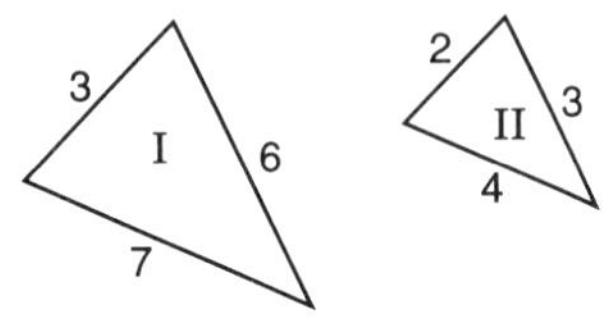

H.

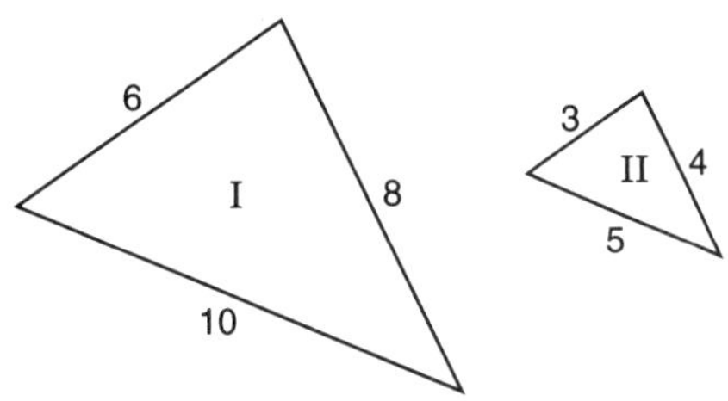

I.

9. By what method is $\triangle ABC \cong \triangle ADC$?

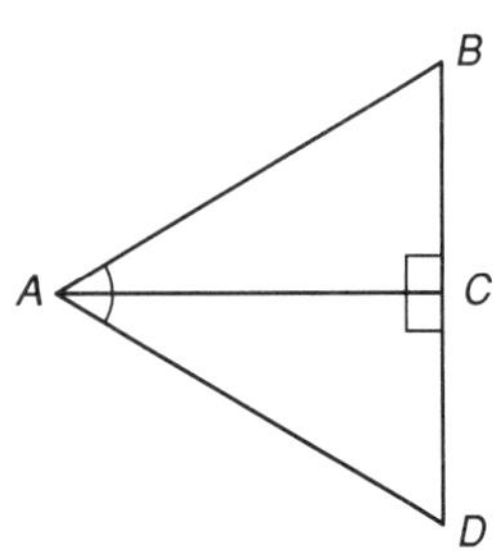

A. angle-angle

B. angle-side-angle

C. side-angle-side

D. angle-angle-side

Think, Solve, Explain: Extended-Response Problem

10. Simon is trying to determine the distance across a pond from point A to point C, as shown in the diagram. He draws triangles with $\overline{BC} \parallel \overline{DE}$.

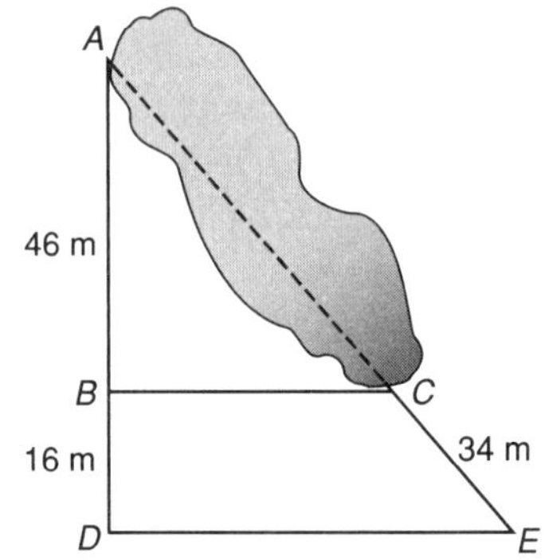

Part A Explain in words why $\triangle ABC \sim \triangle ADE$.

Part B Write a proportion that can be used to find the distance between A and C.

Part C Solve the proportion to find the distance AC in meters.

Multiple-Choice Problems

11. Points R, S, T, and W are located on the number line below. Which point represents a number that, when cubed, results in a number smaller than itself?

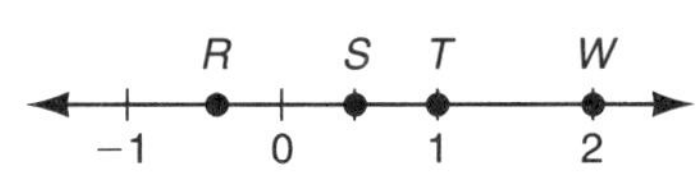

F. R

G. S

H. T

I. W

12. The dimensions of a right triangle are shown below. The ratio of the base of the triangle to its height is $\frac{5\sqrt{2}}{5\sqrt{10}}$.

Which of the following is equivalent to $\frac{5\sqrt{2}}{5\sqrt{10}}$?

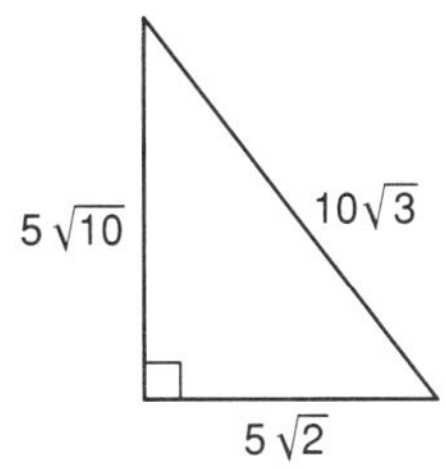

A. $\frac{1}{5}$

B. $\frac{5}{\sqrt{5}}$

C. $\sqrt{5}$

D. $\frac{\sqrt{5}}{5}$

13. Which of the following is NOT equivalent to 8^4?

F. 2^{12}

G. 4^6

H. 16^3

I. 32^2

14. In the orange grove shown below there are 20 trees in the shaded area. If each tree produces approximately 125 oranges, approximately how many oranges can the owner of the grove expect this season?

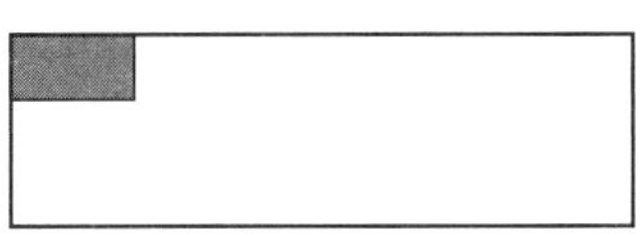

A. 1,875 oranges

B. 2,500 oranges

C. 37,500 oranges

D. 75,000 oranges

15. How many degrees are in the arc labeled x?

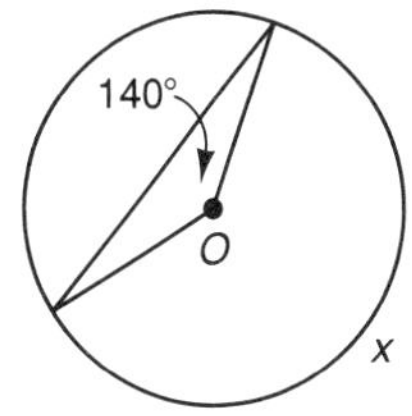

F. 70° **H.** 220°

G. 140° **I.** 280°

16. Christchurch, New Zealand, is located at 43.5° S latitude and Pretoria, South Africa, is located at 25.75° S latitude. To find the approximate airline distance between the two cities, multiply the angle measure between the cities by 70. What is the approximate airline distance between the two cities?

A. 1,240 miles **C.** 2,480 miles

B. 2,425 miles **D.** 4,850 miles

Think, Solve, Explain: Short-Response Problem

17. The figure below is composed of a right triangle and three squares. What is the area of the figure? Show all work.

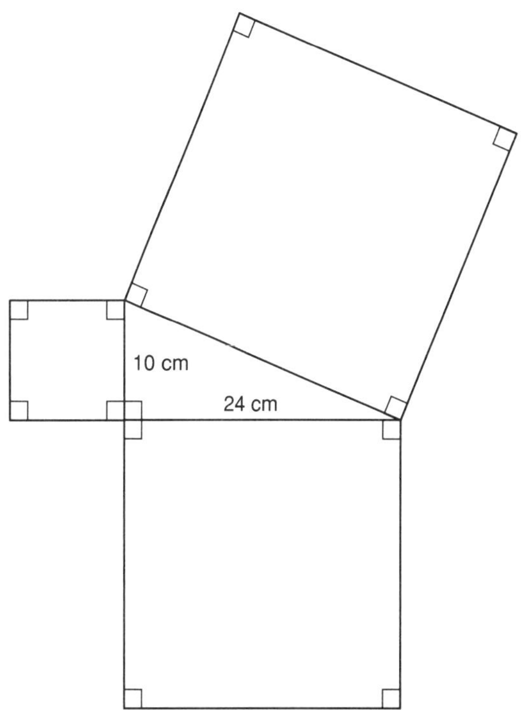

Gridded-Response Problems

18. Kathi's Coat Shop has all its winter coats on sale for 40% off. Elizabeth buys a coat that is originally marked at \$120. After calculating the sale price, Elizabeth gives the clerk a 15% off coupon that she received in the mail. What will she have to pay for the coat?

19. A cylinder with a height of 7 centimeters has a volume of 352 cubic centimeters. What is the total surface area of the cylinder to the nearest centimeter?

20. Tamara buys 8 one-liter bottles of soda for a party of 16 people. How many milliliters of soda, on average, can each person drink?

21. What is the next number in the geometric sequence $\frac{27}{2}, 9, 6, 4, \frac{8}{3}, \ldots$?

22. The system of inequalities and its graph are shown below.

$y < 2x - 4$

$y > \frac{1}{2}x + \frac{1}{2}$

$x < 5$

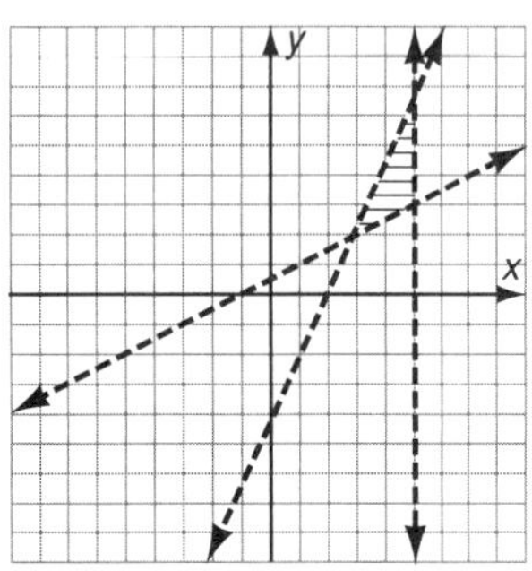

When $x = 4$, what is the integer value of the y-coordinate that satisfies this system?

Multiple-Choice Problems

23. Which of the following shows the relationship of speed to time of a person climbing a hill at a steady pace then starting to run down the hill?

F.

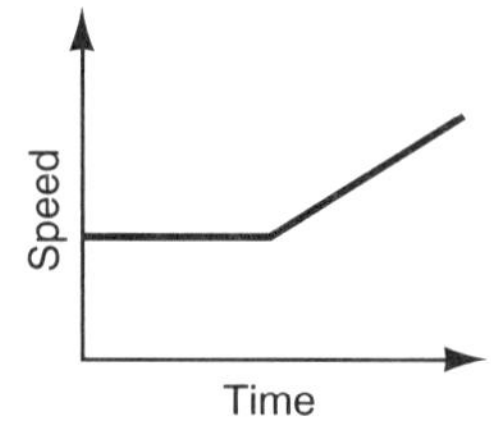

G.

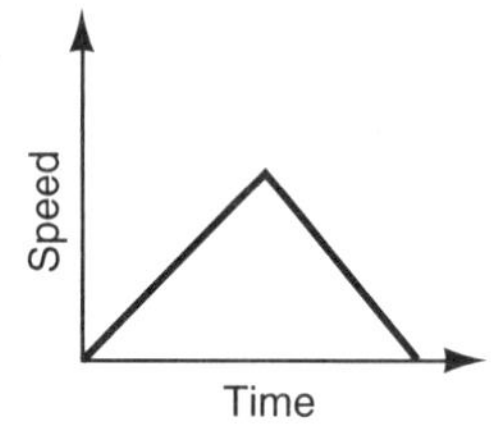

H.

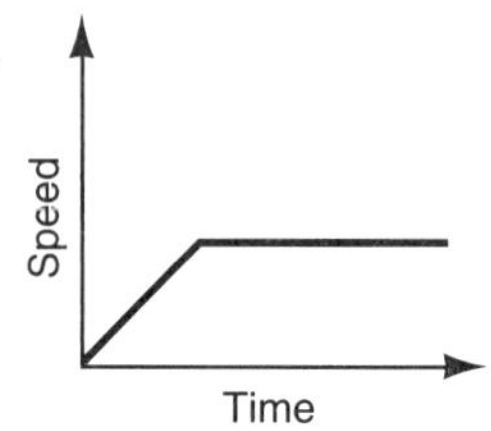

I.

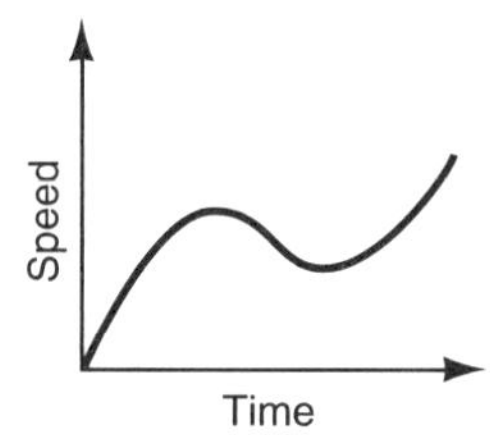

24. Fred has a model of a square pyramid with a base edge of 4 inches and a height of 4 inches. June also has a square pyramid with a height of 4 inches, but its volume is 4 times that of Fred's model. What is the length of the base edge of June's model pyramid?

A. 6 inches

B. 8 inches

C. 12 inches

D. 16 inches

Think, Solve, Explain: Short-Response Problem

25. Carrie bought 2 pairs of shoes and 3 pairs of pants for \$180. Her friend Larue bought the same types, only she bought 3 pairs of shoes and 4 pairs of pants for \$255. This problem can be represented with the following equations where s is the price of a pair of shoes and p is the price of a pair of pants.

$2s + 3p = \$180$

$3s + 4p = \$255$

Show all work necessary to find the price of 1 pair of shoes and 1 pair of pants.

Multiple-Choice Problems

26. The mean for the numbers 1, 1, 1, 1, 2, 6 is 2 and the mode and median are both 1. If the 2 in the data is replaced with 4, which of the following will change?

F. mean

G. median

H. mode

I. range

27. The histogram below displays the heights, in centimeters, of a group of students. How many students are taller than 159 centimeters?

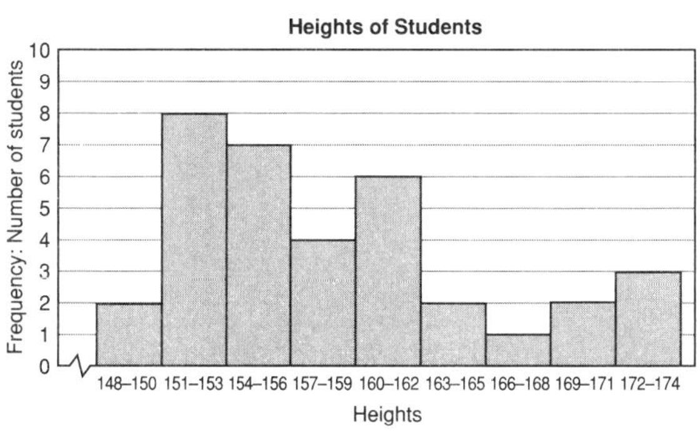

A. 5

B. 8

C. 9

D. 14

28. Using the graph below, which month had the highest temperature and least rainfall?

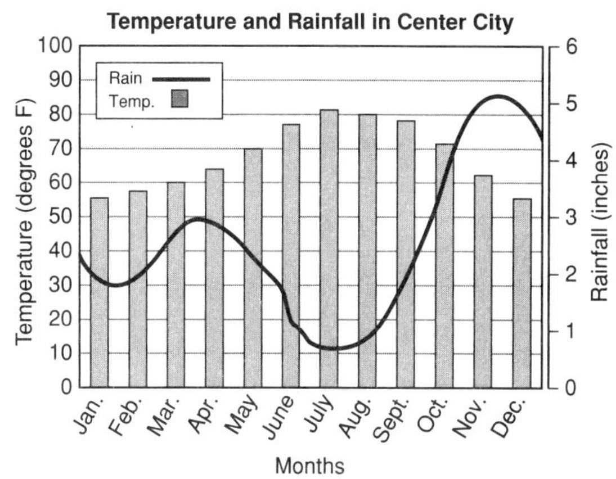

F. January

G. June

H. July

I. December

Gridded-Response Problems

29. A 10-sided die is tossed two times. What is the probability that it will land on 10 both times?

30. The stem-and-leaf plot below displays the grades on a 100-point test. In order to take the next test, a student must have a passing score of 75. How many students can move on to the next test?

Stem	Leaf
10	0 0
9	2 5 7 8 9
8	2 3 5 5 5 7
7	0 0 5 6 8 8
6	0 1 1 3 8
5	2 4 9
4	4

Key: 8|2 = 82

PART II

Multiple-Choice Problems

1. Which of the following relations is an example of a function?

A. {(2, 3), (3, 2), (4, 3)}

B. {(2, 5), (3, 6), (2, 7)}

C. {(4, 1), (1, 4), (1, 5)}

D. {(7, 2), (7, 3), (8, 3)}

2. The table below shows an object's velocity in feet per second. What is its velocity after 10 seconds?

1	32
2	64
3	96
4	128
5	160

F. 192 feet per second

G. 256 feet per second

H. 320 feet per second

I. 352 feet per second

3. Koji wants to find the number of days until his birthday. He made the following table:

September: 30 days
October: 31 days
November: 30 days
December: 31 days
January: 31 days
February: 10 days

Which of the following could be used to find the number of days until his birthday?

A. $3 \cdot 30 + 2 \cdot 31 + 10$

B. $3 \cdot 30 + 3 \cdot 31 + 10$

C. $2 \cdot 30 + 2 \cdot 31 + 10$

D. $2 \cdot 30 + 3 \cdot 31 + 10$

Gridded-Response Problems

4. What is the sum of the first seven terms in the geometric sequence 8, −4, 2, $-1, \frac{1}{2}, \ldots$?

5. Mr. O'Malley, the plumber, gave the Bowman family an estimate to repair the leak in the kitchen. His rate is $52 per hour and the parts will cost $220. His estimate for the job is $454. How many hours will it take him to complete the job?

6. Forrest drove his car at an average of 58 miles per hour for $2\frac{1}{2}$ hours. How far did he drive?

Multiple-Choice Problems

7. A square with a side of 7 units has the same perimeter as a rectangle with a length of 5 units. What is the width of the rectangle?

F. 7 units

G. 8 units

H. 9 units

I. 10 units

8. What is the perimeter of the figure shown below?

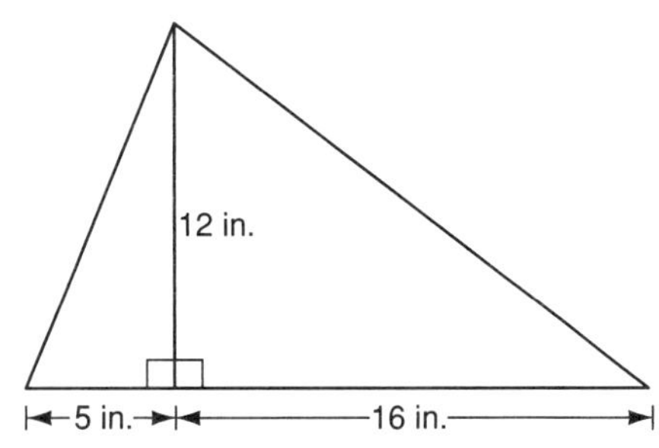

A. 47 inches

B. 54 inches

C. 66 inches

D. 78 inches

9. During a 16-hour period, Angie drank 2 gallons of water. On average, how many ounces of water did she drink in each 30-minute interval?

F. 4 ounces **H.** 12 ounces

G. 8 ounces **I.** 16 ounces

Think, Solve, Explain: Extended-Response Problem

10. Mr. Sanchez wants to purchase a digital camera. He made a scatter plot to show the price and the consumer rating of several cameras.

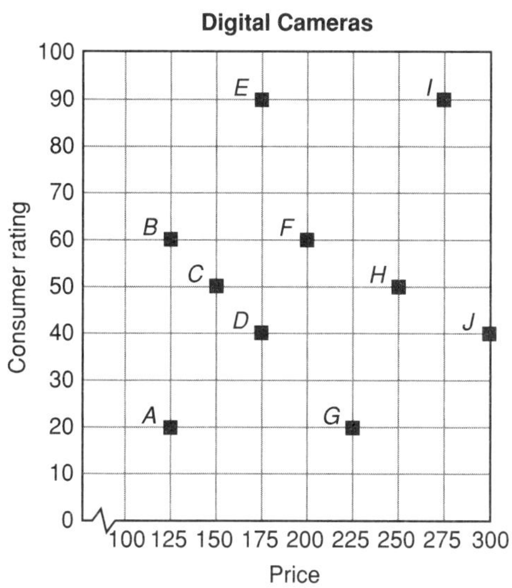

Part A What is the mean price of the cameras? Show all calculations.

Part B What is the mean consumer rating of the cameras? Show all calculations.

Part C Which camera is closest to both means?

Part D Which camera is the best buy? Explain your reasoning.

Multiple-Choice Problems

11. Between what two numbers does 1.34×10^{-2} lie?

A. -2 and -1 **C.** 13 and 14

B. 0 and 1 **D.** 133 and 135

12. Burlines marks their clothes up 140% over cost. How much profit is made on a dress that sells for \$133?

F. \$7 **H.** \$38

G. \$27 **I.** \$45

13. In March, 205 cars were sold at Kenston Auto Dealers. In April, 180 cars were sold. What is the percent of decrease in the number of cars sold?

A. 12.2%

B. 13.9%

C. 87.9%

D. 102.5%

14. Which of the following expressions is equivalent to $6(2a + 3b - 5) - 4(3a - 8b + 1)$?

F. $-14b - 29$

G. $50b - 29$

H. $-14b - 26$

I. $50b - 34$

15. Line segment AB is a median of triangle ACD. Which of the following statements is true?

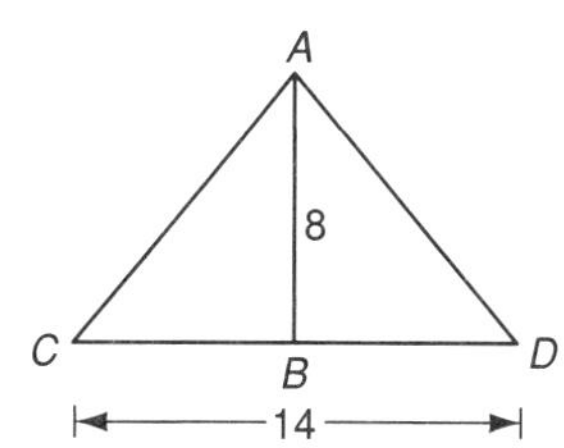

A. $\angle C = \angle D$

B. $\overline{CB}$ bisects $\angle A$

C. $\overline{CD}$ is perpendicular to $\overline{AB}$

D. $\overline{CB} = 7$ units

16. Which letter has horizontal and vertical line symmetry?

F. H

G. M

H. A

I. B

17. A rotation of 240° counterclockwise will map point D onto which point?

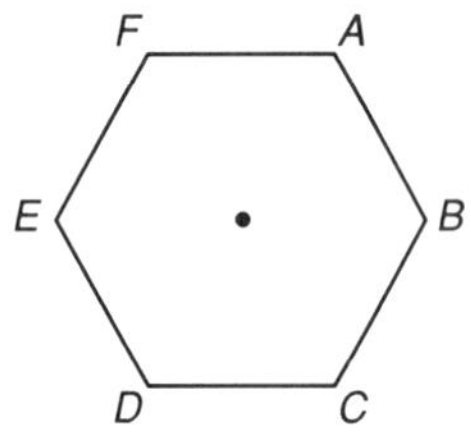

A. point A

B. point B

C. point E

D. point F

Think, Solve, Explain: Short-Response Problem

18. The Crisafis' oil storage tank is in the shape of a right circular cylinder with a height of 5 feet and a diameter of 6 feet. They want to buy a larger tank. They have two choices: one with the same height and a diameter that is 4 feet longer, or one with the same diameter and a height that is 4 feet longer. Which one will be closer to having a volume three times their present tank? Show all work.

Gridded-Response Problems

19. To increase the number 1.3×10^3 to 1,300,000,000, the number must be multiplied by what power of 10?

20. Quadrilateral $ABCD$ is a parallelogram. How many degrees are in $\angle DAB$?

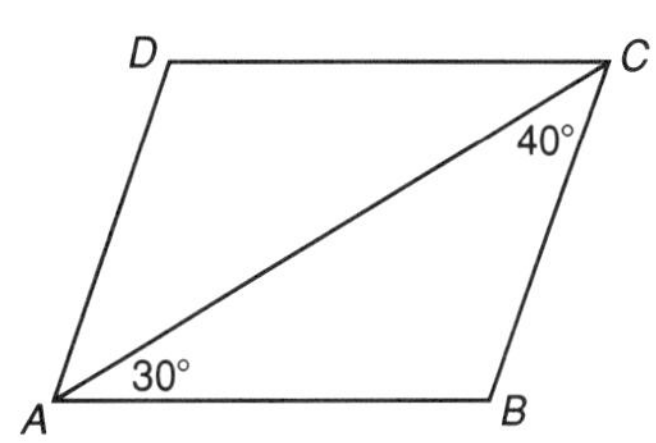

21. What is the distance between point A and point B? Round your answer to the nearest tenth.

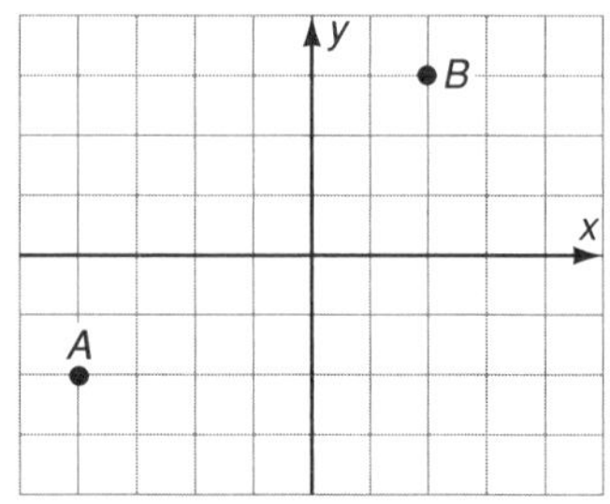

22. What is the y-coordinate of the midpoint of the segment joining A (3, 5) and B (7, 12)?

Think, Solve, Explain: Short-Response Problem

23. *Part A* Plot the points A (2, 4), B (2, −5), and C (−4, −5)

Part B What is the area of triangle ABC?

Part C Triangle ABC is translated so that point A, now known as A', is at the origin. What are the coordinates of point B' and point C'?

Part D What is the area of triangle $A'B'C'$?

Multiple-Choice Problems

24. Members of the Animal Rescue League conducted a survey at the local pet store. Of the 65 people surveyed, 59 of them had at least one pet at home. On their Web site, they reported that 90% of the people in their town had at least one pet at home. What can be said about their conclusion?

F. The conclusion is valid.

G. The conclusion is not valid because the sample size is too small.

H. The conclusion is not valid because of the location of the survey.

I. The conclusion is not valid because it should have been conducted by a professional company.

25. A new club is being formed. There are 5 members, 2 girls and 3 boys. They are going to elect a president, secretary, and treasurer. No member can hold more than one office. Suppose that the president must be a girl, how many different ways can the offices be filled?

A. 10 ways

B. 12 ways

C. 24 ways

D. 60 ways

26. What is the approximate circumference of a circle that has an area of 1.44π square centimeters?

F. 9.0 centimeters

G. 7.5 centimeters

H. 3.8 centimeters

I. 2.9 centimeters

27. A line through point C parallel to $\overline{AB}$ will pass through which of the following points?

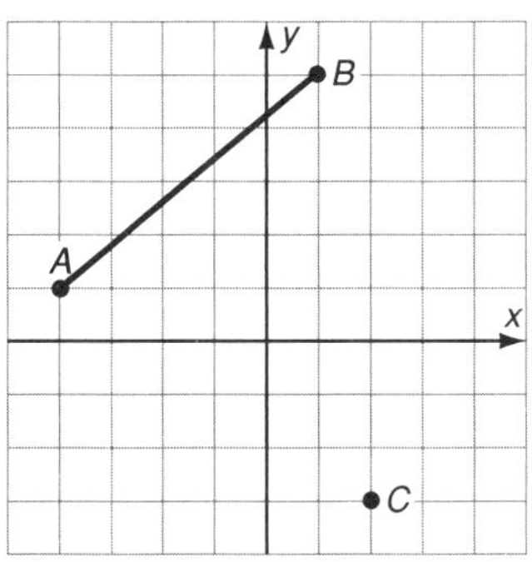

A. (5, 3)

B. (7, 0)

C. (5, 2)

D. (7, 1)

Gridded-Response Problems

28. Use the circle graph below. If the company spends $400,000, how much is budgeted for utilities?

Company Expenses

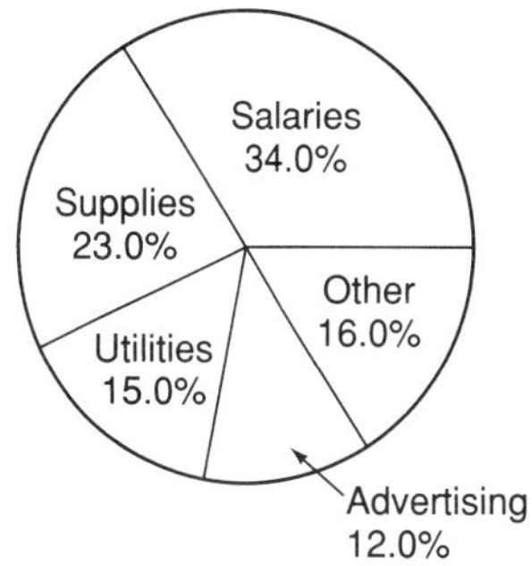

29. At the end of the day, the movie theater owner looked over the breakdown of ticket sales for 511 people that totaled $4,341.

Ticket type	Unit cost	Number of tickets sold
Adult	$9	A
Children	$6	C
Total		511

The following equations represent the information from the table:

$A + C = 511$

$9A + 6C = 4{,}341$

How many children attended the movie that day?

30. What is the mode of the data on the stem-and-leaf plot below?

Stem	Leaf
3	2 3 5
4	1 2 2 3
5	1 7 8

Key: 3|2 = 320

ANSWER SHEET: TEST 4

NOTE: When you answer performance tasks on your own paper, use graph paper if necessary.

PART I

1

2

3 Ⓐ Ⓑ Ⓒ Ⓓ

4 Ⓕ Ⓖ Ⓗ Ⓘ

5 Ⓐ Ⓑ Ⓒ Ⓓ

6 Ⓕ Ⓖ Ⓗ Ⓘ

7 Use your own paper.

8

9

10

11

12 Ⓐ Ⓑ Ⓒ Ⓓ

13 Ⓕ Ⓖ Ⓗ Ⓘ

14 Ⓐ Ⓑ Ⓒ Ⓓ

15

16

17 Use your own paper.

18 Ⓕ Ⓖ Ⓗ Ⓘ

19 Ⓐ Ⓑ Ⓒ Ⓓ

20 Ⓕ Ⓖ Ⓗ Ⓘ

21 Ⓐ Ⓑ Ⓒ Ⓓ

22 Use your own paper.

23

24

25

26 Ⓕ Ⓖ Ⓗ Ⓘ

27 Ⓐ Ⓑ Ⓒ Ⓓ

28 Ⓕ Ⓖ Ⓗ Ⓘ

29 Ⓐ Ⓑ Ⓒ Ⓓ

30 Ⓕ Ⓖ Ⓗ Ⓘ

PART II

1 Ⓐ Ⓑ Ⓒ Ⓓ

2 Ⓕ Ⓖ Ⓗ Ⓘ

3 Ⓐ Ⓑ Ⓒ Ⓓ

4 Ⓕ Ⓖ Ⓗ Ⓘ

5 Ⓐ Ⓑ Ⓒ Ⓓ

6 Ⓕ Ⓖ Ⓗ Ⓘ

7 Use your own paper.

8

9

10

11

12

13

14

15 Ⓐ Ⓑ Ⓒ Ⓓ

16 Ⓕ Ⓖ Ⓗ Ⓘ

17 Ⓐ Ⓑ Ⓒ Ⓓ

18 Use your own paper.

19 Ⓕ Ⓖ Ⓗ Ⓘ

20 Ⓐ Ⓑ Ⓒ Ⓓ

21 Ⓕ Ⓖ Ⓗ Ⓘ

22 Ⓐ Ⓑ Ⓒ Ⓓ

23 Use your own paper.

24

25

26

27 Ⓕ Ⓖ Ⓗ Ⓘ

28 Ⓐ Ⓑ Ⓒ Ⓓ

29 Ⓕ Ⓖ Ⓗ Ⓘ

30 Ⓐ Ⓑ Ⓒ Ⓓ

TEST 4

PART I

Gridded-Response Problems

1. On January 1st, Jeremy put \$3 in his safe. At the beginning of each month, he put twice as much in the safe as he put in the month before. By the end of June, how much money had he put in the safe?

2. The veterinarian prescribed pills for Lauren's cat. The first week the cat was given a total of 15 pills. Each week the total number of pills was decreased by 3. How many pills will the cat be given during the fifth week?

Multiple-Choice Problems

3. Which of the following represents the sentence: Six times a number decreased by ten is the same as twelve?

   A. $x + 6 - 10 = 12$

   B. $6x - 10 = 12$

   C. $10 - 6x = 12$

   D. $\frac{1}{6}x - 10 = 12$

4. Which of the following relations is NOT a function?

   F. $\{(2, 3), (1, 7), (6, 4), (8, 3)\}$

   G. $\{(1, 2), (2, 2), (3, 2), (4, 2)\}$

   H. $\{(3, 3), (4, 4), (5, 5), (6, 6)\}$

   I. $\{(4, 5), (4, 0), (4, 1), (4, 4)\}$

5. Which one of the following coordinates lie in the solution set of the system of inequalities shown below?

   $x + y < 3$

   $x > 1$

   $y < 4$

   A. (2, 0) **C.** (2, 4)

   B. (1, 3) **D.** (2, 3)

6. In square $ABCD$, $\angle CAB = 3x - 9$. What is the value of x?

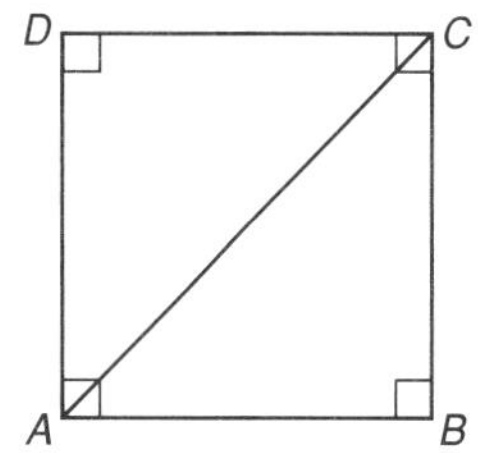

   F. 12

   G. 18

   H. 45

   I. 90

Think, Solve, Explain: Short-Response Problem

7. Bud is building a cabin that is rectangular in shape. Originally, the length was to be 57 feet and the width 46 feet. The builder suggested that he increase the length by 13 feet and increase the width by 16 feet.

   Part A What is the area, in square feet, of the original plan?

   Part B What is the area of the new plan with the suggested additions?

   Part C What is the percent of increase? Round your answer to the nearest whole percent.

   Be sure to show all work necessary to find the answers.

Gridded-Response Problems

8. In the figure below, side AB is parallel to side CD. Triangle ABE is an equilateral triangle. What is the measure of $\angle BCD$?

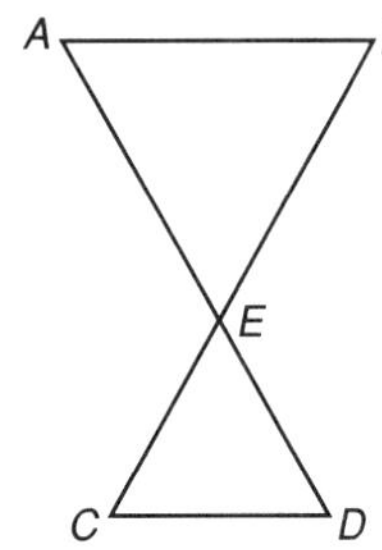

9. In the drawing below, triangle ABC is similar to triangle DEF. What is the length of side DF?

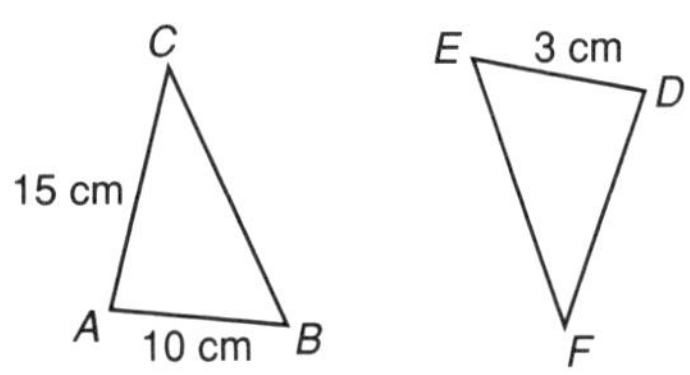

10. In the drawing below, $\overline{AB}$ is reflected over the x-axis. What is the y-coordinate of A'?

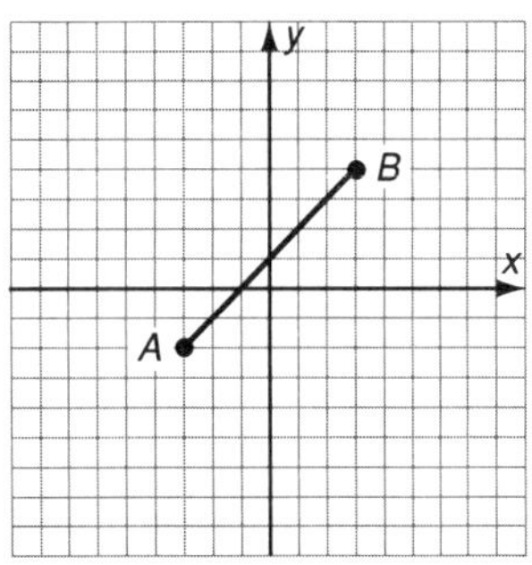

11. The geometric mean between 16 and x is 12. What is the value of x?

Multiple-Choice Problems

12. What is the measure of arc x shown in the diagram below?

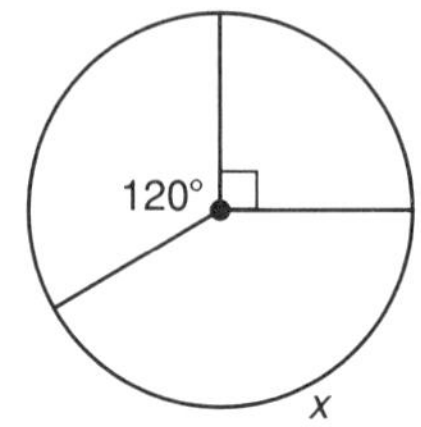

A. 90° **C.** 150°

B. 120° **D.** 180°

13. If it takes $\frac{1}{2}$ pint of strawberries to make 1 jar of strawberry jam, how many quarts of strawberries should be purchased to make 50 jars of jam?

F. 6.25 quarts

G. 12.5 quarts

H. 25 quarts

I. 50 quarts

14. Two poles, 32 feet and 50 feet high, are 70 feet apart. What is the approximate distance from the top of one pole to the top of the other pole?

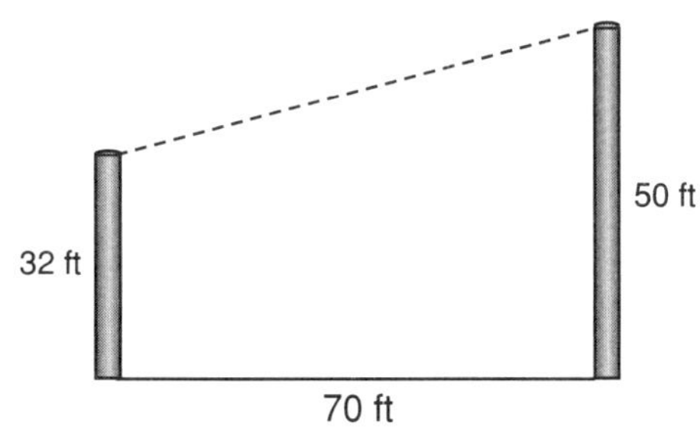

A. 59.36 feet

B. 72.28 feet

C. 77.92 feet

D. 86.02 feet

Gridded-Response Problems

15. In the figure below, what is the area of triangle ABC?

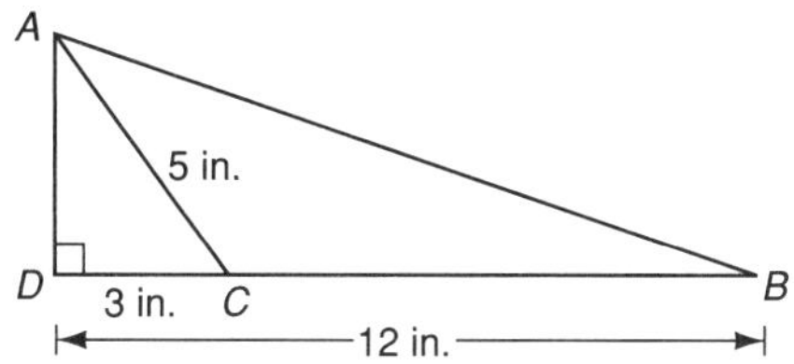

16. A rectangular wading pool is 64 inches long, 54 inches wide, and 12 inches deep. A circular wading pool has a diameter of 60 inches and is 10 inches deep. How many more cubic inches of water does the rectangular pool hold than the circular pool? Use $\pi \approx 3.14$.

Think, Solve, Explain: Extended-Response Problem

17. Single-track recordings are an important part of music sales. The table below shows the percentages of yearly music sales that were made by single-track recordings.

Year	Percent	Year	Percent
1989	10.7	1996	9.3
1990	8.7	1997	9.3
1991	8.8	1998	6.8
1992	7.5	1999	5.4
1993	9.2	2000	2.5
1994	7.4	2001	2.4
1995	7.5	2002	1.9

Part A On a graph label the y-axis from 0 to 11 in intervals of 0.5 unit. Label the x-axis in years from 1989 to 2002. Plot and connect the points for the data in the table to make a line graph.

Part B Use the graph from Part A to find the percentage of singles that you would predict to be sold in 2004.

Part C In 2004, 2.4% of the sales were singles. Explain why the prediction you made in Part B differs from the actual sales.

Multiple-Choice Problems

18. Megan is 16 years old. Which of the following represents her age y years ago?

F. $16 - y$

G. $y - 16$

H. $16y$

I. $\frac{16}{y}$

19. Which of the following is NOT equivalent to 8^3?

A. 2^9

B. $\frac{4^4}{4}$

C. $\frac{4^5}{2}$

D. $\frac{4^6}{8}$

20. Given the values $a = 3$, $b = -2$, $c = \frac{1}{3}$, $x = 1$, and $y = -4$, what is the value of $\frac{a^2 - 2b + 3c}{x^2 - y^2}$?

F. $\frac{14}{17}$

G. $\frac{6}{17}$

H. $-\frac{11}{15}$

I. $-\frac{14}{15}$

21. In 2002, the number of students attending Western Cove High School was 1,727. In 2004, the number of students had increased to 3,800. What is the percent of increase in the student population at Western Cove High School?

A. 0.546% **C.** 54.6%

B. 1.2% **D.** 120%

Think, Solve, Explain: Short-Response Problem

22. Use the coordinate graph below.

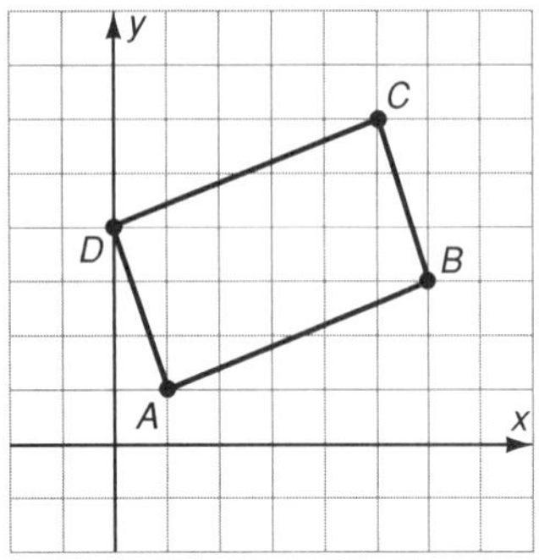

Part A Find the slopes of line segments *AB*, *BC*, *CD*, and *DA*.

Part B What kind of quadrilateral is *ABCD*? As a part of your reasoning, use the slopes you found.

Gridded-Response Problems

23. The metal fabrication plant made 1,328 units last week. This week they produced 1,660 units. What is the percent of increase in the number of units produced this week over the number produced last week?

24. Given the equation $-2(3x - 4) = 2x$, what is the value of x?

25. In a class of 18 students, 7 are boys. Choosing at random, what are the odds that a girl will NOT be chosen first to answer a question?

Multiple-Choice Problems

26. What is the slope of the line parallel to the line containing the points (3, 2) and (1, 1)?

F. -2

G. $-\frac{1}{2}$

H. $\frac{1}{2}$

I. 2

27. The rectangular solids shown below are similar. The volume of solid I is 125 cubic feet and the volume of solid II is 64 cubic feet. If the length of solid I is 15 feet, what is the length of solid II?

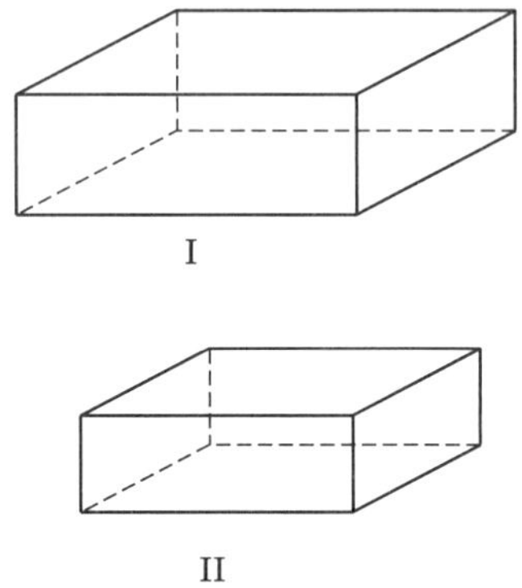

A. 2 feet **C.** 12 feet

B. 7.68 feet **D.** 18.75 feet

28. What is the median of the following data: 6, 2, 0, 5, 8, 2, 2, 5, 7, 2, 5?

F. 2 **H.** 5

G. 4 **I.** 8

29. A die and a coin are tossed simultaneously. What is the probability of getting an odd number and a tail?

A. $\frac{1}{6}$ **C.** $\frac{1}{3}$

B. $\frac{1}{4}$ **D.** $\frac{1}{2}$

30. A famous sonnet has a verb in 12 of its 15 lines, an adjective in 7 lines, and both in 5 lines as shown in the Venn diagram below. How many lines have a verb but no adjective?

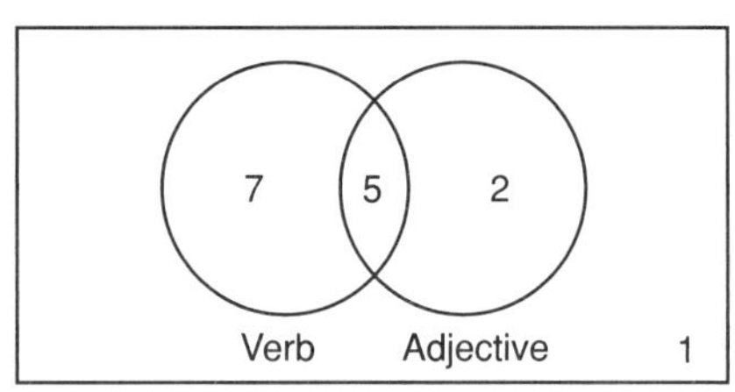

F. 2 lines **H.** 7 lines

G. 3 lines **I.** 12 lines

PART II

Multiple-Choice Problems

1. Each day the number of grams of bacteria triples in a biology experiment. If the scientist began with 5 grams of bacteria, how many grams could be found after 4 days?

A. 20 grams

B. 25 grams

C. 135 grams

D. 625 grams

2. Given $y = 2x^3 + 3$, what is the value of y when $x = -1$?

F. -3

G. 1

H. 5

I. 9

3. What is the solution for the following system of equations?

$2x + y = -3$

$2x - 3y = 17$

A. $(10, -23)$

B. $(-6.5, 10)$

C. $(1, -5)$

D. $(2, -7)$

4. A pair of dice is rolled and the numbers are noted. What is the probability that at least one die shows a 4?

F. $\frac{4}{36}$

G. $\frac{10}{36}$

H. $\frac{11}{36}$

I. $\frac{12}{36}$

5. The height, in centimeters, of the boys and girls in Dr. Burke's class are displayed in the back-to-back stem-and-leaf plot shown below. What conclusion can be made from the displayed data?

Girls		Boys
9	12	1 2 3
8	13	5 7
7 8 8	14	1 2 4 5
1 2 5	15	2 3
3 5 6 7	16	

Key: 12|2 = 122 centimeters

A. There are more boys than girls in the class.

B. The median height of the boys is higher than the median height of the girls.

C. On average, the girls are taller than the boys.

D. The range is the same for the girls' heights and the boys' heights.

6. Tom Bolden, the plumber, created a graph to explain his charges to his customers. His charges are based on time plus cost of materials. How much will it cost Ms. Matrinko to have Tom spend $2\frac{1}{2}$ hours installing a $150 sink?

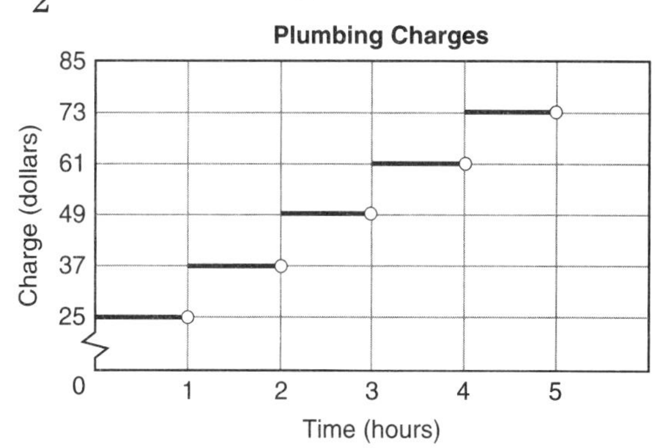

F. $150.00

G. $199.00

H. $211.00

I. $212.50

Think, Solve, Explain: Short-Response Problem

7. In order to work for the Blue Bird Airlines, a person's height must be greater than or equal to 62 inches but less than 72 inches.

Part A Write an inequality or pair of inequalities to represent the airline's requirements.

Part B Sarita's height is 5 feet 2 inches. Should she apply for a job with this airline? Why or why not?

Gridded-Response Problems

8. The table below shows the scores of students in a math competition. The mean score is 34. What is the value of p?

Score	10	20	30	40	50
Number of students	1	2	5	p	3

9. There are 3 blue and 5 red marbles in a jar. Without looking, one marble is taken out of the jar. Without replacing it, another marble is taken out of the jar. What is the probability that the first marble is blue and the second marble is red?

10. Marcus works 8 hours a week and earns \$42. Starting next week, he will earn \$54. How much more money will he be making per hour?

11. The first term of an arithmetic sequence is 6 and the common difference is 3. What is the sum of the first five terms of this sequence?

12. In the figures below, trapezoid $ABCD$ is similar to trapezoid $EFGH$. If $CD = 12$, $AB = 9$, and $GH = 9$ what does EF equal?

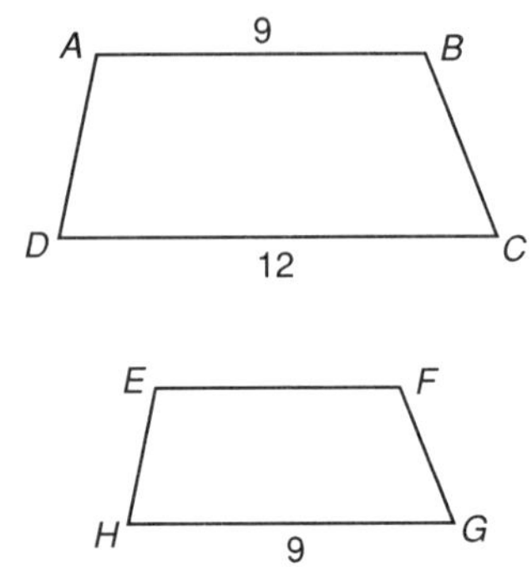

13. What is the sum of the interior angles of a polygon with seven sides?

14. In isosceles right triangle ABC, the hypotenuse BC equals $7\sqrt{2}$. What is the length of leg AC?

Multiple-Choice Problems

15. Which of the following line segments is the altitude of triangle ABC shown below?

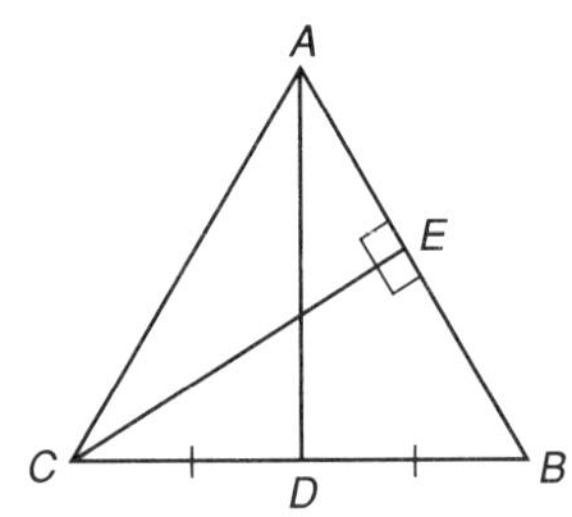

A. $\overline{AD}$

B. $\overline{CD}$

C. $\overline{AE}$

D. $\overline{CE}$

16. The map of an area has a scale of 1 inch = 16 kilometers. On the map the distance from Jefferson to Lincoln City is approximately 5.5 inches. What is the actual distance between the cities?

F. 0.34 kilometer

G. 2.91 kilometers

H. 88 kilometers

I. 880 kilometers

17. In the square pyramid shown below, a plane is passed through the pyramid 15 inches above the base and parallel to the base. What is the area of the shaded cross section?

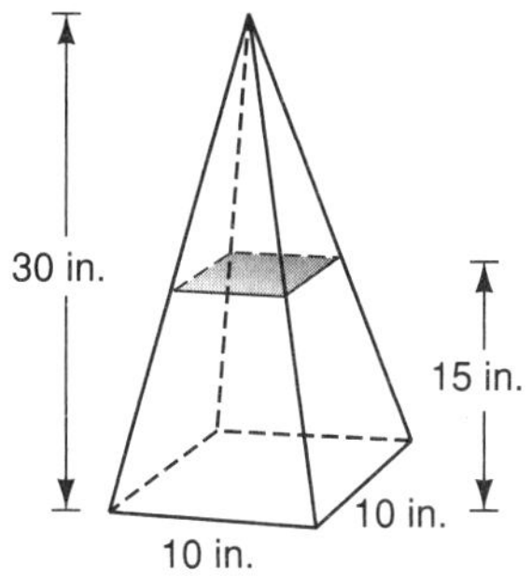

A. 20 square inches

B. 25 square inches

C. 40 square inches

D. 100 square inches

Think, Solve, Explain: Extended-Response Problem

18. Oscar wants to find the width (ST) of the Kissimmee River that is near his house. He marks off lengths of XY perpendicular to XZ and ZS perpendicular to ST.

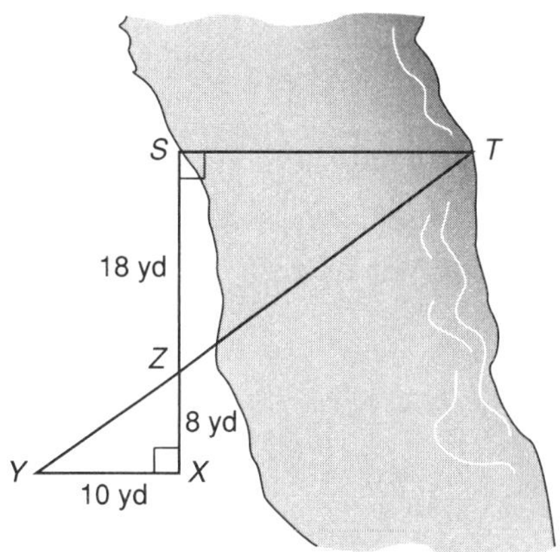

Part A Explain in words why $\triangle XZY \sim \triangle SZT$.

Part B Set up a proportion to find the width of the Kissimmee River.

Part C Solve to find the width of the river. Show all necessary work.

Multiple-Choice Problems

19. Tamara and Sandee are playing a board game than involves rolling a die and moving from space to space on the board. When an odd number is rolled, you move forward that number of spaces. When an even number is rolled, you move back that number of spaces. The first eight times Tamara rolled the following numbers: 5, 2, 3, 4, 1, 5, 2, and 2. Sandee rolled the following numbers: 3, 1, 2, 5, 3, 2, 1, and 3. What are their positions on the board after these eight turns?

F. Tamara is ahead by 8.

G. Sandee is ahead by 8.

H. Tamara is ahead by 3.

I. Sandee is ahead by 3.

20. Which of the following numbers is the greatest?

A. $\sqrt{156}$ **C.** 0.126×10^2

B. $\frac{156}{12}$ **D.** 12% of 156

21. To raise money to purchase new uniforms, the band held a concert. The people were seated on the grass as shown below. In one block of the grid, 50 people were counted. If the people were evenly distributed throughout the area, approximately how many people attended the concert?

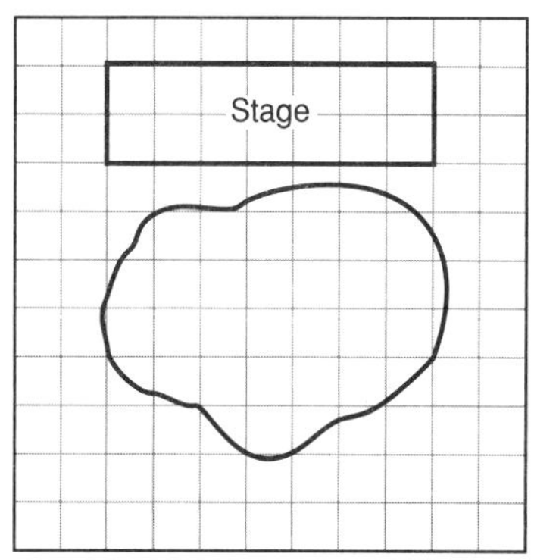

F. 1,000

G. 1,300

H. 1,500

I. 2,100

22. The Perry family is taking a 2,000-mile trip to visit relatives. The first four days they drove the following number of miles: 350, 426, 550, and 360. At the end of four days, approximately what percent of the trip was completed?

A. 16%

B. 31%

C. 69%

D. 84%

Think, Solve, Explain: Short-Response Problem

23. A sphere is inscribed in a cylinder as shown in the diagram below. The cylinder has a radius of 6 inches and a height of 12 inches. How many more cubic inches are in the cylinder than in the sphere? Use $\pi \approx 3.14$. Show all necessary work.

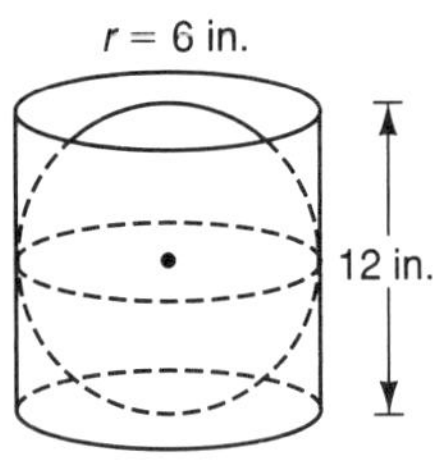

Gridded-Response Problems

24. To reduce 6.2×10^9 to 31, you must divide by 2×10^n. What is the value of n?

25. In the final 4 minutes of the game, a basketball player made 6 three-point baskets and 1 two-point basket. What was his average points per minute played?

26. There are 1.15 statute miles in a nautical mile. The distance from Manila to Shanghai is 1,338 statute miles. The distance from New York City to Havana is 1,227 nautical miles. How many nautical miles longer is the distance from New York City to Havana than from Manila to Shanghai? Round your answer to the nearest hundredth.

Multiple-Choice Problems

27. A piece of cake has been cut and removed as shown in the drawing below. What is the approximate length of the remaining arc x?

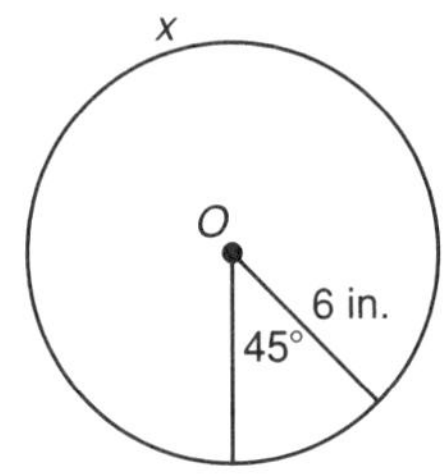

F. 4.7 inches

G. 9.4 inches

H. 16.5 inches

I. 33 inches

28. Milan, Italy, is located at 45.5° N latitude and Nairobi, Kenya, is located at 1.25° S latitude. To find the approximate airline distance between the two cities, multiply the angle measure between them by 70. What is the approximate airline distance between Milan and Nairobi?

A. 3,080 miles

B. 3,100 miles

C. 3,270 miles

D. 3,360 miles

29. Given that $-2(x+4) > 10$, what is the solution set of x?

F. $x < -1$

G. $x > -1$

H. $x < -9$

I. $x > -9$

30. Which of the following is a graph of the inequality $2x - y \leq 5$?

A.

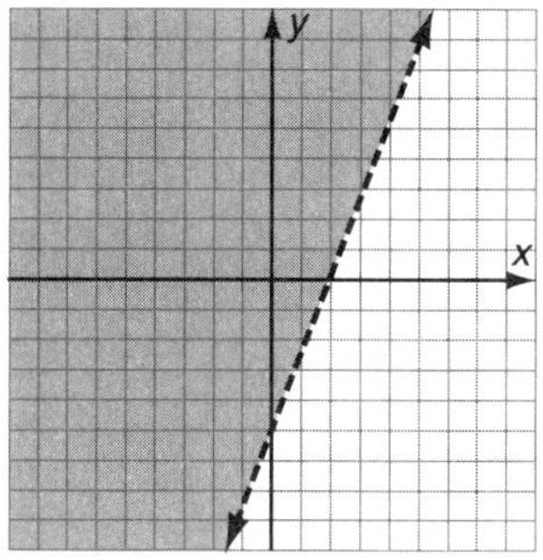

B.

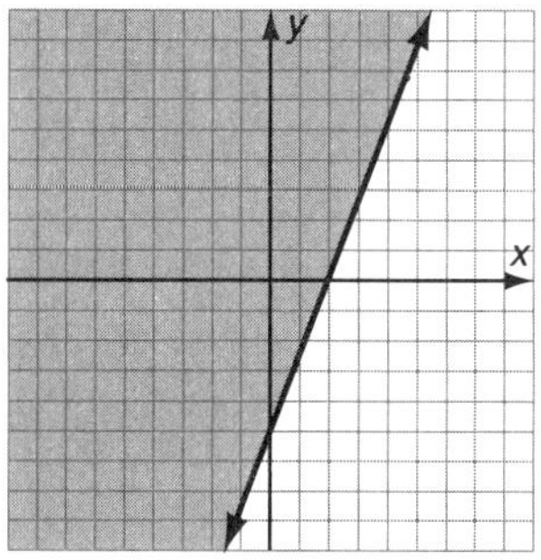

C.

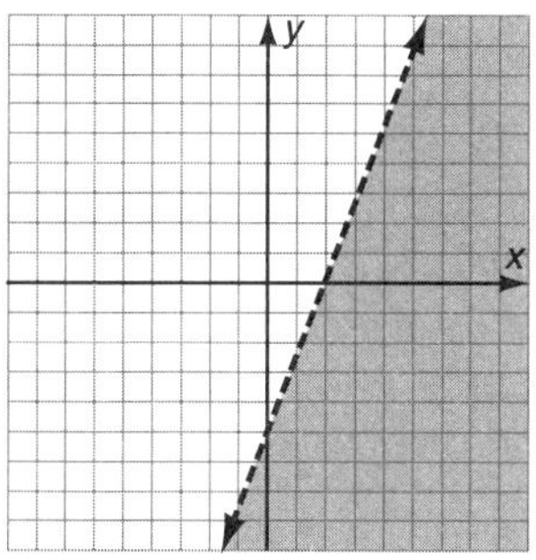

D.

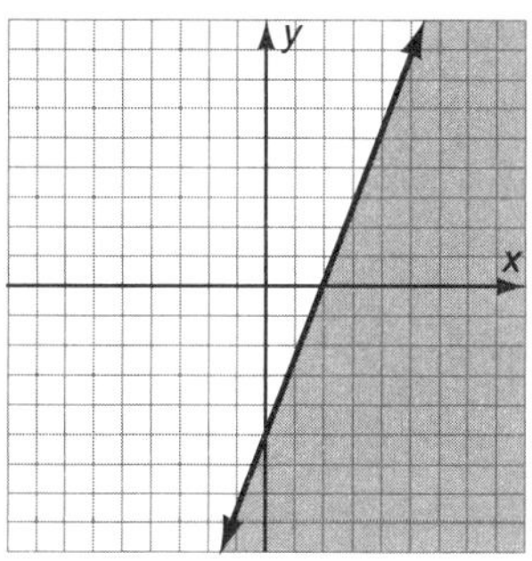

Appendix 1: Practice Exercises

ADDITION AND SUBTRACTION OF INTEGERS

REMEMBER: (1) To add two numbers with *like signs*, add the absolute values of the numbers. The sum has the sign of the numbers being added. (2) To add two numbers with *unlike signs*, subtract their absolute values. The sum has the sign of the number with the greater absolute value. (3) When you add zero to any number, the sum is the number itself. (4) To subtract signed numbers, change the sign of the number being subtracted and follow the rules for addition. (5) When you subtract zero from a number, the difference is the number itself.

PRACTICE

1. $-3 - 2 =$

2. $-5 + 3 =$

3. $-12 - 3 =$

4. $-157 + 108 =$

5. $-7 + 6 =$

6. $-25 + 17 =$

7. $4 - 7 =$

8. $0 - 12 =$

9. $78 + 42 =$

10. $-15 - 9 =$

11. $-16 + 14 =$

12. $8 - 9 =$

13. $-16 - 0 =$

14. $13 - 19 =$

15. $-12 - 12 =$

16. $-7 + 9 =$

17. $24 - 19 - 35 - 6 =$

18. $23 + 3 - 6 - 15 =$

19. $-3 + 8 - 5 + 35 =$

20. $-55 - 11 + 65 + 1 =$

MULTIPLICATION AND DIVISION OF INTEGERS

REMEMBER: (1) When you multiply two numbers with *like signs*, the product is positive. (2) When you multiply two numbers with *unlike signs*, the product is negative. (3) When you divide two numbers with *like signs*, the quotient is positive. (4) When you divide two numbers with *unlike signs*, the quotient is negative.

PRACTICE

1. $-3 \bullet 3 =$
2. $-7(-1) =$
3. $(-11)(1) =$
4. $-51 \div 3 =$
5. $90 \div -15 =$
6. $\dfrac{-120}{-8} =$
7. $(-4)(8) =$
8. $7 \bullet 0 =$
9. $64 \div 32 =$
10. $0 \div (-4) =$
11. $9(-5) =$
12. $\dfrac{18}{-6} =$
13. $-6 \bullet (-5) =$
14. $0 \bullet (-5) =$
15. $18 \div 0 =$
16. $20 \div 2 =$
17. $(-2)(6) =$
18. $\dfrac{-24}{4} =$
19. $-64 \div (-16) =$
20. $8 \bullet (-6) =$

FRACTIONS

REMEMBER: Before adding or subtracting fractions, you may need to find a common denominator and write equivalent fractions. This step is *not* necessary for multiplication or division of fractions. (1) To *add* fractions, add the numerators and keep the common denominator the same. Reduce the sum if possible. (2) To *subtract* fractions, subtract the numerators and keep the common denominator the same. Reduce the difference if possible. (3) To *multiply* fractions, multiply numerators and multiply denominators. Reduce the product if possible. (4) To *divide* fractions, invert the divisor and then multiply. Reduce the result if possible.

PRACTICE

Items 1–6: Write each fraction in lowest terms.

1. $\dfrac{2}{8} =$
2. $\dfrac{4}{10} =$
3. $\dfrac{4}{6} =$
4. $\dfrac{6}{24} =$
5. $\dfrac{4}{12} =$
6. $\dfrac{8}{12} =$

Items 7–12: Write each mixed number as an improper fraction.

7. $4\dfrac{5}{8} =$
8. $3\dfrac{2}{3} =$
9. $1\dfrac{1}{2} =$
10. $6\dfrac{7}{8} =$
11. $4\dfrac{3}{4} =$
12. $7\dfrac{3}{5} =$

Items 13–18: Write each improper fraction as a mixed number.

13. $\frac{12}{7} =$

14. $\frac{16}{5} =$

15. $\frac{9}{2} =$

16. $\frac{7}{3} =$

17. $\frac{8}{5} =$

18. $\frac{4}{2} =$

Items 19–24: Fill in the missing part of the equivalent fraction.

19. $\frac{5}{8} = \frac{?}{16}$

20. $\frac{3}{4} = \frac{?}{24}$

21. $\frac{1}{5} = \frac{?}{10}$

22. $\frac{1}{4} = \frac{3}{?}$

23. $\frac{3}{4} = \frac{12}{?}$

24. $\frac{7}{8} = \frac{21}{?}$

Items 25–39: Perform the indicated operation (addition or subtraction).

25. $\frac{1}{8} + \frac{3}{8} =$

26. $\frac{3}{8} + \frac{1}{2} =$

27. $\frac{7}{12} + \frac{2}{3} =$

28. $\frac{11}{16} + \frac{5}{12} =$

29. $\frac{4}{9} - \frac{1}{9} =$

30. $\frac{5}{6} - \frac{2}{3} =$

31. $\frac{9}{10} - \frac{1}{4} =$

32. $\frac{2}{3} - \frac{1}{7} =$

33. $\begin{array}{r} 10\frac{7}{10} \\ +\ 2\frac{4}{10} \\ \hline \end{array}$

34. $\begin{array}{r} 3\frac{3}{8} \\ +\ 5\frac{5}{6} \\ \hline \end{array}$

35. $\begin{array}{r} 12\frac{4}{5} \\ +\ 9\frac{1}{4} \\ \hline \end{array}$

36. $\begin{array}{r} 8\frac{7}{9} \\ -\ 4\frac{2}{3} \\ \hline \end{array}$

37. $\begin{array}{r} 5 \\ -\ 2\frac{3}{4} \\ \hline \end{array}$

38. $\begin{array}{r} 3\frac{1}{6} \\ -\ 1\frac{5}{8} \\ \hline \end{array}$

39. $\begin{array}{r} 9\frac{7}{8} \\ -\ 4\frac{3}{8} \\ \hline \end{array}$

Items 40–47: Multiply.

40. $\frac{2}{3} \times \frac{3}{5} =$

41. $\frac{3}{8} \times \frac{4}{5} =$

42. $6 \times \frac{2}{3} =$

43. $\frac{5}{12} \times \frac{6}{15} =$

44. $3\frac{1}{3} \times 2\frac{4}{5} =$

45. $4\frac{1}{2} \times 1\frac{1}{3} =$

46. $5\frac{3}{4} \times 3 =$

47. $2\frac{5}{8} \times 1\frac{3}{7} =$

Items 48–55: Divide.

48. $\frac{5}{6} \div \frac{15}{16} =$

49. $\frac{3}{8} \div \frac{27}{32} =$

50. $\frac{24}{28} \div \frac{6}{7} =$

51. $\frac{3}{4} \div \frac{5}{2} =$

52. $\frac{3}{4} \div \frac{1}{8} =$

53. $\frac{1}{3} \div 2 =$

54. $1\frac{1}{2} \div \frac{4}{9} =$

55. $2\frac{1}{2} \div 1\frac{1}{4} =$

PERCENTS, DECIMALS, AND FRACTIONS

REMEMBER: (1) To change a decimal to a fraction, use the appropriate power of 10 for the denominator. Then reduce the fraction, if possible. (2) To change a fraction to a decimal, divide the numerator by the denominator. (3) To change a decimal to a percent, move the decimal point two places to the right and use the word *percent* or the sign %.

PRACTICE ✔

Complete the table.

Fraction	Decimal	Percent
$\frac{1}{2}$		
	0.25	
		75%
	0.2	
$\frac{2}{3}$		
	0.4	
		$33\frac{1}{3}\%$
	0.1	
$\frac{3}{5}$		
	0.3	
		70%
	0.9	

ROUNDING

REMEMBER: To round a number, use this procedure. (1) Decide to what place you will round, and underline the number in that place. (2) If the number to the right of the underlined number is 5 or greater, add 1 to the underlined number and substitute zeros for all the numbers to the right of it. (3) If the number to the right of the underlined number is less than 5, leave the underlined number as is and substitute zeros for all the numbers to the right of it.

PRACTICE

Items 1–3: Round to the nearest thousand.

1. 9,654

2. 14,585

3. 108,277

Items 4–6: Round to the nearest hundredth.

4. 5.396

5. 27.933

6. $94.857

ORDER OF OPERATIONS

REMEMBER: *First* do operations within parentheses or within braces or brackets. *Second*, simplify exponents. That is, raise bases to powers. *Third*, do multiplication and division from left to right. *Fourth*, do addition and subtraction from left to right.

PRACTICE

Simplify.

1. $6(6 + 8) - 32 \div 16$

2. $11 + 2(6 + 4) - 3 \bullet 2^2$

3. $3 + 48 - 16 - 35 \div 7$

4. $(2 \bullet 3 - 12 \div 3) \div 2$

5. $\dfrac{3 \bullet 2^2}{2^3 + 4}$

6. $[3(4 - 1) + 2] \div (20 + 2)$

7. $\dfrac{24 - 8 \bullet 2}{3^3 - 3}$

8. $5[2 + 5(4) - 2^2] - 4^2$

9. $\dfrac{30 - 2 \bullet 4}{14 + 2^2}$

10. $21 - 15 \div 3 + 8 \bullet 4$

CONVERSIONS

REMEMBER: (1) To convert from a smaller unit to a larger unit, *divide* by a conversion factor. (2) To convert from a larger unit to a smaller unit, *multiply* by a conversion factor. (3) In the metric system, the conversion factor is a power of 10.

PRACTICE

Convert, using the conversion chart in the Mathematics Reference Sheet on page 12.

1. 36 inches = ? feet

2. 1 quart = ? cups

3. 5 pounds = ? ounces

4. 2 meters = ? centimeters

5. 3 liters = ? milliliters

6. 2 kilograms = ? grams

7. 9 feet = ? yards

8. 16 cups = ? pints

9. 40 ounces = ? pounds

10. 600 centimeters = ? meters

11. 4000 milliliters = ? liters

12. 2500 milligrams = ? grams

ALGEBRAIC EXPRESSIONS AND EQUATIONS

REMEMBER: To solve verbal problems algebraically, write the given information as an expression or an equation.

PRACTICE

Items 1–8: Write each phrase as an algebraic expression.

1. two times y

2. three less than a

3. five more than p

4. the product of x and seven

5. the quotient when c is divided by d

6. three-fourths of z

7. c less than four

8. four less than one-half of p

Items 9–13: Write each sentence as an algebraic equation.

9. The product of six and a number y is forty-eight.

10. Eight more than x is thirty.

11. Seven times a number h is twenty-eight.

12. Three times the sum of x and two is fifteen.

13. Fourteen less than the product of x and thirteen is twelve.

SOLVING EQUATIONS

REMEMBER: To solve an equation, "undo." (1) *First* undo addition by subtraction and undo subtraction by addition. (2) *Second*, undo multiplication by division and undo division by multiplication. (3) What is done on one side of the equals sign must also be done on the other side of the equals sign.

PRACTICE

Solve for each variable.

1. $x + 2 = 10$

2. $3x = 12$

3. $\frac{1}{2}x = 4$

4. $y - 3 = 4$

5. $5 + y = 9$

6. $\frac{x}{2} = 7$

7. $\frac{2}{3}y = 8$

8. $v - 8 = 2$

9. $6x = 24$

10. $k + 1 = 7$

11. $\frac{x}{4} = 6$

12. $t - 7 = 2$

13. $\frac{1}{4}x = 7$

14. $9x = 36$

15. $7 + w = 15$

16. $\frac{y}{3} = 9$

17. $c - 1 = 8$

18. $5x = 20$

19. $\frac{w}{7} = 1$

20. $p + 9 = 10$

21. $\frac{3}{5}k = 9$

22. $2x = 30$

23. $\frac{y}{6} = 11$

24. $g - 4 = 10$

25. $\frac{7}{10}k = 14$

26. $5y = 25$

SOLVING PROPORTIONS

REMEMBER: Some algebraic equations are proportions that can be solved by cross multiplication.

PRACTICE ✔

Solve each proportion for x.

1. $\frac{x}{20} = \frac{1}{10}$

2. $\frac{x}{12} = \frac{6}{8}$

3. $\frac{4}{x} = \frac{16}{28}$

4. $\frac{5}{x} = \frac{1}{3}$

5. $\frac{3}{10} = \frac{x}{40}$

6. $\frac{2}{3} = \frac{x}{15}$

7. $\frac{7}{9} = \frac{42}{x}$

8. $\frac{5}{18} = \frac{10}{x}$

9. $\frac{9}{10} = \frac{27}{x}$

10. $\frac{4}{9} = \frac{8}{x}$

11. $\frac{24}{35} = \frac{x}{70}$

12. $\frac{3}{4} = \frac{9}{x}$

13. $\frac{x}{8} = \frac{5}{2}$

14. $\frac{16}{x} = \frac{8}{15}$

15. $\frac{2}{3} = \frac{30}{x}$

16. $\frac{3}{x} = \frac{21}{35}$

17. $\frac{4}{7} = \frac{12}{x}$

18. $\frac{3}{5} = \frac{x}{15}$

Appendix 2: Calculator Practice

BASIC OPERATIONS

REMEMBER: (1) Some calculators are programmed to follow the order of operations, but the FCAT calculator is NOT programmed this way. To use the FCAT calculator, do one operation at a time. (2) Order of operations: *First* do operations within parentheses, braces, or brackets. *Second*, raise bases to powers. *Third*, do multiplication and division from left to right. *Fourth*, do addition and subtraction from left to right.

NOTE: Use the ± key to change numbers from positive to negative.

PRACTICE

1. $2 + 3$

2. $5 + 24$

3. $5.3 + 7.89$

4. $\$43.78 + \$25.89$

5. $5 - 2$

6. $35 - 29$

7. $896 - 23$

8. $\$23.15 - \$13.78$

9. 7×15

10. 8.4×2.1

11. 0.8×1.45

12. $35 \div 7$

13. $16.4 \div 1.2$

14. $256 \div 25$

15. $\$56 \div \$7$

16. $8 + 2 \times 6$

17. $15 \div 5 + 2$

18. $5 \times 4 - 3$

19. $-2 - 32 \bullet 2$

20. $-120 \div (-30)$

21. $-7.25 \div 5 + 1.5$

22. $369 \div (-9 \bullet 5)$

23. $(-22 - 62) \div (-4)$

24. $(-6)(-5)(-9)$

25. $-52 + 27 + (-33)$

PERCENTS

REMEMBER: There is more than one way to use a calculator for percents. (1) You can change the percent to a decimal and then multiply. (2) You can also use the calculator's percent key.

EXAMPLE: 25% of 225

Keystrokes	Display
[.] [2] [5] [×] [2] [2] [5] [=]	56.25

Or:

Keystrokes	Display
[2] [2] [5] [×] [2] [5] [%]	56.25

(Do not press =.)

PRACTICE ✔

1. 12% of 48

2. 50% of 95

3. 85% of 195

4. 0.75% of 50

5. 150% of 32

6. 125% of 57

PERCENT AND ADDITION

REMEMBER: For some problems, you need to find the percentage of a number and then add it to the number.

EXAMPLE: A dress costs $55.99 plus 6 percent sales tax. Find the cost.

Keystrokes	Display
[5] [5] [.] [9] [9] [×] [.] [0] [6] [+] [=]	59.3494

Or:

Keystrokes	Display
[5] [5] [.] [9] [9] [×] [1] [.] [0] [6] [=]	59.3494

(Multiply by 1 + the percent.)

In both cases, round to the nearest cent: $59.35

PRACTICE

1. $250 plus 7% tax
2. $850 plus 3% interest
3. 99¢ plus $7\frac{1}{2}$% tax
4. $6.85 plus 6% tax
5. $10,000 plus 4% interest
6. $25.95 plus $7\frac{3}{4}$% tax

PERCENT AND SUBTRACTION

REMEMBER: For some problems, you need to find the percentage of a number and then subtract it from the number.

EXAMPLE: A television is marked 25 percent off the regular price, $589. Find the sale price.

Keystrokes	Display
[5] [8] [9] [×] [.] [2] [5] [=]	147.25
[5] [8] [9] [−] [1] [4] [7] [.] [2] [5] [=]	441.75

Or:

Keystrokes	Display
[5] [8] [9] [×] [.] [2] [5] [=] [−] [=]	441.75

Or:

Keystrokes	Display
[5] [8] [9] [×] [.] [7] [5] [=]	441.75

(Multiply by 1 − the percent.)

PRACTICE

Find the discount and the sale price.

1. 50% off $650
2. 33% off $966
3. 25% off 99¢
4. 40% off $56
5. 60% off $285
6. 20% off $12.50

SQUARE ROOTS

REMEMBER: To take the square root of a number on your calculator, key in the number *before* using the square root key.

PRACTICE ✓

1. $\sqrt{36}$
2. $\sqrt{225}$
3. $\sqrt{8}$
4. $\sqrt{121}$
5. $\sqrt{96}$
6. $\sqrt{49}$

EXPONENTS

REMEMBER: (1) To find the value of a number raised to the second power (exponent is ^{2}), multiply the number times itself. (2) To find the value of a number raised to the third power (exponent is ^{3}), multiply the number times itself, then times itself again. (3) Continue this pattern for greater exponents.

PRACTICE ✓

1. 5^2
2. 3^3
3. 9^2
4. 2^4
5. 12^3

Index

D

E

F

Q

R

S

T